Progress in Mathematics

Volume 223

Fuensanta Andreu-Vaillo
Vicent Caselles
José M. Mazón

Parabolic Quasilinear Equations Minimizing Linear Growth Functionals

Springer Basel AG

Authors:

Fuensanta Andreu-Vaillo
José M. Mazón
Departamento de Análisis Matemático
Universitat de Valencia
Dr. Moliner 50
46100 Burjassot (Valencia)
Spain
e-mail: Fuensanta.Andreu@uv.es
mason@uv.es

Vicent Caselles
Departamento de Tecnología
Universitat Pompeu Fabra
Passeig de Circumvalació, 8
08003 Barcelona
Spain
e-mail: vicent.caselles@upf.edu

2000 Mathematics Subject Classification 35K55, 47H06, 47H20, 65M06, 68U10

A CIP catalogue record for this book is available from the Library of Congress, Washington D.C., USA

Bibliographic information published by Die Deutsche Bibliothek
Die Deutsche Bibliothek lists this publication in the Deutsche Nationalbibliografie; detailed bibliographic data is available in the Internet at <http://dnb.ddb.de>.

ISBN 978-3-0348-9624-5 ISBN 978-3-0348-7928-6 (eBook)
DOI 10.1007/978-3-0348-7928-6

Originally published by Birkhäuser Verlag, Basel, Switzerland in 2004
Softcover reprint of the hardcover 1st edition 2004
Printed on acid-free paper produced of chlorine-free pulp. TCF ∞

ISBN 978-3-0348-9624-5

9 8 7 6 5 4 3 2 1 www.birkhauser-science.com

Ferran Sunyer i Balaguer (1912–1967) was a self-taught Catalan mathematician who, in spite of a serious physical disability, was very active in research in classical mathematical analysis, an area in which he acquired international recognition. His heirs created the Fundació Ferran Sunyer i Balaguer inside the Institut d'Estudis Catalans to honor the memory of Ferran Sunyer i Balaguer and to promote mathematical research.

Each year, the Fundació Ferran Sunyer i Balaguer and the Institut d'Estudis Catalans award an international research prize for a mathematical monograph of expository nature. The prize-winning monographs are published in this series. Details about the prize and the Fundació Ferran Sunyer i Balaguer can be found at

http://www.crm.es/FerranSunyerBalaguer/ffsb.htm

This book has been awarded the Ferran Sunyer i Balaguer 2003 prize.

INSTITVT D'ESTVDIS CATALANS MCMVII

The members of the scientific commitee of the 2003 prize were:

Hyman Bass
University of Michigan

Antonio Córdoba
Universidad Autónoma de Madrid

Warren Dicks
Universitat Autònoma de Barcelona

Paul Malliavin
Université de Paris VI

Joseph Oesterlé
Université de Paris VI

Ferran Sunyer i Balaguer Prize winners:

1992 Alexander Lubotzky
Discrete Groups, Expanding Graphs and Invariant Measures, PM 125

1993 Klaus Schmidt
Dynamical Systems of Algebraic Origin, PM 128

1994 The scientific committee decided not to award the prize

1995 As of this year, the prizes bear the year in which they are awarded, rather than the previous year in which they were announced

1996 V. Kumar Murty and M. Ram Murty
Non-vanishing of L-Functions and Applications, PM 157

1997 A. Böttcher and Y.I. Karlovich
Carleson Curves, Muckenhoupt Weights, and Toeplitz Operators, PM 154

1998 Juan J. Morales-Ruiz
Differential Galois Theory and Non-integrability of Hamiltonian Systems, PM 179

1999 Patrick Dehornoy
Braids and Self-Distributivity, PM 192

2000 Juan-Pablo Ortega and Tudor Ratiu
Hamiltonian Singular Reduction, PM 222

2001 Martin Golubitsky and Ian Stewart
The Symmetry Perspective, PM 200

2002 André Unterberger
Automorphic Pseudodifferential Analysis and Higher Level Weyl Calculi, PM 209

Alexander Lubotzky and Dan Segal
Subgroup Growth, PM 212

Contents

Preface

Our goal in this monograph is to present general existence and uniqueness results for quasilinear parabolic equations whose operator is, in divergence form, the subdifferential of a Lagrangian which is convex in $|\nabla u|$ and has linear growth as $|\nabla u| \to \infty$. We devote particular attention to the case of the minimizing *total variation flow* for which we study the Neumann, Dirichlet and Cauchy problem in $\mathbb{R}^N$ together with the main qualitative properties of its evolution. This kind of problem appears in different contexts: image processing, faceted crystal growth, continuum mechanics, etc. Motivated by the use of the *total variation model in image restoration*, we started our study of the *minimizing total variation* (TV) *flow* in collaboration with C. Ballester, by studying the corresponding Neumann and Dirichlet problems [13], [14]. Later, in a joint paper with J. I. Diaz [15] we studied the asymptotic behaviour of the solutions of these problems. This study was continued in [34] where some extinction profiles were identified. In particular, this provided some explicit solutions of the denoising problem in image processing. The techniques developed for the total variation flow were extended to cover the case of general convex Lagrangians with linear growth rate in the modulus of the gradient, providing a general existence and uniqueness result in this case [16],[17]. Energy functionals with linear growth appear in different contexts, two classical examples being the nonparametric area integrand $f(\xi) = \sqrt{1 + \|\xi\|^2}$, which is associated with the time-dependent minimal surface equation, and the Hencky model in plasticity.

Let us summarize the contents of this book.

Chapter 1 is devoted to the study of the variational approach to image restoration based on total variation minimization subject to the constraints given by the image acquisition model. We review the model initially introduced by L. Rudin, S. Osher and E. Fatemi [175] which had, on one hand, a strong influence in the development of variational models in image denoising and restoration, and, on the other, pioneered the use of the BV model in image processing. The chapter contains the proof of the Chambolle–Lions theorem proving that the constraints can be incorporated by means of a Lagrange multiplier, thus justifying the usual numerical approach to the problem. Then we interpret the corresponding Euler–Lagrange equation in terms of partial differential equations by means of the PDE characterization of the subdifferential of the total variation. This result follows as a consequence of the results in [13] and has been presented in [48]. The approach we present here is a simple and direct approach to the characterization of the subdifferential of positively 1-homogeneous convex functionals of the gradient due to F. Alter in his unpublished work [3]. Then we display a few experiments on image restoration obtained with this model. The chapter also contains a review of the main numerical methods used in the variational approach to image restoration. We apologize in advance for any missing work.

In Chapter 2 we study the Neumann problem for the minimizing total variation flow. First we present the main existence and uniqueness results for this problem, which are essentially taken from [13]. Due to the homogeneity of the operator associated with the problem in L^p for any $p \geq 1$ we prove that the semigroup solutions are strong solutions. This, combined with the regularity results for quasi-minimizers of the perimeter, permits us to prove a regularizing effect on the level lines of the solution, a result which also holds for the solution of the restoration problem. The chapter also contains a proof that solutions of the Neumann problem stabilize as $t \to \infty$ by converging to the mean value of the initial datum.

The Cauchy problem for the total variation flow is studied in Chapter 3. The purpose of this chapter is to prove existence and uniqueness of entropy solutions for initial data in $L^1_{loc}(\mathbb{R}^N)$. This will enable us to study in later chapters the main features of the flow in $\mathbb{R}^N$, thus, dismissing the effect of boundary conditions. First, we study the flow in $L^2(\mathbb{R}^N)$. In Section 2 we prove uniqueness of entropy solutions for initial data in $L^1_{loc}(\mathbb{R}^N)$, using Kruzhkov's method of doubling variables. Then we prove existence for initial data in $L^1_{loc}(\mathbb{R}^N)$. We end up with the study of the time regularity of solutions.

Chapter 4 is devoted to a study of the asymptotic behaviour and qualitative properties of the solutions of the total variation flow in $\mathbb{R}^N$. We start by describing some numerically observed features of the flow, namely that local maxima (resp. minima) immediately decrease (resp. increase) and produce flat zones in the solution. For that we shall need some radially symmetric explicit solutions of the flow. We also note that the length of the level curves of the solutions is a decreasing function of time. Our next purpose will be to describe the extinction profile (the solution has a finite extinction time) of compactly supported solutions. This behaviour is described by a function which is the solution of an eigenvalue problem for the operator $-\text{div}\left(\frac{Du}{|Du|}\right)$. The rest of the chapter is devoted to the study of explicit solutions of this eigenvalue problem in the plane. In the radial case, positive solutions can be fully characterized. Then we look for characteristic functions which are solutions of it. This permits characterization of the bounded sets of finite perimeter $\Omega \subset \mathbb{R}^2$ for which the function $u(t,x) = (1 - \frac{\text{Per}(\Omega)}{|\Omega|}t)^+ \chi_\Omega(x)$ is an entropy solution of the minimizing total variation flow in $\mathbb{R}^2$. As an important by-product of the eigenvalue problem, one can obtain explicit solutions of the Rudin–Osher–Fatemi image denoising model. The results of this chapter have been taken from [13], [15], [34].

Chapter 5 is concerned with the Dirichlet problem for the total variation flow. In this case, the homogeneity of the operator is lost, and the notion of entropy solution in the sense of Kruzhkov is required to obtain a uniqueness result. Existence and time regularity of entropy solutions follow from the usual semigroup theory approach. The techniques introduced in this chapter will be the basis for results in the next two chapters dealing with more general operators. The presentation of this chapter is based on [14].

The next two chapters are devoted to a study of the Dirichlet problem for quasilinear parabolic equations whose operator is, in divergence form, the subdifferential of a Lagrangian which is convex and has linear growth in the magnitude of the gradient. More precisely, we study the Dirichlet problem in a bounded domain Ω with boundary datum $\varphi \in L^1(\partial\Omega)$, for the differential operator $-\mathrm{div}\,\mathbf{a}(x, Du)$, where $\mathbf{a}(x,\xi) = \nabla_\xi f(x,\xi)$, f being a convex function of ξ with linear growth as $\|\xi\| \to \infty$. The regularity assumptions we need to impose on the Lagrangian f exclude the total variation flow, i.e., the case $f(\xi) = \|\xi\|$, which was studied in Chapter 5; but we include many examples relevant in applications, like the nonparametric area integrand and Hencky plasticity. In Chapter 6 we prove existence and uniqueness of strong solutions in $L^2(\Omega)$ using the theory of nonlinear semigroups generated by subdifferential operators. Now, to get the full strength of the abstract result derived from semigroup theory, we need to characterize the subdifferential of the energy functional associated with the problem. In Chapter 7 we prove existence and uniqueness of entropy solutions for data in $L^1(\Omega)$. Existence follows by means of Crandall–Ligget's semigroup generation theorem, while uniqueness is proved using again Kruzhkov's method of doubling variables. The results of these two chapters are essentially taken from [16] and [17], respectively.

The book finishes with three appendices in which we outline some of the main tools used in the above chapters. In the first one (Appendix A) we present without proofs the main results of nonlinear semigroup theory which is the main tool used in this text to prove existence of solutions. Due to the linear growth of the energy functionals associated with the problems studied in this monograph, the natural energy space to study them is the space of functions of bounded variation. In Appendix B we outline some of the main points of the theory of functions of bounded variation used in the previous chapters. Finally, following G. Anzelloti's paper [25], Appendix C is devoted to the main results about pairings between measures and bounded measurable functions, one of the fundamental tools of the text.

It is a pleasure to acknowledge here the debt we owe to our coauthors, namely C. Ballester, G Bellettini, J.I. Diaz and M. Novaga. This monograph could not have been written without their contribution. We would like to thank also F. Alter for permitting us to reproduce his unpublished work [3]. We are also indebted to

B. Rougé and the CNES for stimulating discussions about the restoration problem which gave us a better understanding of it, and for his kind permission to reproduce the images of Chapter 1. We thank M. Bertalmio, A. Solé and B. Rougé for providing us these experiments. Finally we are indebted with L. Rudin from Cognitech Inc. for stimulating us to work on the theoretical analysis of the total variation restoration problem which motivated the subsequent work. Thanks should also be extended to many colleagues with whom we have shared their views on image processing and PDEs, among them we would like to thank Ph. Bénilan, J. Blat, A. Chambolle, P.L. Lions, F. Malgouyres, L. Moisan, S. Moll, J.M. Morel, P. Mulet, S. Osher, G. Sapiro, S. Segura, J. Toledo and J.L. Vázquez.

Last but not least, the first and third authors acknowledge partial support by the Spanish DGICYT, Project PB98-1442, the PNPGC, Project BFM2002-01145 and the RTN Programme of the EC "Nonlinear Partial Differential Equations Describing Front Propogation and other Singular Phenomena", reference HPRN-CT-2002-00274. The second author acknowledges partial support by the Departament d'Universitats, Recerca i Societat de la Informació de la Generalitat de Catalunya, by PNPGC project, reference BFM2000-0962-C02-01, by a CNES project, and, in previous stages of this work by the TMR European Project "Viscosity Solutions and their Applications", reference FMRX-CT98-0234.

Barcelona and Valencia, December 2002

Chapter 1

Total Variation Based Image Restoration

1.1 Introduction

1.1.1 The Image Model

For the purpose of image restoration the process of image formation can be modeled in a first approximation by the formula [207]

$$u_d = Q\{\Pi(k * u) + n\}, \tag{1.1}$$

where u represents the photonic flux, k is the point spread function of the optical-captor joint apparatus, Π is a sampling operator, i.e., a Dirac comb supported by the centers of the matrix of digital sensors, n represents a random perturbation due to photonic or electronic noise, and Q is a uniform quantization operator mapping $\mathbb{R}$ to a discrete interval of values, typically $[0, 255]$.

The point spread function of the optical-captor apparatus. The optical-captor system is modeled by a convolution operator whose kernel k is called its point spread function. Indeed, both the optical system and the captor can be considered as linear and translation invariant systems, and, therefore, each of them is modeled by a convolution operator. The convolution kernel k of the joint system formed by the optics and the captor is thus the convolution of the point spread functions of both separated systems.

In CCD arrays, each detector is a flux integrator (which counts the number of photons arriving to it). Thus, its point spread function is the normalized characteristic function of a square (supposing that each detector has this geometry) $[-\frac{p}{2}, \frac{p}{2}] \times [\frac{p}{2}, \frac{p}{2}]$, i.e.,

$$k_{det}(x, y) = \frac{1}{p^2}\chi_{[-\frac{p}{2},\frac{p}{2}]\times[\frac{p}{2},\frac{p}{2}]}.$$

Its corresponding Fourier transform, also called the *modulated transfer function* of the system, is then

$$MTF_{det}(\xi_1, \xi_2) = \text{sinc}(p\xi_1)\text{sinc}(p\xi_2),$$

where

$$\text{sinc}(x) = \frac{\sin \pi x}{\pi x}.$$

We note that we are using the Fourier transform in the form

$$F(f)(\xi) = \hat{f}(\xi) = \int_{-\infty}^{+\infty} f(y) e^{-2\pi i \xi y}\, dy. \tag{1.2}$$

The optical system has essentially two effects on the image: it projects images of the objects from the object plane to the image plane and degrades them. The degradation of the image due to the optical system makes that a light point source loses definition and appears as a blurred (small) region. This effect can be explained by the wave nature of light and its diffraction theory. We shall discard other degradation effects due to imperfections of optical systems such as lens aberrations [22]. Thus our main source of degradation will be the diffraction of the light when passing through a finite aperture: those systems are called diffraction limited systems.

A light source is called coherent if it emits light with a definite wavelength. If the emitted light is a mixture of wavelengths we say that the source is incoherent. Let us also recall that intensity of the light is given by the square of the electromagnetic field (a solution of Maxwell's equations). These two remarks will be taken into account to obtain the equations relating the electromagnetic field with the intensity field measured by the sensors.

Since we are assuming that the optical system is linear and translation invariant we know that it can be modeled by a convolution operator. Indeed, if the system is linear and translation invariant, it suffices to know the response of the system to a light point source located at the origin, which is modeled by a Dirac delta function δ, since any other light distribution could be approximated (in a weak topology) by superpositions of Dirac functions. The convolution kernel is, thus, the result of the system acting on δ.

We assume that the lens is located in an open bounded region A of a plane. The point spread function $h(x, y)$ in case of a monocromatic wave is approximately given, modulo a phase factor, by the Fourier transform of the characteristic function of the lens aperture:

$$h(x, y) = (\text{phase}) F(\chi_A(\lambda d_i \cdot, \lambda d_i \cdot),$$

where λ represents the wavelength of the light and d_i the distance from the lens to the image plane. This formula is obtained from the Maxwell equations using Kirchoff's scalar theory of diffraction and the Fraunhofer assumptions: the diffraction aperture in the screen is small compared to the distances d_i and R' from the aperture to the image plane and light source, respectively. For a detailed description of the theory of diffraction we refer to [184]. Intuitively, as the light wave arrives, each point of the aperture becomes the source of a spherical wave propagating to the image plane. After considering the above approximation, in particular that d_i is large compared to the dimensions of the aperture and the image, the point spread function h is given by ([184],[63])

$$h(x,y) = \frac{\lambda}{i} e^{\frac{ik}{2d_i}(x^2+y^2)} \int_{-\infty}^{\infty}\int_{-\infty}^{\infty} \chi_A(\lambda d_i x', \lambda d_i y') e^{-2\pi i(xx'+yy')}\, dx'\, dy' \qquad (1.3)$$

where $k = \frac{2\pi}{\lambda}$.

If we measure the light intensity, i.e., the square of the electromagnetic field, and we assume that the system is linear and translation invariant, the formula relating the light intensity emitted by the object I_0 and the light intensity measured by the optical system I is

$$I(x,y) = \int_{-\infty}^{\infty}\int_{-\infty}^{\infty} |h(x-x_0, y-y_0)|^2 I_0(M^{-1}x_0, M^{-1}y_0)\, dx_0\, dy_0, \qquad (1.4)$$

where M is the magnification factor, i.e., the quotient of the distance between two points of the image plane and the corresponding points in the scene, which is given by

$$M = \frac{d_i}{-z_0},$$

z_0 being the distance between the object plane and the plane of the aperture, where the origin of our coordinate system is located.

We shall write $k_{opt}(x,y) = |h(x,y)|^2$ and we call it the point spread function of the optical system. In case of a circular aperture of diameter D and incoherent light source centered around a wavelength λ, the point spread function k_{opt} is given by

$$k_{opt}(x) = \left(2 \frac{J_1(\pi \frac{r}{r_0})}{\pi \frac{r}{r_0}} \right)^2 \qquad (1.5)$$

where $J_1(r)$ is the Bessel function of first class and order 1, r is the radial distance computed in the image plane and

$$r_0 = \frac{\lambda d_i}{D}. \qquad (1.6)$$

If the aperture is a square $[-a,a] \times [-b,b]$, then k_{opt} is given by

$$k_{opt}(x_1,x_2) = \frac{\sin^2\left(\pi \frac{x_1}{x_{01}}\right) \sin^2\left(\pi \frac{x_2}{x_{02}}\right)}{\left(\pi \frac{x_1}{x_{01}}\right)^2 \left(\pi \frac{x_2}{x_{02}}\right)^2} \qquad (1.7)$$

where $x_{01} = \frac{\lambda d_i}{2a}$, $x_{02} = \frac{\lambda d_i}{2b}$.

The point spread function of the joint optical-captor system is the convolution of the point spread functions of both systems, i.e.,

$$k = k_{opt} * k_{det}.$$

In terms of its Fourier transforms, the modulated transforms of the optical system and detector, we have

$$MTF = MTF_{opt} MTV_{det}.$$

Noise. We shall describe the typical noise in case of a CCD array. Light is constituted by photons (quanta of light) and those photons are counted by the detector. Typically, the sensor registers light intensity by transforming the number of photons which arrive to it into an electric charge, counting the electrons which the photons take out of the atoms. This is a process of a quantum nature and therefore there are random fluctuations in the number of photons and photoelectrons on the photoactive surface of the detector. To this source of noise we have to add the thermal fluctuations of the circuits that acquire and process the signal from the detector's photoactive surface. This random thermal noise is usually described by a zero-mean white Gaussian process. The photoelectric fluctuations are more complex to describe: for low light levels, photoelectric emission is governed by Bose–Einstein statistics, which can be approximated by a Poisson distribution whose standard deviation is equal to the square root of the mean; for high light levels, the number of photoelectrons emitted (which follows a Poisson distribution) can be approximated by a Gaussian distribution which, being the limit of a Poisson process, inherits the relation between its standard deviation and its mean [22]. In a first approximation this noise is considered as spatially uncorrelated with a uniform power spectrum, thus a white noise. Finally, both sources of noise are assumed to be independent.

Taken together, both sources of noise are approximated by a Gaussian white noise, which is represented in the basic equation (1.1) by the noise term n. The average signal to noise ratio, called the SNR, can be estimated by the quotient between the signals average and the square root of the variance of the signal.

The detailed description of the noise requires a knowledge of the precise system of image acquisition. More details in the case of satellite images can be found in [172] and references therein.

The processes of image transmission and register generate other types of noise like the loss of some values or a change of the intensity value proportional to it. This could be modeled with a term η in the equation $u_d = Q\{\Pi(k * u) + n\} \cdot \eta$.

1.1.2 Image Restoration

We suppose that our image (or data) u_d is a function defined on a bounded and piecewise smooth open set D of $\mathbb{R}^N$ — typically a rectangle in $\mathbb{R}^2$. From our discussion above, generally, the degradation of the image occurs during image acquisition and can be modeled by a linear and translation invariant blur and additive noise. The equation relating u to u_d can be written as

$$u_d = Ku + n, \tag{1.8}$$

where K is a convolution operator with impulse response k, i.e., $Ku = k * u$, and n is an additive white noise of standard deviation σ. In practice, the noise can be considered as Gaussian.

The problem of recovering u from u_d is ill posed. First, the blurring operator need not be invertible. Second, if the inverse operator K^{-1} exists, applying it to both sides of (1.8) we obtain

$$K^{-1}u_d = u + K^{-1}n. \tag{1.9}$$

Writing $K^{-1}n$ in the Fourier domain, we have

$$K^{-1}n = \left(\frac{\hat{n}}{\hat{k}}\right)^{\vee}$$

where $\hat{f}$ denotes the Fourier transform of f and $f^{\vee}$ denotes the inverse Fourier transform. From this equation, we see that the noise might blow up at the frequencies for which $\hat{k}$ vanishes or becomes small.

Several methods have been proposed to recover u. Most of them can be classified as regularization methods which may take into account statistical properties (Wiener filters), information theoretic properties ([91]), a priori geometric models ([175]) or the functional analytic behaviour of the image given in terms of its wavelet coefficients ([105],[104]).

In case we know nothing about the noise we can set up the restoration problem as a least squares minimization. In this case we consider u and u_d to be deterministic. Then, to obtain the estimate of u from (1.8), we minimize the criterion

$$J(u) = \| Ku - u_d \|_2^2$$

which (assuming that K^tK is invertible) gives an estimate of u in terms of the pseudo-inverse of u_d, i.e.,

$$u^+ = (K^tK)^{-1}K^tu_d,$$

where K^t is the adjoint of K. This is the linear algebraic approach to restoration. As it is well known [22] this estimate of u amplifies the noise due to the ill-conditioning of the operator K.

The typical strategy to solve this ill-conditioning is regularization. Then the solution of (1.8) is estimated by minimizing a functional

$$J_\lambda(u) = \| Ku - u_d \|_2^2 + \gamma \| Qu \|_2^2, \tag{1.10}$$

which yields the estimate

$$u_\gamma = (K^t K + \gamma Q^t Q)^{-1} K^t u_d, \tag{1.11}$$

Q being a regularization operator. Observe that to obtain u_γ we have to solve a system of linear equations. The role of Q is, on one hand, to move the small eigenvalues of K away from zero while leaving the large eigenvalues unchanged, and, on the other hand, to incorporate the a priori (smoothness) knowledge that we have on u.

If we treat u and n as random vectors and we select $Q = R_f^{-1/2} R_n^{1/2}$ with R_f and R_n the image and noise covariance matrices, then (1.11) corresponds to the parametric Wiener filter [63]. When $\lambda = 1$ this corresponds to the Wiener filter that minimizes the mean square error between the original and restored images.

The first regularization method consisted in choosing between all possible solutions of (1.9) the one which minimized the Sobolev (semi) norm of u

$$\int_D |Du|^2 \, dx,$$

which corresponds to the case $Qu = \nabla u$. Then the solution of (1.10) given by (1.11) in the Fourier domain is given by

$$\hat{u} = \frac{\overline{\hat{k}}}{|\hat{k}|^2 + 4\gamma\pi^2|\xi|^2} \hat{u}_d.$$

From the above formula we see that high frequencies of u_d (hence, the noise) are attenuated by the smoothness constraint. This was an important step, but the results were not satisfactory, mainly due to the inability of the previous functional to resolve discontinuities (edges) and oscillatory textured patterns. The smoothness constraint is too restrictive. Indeed, functions in $W^{1,2}(D)$ cannot have discontinuities along rectifiable curves. These observations motivated the introduction of total variation in image restoration models by L. Rudin, S. Osher and E. Fatemi in their seminal work [175]. The a priori hypothesis is that functions of bounded variation (the BV model) [10], [110], [209]) are a reasonable functional model for many problems in image processing, in particular, for restoration problems ([173], [175]). Typically, functions of bounded variation have discontinuities along rectifiable curves, being continuous in some sense (in the measure theoretic sense) away from discontinuities. The discontinuities could be identified with edges. The ability of this functional to describe textures is less clear; some textures can be recovered, but up to a certain scale of oscillation. An interesting experimental discussion of the adequacy of the BV-model to describe real images can be seen in [5], [126].

On the basis of the BV-model, Rudin–Osher–Fatemi [175] proposed to solve the following constrained minimization problem:

$$\begin{aligned}&\text{Minimize} \int_D |Du|\,dx \\ &\text{with} \quad \int_D Ku = \int_D u_d, \quad \int_D |Ku - u_d|^2\,dx = \sigma^2|D|.\end{aligned} \tag{1.12}$$

The first constraint corresponds to the assumption that the noise has zero mean, and the second that its standard deviation is σ. The constraints are a way to incorporate the image acquisition model given in terms of equation (1.8). Under some assumptions on u_d (($H3$) in Section 1.2), the constraint

$$\int_D |Ku - u_d|^2\,dx = \sigma^2|D| \tag{1.13}$$

is equivalent to the constraint

$$\int_D |Ku - u_d|^2\,dx \le \sigma^2|D|,$$

which amounts to saying that σ is an upper bound of the standard deviation of n ([64]). Moreover, assuming that $K1 = 1$ (assumption ($H2$) in Section 1.2), the constraint $\int_D Ku = \int_D u_d$ is automatically satisfied [64].

In practice, the above problem (1.12) is solved via the following unconstrained minimization problem:

$$\text{Minimize} \int_D |Du|\,dx + \lambda \int_D |Ku - u_d|^2\,dx \tag{1.14}$$

for some Lagrange multiplier λ. The constraint has been introduced as a penalization term. The regularization parameter λ controls the trade-off between the goodness of fit of the constraint and the smoothness term given by the total variation. In this formulation, a methodology is required for a correct choice of λ. In [175], Rudin–Osher–Fatemi used the gradient projection method of Rosen ([171]) which leads to the gradient descent PDE associated to the problem (1.14) and updated λ so that the constraint (1.13) is satisfied. The analysis of such an algorithm was initiated in [146]. The most successful analysis of the connections between (1.12) and (1.14) was given by A. Chambolle and P.L. Lions in [64]. Indeed, they proved that both problems are equivalent for some positive value of the Lagrange multiplier λ. We shall reproduce their analysis in Section 1.2.

A different approach was taken in [186], [187], and [185], where the regularization parameter is scale and space adaptative. Indeed, the parameter λ is taken to be x dependent and is written in front of the total variation term

$$\int_D \lambda(x)|Du|.$$

For edge dependent adaptive restoration, $\lambda(x)$ was taken essentially as proportional to $\frac{1}{\epsilon+|Du_e(x)|}$, where u_e is an estimated version of the restored image obtained by a previous total variation restoration [185]. A similar functional in form, but not in purpose, has been used in [155] to control the ringing when extrapolating the spectrum of the image.

To solve (1.14) one formally computes the Euler–Lagrange equation and solves it with Neumann boundary conditions, which amounts to a reflection of the image across the boundary of D. We shall compute in Section 1.3 the differential equation satisfied by the minimum of (1.14). Many numerical methods have been proposed to solve this equation in practice, e.g., [175], [64], [201, 202], [71, 72], [108] (see also [162] for an interesting analysis of the features of most numerical methods, explaining in particular the staircasing effect). Some of them are explicit and based on gradient descent with constant or variable time step and some of them are implicit. We shall briefly review them in Sections 1.4 and 1.5.

Chapters 2 and 4 can be understood as an analysis of the gradient descent flow corresponding to the functional

$$\int_D |Du|,$$

i.e., the evolution problem

$$\frac{\partial u}{\partial t} = \operatorname{div}\left(\frac{Du}{|Du|}\right) \qquad \text{in }]0,\infty[\times D, \tag{1.15}$$

with $u(0,x) = u_0(x)$, $x \in D$ under the Neumann boundary conditions or in the whole space. The explicit solutions computed in Chapter 4, which include explicit solutions of the denoising problem, together with the qualitative properties of the flow give some information on the behaviour of (1.15) when minimizing the total variation. In particular, its behaviour is in contrast with the behaviour of the solutions of the mean curvature motion which also diminishes the total variation of its solutions.

1.2 Equivalence between Constrained and Unconstrained Restoration

As we already mentioned in the introduction to this section, L. Rudin, S. Osher and E. Fatemi [175] proposed to solve the restoration problem by minimizing the total variation under constraints:

$$\begin{aligned} &\text{Minimize} \int_\Omega |Du|\,dx \\ &\text{with} \quad \int_\Omega Ku = \int_\Omega u_d, \quad \int_\Omega |Ku - u_d|^2\,dx = \sigma^2|\Omega|, \end{aligned} \tag{1.16}$$

Ω being the image domain. The first constraint corresponds to the assumption that the noise has zero mean and the second that its standard deviation is σ.

In practice, the above problem is solved via the following unconstrained minimization problem:

$$\text{Minimize} \int_\Omega |Du|\,dx + \frac{\lambda}{2}\int_\Omega |Ku - u_d|^2\,dx \tag{1.17}$$

for some Lagrange multiplier $\lambda \geq 0$. A. Chambolle and P.L. Lions [64] proved that, under some assumptions on K, there is a particular value of the Lagrange multiplier for which both problems are equivalent. Let us give the proof of their result.

First we need to state some assumptions on the data of the problem. For simplicity we shall only consider the case $N = 2$, and assume that Ω is a Lipschitz domain in $\mathbb{R}^2$. Recall that in that case $BV(\Omega) \subseteq L^2(\Omega)$. To simplify our notation we shall assume that

$$|\Omega| = 1.$$

Our set of assumptions is:

(H1) K is a continuous and linear operator in $L^2(\Omega)$.

(H2) $K1 = 1$.

(H3) $\left\| u_d - \int_\Omega u_d \right\| \geq \sigma^2.$

Typically K is a convolution operator and assumption $(H1)$ is satisfied. Assumption $(H2)$ expresses the fact that the optical system preserves the light energy. According to $(H3)$, the data u_d has variance greater than σ^2. Implicit in the treatment will be the fact that n is an oscillatory function which represents a white noise added to the clean image. We assume that n has zero mean and variance σ^2.

Theorem 1.1. *Assume that $(H1, H2, H3)$ hold. Assume that $u_d \in X$, where X is the closure of $K(BV(\Omega))$. Then* (1.16) *has a solution $u \in BV(\Omega)$, and Ku is unique. Moreover, problem* (1.16) *is equivalent to* (1.17) *for a unique (if $\sigma < \|u_d - \int_\Omega u_d\|$) and nonnegative Lagrange multiplier λ. If K is injective, then the solution of both problems is unique.*

Before going into the proof, we shall observe that we may assume that $\int_\Omega u_d = 0$. Indeed, if $\overline{u_d} = u_d - \int_\Omega u_d$, then, due to the assumption $(H2)$, we have that

$$u \text{ is a solution of (1.16) (resp. (1.17))}$$

$$\Leftrightarrow$$

$$\overline{u} = u - \int_\Omega u \text{ is a solution of (1.16) (resp. (1.17)) with } \overline{u_d} \text{ instead of } u_d.$$

Thus, in the rest of the subsection we assume that $\int_\Omega u_d = 0$.

Let us define the functional $\Phi : L^2(\Omega) \to (-\infty, +\infty]$ by

$$\Phi(u) = \begin{cases} \displaystyle\int_\Omega \|Du\| & \text{if } u \in BV(\Omega), \\ +\infty & \text{if } u \in L^2(\Omega) \setminus BV(\Omega). \end{cases} \tag{1.18}$$

Proof. Let u_n be a minimizing sequence for (1.16). As $\int_\Omega \|Du\| + \|Ku\|_2$ is equivalent to the BV-norm, by Poincaré inequality ([209]), $\{u_n\}$ is bounded in $BV(\Omega)$. Thus we can assume that u_n converges weakly in $L^2(\Omega)$ to u (strongly in $L^p(\Omega)$ for all $p < 2$), and Du_n converges weakly as a measure to Du. Since K is a linear continuous map in $L^2(\Omega)$, Ku_n converges weakly in $L^2(\Omega)$ to Ku. We have

$$\begin{aligned} \Phi(u) &\leq \liminf_{n\to\infty} \Phi(u_n), \\ \int_\Omega Ku &= \lim_{n\to\infty} \int_\Omega Ku_n = 0, \\ \|Ku - u_d\|_2 &\leq \lim_{n\to\infty} \|Ku_n - u_d\|_2. \end{aligned}$$

Consider now the function $f(t) = \|tKu - u_d\|_2$ for $t \in [0,1]$. As $f(t)$ is a continuous function, $f(1) \leq \sigma$ and $f(0) \geq \sigma$, there exists some $t \in [0,1]$ such that $f(t) = \sigma$. The function $u' = tu$ satisfies $\int_\Omega u' = 0$ and $\|Ku' - u_d\|_2 = \sigma$. If $t = 0$, we have $\|u_d\|_2 = \sigma$. In that case we may take $u = 0$ as the solution to our problem. Now it is easy to check that if $t \in (0,1)$, then

$$\Phi(u') = t\Phi(u) < \Phi(u) \leq \liminf_{n\to\infty} \Phi(u_n),$$

unless $Du = 0$. This contradicts the fact that $\liminf_{n\to\infty} \Phi(u_n)$ is the infimum of our problem, unless $Du = 0$, in which case $u = 0$ is a solution and $\|u_d\|_2 = \sigma$. Hence we may assume that $t = 1$ and $\|Ku - u_d\|_2 = \sigma$.

Observe that, if we drop the constraint

$$\int_\Omega Ku = \int_\Omega u_d, \tag{1.19}$$

then the above proof shows that for any $\sigma \leq \|u_d\|_2$ we can find a minimizer $\overline{u}$ of H satisfying $\|K\overline{u} - u_d\|_2 = \sigma$. Moreover we necessarily have that

$$\min_{c\in\mathbb{R}} \|K(\overline{u} + c) - u_d\|_2 = \sigma = \|K\overline{u} - u_d\|_2. \tag{1.20}$$

Indeed, if for some $c \in \mathbb{R}$ we have $\|K(\overline{u} + c) - u_d\|_2 < \|K\overline{u} - u_d\| = \sigma$, then the function $g(t) = \|tK(\overline{u} + c) - u_d\|_2$ is continuous and satisfies $g(1) < \sigma$, $g(0) \geq 0$.

Then there exists some $\bar{t} \in [0,1)$ such that $g(\bar{t}) = \sigma$. Let $v = \bar{t}(\overline{u}+c)$. By definition $\|Kv - u_d\|_2 = \sigma$, and we also have

$$\Phi(v) = \bar{t}\Phi(\overline{u}+c) = \bar{t}\Phi(\overline{u}) < \Phi(\overline{u})$$

which is a contradiction implying that (1.20) must hold, unless $\Phi(\overline{u}) = 0$. If $\Phi(\overline{u}) = 0$, then $\overline{u} = \beta$ is a constant. Then, using $(H2)$ we have $\sigma = \|K\overline{u} - u_d\|_2 = \|\beta - u_d\|_2$. Thus

$$\sigma^2 = \int_\Omega (\beta - u_d)^2 = \beta^2 - 2\beta \int_\Omega u_d + \int_\Omega u_d^2 = \beta^2 + \int_\Omega u_d^2 \geq \beta^2 + \sigma^2,$$

which implies that $\beta = 0$. Notice that in this case (1.20) holds. In any case, (1.20) holds. Since $c = 0$ is a minimum of the function $q(c) = \|K(\overline{u}+c) - u_d\|_2$, then $q'(0) = 0$, which implies that

$$\langle K1, K\overline{u} - u_d\rangle = 0.$$

Therefore, assumption $(H2)$ automatically ensures that the minimizer $\overline{u}$ satisfies (1.19). Thus we may forget about the constraint (1.19).

Since we are assuming $(H3)$, using $f(t)$ we prove that the minimum of Φ in the set $\{u : \|Ku - u_d\|_2 \leq \sigma\}$ is reached for some u with $\|Ku - u_d\|_2 = \sigma$, that satisfies (1.19). Therefore, the problem (1.16) is equivalent to the constrained minimization problem

$$\begin{aligned} &\text{Minimize} \int_\Omega \|Du\|\, dx \\ &\text{with} \quad \int_\Omega |Ku - u_d|^2\, dx \leq \sigma^2|\Omega|, \end{aligned} \tag{1.21}$$

in which the constraint is convex.

Now, we can prove the uniqueness of Ku. If both u and v are solutions of (1.16), then $Ku = Kv$. Actually, we have $\Phi(\frac{u+v}{2}) \leq \frac{1}{2}(\Phi(u) + \Phi(v)) = \min \Phi$ and $\|K(\frac{u+v}{2}) - u_d\|_2 \leq \sigma$, with equality if and only if $Ku = Kv$. To prove this last assertion, notice that if $\|K(\frac{u+v}{2}) - u_d\|_2 = \sigma$, then

$$\left\|\frac{1}{2}(Ku - u_d) + \frac{1}{2}(Kv - u_d)\right\|_2 = \sigma = \left\|\frac{1}{2}(Ku - u_d)\right\|_2 + \left\|\frac{1}{2}(Kv - u_d)\right\|_2$$

and the equality holds in the triangle inequality, which implies that $\frac{1}{2}(Ku - u_d)$ and $\frac{1}{2}(Ku - u_d)$ must be colinear. Having the same norm, then $Ku - u_d = Kv - u_d$ and, therefore, $Ku = Kv$. Now, observe that we cannot have $\|K(\frac{u+v}{2}) - u_d\|_2 < \sigma$. If that would be the case, we would define $g(t) = \|tK(\frac{u+v}{2}) - u_d\|_2$ and arguing as above we deduce either that there is some $\bar{t} \in [0,1)$ for which $\bar{t}\frac{u+v}{2}$ has an energy below the infimum of (1.21), which is a contradiction (and thus $Ku = Kv$ holds), or $u = v = 0$, in which case also $Ku = Kv$.

To continue with the proof, we need the following result. □

Proposition 1.2. *If u is a solution of* (1.12), *then there is some $\lambda \geq 0$ such that*

$$-\lambda K^t(Ku - u_d) \in \partial\Phi(u). \tag{1.22}$$

Before proving it, let us recall the following result whose proof can be found in [58], Corollary 2.11.

Proposition 1.3. *Suppose that φ and ψ are two convex, lower semi-continuous and proper functions defined on a Hilbert space. If $D(\varphi) \cap Int(D(\psi)) \neq \emptyset$, then*

$$\partial(\varphi + \psi) = \partial\varphi + \partial\psi.$$

Proof of Proposition 1.2. Denote by $\overline{B}$ the closed unit ball of $L^2(\Omega)$. Set

$$G(u) = \chi_{u_d + \sigma\overline{B}} = \begin{cases} +\infty & \text{if} \quad u \notin u_d + \sigma\overline{B} \Leftrightarrow \|u - u_d\|_2 > \sigma \\ \\ +0 & \text{if} \quad u \in u_d + \sigma\overline{B} \Leftrightarrow \|u - u_d\|_2 \leq \sigma. \end{cases} \tag{1.23}$$

Φ and G are convex lower semi-continuous functions and problem (1.21) is equivalent to minimizing $\Phi(u) + G(Ku)$. We have $D(\Phi) = \{u : \Phi(u) < \infty\} = BV(\Omega)$ and $D(G) = \{u : G(u) < \infty\} = u_d + \sigma\overline{B}$, and, as we assumed that $u_d \in \overline{K(D(\Phi))}$, there exists $\tilde{u} \in D(\Phi)$ with $\|K\tilde{u} - u_d\| < \frac{\sigma}{2}$. Observe that $\tilde{u} \in D(\Phi) \cap \operatorname{Int}(D(G \circ K))$, and, therefore, by Proposition 1.3, we have for all u

$$\partial(\Phi + G \circ K)(u) = \partial\Phi(u) + \partial(G \circ K)(u).$$

Moreover, as G is continuous at $K\tilde{u}$, we have for all u,

$$\partial(G \circ K)(u) = K^t \partial G(Ku)$$

with $\partial G(u) = \{0\}$ if $\|u - u_d\|_2 < \sigma$ and $\partial G(u) = \{\lambda(u - u_d) : \lambda \geq 0\}$ if $\|u - u_d\|_2 = \sigma$ ([109], Proposition 5.7). In particular,

$$\partial(\Phi + G \circ K)(u) = \partial\Phi(u) + K^t \partial G(Ku).$$

If u is a solution of (1.16) and thus of (1.21), then $0 \in \partial(\Phi + G \circ K)(u)$. As any solution of (1.16) satisfies $\|Ku - u_d\|_2 = \sigma$, this shows that

$$\exists\lambda \geq 0, 0 \in \partial\Phi(u) + \lambda K^t(Ku - u_d) \Leftrightarrow \exists\lambda \geq 0, -\lambda K^t(Ku - u_d) \in \partial\Phi(u). \qquad \square$$

Note that the last result implies that for this $\lambda \geq 0$, u is a minimizer of the convex functional $\Phi(u) + \frac{\lambda}{2}\|Ku - u_d\|_2$ which is the functional corresponding to problem (1.17). Conversely, a minimizer u of this functional is obviously a solution of (1.16) for the particular value $\sigma = \|Ku - u_d\|_2$. This establishes the equivalence between problems (1.16) (with $0 < \sigma \leq \|u_d - \int_\Omega u_d\|_2$) and (1.17) (with $\lambda \geq 0$). The only assertion of Theorem 1.1 remaining to be proved is that the correspondence between σ and λ is unique as soon as $\|u_d - \int_\Omega u_d\|_2 > \sigma$. This will be the content of Lemma 1.4.

First of all observe that when $\lambda \geq 0$, problem (1.17) has a solution $u^\lambda \in BV(\Omega)$ which is unique as soon as K is injective. Notice that the solutions of (1.17) for $\lambda > 0$ satisfy the constraint (1.19). Thus as $\lambda \to 0$, u^λ converges to a solution of (1.17) for $\lambda = 0$ that satisfies (1.19). Thus, for $\lambda = 0$ we need to add explicitly the condition (1.19), otherwise any constant would be a solution of the problem.

Because of the strict convexity in Ku of the term $\|Ku - u_d\|_2$, it is straightforward to check that Ku^λ is unique, even if u^λ is not. This implies that the function $\sigma(\lambda) = \|Ku^\lambda - u_d\|_2$ is well defined. The behaviour of $\sigma(\lambda)$ is described in the following lemma.

Lemma 1.4. *The function $\sigma(\lambda)$ is a nondecreasing and continuous function. It maps $[0,\infty)$ onto $(0, \|u_d - \int_\Omega u_d\|_2]$. Moreover, there exits $\underline{\lambda} \geq 0$ such that $\sigma(\lambda)$ is strictly decreasing over $[\underline{\lambda}, +\infty)$, and $\sigma(\lambda) = \|u_d - \int_\Omega u_d\|_2$ if $0 \leq \lambda \leq \underline{\lambda}$.*

For convenience, we shall prove it at the end of the next section. With it, the proof of Theorem 1.1 is concluded.

1.3 The Partial Differential Equation Satisfied by the Minimum of (1.17)

Our task will be to give a sense to (1.22) as a partial differential equation, describing the subdifferential of Φ in a distributional sense. To be precise we should not say distributional sense since the test functions will be functions in $BV(\Omega)$. The proof of Proposition 1.10 can be found in [13],[14], [48],[79], [198]. We shall give a simple approach to the characterization of $\partial\Phi$ which is due to F. Alter [3].

Let E be a normed space, and let E^* be its dual space. Let $\Psi : E \to [0,\infty]$ be any function. Let us define $\tilde{\Psi} : E^* \to [0,\infty]$ by

$$\tilde{\Psi}(x^*) = \sup\left\{ \frac{\langle x^*, y\rangle}{\Psi(y)} \ : \ y \in E \right\} \tag{1.24}$$

with the convention that $\frac{0}{0} = 0$, $\frac{0}{\infty} = 0$. Note that $\tilde{\Psi}(x^*) \geq 0$, for any $x^* \in E^*$. Note also that the supremum is attained on the set of $y \in E^*$ such that $\langle x^*, y\rangle \geq 0$. Note also that we have the following Cauchy–Schwartz inequality

$$\langle x^*, y\rangle \leq \tilde{\Psi}(x^*)\Psi(y) \quad \text{if } \Psi(y) > 0.$$

The following lemma is a simple consequence of the above definition.

Lemma 1.5. *Let $\Psi_1, \Psi_2 : E \to [0,\infty]$. If $\Psi_1 \leq \Psi_2$, then $\tilde{\Psi}_2 \leq \tilde{\Psi}_1$.*

Proposition 1.6. *If Ψ is convex, lower semi-continuous and positive homogeneous of degree* 1, *then $\tilde{\tilde{\Psi}}|E = \Psi$.*

Proof. Since $\frac{\langle y^*, x\rangle}{\Psi(x)} \le \tilde{\Psi}(y^*)$ for any $x \in E$, $y^* \in E^*$, we also have that $\frac{\langle y^*, x\rangle}{\tilde{\Psi}(y^*)} \le \Psi(x)$ for any $x \in E$, $y^* \in E^*$. This implies that $\tilde{\tilde{\Psi}}(x) \le \Psi(x)$ for any $x \in E$. Assume that there is some $x \in E$ and $\epsilon > 0$ such that $\tilde{\tilde{\Psi}}(x) + \epsilon < \Psi(x)$, hence, in particular, $\Psi(x) > 0$ and $\tilde{\tilde{\Psi}}(x) < \infty$. Using Hahn–Banach's theorem, there is a linear form $y^* \in E^*$ separating x from the closed convex set $C := \{z \in E : \Psi(z) \le \tilde{\tilde{\Psi}}(x) + \epsilon\}$. Since $0 \in C$ we may even assume that $\langle y^*, x\rangle = 1$ and $\langle y^*, z\rangle \le \alpha < 1$ for any $z \in C$. Note that, from the definition of $\tilde{\tilde{\Psi}}$, we have

$$\tilde{\tilde{\Psi}}(x) \ge \frac{1}{\tilde{\Psi}(y^*)}. \tag{1.25}$$

Let us prove that $\tilde{\Psi}(y^*) \le \frac{1}{\tilde{\tilde{\Psi}}(x)+\epsilon}$. For that it will be sufficient to prove that

$$\frac{\langle y^*, z\rangle}{\Psi(z)} \le \frac{1}{\tilde{\tilde{\Psi}}(x) + \epsilon} \tag{1.26}$$

for any $z \in E$ such that $\langle y^*, z\rangle \ge 0$. Let $z \in E$, $\langle y^*, z\rangle \ge 0$. If $\Psi(z) = \infty$, then (1.26) holds. If $\Psi(z) = 0$, then also $\Psi(tz) = 0$ for any $t \ge 0$. Hence $tz \in C$ for all $t \ge 0$, and we have that $0 \le \langle y^*, tz\rangle \le 1$ for all $t \ge 0$. Thus $\langle y^*, z\rangle = 0$, and, therefore, (1.26) holds. Finally, assume that $0 < \Psi(z) < \infty$. Let $t > 0$ be such that $\Psi(tz) = \tilde{\tilde{\Psi}}(x) + \epsilon$. Using that $tz \in C$, we have

$$\frac{\langle y^*, z\rangle}{\Psi(z)} = \frac{\langle y^*, tz\rangle}{\Psi(tz)} \le \frac{1}{\tilde{\tilde{\Psi}}(x) + \epsilon}.$$

Both (1.25) and (1.26) give a contradiction. We conclude that $\tilde{\tilde{\Psi}}(x) = \Psi(x)$ for any $x \in E$. □

Lemma 1.7. *Assume that Ψ is convex, lower semi-continuous and positive homogeneous of degree* 1. *If $u \in D(\partial\Psi)$ and $v^* \in \partial\Psi(u)$, then $\langle v^*, u\rangle = \Psi(u)$.*

Proof. Indeed, if $v^* \in \partial\Psi(u)$, then

$$\langle v^*, x - u\rangle \le \Psi(x) - \Psi(u), \quad \text{for all } x \in E.$$

To obtain the result it suffices to take $x = 0$ and $x = 2u$ in the above inequality. □

Theorem 1.8. *Assume that Ψ is convex, lower semi-continuous and positive homogeneous of degree* 1. *Then $v^* \in \partial\Psi(u)$ if and only if $\tilde{\Psi}(v^*) \le 1$ and $\langle v^*, u\rangle = \Psi(u)$ (hence, $\tilde{\Psi}(v^*) = 1$ if $\Psi(u) > 0$).*

Proof. When $\langle v^*, u\rangle = \Psi(u)$, condition $v^* \in \partial\Psi(u)$ may be written as $\langle v^*, x\rangle \leq \Psi(x)$ for all $x \in E$, which is equivalent to $\tilde{\Psi}(v^*) \leq 1$. □

Let Ω be a bounded domain in $\mathbb{R}^N$ with Lipschitz boundary. Let us consider the space (see Appendix C)

$$X(\Omega)_2 := \left\{z \in L^\infty(\Omega, \mathbb{R}^N) \ : \ \operatorname{div}(z) \in L^2(\Omega)\right\}.$$

Let us define, for $v \in L^2(\Omega)$,

$$\Psi(v) = \inf\left\{\| z \|_\infty : \ z \in X(\Omega)_2, \, v = -\operatorname{div}(z) \text{ in } \mathcal{D}'(\Omega), \, [z,\nu] = 0\right\}, \tag{1.27}$$

where ν denotes the outward unit normal to $\partial\Omega$ and $[z,\nu]$ is the trace of the normal component of z (see Appendix C).

Observe that Ψ is convex, lower semi-continuous and positive homogeneous of degree 1. Moreover, it is easy to see that, if $\Psi(v) < \infty$, the infimum in (1.27) is attained, i.e., there is some $z \in X(\Omega)_2$ such that $v = -\operatorname{div}(z)$ in $\mathcal{D}'(\Omega)$, $[z,\nu] = 0$ and $\Psi(v) = \|z\|_\infty$.

Proposition 1.9. *We have that* $\Psi = \tilde{\Phi}$.

Proof. Let $v \in L^2(\Omega)$. If $\Psi(v) = \infty$, then we have $\tilde{\Phi}(v) \leq \Psi(v)$. Thus, we may assume that $\Psi(v) < \infty$. Let $z \in X(\Omega)_2$ be such that $v = -\operatorname{div}(z)$ and $[z,\nu] = 0$. Then

$$\int_\Omega vu \, dx = \int_\Omega (z, Du) \leq \| z \|_\infty \Phi(u) \quad \text{for all } u \in BV(\Omega) \cap L^2(\Omega).$$

Taking supremums in u we obtain $\tilde{\Phi}(v) \leq \| z \|_\infty$. Now, taking infimums in z, we obtain $\tilde{\Phi}(v) \leq \Psi(v)$.

To prove the opposite inequality, let us denote

$$D = \left\{\operatorname{div}(z) \ : \ z \in C_0^\infty(\Omega, \mathbb{R}^N)\right\}.$$

Then

$$\begin{aligned}
\sup_{v \in L^2} \frac{\int_\Omega uv\, dx}{\Psi(v)} \ &\geq \ \sup_{v \in D} \frac{\int_\Omega uv\, dx}{\Psi(v)} \geq \sup_{v \in D, \Psi(v) < \infty} \frac{\int_\Omega uv\, dx}{\Psi(v)} \\
&\geq \sup_{z \in C_0^\infty(\Omega, \mathbb{R}^N)} \frac{-\int_\Omega u\operatorname{div}(z)\, dx}{\| z \|_\infty} \geq \Phi(u).
\end{aligned}$$

Thus, $\Phi \leq \tilde{\Psi}$. This implies that $\tilde{\tilde{\Psi}} \leq \tilde{\Phi}$, and, using Proposition 1.6, we obtain that $\Psi \leq \tilde{\Phi}$. □

Proposition 1.10. *Let $u, v \in L^2(\Omega)$, $u \in BV(\Omega)$. The following assertions are equivalent:*

(a) $v \in \partial\Phi(u)$,

(b) *we have*

$$\int_\Omega vu\, dx = \Phi(u), \tag{1.28}$$

$$\exists z \in X(\Omega)_2,\ \|z\|_\infty \leq 1, \text{ such that } v = -\text{div}(z) \quad \text{in } \mathcal{D}'(\Omega), \tag{1.29}$$

and

$$[z, \nu] = 0 \quad \text{on } \partial\Omega. \tag{1.30}$$

(c) (1.29) *and* (1.30) *hold and*

$$\int_\Omega (z, Du) = \int_\Omega \|Du\|. \tag{1.31}$$

Proof. By Theorem 1.8, we have that $v \in \partial\Phi(u)$ if and only if $\tilde{\Phi}(v) \leq 1$ and $\int_\Omega vu\, dx = \Phi(u)$. Since $\tilde{\Phi} = \Psi$, the equivalence of (*a*) and (*b*) follows from the definition of Ψ and the observation following it. If (*b*) holds, integrating by parts in (1.28) and using (1.30) we obtain (1.31). The converse implication follows in the same way. □

Thus, according to the above results, the equation $v = -\text{div}\left(\frac{Du}{|Du|}\right)$ must be understood in the following sense: there exists a vector field $z \in X(\Omega)_2$ with $\|z\|_\infty \leq 1$, such that $v = -\text{div}(z)$ in $\mathcal{D}'(\Omega)$, satisfying (1.31). The boundary condition will be written in the form $[z, \nu] = 0$ in $\partial\Omega$. Hence the minimum u of (1.17) is a solution of

$$\begin{cases} -\text{div}\left(\dfrac{Du}{|Du|}\right) + \lambda K^t(Ku - u_d) = 0 & \text{in } \Omega, \\ [z, \nu] = 0 & \text{on } \partial\Omega. \end{cases} \tag{1.32}$$

Proof of Lemma 1.4. Let $\lambda > \mu \geq 0$. We have

$$\int_\Omega \|Du^\lambda\| + \frac{\lambda}{2}\|Ku^\lambda - u_d\|^2 \leq \int_\Omega \|Du^\mu\| + \frac{\lambda}{2}\|Ku^\mu - u_d\|^2 \tag{1.33}$$

and

$$\int_\Omega \|Du^\mu\| + \frac{\mu}{2}\|Ku^\mu - u_d\|^2 \leq \int_\Omega \|Du^\lambda\| + \frac{\mu}{2}\|Ku^\lambda - u_d\|^2. \tag{1.34}$$

Combining both inequalities, we get

$$(\lambda - \mu)\sigma(\lambda)^2 \leq (\lambda - \mu)\sigma(\mu)^2$$

and this shows that $\sigma(\cdot)$ is not increasing.

Notice that $u^0 = \int_\Omega u_d$ and $\sigma(0) = \|u_d - \int_\Omega u_d\|$. Let us observe that $\sigma(\lambda) \to 0+$ as $\lambda \to \infty$. Indeed, by the first assertion of Theorem 1.1, for any $\sigma \in (0, \|u_d - \int_\Omega u_d\|]$, problem (1.16) has a solution u which satisfies (1.22) for some $\lambda \geq 0$, and, therefore, u is a minimizer of (1.17). In other words, $\sigma(\lambda) = \sigma$. This proves that $\sigma(\lambda) \to 0+$ as $\lambda \to \infty$. Notice that the continuity of $\sigma(\lambda)$ also follows.

To prove that $\sigma(\cdot)$ is strictly decreasing, assume on the contrary that there exist $\lambda < \mu$ such that $\sigma(\lambda) = \sigma(\mu)$. Equations (1.33) and (1.34) show this time that $\Phi(u^\lambda) = \Phi(u^\mu)$ and, in fact, that u^λ is a solution of (1.17) for any $\lambda' \in [\lambda, \mu]$. This means that

$$\forall \lambda' \in [\lambda, \mu], \quad -\lambda' K^t \left(K u^\lambda - u_d\right) \in \partial\Phi(u^\lambda), \tag{1.35}$$

or in other words, u^λ satisfies the equation

$$\forall \lambda' \in [\lambda, \mu], \quad -\mathrm{div}\left(\frac{Du^\lambda}{|Du^\lambda|}\right) + \lambda' K^t \left(K u^\lambda - u_d\right) = 0, \tag{1.36}$$

and the Neumann boundary conditions. Multiplying (1.36) by u^λ and integrating by parts, we deduce that

$$\forall \lambda' \in [\lambda, \mu], \quad \int_\Omega \|Du^\lambda\| = -\lambda' \langle K u^\lambda - u_d, K u^\lambda \rangle \tag{1.37}$$

which implies that $\int_\Omega \|Du^\lambda\| = 0$ and, therefore, $u^\lambda = \int_\Omega u_d$. Hence $\sigma(\lambda) = \|u_d - \int_\Omega u_d\|_2$.

We have established that if u is a solution of (1.17) for both λ and μ, $\lambda < \mu$, then $\Phi(u) = 0$. The consequence of this fact is that $\sigma(\lambda)$ has to be strictly decreasing, except possibly on $[0, \underline{\lambda}]$ for some $\underline{\lambda} \geq 0$, where it takes the value $\|u_d - \int_\Omega u_d\|_2$. □

Let us finally observe that it is possible to have $\underline{\lambda} > 0$ in the last lemma. Indeed, as in Lemma 3.3, 0 is a solution of (1.17) if and only if there exists a vector field $\xi \in L^\infty(\Omega, \mathbb{R}^2)$ with $\|\xi\|_\infty \leq 1$ such that

$$-\mathrm{div}(\xi) + \lambda K^t u_d = 0.$$

Let $V := \{\mathrm{div}(\xi) : \xi \in L^\infty(\Omega, \mathbb{R}^2),\ \|\xi\|_\infty \leq 1\}$. Then, $\underline{\lambda}$ may be defined as

$$\underline{\lambda} = \max\{\lambda : \lambda K^t u_d \in V\}.$$

Now, let f be any smooth function defined on $\partial\Omega$ and let v be the solution of

$$\begin{cases} \Delta v = K^t u_d & \text{in} \quad \Omega, \\ v = f & \text{in} \quad \partial\Omega. \end{cases} \tag{1.38}$$

If $K^t u_d \in L^p(\Omega)$ with $p > 2$, then $v \in W^{2,p}(\Omega) \subseteq C^1(\overline{\Omega})$. Consequently we obtain

$$\underline{\lambda} \geq \frac{1}{\|\nabla v\|_\infty}.$$

1.4 Algorithm and Numerical Experiments

1.4.1 Description of the Numerical Algorithm

Let us describe the algorithm we have used in our numerical experiments. In the continuous domain we assume that the image u is defined on the square $\Omega = [0,1] \times [0,1]$ and the restoration functional is

$$\text{Minimize}_{u \in BV(\Omega)} \int_\Omega \|Du\|\, dx + \frac{\lambda}{2} \int_\Omega |Ku - u_d|^2\, dx, \tag{1.39}$$

where the convolution kernel K and the data u_d are known. The corresponding Euler–Lagrange equation is (1.32). Since the total variation is not differentiable at 0, we need to approximate it by a differentiable functional. The usual approximation of the energy functional (1.39) is

$$\text{Minimize} \int_\Omega \sqrt{\epsilon^2 + \|Du\|^2}\, dx + \frac{\lambda}{2} \int_\Omega |Ku - u_d|^2\, dx, \quad \epsilon > 0, \tag{1.40}$$

whose associated Euler–Lagrange equation is

$$\begin{cases} -\operatorname{div}\left(\dfrac{Du}{\sqrt{\epsilon^2 + |Du|^2}} \right) + \lambda K^t(Ku - u_d) = 0 & \text{in } \Omega, \\[2ex] \dfrac{Du}{\sqrt{\epsilon^2 + |Du|^2}} \cdot \nu = 0 & \text{on } \partial\Omega. \end{cases} \tag{1.41}$$

For later use, let us denote the energy in (1.40) by $E_\epsilon(u)$. We note that the nondifferentiability at 0 of the total variation term is at the origin of the staircasing effect (ramps are transformed in stairs) which is visible in the numerical experiments performed with (1.39) ([162]).

To proceed with the discrete numerical algorithm, we assume that we have an image u defined on $\{0, 1, ..., N-1\} \times \{0, 1, ..., N-1\}$. We replace the gradient by a discrete approximation: we shall use the notation

$$\nabla^{+,+} u = (\nabla_x^+ u, \nabla_y^+ u), \quad \nabla^{+,-} u = (\nabla_x^+ u, \nabla_y^- u),$$
$$\nabla^{-,+} u = (\nabla_x^- u, \nabla_y^+ u), \quad \nabla^{-,-} u = (\nabla_x^- u, \nabla_y^- u)$$

where

$$\nabla_x^+ u(i,j) = u(i+1,j) - u(i,j), \quad \nabla_x^- u(i,j) = u(i,j) - u(i-1,j),$$
$$\nabla_y^+ u(i,j) = u(i,j+1) - u(i,j), \quad \nabla_y^- u(i,j) = u(i,j) - u(i,j-1).$$

In principle, we could use any of the approximations

$$\sum_{i,j=0}^{N-1} \sqrt{\epsilon^2 + |\nabla^{\pm,\pm} u(i,j)|^2} + \frac{\lambda}{2} \sum_{i,j=0}^{N-1} |Ku(i,j) - u_d(i,j)|^2,$$

but we observe that using only one of the approximations produces some artifacts due to the interaction of the approximation with the data, in particular, with the noise. Thus the best is either to use alternatively all of them, or to use the discrete functional

$$E_\epsilon(u) := \frac{1}{4} \sum_{\alpha,\beta=+,-} \sum_{i,j=0}^{N-1} \sqrt{\epsilon^2 + |\nabla^{\alpha,\beta} u(i,j)|^2} + \frac{\lambda}{2} \sum_{i,j=0}^{N-1} |Ku(i,j) - u_d(i,j)|^2. \quad (1.42)$$

Note that the dual operators to $\nabla^{+,+}$, $\nabla^{+,-}$, $\nabla^{-,+}$, $\nabla^{-,-}$ are, respectively, the operators $\mathrm{div}^{-,-}$, $\mathrm{div}^{-,+}$, $\mathrm{div}^{+,-}$, $\mathrm{div}^{+,+}$. Applying the gradient descent method, to minimize (1.42) we iteratively actualize the solution using the equation

$$u^{n+1} = u^n + \tfrac{\Delta t}{4} \sum_{\alpha,\beta=+,-} \mathrm{div}^{\alpha *,\beta *} \left(\frac{\nabla^{\alpha,\beta} u^n}{\sqrt{\epsilon^2 + |\nabla^{\alpha,\beta} u^n|^2}} \right) - \Delta t \lambda K^t (K u^n - u_d), \quad (1.43)$$

where $div^{\alpha *,\beta *}$ denotes the dual operator of $\nabla^{\alpha,\beta}$. In practice, it is very important to guarantee that the energy decreases along the evolution, i.e., that

$$E_\epsilon(u^{n+1}) \leq E_\epsilon(u^n). \quad (1.44)$$

For that we have to control the time increment Δt. Indeed, as in [155], at each iteration, we only accept Δt if (1.44) holds. We have the possibility to choose Δt so that the energy has the largest decrease in the direction of the energy gradient. If $\nabla E_\epsilon(u)$ denotes the energy gradient which is given in the right-hand side of (1.43), then Δt can be chosen as a solution of

$$\min_{s>0} E_\epsilon(u^n - s\nabla E_\epsilon(u^n)). \quad (1.45)$$

In practice one observes that long term decreasing of the energy is favored not by the optimal choice of Δt but by choosing a constant value of Δt and decreasing it when (1.44) is not satisfied ([155]). Many authors have proposed and studied numerical methods to minimize (1.39) [175], [64], [201, 202], [71, 72], [108], [152], some of them being implicit methods. Indeed, for this problem, implicit methods are accurate and well founded to obtain convergence to the solution, but are time consuming. That reason caused us to choose an explicit method to solve (1.39).

In case that we want to compute the parameter λ which satisfies the constraint (1.13) we may use Uzawa's method [80] in order to update the parameter λ. The algorithm is:

(i) Initially, take $\lambda > 0$ small enough so that

$$Q(u^\lambda) := \frac{1}{|\Omega|} \int_\Omega (Ku^\lambda - u_d)^2 > \sigma^2. \quad (1.46)$$

(ii) For each value $\lambda > 0$, we solve iteratively (1.43), until we reach the asymptotic state u^λ.

(iii) Recompute $\lambda = \max(\lambda + \rho(Q(u^\lambda) - \sigma^2), 0)$ (with $\rho > 0$ small enough) and iterate (ii) and (iii) until λ does not change significantly.

Observe that λ will be increased (the constraint is enforced) if the empirical value $Q(u^\lambda)$ is above the expected value σ^2 and decreased (the constraint is relaxed) if it is below. For a detailed study of the problem of computing the value of the Lagrange multiplier λ satisfying (1.13) we refer to [200].

1.4.2 Description of the Data and Experiments

In case of satellite SPOT images, the operator K is well approximated by a convolution kernel k, called the impulse response of the system. Let us recall the form of k in case of satellite SPOT images. The Fourier transform $\hat{k}$ of k, called the *modulated transfer function* (MTF), is the product of three MTFs corresponding to the imperfection of the optics, the imperfection and the size of the CCD detectors, and the motion of the satellite [172].

For simplicity, we shall consider a model of degradations which corresponds to the satellite SPOT 5. The Fourier transform of the impulse response k is supported on $[-\frac{1}{2}, \frac{1}{2}]$. The convolution kernel is given by

$$\hat{k}(\xi, \eta) = e^{-2\gamma_\xi|\xi| - 2\gamma_\eta|\eta|} \left(\frac{\sin(2\pi\xi)}{2\pi\xi}\right)\left(\frac{\sin(2\pi\eta)}{2\pi\eta}\right)\left(\frac{\sin(\pi\xi)}{\pi\xi}\right), \tag{1.47}$$

$\xi, \eta \in [-\frac{1}{2}, \frac{1}{2}]$, where $\gamma_\xi = 1.505$, $\gamma_\eta = 1.412$. Observe that the kernel vanishes only at the boundary of the Fourier domain, hence K is injective. The case of other models corresponding to kernels which vanish at intermediate frequencies was also considered in [108], [48].

We have assumed that the noise is a white Gaussian noise of standard deviation σ. Following [172], the signal to noise ratio, or SNR, can be approximated by the quotient between the average of the signal and the standard deviation of the signal for the average luminance

$$SNR = \frac{\text{Average}}{\sigma}.$$

Thus $\sigma = \frac{\text{Average}}{SNR}$. In practice, for SPOT images, the values of $SNR = 50, 100$ are realistic and give rise to the same difficulty as the real noise.

The restored images will be compared to a reference image. The blurred images have been created artificially from an airplane image. But, in order to take into account all the degradations suffered by the images, including aliasing, the airplane image is over-sampled with respect to the blurred observed image. Consequently, the restored image cannot be directly compared to the sharp airplane image, since they have not the same scale of sampling. Thus, the reference image is constructed from the over-sampled airplane image by convolving it with a prolate function followed by a sub-sampling [172, 108]. The reference image has an optimal resolution without aliasing and is the best we can hope to obtain from the degraded ones. The degraded images are constructed following the models described above. Figure 1.1 displays the reference image. Figure 1.2 displays the degraded image constructed using the convolution kernel of model (1.47) and a noise of standard deviation $\sigma = 1$, which corresponds to an $SNR = 50$.

Figure 1.3 displays the result of restoring Figure 1.2 using (1.43) with λ computed automatically. Figure 1.4 displays some details of it. We may conclude that total variation based image restoration is able to recover a good estimate of the original uncorrupted image, at least for Gaussian noise. The results for the restoration problem show that the method is able to recover the edges of the image, and also some small details but has more difficulties to recover some oscillating patterns, as textures, and has the inconvenience of somewhat enhancing the noise when the parameter is chosen so that texture is not eliminated.

Figure 1.1: Reference satellite image.

Figure 1.2: Degraded image using the convolution kernel of model (1.47) and white noise of standard deviation $\sigma = 1$.

Figure 1.3: Restoration of the degraded image of Fig. 1.2

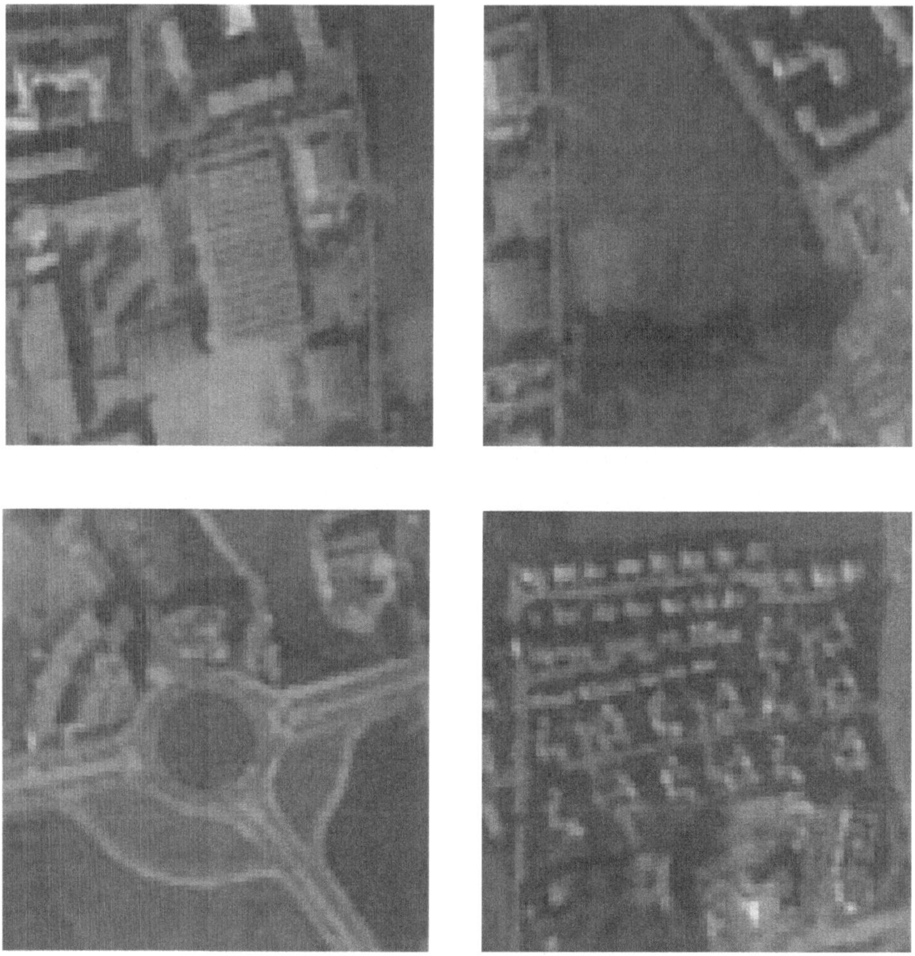

Figure 1.4: Four different zoomed details of the restored image of Fig. 1.3. Top left, textured area, top right, more flat zone. Bottom left, detail of the road, bottom right, detail of another textured zone.

1.5 Review of Numerical Methods

Total variation image restoration was initially proposed by L. Rudin, S. Osher and E. Fatemi in their seminal paper [175]. The authors proposed to choose from all functions satisfying the constraints given by the image acquisition model the one which minimizes total variation. They used a steepest descent method to solve (1.41). The Lagrange multiplier λ was updated at each iteration using the gradient-projection method of Rosen [171]. The mathematical analysis of this method was pursued in [146], which, remains unpublished.

The introduction of total variation as a smoothness constraint was motivated by the choice of BV functions as the underlying functional model for images. The question of which is the right functional space to describe images has motivated much research [5], [126], [9], [151], [153], [159]. From a practical point of view, the answer may depend on the application. Thus, BV-functions have been used in image denoising and restoration [175], [153], [82], [81], segmentation and edge detection [87], [86] (see also [158], [156]), and inspired many other applications (see the references). Upper semi-continuous functions constitute the right function lattice to develop mathematical morphology [180, 181]. The use of Besov spaces is common in image compression [92], [65], [81], [102, 103, 104, 105, 106, 107], [151], [153].

Since the steepest descent method used in our experiments has a linear convergence rate (see [161]) many authors have proposed different algorithms to solve (1.32). Our purpose in this section is to review their work.

C.R. Vogel and M.E. Oman proposed a lagged diffusivity fixed point iteration to solve (1.32) in case of the denoising problem [201] or in the general restoration problem [202, 203, 204]. For that they use the approximating functional (1.40) whose Euler–Lagrange equation is (1.41). Writing

$$L(u)v = -\text{div}\left(\frac{Dv}{\sqrt{|Du|^2+\epsilon^2}}\right),$$

we may write (1.41) as

$$\lambda K^*(Ku - u_d) + L(u)u = 0 \tag{1.48}$$

supplemented with Neumann boundary conditions. Then Vogel and Oman propose to solve (1.48) by means of the fixed point iteration

$$[K^*K + \frac{1}{\lambda}L(u^m)]u^{m+1} = K^*u_d, \quad m = 0, 1, ... \tag{1.49}$$

In practice one starts with $u^0 = u_d$. This system is solved using a preconditioned conjugate gradient method PGC and it is described in detail in [202, 203, 204, 205]. Even if the rate of convergence is linear, the authors have found it to be quite rapid in practice [202, 203, 204]. The analysis of the convergence of this numerical scheme for the denoising problem was studied in [100] proving that the algorithm is locally convergent with a linear rate of convergence. The lagged diffusivity method can be viewed as a special case of the half quadratic regularization of D. Geman and G. Reynolds [116] and the ARTUR scheme of Charbonnier, Blanc-Feraud, Aubert and Barlaud [77] and in this more general context there exist several proofs of its convergence [116, 77, 64, 70]. Again, in the case of the denoising problem, a Newton type method was proposed in [201] which had the inconvenience of having a small domain of convergence when ϵ is small. This was later improved by T.F. Chan, H.M. Zhou and R.H. Chan in [67] by using a continuation method for choosing ϵ. A theoretical analysis of the existence, uniqueness and stability under several perturbations of solutions of (1.17) was given in [1]. A full account of some of the numerical approaches to total variation regularization, including methods to select the parameter λ, can be found in the book [200].

To improve the rate of convergence T. Chan, G. Golub and P. Mulet proposed a linearization of (1.41) based on a dual variable [71, 72, 68]. The idea is to introduce

$$w_\epsilon = \frac{Du}{\sqrt{\epsilon^2 + |Du|^2}}$$

as a new variable and replace (1.41) by the following system of equations in the (u, w) variables:

$$\begin{cases} -\operatorname{div} w_\epsilon + \lambda K^t(Ku - u_d) = 0 & \text{in } \Omega, \\ w_\epsilon \sqrt{\epsilon^2 + |Du|^2} - Du = 0 & \text{in } \Omega, \\ w_\epsilon \cdot \nu = 0 & \text{on } \partial\Omega. \end{cases} \tag{1.50}$$

This system of equations can be obtained if we write the dual problem, and the solution for both problems, primal and dual, is given by (1.50). Indeed, for simplicity, let us consider the case where $\epsilon = 0$ and write w instead of w_0. If

$$\Phi(u, w) = \int_\Omega u \operatorname{div}(w) + \frac{\lambda}{2}(Ku - u_d)^2 \, dxdy,$$

then

$$\min_u \Phi(u) + \frac{\lambda}{2} \int_\Omega (Ku - u_d)^2 \, dxdy = \min_u \min_{w \in \mathcal{V}} \Phi(u, w),$$

where $\mathcal{V} = \{w \in X_2(\Omega) : \| w \|_\infty \leq 1\}$. By using arguments of convex programming [109] we have

$$\min_u \min_{w \in \mathcal{V}} \Phi(u, w) = \Phi(u^*, w^*) = \min_{w \in \mathcal{V}} \min_u \Phi(u, w) \tag{1.51}$$

and w^* is the solution of the dual problem

$$\sup_{w \in \mathcal{V}} \Psi(w),$$

where

$$\Psi(w) = \min_u \Phi(u, w).$$

The solution (u^*, w^*) of (1.51) is given by the solution of (1.50) ([71]). The authors solve (1.50) using an approximate Newton's method [71, 72, 68] and show that it is quadratically convergent.

Let us mention the efforts by Strong and Chan to define scale dependent total variation denoising [185], [186], [187] which is directed to find a relation between the size of the objects and the scale of regularization. To construct an effective method for $2D$ images they proposed a model

$$\int_\Omega \lambda(x)\|Du\| + \int_\Omega |u - u_d|^2 \, dx,$$

where $\lambda(x)$ was taken essentially as proportional to $\frac{1}{\epsilon + |Du_e(x)|}$ and u_e is an estimated version of the restored image obtained by a previous total variation restoration [185].

In [116] D. Geman and G. Reynolds proposed to minimize the functional

$$J(u) = \sum_{i,j} [\varphi(\nabla_x^+ u(i,j)) + \varphi(\nabla_y^+ u(i,j))] + \lambda \parallel Ku - u_d \parallel_2^2, \tag{1.52}$$

where the function $\varphi : [0, \infty) \to \mathbb{R}$ was chosen such that $\varphi(\sqrt{t})$ is concave, $\varphi(0) = -1$, and $\lim_{t \to +\infty} \varphi(t) = 0$. These conditions allow us to reconstruct sharp transitions between distinct regions of the image. Moreover to minimize (1.52) the authors proposed a half-quadratic regularization method [116, 117]. According to the authors [116, 117], the basic idea is to introduce a new objective function which, although defined over a extended domain, has the same minimum in u as J and can be manipulated with linear algebraic methods. For that, the authors constructed a function ψ defined on an interval $(0, M]$ such that

$$\varphi(t) = \inf_{0 < w \leq M} (wt^2 + \psi(w)) \qquad \forall t \geq 0, \tag{1.53}$$

which permitted definition of a functional of the form

$$J^*(u, b_x, b_y) =$$

$$\sum_{i,j} [\varphi^*(\nabla_x^+ u(i,j), b_x(i,j)) + \varphi^*(\nabla_y^+ u(i,j), b_y(i,j))] + \lambda \parallel Ku - u_d \parallel_2^2,$$

where $\varphi^*(t, w) = wt^2 + \psi(w)$, which satisfies

$$J(u) = \inf_{(b_x, b_y)} J^*(u, b_x, b_y).$$

In [77] P. Charbonnier, L. Blanc-Feraud, G. Aubert and M. Barlaud studied the conditions that φ must satisfy to have an edge preserving regularization method.

Moreover, under some assumptions on φ (classified as basic assumptions: (a) $\varphi(t) \geq 0\ \forall t \geq 0$, with $\varphi(0) = 0$, (b) $\varphi(t) = \varphi(-t)$, (c) φ is continuously differentiable, (d) $\varphi'(t) \geq 0\ \forall t \geq 0$, and edge preservation assumptions : (e) $\frac{\varphi'(t)}{2t}$ is continuous and strictly decreasing on $[0,\infty)$, (f) $\lim_{t\to+\infty} \frac{\varphi'(t)}{2t} = 0$, (g) $\lim_{t\to 0+} \frac{\varphi'(t)}{2t} = M$, $0 < M < \infty$) the authors proved that φ could be written in the form (1.53) with $\psi : (0, M] \to [0,\infty)$ being a strictly convex and decreasing function. This implies that J^* is convex in the b variables when u is fixed and the authors proved that the minimum is explicitly given by

$$b_x(i,j) = \frac{\varphi'(\nabla_x^+(i,j))}{2\nabla_x^+(i,j)} \quad \text{and} \quad b_y(i,j) = \frac{\varphi'(\nabla_y^+(i,j))}{2\nabla_y^+(i,j)}.$$

Then the algorithm consists in iteratively computing until convergence

$$(b_x^{n+1}, b_y^{n+1}) = \operatorname{argmin}_{(b_x,b_y)} J^*(u^n, b_x, b_y),$$

$$u^{n+1} = \operatorname{argmin}_u J^*(u, b_x^{n+1}, b_y^{n+1}),$$

starting with $u^0 = 0$. The above algorithm can also be applied to functions φ which are nonconvex, e.g., $\varphi(s) = \frac{s^2}{1+s^2}$, $\varphi(s) = \log(1+s^2)$, but its convergence is only proved when φ is convex [77]. The book by G. Aubert and P. Kornprobst [28] contains a full account of this method together with other interesting applications of BV functions to image processing, in particular, to optical flow computation or sequence segmentation. Other related references are [50], [190], [78], [176].

A. Chambolle and P.L. Lions studied in [64] the equivalence between the constrained problem (1.16) and the unconstrained one (1.17) and they proved Theorem 1.1. Moreover, they also studied the convergence of the relaxation algorithm introduced in [77] and inspired by the work of Geman and Reynolds [116]. Let us describe in detail the algorithm when applied to total variation. Let

$$\Phi_\epsilon(x) := \begin{cases} \dfrac{1}{2\epsilon}x^2 & \text{if } |x| \leq \epsilon, \\[2ex] |x| - \dfrac{\epsilon}{2} & \text{if } \epsilon \leq |x| \leq \frac{1}{\epsilon}, \\[2ex] \dfrac{\epsilon}{2}x^2 + \dfrac{1}{2}\left(\dfrac{1}{\epsilon} - \epsilon\right) & \text{if } |x| \geq \frac{1}{\epsilon}, \end{cases} \tag{1.54}$$

and consider the problem

$$\text{Minimize}_{u\in W^{1,2}(\Omega)} \int_\Omega \Phi_\epsilon(|\nabla u|) + \frac{\lambda}{2}\int_\Omega |u - u_d|^2\, dx. \tag{1.55}$$

Consider the functional

$$E(u,v) = \int_\Omega v|\nabla u| + \int_\Omega \frac{1}{v} + \frac{\lambda}{2}\int_\Omega |u-u_d|^2\, dx,$$

where $u \in W^{1,2}(\Omega)$ and $v \in L^2(\Omega)$, $\epsilon \leq v \leq \frac{1}{\epsilon}$. Starting from any u^1 and v^1 (for instance $v^1 = 1$), we construct iteratively

$$u^{n+1} = \text{argmin}_{u \in W^{1,2}(\Omega)} E(u, v^n),$$

$$v^{n+1} = \text{argmin}_{\epsilon \leq v \leq \frac{1}{\epsilon}} E(u^{n+1}, v) = \epsilon \vee \frac{1}{|\nabla u^{n+1}|} \wedge \frac{1}{\epsilon}.$$

The sequence u^n converges strongly in $L^2(\Omega)$ and weakly in $W^{1,2}(\Omega)$ to the minimum of (1.55) [64].

E. Casas, K. Kunisch and C. Pola [62] studied problem (1.17) and its numerical approximation. They described the subdifferential of Φ in the BV space. G. Chavent and K. Kunisch [79] studied the minimization problem

$$\text{Minimize}_{u \in BV(\Omega) \cap L^2(\Omega)} \frac{1}{2} \parallel Ku - u_d \parallel_2^2 + \frac{\alpha}{2} \parallel u \parallel_2^2 + \beta \int_\Omega \|Du\|,$$

$\alpha \geq 0$, $\beta > 0$, and proved existence, uniqueness and stability with respect to perturbations of the data. They also obtained a characterization of the subdifferential similar to the one presented here. Let us also mention the interesting numerical approach proposed in [136] based in an augmented Lagrangian technique to solve the nonsmooth convex minimization problem for the functional

$$\int_\Omega \left(\frac{\mu}{2} |\nabla u|^2 + g|\nabla u| \right) dx + \frac{1}{2} \int_\Omega |u - u_d|^2 \, dx,$$

where $\mu, g > 0$. The condition $\mu > 0$ is only required for the analysis of the above problem in the infinite dimensional case. In the corresponding discrete case, the analysis of the convergence of their numerical algorithm covers also the case $\mu = 0$, which amounts to total variation denoising. In their algorithm, the vector field z which represents $\frac{Du}{|Du|}$ in Proposition 1.10 is introduced as a Lagrangian or dual variable $z(x)$ and the system satisfied by (u, λ) is solved. Their numerical method is different from the ones described above, an active set strategy based on the first-order augmented Lagrangian of the dual variable is employed [136]. A very interesting algorithm to solve the denoising problem based on the study of the dual problem (which amounts to computing the vector field z described in Proposition 1.10) has been proposed by A. Chambolle [66].

L. Vese [198] studied problem (1.17) (indeed, a more general version with $\int_\Omega \varphi(|Du|)$ in place of $\int_\Omega |Du|$ with φ convex and $\frac{\varphi(r)}{r} \to c \in (0, \infty)$ as $r \to +\infty$) from the theoretical and numerical point of view. She proved existence and uniqueness results and gave a characterization of $\partial\Phi$ similar to the one given in Proposition 1.10. She also studied the convergence of a numerical approximation to the solution of (1.17), in which the functional is first regularized and the solution of the regularized functional is approximated by half-quadratic regularization [198].

M. Nikolova [162] studied the discrete version of functionals of the form (1.52) for smooth and nonsmooth potential functions φ. She proved that to recover strongly homogeneous zones in an image which has been corrupted by noise or to preserve them in case of small variations of the data, a necessary and sufficient condition (modulo some other technical conditions) is that φ is nonsmooth at 0.

Y. Meyer [153] interpreted the total variation denoising model

$$\text{Minimize} \int_\Omega \|Du\| \, dx + \frac{\lambda}{2} \int_\Omega |u - u_d|^2 \, dx \tag{1.56}$$

in the more general context of what he called $u + v$ models. In this context the given image u_d is decomposed as the sum of two components $u(x) + v(x)$. The first component $u(x)$ contains the main geometric structures of the image and it models the objects present in it. The second component $v(x)$ contains both the textured parts and the noise. In the case of the TV denoising model (1.56), we are decomposing $u_d(x)$ as

$$u_d(x) = u - \frac{1}{\lambda} \operatorname{div} \left(\frac{Du}{|Du|} \right),$$

thus $v(x) = \frac{1}{\lambda} \operatorname{div} \left(\frac{Du}{|Du|} \right)$. The question arises if this is a reasonable $u + v$ model. We refer to [153] for a more detailed study of the model in this context (other $u + v$ models inspired in Meyer's work can be found in [199, 166]). Let us finally mention the following result which can be found in [153].

Theorem 1.11. *Let $\Omega = \mathbb{R}^2$. If $f \in L^2(\mathbb{R}^2)$ and (u, v) is a minimizer of (1.56), then all wavelet coefficients (with respect to an orthonormal wavelet basis) of v are less than $1/\lambda$. Conversely if a soft thresholding is applied to u_d with threshold $10/\lambda$, then the resulting function $\widehat{u_d}$ is a good substitute for being the optimal u. Indeed $u_d = \widehat{u_d} + R$, where $\lambda \parallel R \parallel_2^2 \leq C\omega_\lambda(u_d)$, C is a positive constant which does not depend on λ and $\omega_\lambda(u_d)$ denotes the minimum value of (1.56).*

Besides the above references, other extensions and applications of the idea of Total Variation have been proposed by many authors: satellite image deblurring with spectrum interpolation ([108]), image restoration combining Total Variation and wavelets ([149, 150]), edge direction preserving image zooming ([148]), anti-ringing deconvolution ([155]), blind deconvolution ([73] and its references), denoising of color or vector valued images ([51], [52]), variational restoration of non-flat image features ([75], [76], [188], [189], [165]), Total Variation denoising with local constraints ([53], [48]), etc.

Chapter 2

The Neumann Problem for the Total Variation Flow

2.1 Introduction

This chapter is devoted to prove existence and uniqueness of solutions for the minimizing total variation flow with Neumann boundary conditions, namely

$$\begin{cases} \dfrac{\partial u}{\partial t} = \operatorname{div}\left(\dfrac{Du}{|Du|}\right) & \text{in } Q = (0,\infty) \times \Omega, \\[2ex] \dfrac{\partial u}{\partial \eta} = 0 & \text{on } S = (0,\infty) \times \partial\Omega, \\[2ex] u(0,x) = u_0(x) & \text{in } x \in \Omega, \end{cases} \tag{2.1}$$

where Ω is a bounded set in $\mathbb{R}^N$ with Lipschitz continuous boundary $\partial\Omega$ and $u_0 \in L^1(\Omega)$. As we saw in the previous chapter, this partial differential equation appears when one uses the steepest descent method to minimize the total variation, a method introduced by L. Rudin, S. Osher and E. Fatemi ([174], [175]) in the context of image denoising and reconstruction. Then solving (2.1) amounts to regularizing or, in other words, to filtering the initial datum u_0. This filtering process has less destructive effect on the edges than filtering with a Gaussian, i.e., than solving the heat equation with initial condition u_0. In this context the given *image* u_0 is a function defined on a bounded, smooth or piecewise smooth open subset Ω of $\mathbb{R}^N$; typically, Ω will be a rectangle in $\mathbb{R}^2$. As argued in [7], the choice of Neumann boundary conditions is a natural choice in image processing. It corresponds to the reflection of the picture across the boundary and has the advantage of not imposing any value on the boundary and not creating edges on it. When dealing with the deconvolution or reconstruction problem one minimizes the *total variation functional*, i.e., the functional

$$\int_\Omega |Du| \tag{2.2}$$

under some constraints which model the process of image acquisition, including blur and noise ([146], [174], [175], [71], [72], [64], [201],[202]).

In this chapter we shall prove existence and uniqueness of solutions of (2.1) for initial data in $L^1(\Omega)$. To make precise our notion of solution we need the following functional space. By $L^1_w(0,T;BV(\Omega))$ we denote the space of weakly measurable functions $w:[0,T]\to BV(\Omega)$ (i.e., $t\in[0,T]\to\langle w(t),\phi\rangle$ is measurable for every $\phi\in BV(\Omega)^*$) such that $\int_0^T \|w(t)\|<\infty$. Observe that, since $BV(\Omega)$ has a separable predual (see Remark B.7), it follows easily that the map $t\in[0,T]\to\|w(t)\|$ is measurable. We also need the following truncation functions: $T_k(r)=[k-(k-|r|)^+]\mathrm{sign}_0(r)$, $k\geq 0$, $r\in\mathbb{R}$.

If μ is a (possibly vector valued) Radon measure and f is a Borel function, the integration of f with respect to μ will be denoted by $\int f d\mu$. When μ is the Lebesgue measure, the symbol dx will often be omitted.

Our concept of solution is the following

Definition 2.1. A measurable function $u:(0,T)\times\Omega\to\mathbb{R}$ is a *weak solution* of (2.1) in $(0,T)\times\Omega$ if $u\in C([0,T],L^1(\Omega))\cap W^{1,1}_{loc}(0,T;L^1(\Omega))$, $T_k(u)\in L^1_w(0,T;BV(\Omega))$ for all $k>0$ and there exists $z\in L^\infty((0,T)\times\Omega)$ with $\|z\|_\infty\leq 1$, $u_t=\mathrm{div}(z)$ in $\mathcal{D}'((0,T)\times\Omega)$ such that

$$\int_\Omega (T_k(u(t))-w)u_t(t)\,dx\leq\int_\Omega z(t)\cdot\nabla w\,dx-\int_\Omega\|DT_k(u(t))\| \tag{2.3}$$

for every $w\in W^{1,1}(\Omega)\cap L^\infty(\Omega)$ and a.e. on $[0,T]$.

The main result of this chapter is the following:

Theorem 2.2. *Let $u_0\in L^1(\Omega)$. Then there exists a unique weak solution of* (2.1) *in $(0,T)\times\Omega$ for every $T>0$ such that $u(0)=u_0$. Moreover, if $u(t),\hat{u}(t)$ are weak solutions corresponding to initial data $u_0,\hat{u}_0$, respectively, then*

$$\|(u(t)-\hat{u}(t))^+\|_1\leq\|(u_0-\hat{u}_0)^+\|_1 \quad\text{and}\quad \|u(t)-\hat{u}(t)\|_1\leq\|u_0-\hat{u}_0\|_1, \tag{2.4}$$

for all $t\geq 0$.

To prove Theorem 2.2 we shall use the techniques of completely accretive operators and the Crandall–Liggett semigroup generation theorem (see Apendix A). For that, we shall associate a completely accretive operator $\mathcal{A}$ to the formal differential expression $-\mathrm{div}(\frac{Du}{|Du|})$ together with Neumann boundary conditions.

Then, using Crandall–Liggett's semigroup generation theorem we conclude that the abstract Cauchy problem in $L^1(\Omega)$

$$\begin{cases} \dfrac{du}{dt} + \mathcal{A}u \ni 0, \\ \\ u(0) = u_0 \end{cases} \tag{2.5}$$

has a unique strong solution $u \in C([0,T], L^1(\Omega)) \cap W^{1,1}_{loc}(0,T; L^1(\Omega))$ $(\forall T > 0)$ with initial datum $u(0) = u_0$. In Section 2.4 we shall prove that strong solutions of (2.5) coincide with weak solutions of (2.1). The chapter finishes with the study of the asymptotic behaviour of the solutions of problem (2.1).

2.2 Strong Solutions in $L^2(\Omega)$

Consider the energy functional $\Phi : L^2(\Omega) \to (-\infty, +\infty]$ defined by

$$\Phi(u) = \begin{cases} \displaystyle\int_\Omega \|Du\| & \text{if } u \in BV(\Omega) \cap L^2(\Omega), \\ \\ +\infty & \text{if } u \in L^2(\Omega) \setminus BV(\Omega). \end{cases} \tag{2.6}$$

Since the functional Φ is convex, lower semi-continuous and proper, then $\partial\Phi$ is a maximal monotone operator with dense domain, generating a contraction semigroup in $L^2(\Omega)$ (see Appendix A or [58]). Therefore, we have the following result.

Theorem 2.3. *Let $u_0 \in L^2(\Omega)$. Then there exists a unique strong solution in the semigroup sense u of (2.1) in $[0,T]$ for every $T > 0$, i.e., $u \in C([0,T]; L^2(\mathbb{R}^N)) \cap W^{1,2}_{\rm loc}(0,T; L^2(\Omega))$, $u(t) \in D(\partial\Phi)$ a.e. in $t \in [0,T]$ and*

$$-u'(t) \in \partial\Phi(u(t)) \quad \text{a.e. in } t \in [0,T]. \tag{2.7}$$

Moreover, if u and v are the strong solutions of (2.1) corresponding to the initial conditions $u_0, v_0 \in L^2(\Omega)$, then

$$\|u(t) - v(t)\|_2 \le \|u_0 - v_0\|_2 \quad \text{for any } t > 0. \tag{2.8}$$

The semigroup theory immediately provides us with existence and uniqueness results of (2.1) in the semigroup sense. The characterization of $\partial\Phi$ given in Chapter 1, Proposition 1.10 permits us to write Theorem 2.3 in more classical terms. Let us describe the characterization of $\partial\Phi$ so that we can use test functions in $BV(\Omega)$. Later we shall extend the existence and uniqueness results to the case of initial conditions in $L^1(\Omega)$.

Lemma 2.4. *The following assertions are equivalent:*

(a) $(u,v) \in \partial\Phi$;

(b)

$$u \in L^2(\Omega) \cap BV(\Omega), \; v \in L^2(\Omega), \tag{2.9}$$

$$\exists z \in X(\Omega)_2, \; \|z\|_\infty \leq 1, \text{ such that } v = -\text{div}(z) \quad \text{in} \quad \mathcal{D}'(\Omega), \tag{2.10}$$

and

$$\int_\Omega (z, Du) = \int_\Omega \|Du\|, \tag{2.11}$$

$$[z,\nu] = 0 \quad \text{on} \quad \partial\Omega; \tag{2.12}$$

(c) (2.9) *and* (2.10) *hold, and*

$$\int_\Omega (w-u)v\,dx \leq \int_\Omega z\cdot\nabla w\,dx - \int_\Omega \|Du\|, \qquad \forall w \in W^{1,1}(\Omega)\cap L^2(\Omega); \tag{2.13}$$

(d) (2.9) *and* (2.10) *hold, and*

$$\int_\Omega (w-u)v\,dx \leq \int_\Omega (z, Dw) - \int_\Omega \|Du\| \qquad \forall w \in L^2(\Omega)\cap BV(\Omega); \tag{2.14}$$

(e) (2.9) *and* (2.10) *hold, and* (2.14) *holds with the equality instead of the inequality.*

Proof. The equivalence of (a) and (b) has been proved in Proposition 1.10. To obtain (e) from (b) it suffices to multiply both terms of the equation $v = -\text{div}(z)$ by $w-u$, for $w \in L^2(\Omega)\cap BV(\Omega)$ and to integrate by parts using Theorem C.9. It is clear that (e) implies (d), and (d) implies (c). To prove that (b) follows from (d) we choose $w = u$ in (2.14) and we obtain that

$$\int_\Omega \|Du\| \leq \int_\Omega (z, Du) \leq \|z\|_\infty \int_\Omega \|Du\| \leq \int_\Omega \|Du\|.$$

To obtain (2.12) we choose $w = u \pm \varphi$ in (2.14) with $\varphi \in C^\infty(\overline{\Omega})$ and we obtain

$$\pm \int_\Omega v\varphi\,dx \leq \pm \int_\Omega z\cdot D\varphi = -\pm \int_\Omega \text{div}(z)\,\varphi\,dx + \pm \int_{\partial\Omega} [z,\nu]\,\varphi\,d\mathcal{H}^{N-1},$$

which implies (2.12). In order to prove that (c) implies (d), let $w \in BV(\Omega)\cap L^2(\Omega)$. Using Theorem B.3 and Lemma C.8 we know that there exists a sequence $w_n \in C^\infty(\Omega)\cap BV(\Omega)\cap L^2(\Omega)$ such that

$$w_n \to w \quad \text{in} \quad L^2(\Omega) \quad \text{and} \quad \int_\Omega |\nabla w_n|\,dx \to \int_\Omega \|Dw\|.$$

Then

$$\int_\Omega z \cdot \nabla w_n \, dx = -\int_\Omega \operatorname{div}(z)\, w_n \, dx + \int_{\partial\Omega} [z,\nu] w_n \, d\mathcal{H}^{N-1}$$
$$\to -\int_\Omega \operatorname{div}(z)\, w \, dx + \int_{\partial\Omega} [z,\nu] w \, d\mathcal{H}^{N-1} = \int_\Omega (z, Dw).$$

Now, we use w_n as test function in (2.13) and let $n \to \infty$ to obtain (2.14). □

Definition 2.5. We say that $u \in C([0,T]; L^2(\Omega))$ is a *strong solution* of (2.1) if

$$u \in W^{1,2}_{\mathrm{loc}}(0,T; L^2(\Omega)) \cap L^1_w(]0,T[; BV(\Omega)),$$

$u(t) = u_0$, and there exists $z \in L^\infty\left(]0,T[\times\Omega; \mathbb{R}^N\right)$ such that $\|z\|_\infty \le 1$,

$$[z(t),\nu] = 0 \quad \text{in } \partial\Omega, \text{ a.e. } t \in [0,T]$$

satisfying

$$u_t = \operatorname{div}(z) \qquad \text{in } \mathcal{D}'\left(]0,T[\times\Omega\right)$$

and

$$\int_\Omega (u(t) - w) u_t(t) \, dx = \int_\Omega (z(t), Dw) - \int_\Omega \|Du(t)\| \tag{2.15}$$
$$\forall w \in L^2(\Omega) \cap BV(\Omega), \text{ a.e. } t \in [0,T].$$

Obviously, using Lemma 2.4, a strong solution of (2.1) is a strong solution in the sense of semigroups. The converse implication follows along the same lines, except for the measurability of $z(t,x)$. To ensure the joint measurability of z one takes into account that, by Crandall–Liggett's theorem (Theorem A.28), semigroup solutions can be approximated by implicit-in-time discretizations of (2.7), and one constructs a function $z(t,x) \in L^\infty((0,T)\times\Omega)$ satisfying the requirements contained in Definition 2.5. We do not give the details of this proof here, since a similar argument will be used in Section 2.4 to construct solutions for initial data in $L^1(\Omega)$. We have obtained the following result.

Theorem 2.6. *Let $u_0 \in L^2(\Omega)$. Then there exists a unique strong solution u of (2.1) in $[0,T] \times \Omega$ for every $T > 0$. Moreover, if u and v are the strong solutions of (2.1) corresponding to the initial conditions $u_0, v_0 \in L^2(\Omega)$, then*

$$\|u(t) - v(t)\|_2 \le \|u_0 - v_0\|_2 \quad \text{for any } t > 0. \tag{2.16}$$

2.3 The Semigroup Solution in $L^1(\Omega)$

Let us introduce the following operator $\mathcal{A}$ in $L^1(\Omega)$. Let us define the space (see Appendix C)

$$X(\Omega)_1 := \left\{ z \in L^\infty(\Omega, \mathbb{R}^N) \ : \ \operatorname{div}(z) \in L^1(\Omega) \right\}.$$

Definition 2.7. $(u,v) \in \mathcal{A}$ if and only if $u, v \in L^1(\Omega)$, $T_k(u) \in BV(\Omega)$ for all $k > 0$, and there exists $z \in X(\Omega)_1$ with $\|z\|_\infty \le 1$, $v = -\mathrm{div}(z)$ in $\mathcal{D}'(\Omega)$ such that

$$\int_\Omega (w - T_k(u))v\,dx \le \int_\Omega z \cdot \nabla w\,dx - \int_\Omega \|DT_k(u)\|,$$

for all $w \in W^{1,1}(\Omega) \cap L^\infty(\Omega)$ and $k > 0$.

Theorem 2.8. *The operator $\mathcal{A}$ is m-completely accretive in $L^1(\Omega)$ with dense domain. For any $u_0 \in L^1(\Omega)$ the semigroup solution $u(t) = e^{-t\mathcal{A}}u_0$ is a strong solution of*

$$\frac{du}{dt} + \mathcal{A}u \ni 0, \qquad u(0) = u_0. \tag{2.17}$$

To prove Theorem 2.8 we need to use test functions in $BV(\Omega) \cap L^\infty(\Omega)$. The next lemma shows that this is indeed possible.

Lemma 2.9. *We have the following characterization of the operator $\mathcal{A}$:*
$(u,v) \in \mathcal{A}$ if and only if $u, v \in L^1(\Omega)$, $T_k(u) \in BV(\Omega)$ for all $k > 0$, and there exists $z \in X(\Omega)_1$ with $\|z\|_\infty \le 1$, $v = -\mathrm{div}(z)$ in $\mathcal{D}'(\Omega)$ such that

$$\int_\Omega (w - T_k(u))v\,dx \le \int_\Omega (z, Dw) - \int_\Omega \|DT_k(u)\|, \tag{2.18}$$

for all $w \in BV(\Omega) \cap L^\infty(\Omega)$ and $k > 0$.
Moreover, we have that

(i) $\displaystyle\int_\Omega (z, DT_k(u)) = \int_\Omega \|DT_k(u)\|$ for all $k > 0$,

(ii) $\displaystyle\int_\Omega vT_k(u)\,dx = \int_\Omega \|DT_k(u)\|$ for all $k > 0$,

(iii) $\displaystyle\int_\Omega wv\,dx = \int_\Omega (z, Dw)$ for all $w \in BV(\Omega) \cap L^\infty(\Omega)$.

Remark 2.10. As a consequence we also have the following characterization of the operator $\mathcal{A}$.
$(u,v) \in \mathcal{A}$ if and only if $u, v \in L^1(\Omega)$, $T_k(u) \in BV(\Omega)$ for all $k > 0$, and there exists $z \in X(\Omega)_1$ with $\|z\|_\infty \le 1$, $v = -\mathrm{div}(z)$ in $\mathcal{D}'(\Omega)$ such that

$$\int_\Omega (w - T_k(u))v\,dx = \int_\Omega (z, Dw - DT_k(u)) \quad \forall w \in BV(\Omega) \cap L^\infty(\Omega),\ \forall k > 0,$$

or equivalently

$$\int_\Omega (z, DT_k(u)) = \int_\Omega \|DT_k(u)\|, \quad \forall k > 0, \quad \text{and} \quad [z, \nu] = 0 \ \text{ on } \partial\Omega.$$

Proof. The characterization of $\mathcal{A}$ stated in the lemma follows by approximating functions in $BV(\Omega)$ by functions in $W^{1,1}(\Omega)$ as in the equivalence $(c) \iff (d)$ in Lemma 2.4.

Now, taking $w = T_k(u)$ in (2.18) we obtain

$$0 \leq \int_\Omega (z, DT_k(u)) - \int_\Omega \|DT_k(u)\|.$$

Thus,

$$\int_\Omega (z, DT_k(u)) \leq \|z\|_\infty \int_\Omega \|DT_k(u)\| \leq \int_\Omega \|DT_k(u)\| \leq \int_\Omega (z, DT_k(u)),$$

and (i) follows. To prove (ii) we take $w = 0$ in (2.18) to obtain

$$\int_\Omega \|DT_k(u)\| \leq \int_\Omega vT_k(u)\, dx$$

and then $w = 2T_k(u)$ to obtain, using (i),

$$\int_\Omega vT_k(u)\, dx \leq 2\int_\Omega (z, DT_k(u)) - \int_\Omega \|DT_k(u)\| = \int_\Omega \|DT_k(u)\|.$$

Consequently, (ii) holds.

After using (ii) in (2.18) we may write

$$\int_\Omega wv\, dx \leq \int_\Omega (z, Dw)$$

for any $w \in BV(\Omega) \cap L^\infty(\Omega)$. Since the same inequality holds for $-w \in BV(\Omega) \cap L^\infty(\Omega)$, we obtain (iii). □

Remark 2.11. Recall that we denote by $\theta(z, Dw, \cdot)$ the Radon–Nikodym derivative of the measure (z, Dw) with respect to the measure $\|Dw\|$ (see Appendix C). As a consequence of (i) we have that $\theta(z, DT_k(u), x) = 1$ a.e. with respect to the measure $\|DT_k(u)\|$. In case that $z \in C(\Omega, \mathbb{R}^N)$, this implies that

$$z(x) \cdot \frac{DT_k(u)}{\|DT_k(u)\|}(x) = 1, \qquad \|DT_k(u)\|\text{-a.e.},$$

where $\frac{DT_k(u)}{\|DT_k(u)\|}$ denotes the density of $DT_k(u)$ with respect to $\|DT_k(u)\|$ (see Theorem C.14), Heuristically, this amounts to saying that $z = \frac{Du}{\|Du\|}$. When z is not continuous we have that

$$z(x) \cdot \frac{DT_k(u)}{\|DT_k(u)\|}(x) = 1, \qquad \|\nabla T_k(u)\|\text{-a.e.},$$

where $\|\nabla T_k(u)\|$ denotes the absolutely continuous part of $\|DT_k(u)\|$ with respect to the Lebesgue measure in $\mathbb{R}^N$ (see Theorem C.14). In particular, if $u \in W^{1,1}(\Omega) \cap L^\infty(\Omega)$ we have that

$$z(x) \cdot \frac{\nabla u}{\|\nabla u\|}(x) = 1, \qquad \|\nabla u\|\text{-a.e.}$$

Proof of Theorem 2.8. Let $(u, v), (\hat{u}, \hat{v}) \in \mathcal{A}$, $p \in P_0$. We have to prove that

$$\int_\Omega p(u - \hat{u})(v - \hat{v})\, dx \geq 0. \tag{2.19}$$

Let $z, \hat{z} \in X(\Omega)_1$, $\|z\|_\infty \leq 1, \|\hat{z}\|_\infty \leq 1$, be such that $v = -\text{div}(z)$, $\hat{v} = -\text{div}(\hat{z})$ and

$$\int_\Omega (w - T_k(u))v\, dx = \int_\Omega (z, Dw) - \int_\Omega \|DT_k(u)\|, \tag{2.20}$$

$$\int_\Omega (w - T_k(\hat{u}))\hat{v}\, dx = \int_\Omega (\hat{z}, Dw) - \int_\Omega \|DT_k(\hat{u})\|, \tag{2.21}$$

for any $w \in BV(\Omega) \cap L^\infty(\Omega)$ and any $k > 0$. As observed in the previous remark, $\theta(z, DT_k(u), x) = 1$ $\|DT_k(u)\|$ – a.e., and, using Corollary C.7, we obtain that

$$\int_B (z, DT_k(u)) = \int_B \theta(z, DT_k(u), x)\|DT_k(u)\| = \int_B \|DT_k(u)\|,$$

$$\left|\int_B (\hat{z}, DT_k(u))\right| \leq \int_B \|DT_k(u)\|$$

for any Borel set $B \subseteq \Omega$. Similarly,

$$\int_B (\hat{z}, DT_k(\hat{u})) = \int_B \|DT_k(\hat{u})\|,$$

$$\left|\int_B (z, DT_k(\hat{u}))\right| \leq \int_B \|DT_k(\hat{u})\|$$

for any Borel set $B \subseteq \Omega$. It follows that

$$\int_B (z - \hat{z}, D(T_k(u) - T_k(\hat{u}))) \geq 0$$

for any Borel set $B \subseteq \Omega$. This implies that

$$\theta(z - \hat{z}, D(T_k(u) - T_k(\hat{u})), x) \geq 0 \quad \|D(T_k(u) - T_k(\hat{u}))\|\text{-a.e.}.$$

Since, according to Corollary C.16, we have that

$$\theta(z - \hat{z}, Dp(T_k(u) - T_k(\hat{u})), x) = \theta(z - \hat{z}, D(T_k(u) - T_k(\hat{u})), x)$$

a.e. with respect to the measures $\|D(T_k(u) - T_k(\hat{u}))\|$ and $\|Dp(T_k(u) - T_k(\hat{u}))\|$. We conclude that

$$\theta(z - \hat{z}, Dp(T_k(u) - T_k(\hat{u})), x) \geq 0, \quad \|Dp(T_k(u) - T_k(\hat{u}))\|\text{-a.e.} \tag{2.22}$$

Taking $w = T_k(u) + p(T_k(u) - T_k(\hat{u}))$ in (2.20) and $w = T_k(\hat{u}) - p(T_k(u) - T_k(\hat{u}))$ in (2.21), adding both terms, and using (2.22), we obtain

$$\int_\Omega p(T_k(u) - T_k(\hat{u}))(v - \hat{v})\, dx = \int_\Omega (z - \hat{z}, Dp(T_k(u) - T_k(\hat{u})))$$

$$= \int_\Omega \theta(z - \hat{z}, Dp(T_k(u) - T_k(\hat{u})), x)\|Dp(T_k(u) - T_k(\hat{u}))\| \geq 0.$$

The inequality (2.19) follows by letting $k \to \infty$. Therefore $\mathcal{A}$ is completely accretive.

Let us prove $\mathcal{A}$ is closed in $L^1(\Omega)$. Let $(u_n, v_n) \in \mathcal{A}$ be such that $u_n \to u$, $v_n \to v$ in $L^1(\Omega)$ as $n \to \infty$. Since $(u_n, v_n) \in \mathcal{A}$, there exists $z_n \in X(\Omega)_1$, $\|z_n\|_\infty \leq 1$ with $v_n = -\mathrm{div}(z_n)$ in $\mathcal{D}'(\Omega)$ such that

$$\int_\Omega (w - T_k(u_n))v_n\, dx \leq \int_\Omega z_n \cdot \nabla w\, dx - \int_\Omega \|DT_k(u_n)\|, \tag{2.23}$$

for all $w \in W^{1,1}(\Omega) \cap L^\infty(\Omega)$ and $k > 0$. Taking $w = 0$ in (2.23) we have that

$$\|DT_k(u_n)\| \leq \int_\Omega T_k(u_n)v_n\, dx \leq k \sup_n \|v_n\|_1.$$

It follows that $T_k(u) \in BV(\Omega)$. Since $\|z_n\|_\infty \leq 1$ we may assume that $z_n \rightharpoonup z$ in the weak* topology of $L^\infty(\Omega, \mathbb{R}^N)$ with $\|z\|_\infty \leq 1$. Now, letting $n \to \infty$ in (2.23) we obtain that

$$\int_\Omega (w - T_k(u))v\, dx \leq \int_\Omega z \cdot \nabla w\, dx - \int_\Omega \|DT_k(u)\|.$$

Hence $(u, v) \in \mathcal{A}$, and $\mathcal{A}$ is closed.

Let us prove that $\partial\Phi \subset \mathcal{A}$. Having in mind Lemma 2.4 and Remark 2.10, we only need to prove that if $(u, v) \in \partial\Phi$, then

$$\int_\Omega (z, DT_k(u)) = \int_\Omega \|DT_k(u)\|, \qquad \text{for all } k > 0. \tag{2.24}$$

In fact, according to Corollary C.16, we have that

$$\theta(z, DT_k(u), x) = \theta(z, Du, x)$$

a.e. with respect to the measures $\|DT_k(u)\|$ and $\|Du\|$. Now, since

$$\int_\Omega (z, Du) = \int_\Omega \|Du\|,$$

we have $\theta(z, Du, x) = 1$ a.e. with respect to the measure $\|Du\|$. Hence

$$\int_\Omega (z, DT_k(u)) = \int_\Omega \theta(z, DT_k(u), x) \, \|DT_k(u)\| = \int_\Omega \|DT_k(u)\|.$$

Since $\partial\Phi \subset \mathcal{A}$, we have that $R(I + \mathcal{A})$ is dense in $L^1(\Omega)$. Then by Proposition A.42, it follows that $\mathcal{A}$ is m-completely accretive in $L^1(\Omega)$. The density of the domain follows from the density of the domain of $\partial\Phi$. By Crandall–Ligget's theorem, $\mathcal{A}$ generates a contraction semigroup in $L^1(\Omega)$ given by the exponential formula

$$e^{-t\mathcal{A}} u_0 = \lim_{n\to\infty} \left(I + \frac{t}{n} A \right)^{-n} u_0 \quad \text{for any } u_0 \in L^1(\Omega).$$

The function $u(t) = e^{-t\mathcal{A}} u_0$ is a mild solution of

$$\begin{cases} \dfrac{du}{dt} + \mathcal{A}u \ni 0, \\ \\ u(0) = u_0. \end{cases} \tag{2.25}$$

To prove that $u(t)$ is a strong solution of (2.25) we shall use the regularizing effect due to the homogeneity of the operator $\mathcal{A}$. Let us first observe that

$$\text{if } (u, v) \in \mathcal{A} \text{ and } \lambda > 0, \text{ then } (\lambda u, v) \in \mathcal{A}. \tag{2.26}$$

Indeed, let $(u, v) \in \mathcal{A}$ and let $z \in X(\Omega)_1$ with $\|z\|_\infty \leq 1$, $v = -\mathrm{div}(z)$ in $\mathcal{D}'(\Omega)$ satisfying

$$\int_\Omega (w - T_k(u)) v \, dx \leq \int_\Omega (z, Dw) - \int_\Omega \|DT_k(u)\|, \tag{2.27}$$

for all $w \in BV(\Omega) \cap L^\infty(\Omega)$, and $k > 0$. Then, take as test function in (2.27) $w + T_k(u) - T_k(\lambda u)$ instead of $w \in BV(\Omega) \cap L^\infty(\Omega)$ to obtain

$$\int_\Omega (w - T_k(\lambda u)) v \, dx \leq \int_\Omega (z, Dw) - \int_\Omega \|DT_k(\lambda u)\|.$$

In other words, $(\lambda u, v) \in \mathcal{A}$. From (2.26) it follows immediately that

$$\frac{1}{\lambda} (I + \lambda\mu\mathcal{A})^{-1} u_0 = (I + \mu\mathcal{A})^{-1} \left(\frac{1}{\lambda} u_0 \right) \tag{2.28}$$

for any $\lambda, \mu > 0$ and any $u_0 \in L^1(\Omega)$. Iterating (2.28) and taking $\mu = \frac{t}{n}$ we obtain

$$\left(I + \frac{t}{n} A \right)^{-n} \left(\frac{1}{\lambda} u_0 \right) = \frac{1}{\lambda} \left(I + \lambda \frac{t}{n} A \right)^{-n} u_0 \tag{2.29}$$

for any $\lambda > 0$, $n \in \mathbb{N}$ and $u_0 \in L^1(\Omega)$. Writing $S(t) = e^{-t\mathcal{A}}$ and letting $n \to \infty$ in (2.29) we may write

$$S(t)\left(\frac{1}{\lambda}u_0\right) = \frac{1}{\lambda}S(\lambda t)u_0, \tag{2.30}$$

for any $\lambda > 0$ and any $u_0 \in L^1(\Omega)$. Now, let $u_0 \in L^1(\Omega)$ and $u(t) = S(t)u_0$. Since $\mathcal{A}$ is m-completely accretive in $L^1(\Omega)$, $u(t)$ will be a strong solution of (2.25) once we know that $S(t)u_0 \in D(\mathcal{A})$ for all $t > 0$ (Corollary A.46). From the proof of Theorem 4.2 in [44] it is sufficient to prove that, given $t > 0$, for some sequence $t_n \downarrow 0$,

$$\left\{\frac{S(t+t_n)u_0 - S(t)u_0}{t_n}\right\}_{n=1}^{\infty} \quad \text{is weakly convergent in } L^1(\Omega). \tag{2.31}$$

Fix $t > 0$ and let $h > 0$, $\lambda = 1 + \frac{h}{t}$. Using (2.30) we have that

$$\begin{aligned} S(t+h)u_0 - S(t)u_0 &= S(\lambda t)u_0 - S(t)u_0 = \lambda S(t)\left(\frac{1}{\lambda}u_0\right) - S(t)u_0 \\ &= \lambda\left[S(t)\left(\frac{1}{\lambda}u_0\right) - S(t)u_0\right] + (\lambda - 1)S(t)u_0. \end{aligned}$$

From this, it follows that

$$|S(t+h)u_0 - S(t)u_0| \le \lambda\left|S(t)\left(\frac{1}{\lambda}u_0\right) - S(t)u_0\right| + |\lambda - 1||S(t)u_0|. \tag{2.32}$$

The complete accretivity of $\mathcal{A}$ implies that

$$\begin{aligned} &S(t)\left(\frac{1}{\lambda}u_0\right) - S(t)u_0 \ll \frac{1}{\lambda}u_0 - u_0, \\ &S(t)u_0 \ll u_0. \end{aligned}$$

Since $u \ll v$, $u, v \in \mathcal{M}(\Omega)$ implies that $\alpha u \ll \alpha v$, $\alpha > 0$, and $|u| \ll |v|$, the previous relations in turn imply that

$$\begin{aligned} &\lambda\left|S(t)\left(\frac{1}{\lambda}u_0\right) - S(t)u_0\right| \ll (\lambda - 1)|u_0|, \\ &(\lambda - 1)|S(t)u_0| \ll (\lambda - 1)|u_0|. \end{aligned} \tag{2.33}$$

Since the set $\{f \in \mathcal{M}(\Omega) : f \ll (\lambda - 1)|u_0|\}$ is convex we deduce from (2.32) and (2.33) that

$$|S(t+h)u_0 - S(t)u_0| \ll 2(\lambda - 1)|u_0| = 2\frac{h}{t}|u_0|,$$

hence,

$$\frac{|S(t+h)u_0 - S(t)u_0|}{h} \ll \frac{2}{t}|u_0|. \tag{2.34}$$

Now, using Proposition A.39 we conclude that

$$\left\{ \frac{|S(t+h)u_0 - S(t)u_0|}{h} \right\}_{h>0}$$

is weakly compact in $L^1(\Omega)$ and (2.31) holds. Notice also that as a consequence of (2.34) we obtain

$$|u'(t)| \leq \frac{2}{t} |u_0|. \tag{2.35}$$

□

Remark 2.12. Let us mention another proof of the complete accretivity of $\mathcal{A}$. For that we consider the functional $\Gamma : L^1(\Omega) \to (-\infty, +\infty]$ defined by

$$\Gamma(u) = \begin{cases} \int_\Omega |\nabla u| & \text{if } u \in W^{1,1}(\Omega), \\ \\ +\infty & \text{if } u \in W^{1,1}(\Omega). \end{cases} \tag{2.36}$$

Then, using Lemma A.48 we know that the operator $\partial_{L^1(\Omega)}\Gamma$ in $L^1(\Omega)$ defined by

$$(u,v) \in \partial_{L^1(\Omega)}\Gamma \quad \text{if and only if } u \in W^{1,1}(\Omega),\ v \in L^1(\Omega), \text{ and}$$

$$\Gamma(w) \geq \Gamma(u) + \int_\Omega (w-u)v\,dx, \quad \forall w \in L^1(\Omega) \text{ such that } (w-u)v \in L^1(\Omega)$$

is completely accretive in $L^1(\Omega)$. Now, the lower semi-continuous envelope of the functional Γ is the functional Ψ given by

$$\Psi(u) = \begin{cases} \int_\Omega \|Du\| & \text{if } u \in BV(\Omega), \\ \\ +\infty & \text{if } u \in BV(\Omega), \end{cases} \tag{2.37}$$

and, using Theorem A.50, we know that $\overline{\partial_{L^1(\Omega)}\Psi}^{L^1(\Omega)}$ is m-completely accretive in $L^1(\Omega)$. Using that $\mathcal{A}$ is completely accretive and closed in $L^1(\Omega)$, and Proposition A.42 we obtain that $\mathcal{A}= \overline{\partial_{L^1(\Omega)}\Psi}^{L^1(\Omega)}$.

2.4 Existence and Uniqueness of Weak Solutions

Lemma 2.13. *Let $u \in C([0,T], L^1(\Omega)) \cap W^{1,1}_{loc}(0,T; L^1(\Omega))$ be the strong solution of (2.17) with initial condition $u(0) = u_0 \in L^1(\Omega)$. Let $J_k(r) = \int_0^r T_k(s)ds$, $k > 0$. Then*

$$\int_\Omega J_k(u(t))\,dx + \int_0^t \int_\Omega \|DT_k(u(s))\| ds \leq \int_\Omega J_k(u_0)\,dx \tag{2.38}$$

for all $t > 0$ and all $k > 0$.

Proof. Since, a.e. on $[0,T]$, $(u(t), -u_t(t)) \in \mathcal{A}$, for almost all $t \in [0,T]$ there exists $z(t) \in X(\Omega)_1$ with $\|z(t)\|_\infty \leq 1$ such that $u_t(t) = \operatorname{div}(z(t))$ and

$$\int_\Omega (T_k(u(t)) - w)u_t(t)\,dx \leq \int_\Omega z(t)\cdot\nabla w\,dx - \int_\Omega \|DT_k(u(t))\| \tag{2.39}$$

for all $w \in W^{1,1}(\Omega) \cap L^\infty(\Omega)$, and all $k > 0$. Now set $w = 0$ in (2.39) to get

$$\frac{d}{dt}\int_\Omega J_k(u(t))\,dx + \int_\Omega \|DT_k(u(t))\| \leq 0.$$

Integrating this expression we obtain (2.38). □

Lemma 2.14. *Let $u, v \in L^1(\Omega)$, $u \in BV(\Omega) \cap L^\infty(\Omega)$, $z \in X(\Omega)_1$, with $\|z\|_\infty \leq 1$ and $v = -\operatorname{div}(z)$. Suppose that*

$$\int_\Omega (w-u)v\,dx \leq \int_\Omega z\cdot\nabla w\,dx - \int_\Omega \|Du\| \quad \forall w \in W^{1,1}(\Omega)\cap L^\infty(\Omega). \tag{2.40}$$

Then

$$\int_\Omega (w - T_k(u))v\,dx \leq \int_\Omega z\cdot\nabla w\,dx - \int_\Omega \|DT_k(u)\|, \qquad \forall w \in W^{1,1}(\Omega)\cap L^\infty(\Omega),\ \forall k>0. \tag{2.41}$$

Proof. As in Lemma 2.9, we observe that we may use test functions $w \in BV(\Omega) \cap L^\infty(\Omega)$ in (2.40). Let $G_k(r) = r - T_k(r)$. If we set $w = u$ in (2.40) we have that

$$\int_\Omega (z, Du) = \int_\Omega \|Du\|. \tag{2.42}$$

Since

$$\int_\Omega (z, DT_k(u)) \leq \int_\Omega \|DT_k(u)\|, \quad \int_\Omega (z, DG_k(u)) \leq \int_\Omega \|DG_k(u)\|,$$

for any $k > 0$, and, by Proposition B.17, we have

$$\begin{aligned}\|Du\| &= \int_\Omega (z, Du) = \int_\Omega (z, DT_k(u) + DG_k(u)) \\ &\leq \int_\Omega \|DT_k(u)\| + \int_\Omega \|DG_k(u)\| = \int_\Omega \|Du\|,\end{aligned}$$

we obtain

$$\int_\Omega (z, DT_k(u)) = \int_\Omega \|DT_k(u)\|, \quad \int_\Omega (z, DG_k(u)) = \int_\Omega \|DG_k(u)\|.$$

Now, set $w = \varphi + G_k(u)$, $\varphi \in W^{1,1}(\Omega) \cap L^\infty(\Omega)$, in (2.40) to obtain

$$\begin{aligned}\int_\Omega (\varphi - T_k(u))v\,dx &\leq \int_\Omega z \cdot \nabla\varphi\,dx + \int_\Omega (z, DG_k(u)) - \int_\Omega \|Du\| \\ &= \int_\Omega z \cdot \nabla\varphi\,dx - \int_\Omega \|DT_k(u)\|.\end{aligned}$$

□

We shall need the following lemma whose proof is straightforward.

Lemma 2.15. *Let $u \in C([0,T], L^1(\Omega)) \cap W^{1,1}_{loc}(0,T; L^1(\Omega))$, $z \in L^\infty((0,T) \times \Omega)$ with $\|z\|_\infty \leq 1$ and such that*

$$u_t = \mathrm{div}(z) \qquad \text{in } \mathcal{D}'((0,T) \times \Omega).$$

Then for almost all $t \in [0,T]$,

$$u_t(t) = \mathrm{div}(z(t)) \qquad \text{in } \mathcal{D}'(\Omega).$$

Proof of Theorem 2.2. Let $u \in C([0,T], L^1(\Omega)) \cap W^{1,1}_{loc}(0,T; L^1(\Omega))$ be the strong solution of (2.17). Let us assume that $u_0 \in L^\infty(\Omega) \cap D(\mathcal{A})$. By the complete accretivity of $\mathcal{A}$ we know that $\|u(t)\|_\infty \leq \|u_0\|_\infty$ and taking $k > \|u_0\|_\infty$ we conclude by using Lemma 2.13 that $u \in L^1(0,T; BV(\Omega))$. Since $u(t)$ is a strong solution of (2.17), the set K consisting of those values of $t \in [0,T]$ for which either u is not differentiable at t, or t is not a Lebesgue point for u', or $u' + \mathcal{A}u \not\ni 0$, is a null subset of $[0,T]$. Then, since $u' \in L^1(0,T; L^1(\Omega))$, Lemma A.8 guarantees that for each $\epsilon > 0$ there exists a partition $0 = t_0 < t_1 < \cdots < t_{n-1} \leq T < t_n$ with the properties: $t_k \notin K$, $t_k - t_{k-1} < \epsilon$, for $k = 1, \ldots, n$ and

$$\sum_{k=1}^n \int_{t_{k-1}}^{t_k} \|u'(s) - u'(t_k)\|\,ds < \epsilon.$$

If we define u_ϵ as $u_\epsilon(0) = u_0$, $u_\epsilon(t) = u(t_k)$ on $]t_{k-1}, t_k]$, $k = 1, \ldots, n$, then $u_\epsilon \to u$ in $C(0,T; L^1(\Omega))$.

Since $(u(t_k), -u'(t_k)) \in \mathcal{A}$, there exists $z_k \in X(\Omega)_1$, with $u'(t_k) = \mathrm{div}(z_k)$ in $\mathcal{D}'(\Omega)$ such that

$$\int_\Omega (u(t_k) - w)u'(t_k)\,dx \leq \int_\Omega z_k \cdot \nabla w\,dx - \int_\Omega \|Du(t_k)\|$$

for all $w \in W^{1,1}(\Omega) \cap L^\infty(\Omega)$. Thus, if we set $z_\epsilon(t) = z_k$ and $v_\epsilon(t) = u'(t_k)$ on $]t_{k-1}, t_k]$, $k = 1, \ldots, n$, we get

$$\int_0^T \int_\Omega (u_\epsilon(t) - w)\, v_\epsilon(t)\, \varphi\,dxdt \leq \int_0^T \left\{ \int_\Omega z_\epsilon(t) \cdot \nabla w\,dx - \int_\Omega \|Du_\epsilon(t)\| \right\} \varphi(t)\,dt$$

for all $w \in W^{1,1}(\Omega) \cap L^\infty(\Omega)$ and all $\varphi \in C_0^1(0,T)$, $\varphi \geq 0$. Now, letting $\epsilon \to 0^+$, and applying the Vitali convergence theorem, it follows that there exists $z \in L^\infty((0,T) \times \Omega)$ with $\|z\|_\infty \leq 1$ such that $u_t = \mathrm{div}(z)$ in $\mathcal{D}'((0,T) \times \Omega)$ and

$$\int_0^T \int_\Omega (u(t) - w)\, u_t\, \varphi(t)\, dxdt \leq \int_0^T \left\{ \int_\Omega z \cdot \nabla w\, dx - \int_\Omega \|Du(t)\| \right\} \varphi(t)\, dt$$

for all $w \in W^{1,1}(\Omega) \cap L^\infty(\Omega)$ and all $\varphi \in C_0^1(0,T)$, $\varphi \geq 0$. Since

$$(u - w)u_t \in L^1_{loc}(0,T; L^1(\Omega)), \quad \int_\Omega z \cdot \nabla w\, dx - \int_\Omega \|Du(\cdot)\| \in L^1(0,T)$$

it follows that

$$\int_\Omega (u(t) - w)\, u_t(t)\, dx \leq \int_\Omega z(t) \cdot \nabla w\, dx - \int_\Omega \|Du(t)\|$$

for every $w \in W^{1,1}(\Omega) \cap L^\infty(\Omega)$ and a.e. on $[0,T]$. Now, using Lemmas 2.14 and 2.15 we obtain that

$$\int_\Omega (T_k(u(t)) - w)\, u_t(t)\, dx \leq \int_\Omega z(t) \cdot \nabla w\, dx - \int_\Omega \|DT_k(u(t))\| \tag{2.43}$$

for every $w \in W^{1,1}(\Omega) \cap L^\infty(\Omega)$ and a.e. on $[0,T]$. We have shown that $u(t)$ is a weak solution of (2.1).

Now, let $u_0 \in L^1(\Omega)$ and let $u_{0n} \in L^\infty(\Omega) \cap D(\mathcal{A})$. Let u_n, u be the strong solutions of (2.17) with initial data u_{0n}, u_0, respectively. We know that u_n converges to u in $C([0,T], L^1(\Omega))$ and $u \in C([0,T], L^1(\Omega)) \cap W^{1,1}_{loc}(0,T; L^1(\Omega))$. By Lemma 2.13, we have that

$$\int_0^T \int_\Omega \|DT_k(u_n(s))\|\, ds \leq \int_\Omega J_k(u_{0n})\, dx \tag{2.44}$$

for all $T > 0$ and all $k > 0$. It follows that $T_k(u) \in L^1(0,T; BV(\Omega))$ for all $k > 0$. By the previous paragraph, there exist $z_n \in L^\infty((0,T) \times \Omega)$ with $\|z_n\|_\infty \leq 1$ such that $(u_n)_t = \mathrm{div}(z_n)$ in $\mathcal{D}'((0,T) \times \Omega)$ and

$$\begin{aligned} &\int_0^T \int_\Omega (T_k(u_n(t)) - w)\, (u_n)_t\, \varphi(t)\, dxdt \\ &\leq \int_0^T \int_\Omega z_n(t) \cdot \nabla w\, \varphi(t)\, dxdt - \int_0^T \int_\Omega \|DT_k(u_n(t))\| \varphi(t)\, dt \end{aligned}$$

for all $w \in W^{1,1}(\Omega) \cap L^\infty(\Omega)$, all $\varphi \in C_0^1(0,T)$, $\varphi \geq 0$ and all $k > 0$. Write the previous expression in the form

$$\begin{aligned} &\int_0^T \int_\Omega (u_n(t)w - J_k(u_n))\, \varphi'(t)\, dxdt \\ &\leq \int_0^T \int_\Omega z_n(t) \cdot \nabla w\, \varphi(t)\, dxdt - \int_0^T \int_\Omega \|DT_k(u_n(t))\| \varphi(t)\, dt \end{aligned} \tag{2.45}$$

for all $w \in W^{1,1}(\Omega) \cap L^\infty(\Omega)$, all $\varphi \in C_0^1(0,T)$, $\varphi \geq 0$ and all $k > 0$. Modulo a subsequence, we may assume that $z_n \rightharpoonup z$ in the weak* topology of $L^\infty((0,T)\times\Omega)$. Now letting $n \to \infty$ in (2.45) we obtain

$$\int_0^T \int_\Omega (u(t)w - J_k(u))\,\varphi'(t)\,dxdt$$
$$\leq \int_0^T \int_\Omega z(t)\cdot\nabla w\,\varphi(t)\,dxdt - \int_0^T \int_\Omega \|DT_k(u(t))\|\varphi(t)\,dt$$

for all $w \in W^{1,1}(\Omega) \cap L^\infty(\Omega)$, all $\varphi \in C_0^1(0,T)$, $\varphi \geq 0$ and all $k > 0$. Integrating by parts with respect to t in the left-hand side of the above expression we obtain

$$\int_0^T \int_\Omega (T_k(u(t)) - w)\,u_t\,\varphi(t)\,dxdt$$
$$\leq \int_0^T \left\{\int_\Omega z\cdot\nabla w\,dx - \int_\Omega \|DT_k(u(t))\|\right\}\varphi(t)\,dt$$

for all $w \in W^{1,1}(\Omega) \cap L^\infty(\Omega)$, all $\varphi \in C_0^1(0,T)$, $\varphi \geq 0$ and all $k > 0$. Since

$$(T_k(u) - w)u_t \in L^1_{loc}(0,T;L^1(\Omega)), \quad \int_\Omega z\cdot\nabla w\,dx - \|DT_k(u(\cdot))\| \in L^1(0,T)$$

it follows that

$$\int_\Omega (T_k(u(t)) - w)u_t(t)\,dx \leq \int_\Omega z(t)\cdot\nabla w\,dx - \int_\Omega \|DT_k(u(t))\|$$

for all $w \in W^{1,1}(\Omega) \cap L^\infty(\Omega)$ and all $k > 0$, a.e. on $[0,T]$. Finally observe that $u_t = \operatorname{div}(z)$ in $\mathcal{D}'((0,T)\times\Omega)$ and $\|z\|_\infty \leq 1$. We conclude that u is a weak solution of (2.1).

For further reference, let us observe that, according to Lemma 2.9, we also have

$$\int_\Omega (T_k(u(t)) - w)u_t(t)\,dx = \int_\Omega (z(t), Dw) - \int_\Omega \|DT_k(u(t))\| \tag{2.46}$$

for all $w \in BV(\Omega) \cap L^\infty(\Omega)$ and all $k > 0$, a.e. on $[0,T]$.

Let us finally observe that a weak solution of (2.1) is a strong solution of (2.17). Let u be a weak solution of (2.1) in $(0,T)\times\Omega$. Then $u \in C([0,T],L^1(\Omega)) \cap W^{1,1}_{loc}((0,T),L^1(\Omega))$, $T_k(u) \in L^1([0,T],BV(\Omega))$ for all $k > 0$ and there exists $z \in L^\infty((0,T)\times\Omega)$ with $\|z\|_\infty \leq 1$, $u_t = \operatorname{div}(z)$ in $\mathcal{D}'((0,T)\times\Omega)$ such that

$$\int_\Omega (T_k(u(t)) - w)u_t(t)\,dx \leq \int_\Omega z(t)\cdot\nabla w\,dx - \int_\Omega \|DT_k(u(t))\| \tag{2.47}$$

for every $w \in W^{1,1}(\Omega) \cap L^\infty(\Omega)$ and a.e. on $[0,T]$. By Lemma 2.15 we have that a.e. on $[0,T]$, $u_t(t) = \mathrm{div}(z(t))$. Hence a.e. on $[0,T]$, $z(t) \in X(\Omega)_1$ and, since (2.47) holds also a.e. on $[0,T]$, we have that a.e. on $[0,T]$ $(u(t), -u_t(t)) \in \mathcal{A}$, i.e.,

$$u'(t) + \mathcal{A}u(t) \ni 0 \qquad \text{a.e. on } [0,T].$$

Therefore u is a strong solution of (2.17). The uniqueness of weak solutions of (2.1) follows as a consequence of the uniqueness of strong solutions of (2.17). The comparison estimates (2.4) follow from the complete accretivity of $\mathcal{A}$. □

2.5 An L^N-L^∞ Regularizing Effect

Let us first remark that there is no L^1-L^∞ or L^1-L^2 regularizing effect. Indeed, $v(t,x) = \frac{1}{\|x\|^{N/2}} - \frac{t}{\|x\|}$ solves (2.1) in $(0,1) \times B_1(0)$ with initial datum $v_0(x) = \frac{1}{\|x\|^{N/2}}$. Observe that $v(t) \in L^1(B_1(0)) \setminus L^2(B_1(0))$, $0 \leq t < 1$. Obviously, this solution does not satisfy Neumann boundary conditions but it may be used together with a comparison principle to build up a solution $u(t,x)$ of (2.1) which is in $L^1(\Omega) \setminus L^2(\Omega)$. Indeed, given $A \geq 1$, let $u(t,x)$ the solution of the Neumann problem in $(0,\infty) \times B_1(0)$ with $u(0,x) = \frac{A}{\|x\|^{\frac{N}{2}}}$. Since $v_A(t,x) = A$ is a solution of the Neumann problem in $(0,\infty) \times B_1(0)$, by comparison, we have

$$A = v_A(t,x) \leq u(t,x) \quad \text{in} \quad (0,\infty) \times B_1(0).$$

Hence, $v(t,x) \leq u(t,x)$ in the parabolic boundary of $(0,1) \times B_1(0)$. Now, working as in the proof of Proposition 4.10, we get that

$$v(t,x) \leq u(t,x) \qquad \text{for all} \quad (t,x) \in (0,1) \times B_1(0).$$

Therefore, $u(t) \in L^1(B_1(0)) \setminus L^2(B_1(0))$, $0 \leq t < 1$.

Let us prove that if the initial condition is in $L^N(\Omega)$ then the solution is in $L^\infty(\Omega)$ for any $t > 0$.

Theorem 2.16. *Let $u(t)$ be the strong solution of (2.1) such that $u(0) = u_0$. If $u_0 \in L^N(\Omega)$, then $u(t) \in L^\infty(\Omega)$ for any $t > 0$.*

The result will be a consequence of the homogeneity estimate (2.34) and next result ([89]).

Theorem 2.17. *Let Ω be an open bounded set in $\mathbb{R}^N$ with Lipschitz boundary. Let $u \in BV(\Omega)$. Assume that there is $z \in X(\Omega)_N$ with $\|z\|_\infty \leq 1$ such that $(z, Du) = \|Du\|$ in Ω, and $[z,\nu] = 0$ in $\partial\Omega$. Then $u \in L^r(\Omega)$ for all $r < \infty$. If $z \in X(\Omega)_q$ with $q > N$, then $u \in L^\infty(\Omega)$.*

Proof. Let $f \in L^N(\Omega)$ be such that

$$-\mathrm{div}(z) = f \quad \text{in } \Omega. \tag{2.48}$$

Let $1^* = \frac{N}{N-1}$. Multiplying (2.48) by $|T_k(u)|^{1^*-1}T_k(u)$, $k > 0$, and integrating in Ω we obtain

$$\int_\Omega \|D(|T_k(u)|^{1^*-1}T_k(u))\| = \int_\Omega f|T_k(u)|^{1^*-1}T_k(u).$$

Since

$$\left(\int_\Omega |w - w_\Omega|^{1^*}\right)^{\frac{1}{1^*}} \leq C \int_\Omega \|Dw\|$$

for any $w \in BV(\Omega)$ and some constant $C > 0$ (see, (B.2)), we have

$$\left(\int_\Omega \left||T_k(u)|^{1^*-1}T_k(u) - \left(|T_k(u)|^{1^*-1}T_k(u)\right)_\Omega\right|^{1^*}\right)^{\frac{1}{1^*}}$$
$$\leq C \int_\Omega |f||T_k(u)|^{1^*} \leq C\left(m\int_{\Omega\setminus G_m} |T_k(u)|^{1^*} + \int_{G_m} |f||T_k(u)|^{1^*}\right)$$

where

$$G_m = \{x \in \Omega : |f(x)| \geq m\}.$$

Choose m large enough in order that

$$\|f\|_{L^N(G_m)} \leq \frac{1}{2C}.$$

Then, since

$$\left(\int_\Omega \left|\left(|T_k(u)|^{1^*-1}T_k(u)\right)_\Omega\right|^{1^*}\right)^{\frac{1}{1^*}} \leq \frac{1}{|\Omega|^{\frac{1}{N}}}\int_\Omega |T_k(u)|^{1^*},$$

we have

$$\left(\int_\Omega |T_k(u)|^{(1^*)^2}\right)^{\frac{1}{1^*}}$$
$$\leq C\left(m\int_{\Omega\setminus G_m} |T_k(u)|^{1^*} + \frac{1}{2C}\left(\int_{G_m} |T_k(u)|^{(1^*)^2}\right)^{\frac{1}{1^*}}\right) + \frac{1}{|\Omega|^{\frac{1}{N}}}\int_\Omega |T_k(u)|^{1^*}.$$

Letting $k \to \infty$ we obtain that $u \in L^{(1^*)^2}(\Omega)$, with

$$\left(\int_\Omega |u|^{(1^*)^2}\right)^{\frac{1}{1^*}} \leq 2\left(Cm + \frac{1}{|\Omega|^{\frac{1}{N}}}\right)\int_\Omega |u|^{1^*}.$$

Iterating this process one gets that $|u|^{(1^*)^n} \in L^1(\Omega)$ for all $n \geq 1$, and then $u \in L^r(\Omega)$ for all $r < \infty$.

We suppose now that $f \in L^q(\Omega)$ for some $q > N$. Let us multiply (2.48) by $|T_k(u)|^{j-1}T_k(u)$, $k > 0$, and integrate in Ω to obtain

$$\int_\Omega \|D(|T_k(u)|^{j-1}T_k(u))\| = \int_\Omega f|T_k(u)|^{j-1}T_k(u).$$

Let q' be the conjugate exponent of q. Using the Sobolev–Poincaré inequality (B.2) we obtain

$$\left(\int_\Omega \left||T_k(u)|^{j-1}T_k(u) - \left(|T_k(u)|^{j-1}T_k(u)\right)_\Omega\right|^{1^*}\right)^{\frac{1}{1^*}}$$
$$\leq C\int_\Omega \|D(|T_k(u)|^{j-1}T_k(u))\| \leq C\|f\|_q \left(\int_\Omega |T_k(u)|^{jq'}\right)^{\frac{1}{q'}}.$$

Consequently, using Hölder's inequality,

$$\left(\int_\Omega |T_k(u)|^{j1^*}\right)^{\frac{1}{1^*}} \leq C\|f\|_q \left(\int_\Omega |T_k(u)|^{jq'}\right)^{\frac{1}{q'}} + \frac{1}{|\Omega|^{\frac{1}{N}}}\int_\Omega |T_k(u)|^j$$
$$\leq D\left(\int_\Omega |T_k(u)|^{jq'}\right)^{\frac{1}{q'}},$$

where

$$D = C\|f\|_q + |\Omega|^{\frac{1}{q}-\frac{1}{N}}.$$

Taking $j = \frac{1^*}{q'}$ and letting $k \to \infty$ one gets both that $u \in L^{\frac{(1^*)^2}{q'}}(\Omega)$ and $|u|^{\frac{1^*}{q'}} \in BV(\Omega)$ with

$$\left(\int_\Omega \left(|u|^{1^*}\right)^{\frac{1^*}{q'}}\right)^{\frac{1}{1^*}} \leq D\left(\int_\Omega |u|^{1^*}\right)^{\frac{1}{q'}}.$$

Iterating this process we have

$$\left(\int_\Omega \left(|u|^{1^*}\right)^{\left(\frac{1^*}{q'}\right)^n}\right)^{\frac{1}{1^*}} \leq D^{1+\frac{1^*}{q'}+\cdots+\left(\frac{1^*}{q'}\right)^{n-1}} \left(\int_\Omega |u|^{1^*}\right)^{\left(\frac{1^*}{q'}\right)^n\frac{1}{1^*}}.$$

Let us write $K = \frac{q'}{1^*} < 1$, then we have

$$\left(\int_\Omega \left(|u|^{1^*}\right)^{\left(\frac{1^*}{q'}\right)^n}\right)^{\left(\frac{q'}{1^*}\right)^n} \leq D^{K1^*\frac{1-K^n}{1-K}}\int_\Omega |u|^{1^*}.$$

Letting $n \to \infty$ we obtain $u \in L^\infty(\Omega)$ and

$$\| |u|^{1^*} \|_\infty \leq D^{\frac{K1^*}{1-K}} \int_\Omega |u|^{1^*},$$

thus

$$\|u\|_\infty \leq D^{\frac{K}{1-K}} \|u\|_{1^*}. \qquad \square$$

Proof of Theorem 2.16. Let $u_0 \in L^N(\Omega)$. By estimate (2.34) and Lemma 2.15 we have that $u_t(t) \in L^N(\Omega)$ and

$$u_t(t) = \mathrm{div}(z(t)) \qquad \text{in } \mathcal{D}'(\Omega)$$

for almost all $t > 0$. By Theorem 2.17 we obtain that $u(t) \in L^q(\Omega)$ for any $q \in [1, \infty)$ and almost all $t > 0$. Using again Theorem 2.17 we deduce that $u(t) \in L^\infty(\Omega)$ for almost all $t > 0$, hence also for all $t > 0$. $\square$

2.6 Asymptotic Behaviour of Solutions

We start by proving that the mild solutions of problem (2.1) stabilize as $t \to 0$ by converging to a constant function. In order to prove the stabilization theorem we need the orbits to be relatively compact.

Lemma 2.18. *Let $(S(t))_{t\geq 0}$ be the semigroup generated by $\mathcal{A}$. Then, for every $u_0 \in L^1(\Omega)$, the orbit $\gamma(u_0) = \{S(t)u_0 \ : \ t \geq 0\}$ is a relatively compact subset of $L^1(\Omega)$.*

Proof. Let J_λ be the resolvent of $\mathcal{A}$. Then, $J_\lambda(B)$ is a relatively compact subset of $L^1(\Omega)$ if B is a bounded subset of $L^\infty(\Omega)$. In fact, let B a bounded subset of $L^\infty(\Omega)$. Take $\{f_n\}_{n=1}^\infty \subseteq B$ and let $u_n := J_\lambda f_n$. Set $M := \sup_{n\in\mathbb{N}} \|f_n\|_\infty < \infty$. Since $\mathcal{A}$ is m-completely accretive (Theorem 2.8), $\|u_n\|_\infty \leq M$ for every $n \in \mathbb{N}$. Moreover, since $(u_n, \frac{1}{\lambda}(f_n - u_n)) \in \mathcal{A}$, by Lemma 2.9,

$$\int_\Omega \|Du_n\| = \int_\Omega \frac{1}{\lambda}(f_n - u_n)u_n \, dx \leq \frac{1}{\lambda} M^2 \mathcal{L}^N(\Omega) \qquad \text{for all } n \in \mathbb{N}.$$

Thus, $\{u_n \ : \ n \in \mathbb{N}\}$ is a bounded sequence in $BV(\Omega)$, and by Theorem B.21 we have that $\{u_n \ : \ n \in \mathbb{N}\}$ is a relatively compact subset of $L^1(\Omega)$.

Consider first $u_0 \in \mathcal{D}(\mathcal{A}) \cap L^\infty(\Omega)$. Then, since

$$\|S(t)u_0\|_\infty \leq \|u_0\|_\infty \quad \text{for all } t \geq 0,$$

we have that $J_\lambda(\gamma(u_0))$ is a relatively compact subset of $L^1(\Omega)$ for all $\lambda > 0$. Moreover,

$$\|S(t)u_0 - J_\lambda S(t)u_0\|_1 \leq \lambda \inf \{\|v\|_1 \ : \ v \in \mathcal{A}(u_0)\}.$$

Hence, $\gamma(u_0)$ is relatively compact in $L^1(\Omega)$. Finally, since $\mathcal{D}(\mathcal{A}) \cap L^\infty(\Omega)$ is dense in $L^1(\Omega)$, given $u_0 \in L^1(\Omega)$ and $\epsilon > 0$, there exists $v_0 \in \mathcal{D}(\mathcal{A}) \cap L^\infty(\Omega)$ such that $\|u_0 - v_0\|_1 < \epsilon$. Thus we have,

$$\sup_{t\geq 0} \inf_{s\geq 0} \|S(t)u_0 - S(s)v_0\|_1 \leq \sup_{t\geq 0} \|S(t)u_0 - S(t)v_0\|_1 \leq \|u_0 - v_0\|_1 < \epsilon.$$

It follows that $\gamma(u_0)$ is relatively compact in $L^1(\Omega)$. □

We need the following result about the conservation of mass.

Lemma 2.19. *Let $(S(t))_{t\geq 0}$ be the semigroup generated by $\mathcal{A}$. Then, we have conservation of mass, that is,*

$$\int_\Omega S(t)u_0 \, dx = \int_\Omega u_0 \, dx, \qquad \text{for all } \ t \geq 0.$$

Proof. Given $u_0 \in L^1(\Omega)$, let $u(t) = S(t)u_0$. Then, $(u(t), -u'(t)) \in \mathcal{A}$. Hence, taking $w = T_k(u(t)) \pm 1$ as test function in (2.18), we obtain that $\int_\Omega u'(t) = 0$. Consequently, the function $t \mapsto \int_\Omega u(t)$ is constant, and the proof concludes. □

We denote by $\omega(u_0)$ the ω-limit set of u_0, i.e.,

$$\omega(u_0) := \left\{ v \in L^1(\Omega) \ : \ \exists t_n \to +\infty, \ \lim_{k\to\infty} S(t_{n_k})u_0 = v \right\}.$$

Theorem 2.20. *Let $(S(t))_{t\geq 0}$ be the semigroup generated by $\mathcal{A}$. Then*

$$\|S(t)u_0 - (u_0)_\Omega\|_1 \to 0 \quad \text{as} \ \ t \to \infty,$$

where

$$(u_0)_\Omega = \frac{1}{\mathcal{L}^N(\Omega)} \int_\Omega u_0(x) \, dx.$$

Moreover, if $u_0 \in L^\infty(\Omega)$ there exists a constant C, independent of u_0, such that

$$\|S(t)u_0 - (u_0)_\Omega\|_p \leq \frac{C\|u_0\|_2^2}{t} \quad \text{for all } \ t > 0, \ \text{ and } \ 1 \leq p \leq \frac{N}{N-1}.$$

Proof. Suppose first that $u_0 \in L^\infty(\Omega)$. Since $\mathcal{A}$ is completely accretive then $\|u(t)\|_\infty \leq \|u_0\|_\infty$. Using $k > \|u_0\|_\infty$ and letting $t \to \infty$ in (2.38) we have

$$\int_0^\infty \int_\Omega \|DS(\tau)u_0\| \, d\tau \leq \frac{1}{2} \int_\Omega u_0^2 \, dx. \tag{2.49}$$

Thus, there exists a sequence $t_n \to \infty$, such that $\displaystyle\int_\Omega \|DS(t_n)u_0\| \to 0$ as $n \to \infty$. Now by Lemma 2.18, there exists a subsequence (t_{n_k}) such that

$$\lim_{k\to\infty} S(t_{n_k})u_0 = v \in \omega(u_0),$$

and by the lower semi-continuity of the total variation, it follows that

$$\int_\Omega \|Dv\| \leq \liminf_{k\to\infty} \int_\Omega \|DS(t_{n_k})u_0\| = 0.$$

Therefore, v is a constant K, and consequently, $S(t)K = K$ for all $t \geq 0$; since the operators $S(t)$ are contractions we get $\omega(u_0) = \{K\}$ and

$$\lim_{t\to\infty} S(t)u_0 = K.$$

Now, as a consequence of Lemma 2.19, $K = (u_0)_\Omega$ and the proof for the case $u_0 \in L^\infty(\Omega)$ concludes. From the above the same conclusion in the general case $u_0 \in L^1(\Omega)$ is easily obtained.

Finally, suppose $u_0 \in L^\infty(\Omega)$. Then, by (2.49) we have that

$$\int_0^t \int_\Omega \|DS(s)u_0\| \, ds \leq \frac{1}{2}\|u_0\|_2^2 \quad \forall\, t > 0. \tag{2.50}$$

On the other hand, since $(S(s)u_0)_\Omega = (u_0)_\Omega$, by the Poincaré inequality (Theorem B.19), it follows that

$$\|S(s)u_0 - (u_0)_\Omega\|_p = \|S(s)u_0 - (S(s)u_0)_\Omega\|_p \leq M \int_\Omega \|DS(s)u_0\|, \tag{2.51}$$

for all $s > 0$, and $1 \leq p \leq \frac{N}{N-1}$. Then, (2.50) and (2.51) imply that

$$\int_0^t \|S(s)u_0 - (u_0)_\Omega\|_p \, ds \leq \frac{M}{2}\|u_0\|_2^2 \quad \forall\, t > 0. \tag{2.52}$$

Now, since $\mathcal{A}$ is completely accretive and $V(u) = \|u - (u_0)_\Omega\|_p$ is a Lyapunov functional for the semigroup generated by $\mathcal{A}$, using (2.52) we get

$$t\|S(t)u_0 - (u_0)_\Omega\|_p \leq \int_0^t \|S(s)u_0 - (u_0)_\Omega\|_p \, ds \leq \frac{M}{2}\|u_0\|_2^2,$$

concluding the proof. □

Now, we are going to prove, by energy methods, as in [23] (see also the monograph [24]), that in the two dimensional case, in fact, this asymptotic state is reached in finite time.

Theorem 2.21. *Suppose $N = 2$. Let $u_0 \in L^2(\Omega)$ and $u(t,x)$ the unique weak solution of problem (2.1). Then there exists a finite time T_0 such that*

$$u(t) = (u_0)_\Omega = \frac{1}{\mathcal{L}^N(\Omega)} \int_\Omega u_0(x) \, dx \qquad \forall\, t \geq T_0.$$

Proof. Since u is a weak solution of problem (2.1), there exists $z \in L^\infty(Q)$ with $\|z\|_\infty \leq 1$, $u_t = \mathrm{div}(z)$ in $\mathcal{D}'(Q)$ such that

$$\int_\Omega (u(t) - w)u_t(t)\,dx = \int_\Omega (z(t), Dw) - \int_\Omega \|Du(t)\| \tag{2.53}$$

for all $w \in BV(\Omega) \cap L^\infty(\Omega)$. Hence, taking $w = (u_0)_\Omega$ as test function in (2.53), it yields

$$\int_\Omega \big(u(t) - (u_0)_\Omega\big)u_t(t)\,dx = -\int_\Omega \|Du(t)\|.$$

Now, by Poincaré inequality for BV functions (Theorem B.19) and having in mind that we have conservation of mass, we obtain

$$\|u(t) - (u_0)_\Omega\|_2 \leq C\int_\Omega \|Du(t)\|.$$

Thus, we get

$$\frac{1}{2}\frac{d}{dt}\int_\Omega \big(u(t) - (u_0)_\Omega\big)^2\,dx + \frac{1}{C}\|u(t) - (u_0)_\Omega\|_2 \leq 0. \tag{2.54}$$

Therefore, the function

$$y(t) := \int_\Omega \big(u(t) - (u_0)_\Omega\big)^2\,dx$$

satisfies the inequality

$$y'(t) + My(t)^{1/2} \leq 0,$$

from which it follows that there exists $T_0 > 0$ such that $y(t) = 0$ for all $t \geq T_0$. $\square$

By Theorem 2.21, given $u_0 \in L^2(\Omega)$, if $u(t,x)$ is the unique weak solution of problem (2.1), then

$$T^*(u_0) := \inf\{t > 0 \;:\; u(t) = (u_0)_\Omega\} < \infty.$$

The study of the behaviour of $u(t)$ near $T^*(u_0)$ can be carried out as in the case of the Cauchy problem (see Chapter 4). As in that case, before proving the result, lower and upper bounds on the rate of decay of $\|u(t) - (u_0)_\Omega\|_2$ are established.

Lemma 2.22. *Suppose $N = 2$. Let $u_0 \in L^2(\Omega)$ and let $u(t,x)$ be the unique solution of problem* (2.1)*. Then, we have:*

(i) *There exists a constant C_1 independent of the initial data, such that*

$$C_1(T^*(u_0) - t) \leq \|u(t) - (u_0)_\Omega\|_2 \qquad \text{for } 0 \leq t \leq T^*(u_0). \tag{2.55}$$

(ii) *Given* $0 < \tau < T^*(u_0)$, *we have*

$$\|u(t) - (u_0)_\Omega\|_\infty \leq \frac{2\|u_0\|_\infty}{\tau}(T^*(u_0) - t) \qquad \text{for} \quad \tau \leq t \leq T^*(u_0). \quad (2.56)$$

Proof. Note that by Theorem 2.16 we may assume that $u_0 \in L^\infty(\Omega)$.
(i) Working as in the proof of Theorem 2.21, we get

$$\frac{1}{2}\frac{d}{dt}\int_\Omega (u(t) - (u_0)_\Omega)^2 \, dx + C_1\|u(t) - (u_0)_\Omega\|_2 \leq 0.$$

Hence

$$\frac{d}{dt}\left[\left(\int_\Omega |u(t) - (u_0)_\Omega|^2 \, dx\right)^{\frac{1}{2}}\right] + C_1 \leq 0. \quad (2.57)$$

Then, given $0 \leq t \leq T^*(u_0)$, integrating (2.57) from t to $T^*(u_0)$ we obtain (2.55).

(ii) The proof is a consequence of the regularizing effect due to the homogeneity of the operator $\mathcal{A}$. Recall that in (2.35) we have established

$$|u'(t)| \leq \frac{2}{t}|u_0| \quad \text{for almost all } \ t > 0. \quad (2.58)$$

□

As in the case of the Cauchy problem (see Chapter 4) we can prove the following result.

Theorem 2.23. *Suppose* $N = 2$. *Let* $u_0 \in L^2(\Omega)$ *and let* $u(t,x)$ *be the unique weak solution of problem* (2.1). *Let*

$$w(t,x) := \begin{cases} \dfrac{u(t,x) - (u_0)_\Omega}{T^*(u_0) - t} & \text{if } \ 0 \leq t < T^*(u_0), \\[2ex] 0 & \text{if } \ t \geq T^*(u_0). \end{cases}$$

Then, there exists an increasing sequence $t_n \to T^*(u_0)$, *and a solution* $v^* \neq 0$ *of the stationary problem*

$$(S_N) \begin{cases} -\mathrm{div}\left(\dfrac{Dv}{|Dv|}\right) = v & \text{in} \quad \Omega, \\[2ex] \dfrac{\partial v}{\partial \eta} = 0 & \text{on} \quad \partial\Omega, \end{cases}$$

such that

$$\lim_{n\to\infty} w(t_n) = v^* \quad \text{in} \ \ L^p(\Omega)$$

for all $1 \leq p < \infty$.

2.7 Regularity of the Level Lines

Theorem 2.24. *Let $u_0 \in L^N(\Omega)$ and let $u(t)$ be the strong solution of (2.1) corresponding to the initial datum $u(0) = u_0$. Then for any $t > 0$ and for almost $\lambda \in \mathbb{R}$ the reduced boundary $\partial^*[u(t) > \lambda]$ is relatively open in $\partial[u(t) > \lambda]$ and is a hypersurface of class $C^{1,\alpha}$ for any $\alpha < 1$. Moreover the closed set $\Sigma(E_\lambda) = \partial E_\lambda \setminus \partial^* E_\lambda$ is empty if $N < 8$, discrete if $N = 8$ and has Haussdorff dimension not greater than $N - 8$ if $N > 8$. This result can be more precise if $N = 2$, in this case $\partial[u(t) > \lambda]$ is of class $C^{1,1}$.*

By Theorem 2.16 we know that $u(t) \in L^\infty(\Omega)$ for all $t > 0$. Then Theorem 2.24 is a consequence of the following theorem which collects some results that have been proved in the literature. For a proof we refer to [8, 11] and the references therein.

Theorem 2.25. *Let $u \in BV(\Omega)$, $z \in X(\Omega)_p$, $N \leq p \leq \infty$, be such that $(z, Du) = \|Du\|$. Then for almost all levels $\lambda \in \mathbb{R}$, the sets $E_\lambda = [u > \lambda]$ satisfy:*

(i) *If $N < p < \infty$ ($p = \infty$), then the reduced boundary $\partial^* E_\lambda$ is relatively open in ∂E_λ and is a hypersurface of class $C^{1,\alpha}$ for any $\alpha < \frac{p-N}{2p}$ (resp., for any $\alpha < 1$). Moreover the closed set $\Sigma(E_\lambda) = \partial E_\lambda \setminus \partial^* E_\lambda$ is empty if $N < 8$, discrete if $N = 8$ and has Haussdorff dimension not greater than $N - 8$ if $N > 8$.*

(ii) *If $p = N$, there is a closed set $\Sigma(E_\lambda)$ of Haussdorff dimension not greater than $N - 8$ such that $\partial E_\lambda \setminus \Sigma(E_\lambda)$ is an $(N-1)$-dimensional manifold of class $C^{0,\alpha}$ for all $\alpha < 1$.*

If $N = 2$, these results can be more precise. If $p = 2$, then ∂E_λ is locally parameterizable with a bi-Lipschitz map (a Lipschitz map with a Lipschitz inverse). If $p = \infty$, ∂E_λ is of class $C^{1,1}$.

Proof. Let λ be such that $[u > \lambda]$ has finite perimeter in Ω (in particular, for almost every λ). Let $x \in \Omega$, F be a finite perimeter set such that $F \triangle [u > \lambda] \subset\subset B_\rho(x) \subseteq \Omega$. Then

$$\int_{[u>\lambda]\cap\Omega} \operatorname{div}(z) - \int_{F\cap\Omega} \operatorname{div}(z) \leq P(F, \Omega) - P([u > \lambda], \Omega). \tag{2.59}$$

We have

$$P([u > \lambda], \Omega) \leq P(F, \Omega) - \int_{[u>\lambda]\Delta F} \operatorname{div}(z)$$

and, thus, also

$$\begin{aligned} P([u>\lambda], B_\rho(x)) &\le P(F, B_\rho(x)) - \int_{[u>\lambda]\Delta F} \mathrm{div}(z) \\ &\le P(F, B_\rho(x)) + \|\mathrm{div}(z)\|_{L^N(B_\rho(x))} |[u>\lambda]\Delta F|^{\frac{N}{N-1}} \\ &\le P(F, B_\rho(x)) + \omega_N^{\frac{p-N}{N}} \|\mathrm{div}(z)\|_{L^p(B_\rho(x))} \rho^{2\alpha} |[u>\lambda]\Delta F|^{\frac{N}{N-1}} \end{aligned}$$

with $\alpha = \frac{p-N}{2p}$. This permits us to prove that there is a constant $C(N)$ such that

$$|[u>\lambda]\Delta F| \le C(N) P([u>\lambda], B_\rho(x)),$$

hence

$$P([u>\lambda], B_\rho(x)) \le \frac{1}{1-\eta(\rho)} P(F, B_\rho(x)),$$

where $\eta(\rho) = C(N)\omega_N^{\frac{p-N}{N}} \|\mathrm{div}(z)\|_{L^p(B_\rho(x))} \rho^{2\alpha}$. The above inequality may be written as

$$P([u>\lambda], B_\rho(x)) \le (1+\omega(\rho)) P(F, B_\rho(x)),$$

where $\omega(\rho) = \frac{\eta(\rho)}{1-\eta(\rho)}$. In other words, $[u>\lambda]$ is a quasi-minimizer of the perimeter. The study of the regularity of quasi-minimizers of the perimeter can be found in [8], [11] and the references therein. □

Chapter 3

The Total Variation Flow in $\mathbb{R}^N$

The purpose of this chapter is to prove existence and uniqueness of the minimizing total variation flow in $\mathbb{R}^N$

$$\frac{\partial u}{\partial t} = \operatorname{div}\left(\frac{Du}{|Du|}\right) \qquad \text{in }]0,\infty[\times\mathbb{R}^N, \tag{3.1}$$

coupled with the initial condition

$$u(0,x) = u_0(x) \qquad x \in \mathbb{R}^N, \tag{3.2}$$

when $u_0 \in L^1_{\rm loc}(\mathbb{R}^N)$.

3.1 Initial Conditions in $L^2(\mathbb{R}^N)$

Throughout this section, given a (possibly vector-valued) function f depending on space and time, we usually write $f(t)$ to mean the function $f(t,\cdot)$.

Definition 3.1. A function $u \in C([0,T];L^2(\mathbb{R}^N))$ is called a *strong solution* of (3.1) if

$$u \in W^{1,2}_{\rm loc}(0,T;L^2(\mathbb{R}^N)) \cap L^1_w(0,T;BV(\mathbb{R}^N))$$

and there exists $z \in L^\infty\left(]0,T[\times\mathbb{R}^N;\mathbb{R}^N\right)$ with $\|z\|_\infty \le 1$ such that

$$u_t = \operatorname{div}(z) \qquad \text{in } \mathcal{D}'(]0,T[\times\mathbb{R}^N)$$

and

$$\int_{\mathbb{R}^N}(u(t)-w)u_t(t)\,dx = \int_{\mathbb{R}^N}(z(t),Dw) - \int_{\mathbb{R}^N}\|Du(t)\| \tag{3.3}$$

for all $w \in L^2(\mathbb{R}^N) \cap BV(\mathbb{R}^N)$, a.e. $t \in [0,T]$.

The aim of this section is to prove the following result.

Theorem 3.2. *Let $u_0 \in L^2(\mathbb{R}^N)$. Then there exists a unique strong solution u of (3.1), (3.2) in $[0,T] \times \mathbb{R}^N$ for every $T > 0$. Moreover, if u and v are the strong solutions of (3.1) corresponding to the initial conditions $u_0, v_0 \in L^2(\mathbb{R}^N)$, then*

$$\|(u(t) - v(t))^+\|_2 \leq \|(u_0 - v_0)^+\|_2 \quad \text{for any } t > 0. \tag{3.4}$$

Proof. Let us introduce the following multivalued operator $\mathcal{B}$ in $L^2(\mathbb{R}^N)$: a pair of functions (u,v) belongs to the graph of $\mathcal{B}$ if and only if

$$u \in L^2(\mathbb{R}^N) \cap BV(\mathbb{R}^N), \; v \in L^2(\mathbb{R}^N), \tag{3.5}$$

$$\text{there exists } z \in X(\mathbb{R}^N)_2 \text{ with } \|z\|_\infty \leq 1, \text{ such that } v = -\mathrm{div}(z) \tag{3.6}$$

and

$$\int_{\mathbb{R}^N} (w-u)v\,dx \leq \int_{\mathbb{R}^N} z \cdot \nabla w\,dx - \int_{\mathbb{R}^N} \|Du\|, \quad \forall w \in L^2(\mathbb{R}^N) \cap W^{1,1}(\mathbb{R}^N).$$

Let also $\Psi : L^2(\mathbb{R}^N) \to \,]-\infty, +\infty]$ be the functional defined by

$$\Psi(u) := \begin{cases} \displaystyle\int_{\mathbb{R}^N} \|Du\| & \text{if } \; u \in L^2(\mathbb{R}^N) \cap BV(\mathbb{R}^N), \\ +\infty & \text{if } \; u \in L^2(\mathbb{R}^N) \setminus BV(\mathbb{R}^N). \end{cases} \tag{3.7}$$

Since Ψ is convex and lower semi-continuous in $L^2(\mathbb{R}^N)$, its subdifferential $\partial\Psi$ is a maximal monotone operator in $L^2(\mathbb{R}^N)$.

We divide the proof of the theorem into two steps.

Step 1. The following assertions are equivalent:

(a) $(u,v) \in \mathcal{B}$;

(b) (3.5) and (3.6) hold,

and

$$\int_{\mathbb{R}^N} (w-u)v\,dx \leq \int_{\mathbb{R}^N} (z, Dw) - \int_{\mathbb{R}^N} \|Du\| \tag{3.8}$$

for all $w \in L^2(\mathbb{R}^N) \cap BV(\mathbb{R}^N)$;

(c) (3.5) and (3.6) hold, and (3.8) holds with the equality instead of the inequality;

(d) (3.5) and (3.6) hold, and

$$\int_{\mathbb{R}^N} (z, Du) = \int_{\mathbb{R}^N} \|Du\|. \tag{3.9}$$

It is clear that (c) implies (b), and (b) implies (a), while (d) follows from (b) taking $w = u$ in (3.8) and using (C.8). In order to prove that (a) implies (b) it is enough to use Theorem B.3 and Lemma C.8 as in the proof of Lemma 2.9. To obtain (c) from (d) it suffices to multiply both terms of the equation $v = -\mathrm{div}(z)$ by $w - u$, for $w \in L^2(\mathbb{R}^N) \cap BV(\mathbb{R}^N)$, and to integrate by parts using (C.11).

Step 2. We also have $\mathcal{B} = \partial\Psi$. The proof is similar to the one given in Section 2.2 for the Neumann problem and we omit the details.

As a consequence, the semigroup generated by $\mathcal{B}$ coincides with the semigroup generated by $\partial\Psi$ and therefore $u(t,x) = e^{-t\mathcal{B}}u_0(x)$ is a strong solution of

$$u_t + \mathcal{B}u \ni 0,$$

i.e., $u \in W^{1,2}_{\mathrm{loc}}(]0,T[; L^2(\mathbb{R}^N))$ and $-u_t(t) \in \mathcal{B}u(t)$ for almost all $t \in\ ]0,T[$. Then, according to the equivalence proved in Step 1, we have that

$$\int_{\mathbb{R}^N} (u(t) - w)u_t(t)\, dx = \int_{\mathbb{R}^N} (z(t), Dw) - \int_{\mathbb{R}^N} \|Du(t)\| \tag{3.10}$$

for all $w \in L^2(\mathbb{R}^N) \cap BV(\mathbb{R}^N)$ and for almost all $t \in\ ]0,T[$. Now, choosing $w = u - \varphi$, $\varphi \in C_0^\infty(\mathbb{R}^N)$, we see that $u_t(t) = \mathrm{div}(z(t))$ in $\mathcal{D}'(\mathbb{R}^N)$ for almost every $t \in\ ]0,T[$. We deduce that $u_t = \mathrm{div}(z)$ in $\mathcal{D}'(]0,T[\times\mathbb{R}^N)$. We have proved that u is a strong solution of (3.1) in the sense of Definition 3.1.

The contractivity estimate (3.4) of Theorem 3.2 follows as in Theorem 2.2. This concludes the proof of the theorem. □

Given a function $g \in L^2(\mathbb{R}^N) \cap L^N(\mathbb{R}^N)$, we define

$$\|g\|_* := \sup\left\{\left|\int_{\mathbb{R}^N} g(x)u(x)\, dx\right| : u \in L^2(\mathbb{R}^N) \cap BV(\mathbb{R}^N),\ \int_{\mathbb{R}^N} \|Du\| \le 1\right\}.$$

Part (b) of the next lemma gives a characterization of $\mathcal{B}(0)$ which will be useful in Section 4.5 to find vector fields whose divergence is assigned. This part of the lemma was proved in [153] in the context of the analysis of the Rudin–Osher–Fatemi model for image denoising; for the sake of completeness, we shall include its proof.

Lemma 3.3. *Let $f \in L^2(\mathbb{R}^N) \cap L^N(\mathbb{R}^N)$ and $\lambda > 0$. The following assertions hold.*

(a) *The function u is the solution of*

$$\min_{w \in L^2(\mathbb{R}^N) \cap BV(\mathbb{R}^N)} D(w), \quad D(w) := \int_{\mathbb{R}^N} \|Dw\| + \frac{1}{2\lambda}\int_{\mathbb{R}^N} (w - f)^2\, dx \tag{3.11}$$

if and only if there exists $z \in X(\mathbb{R}^N)_2$ satisfying (3.9) with $\|z\|_\infty \le 1$ and $-\lambda\,\mathrm{div}(z) = f - u$.

(b) *The function* $u \equiv 0$ *is the solution of* (3.11) *if and only if* $\|f\|_* \leq \lambda$.

(c) *If* $N = 2$, $\mathcal{B}(0) = \{f \in L^2(\mathbb{R}^2) : \|f\|_* \leq 1\}$.

Proof. (a) Thanks to the strict convexity of D, u is the solution of (3.11) if and only if $0 \in \partial D(u) = \partial\Psi(u) + \frac{1}{\lambda}(u - f) = \mathcal{B}(u) + \frac{1}{\lambda}(u - f)$, where Ψ is defined in (3.7) and the last equality follows from Step 2 in the proof of Theorem 3.2. This means, recalling the definition of $\mathcal{B}$ in the proof of Theorem 3.2, that there exists $z \in X(\mathbb{R}^N)_2$ satisfying (3.9) with $\|z\|_\infty \leq 1$ and $-\lambda \operatorname{div}(z) = f - u$.
(b) The function $u \equiv 0$ is the solution of (3.11) if and only if

$$\int_{\mathbb{R}^N} \|Dv\| + \frac{1}{2\lambda}\int_{\mathbb{R}^N} (v - f)^2\, dx \geq \frac{1}{2\lambda}\int_{\mathbb{R}^N} f^2\, dx \qquad \forall v \in L^2(\mathbb{R}^N) \cap BV(\mathbb{R}^N). \tag{3.12}$$

Replacing v by ϵv (where $\epsilon > 0$), expanding the L^2-norm, dividing by $\epsilon > 0$, and letting $\epsilon \to 0+$ we have

$$\left|\int_{\mathbb{R}^N} f(x)v(x)\, dx\right| \leq \lambda \int_{\mathbb{R}^N} \|Dv\| \qquad \forall v \in L^2(\mathbb{R}^N) \cap BV(\mathbb{R}^N). \tag{3.13}$$

Since (3.13) implies (3.12), we have that (3.12) and (3.13) are equivalent. The assertion follows by observing that (3.13) is equivalent to $\|f\|_* \leq \lambda$.
(c) Let $N = 2$. We have

$$\mathcal{B}(0) = \left\{v \in L^2(\mathbb{R}^2) \;:\; \exists z \in X(\mathbb{R}^2)_2,\ \|z\|_\infty \leq 1, -\operatorname{div}(z) = v\right\}.$$

On the other hand, from (a) and (b) it follows that $\|f\|_* \leq 1$ if and only if there exists $z \in X(\mathbb{R}^2)_2$ with $\|z\|_\infty \leq 1$ and such that $f = -\operatorname{div}(z)$. Then the assertion follows. □

Let us give a heuristic explanation of what the vector field z represents. Condition (3.9) essentially means that z has unit norm and is orthogonal to the level sets of u. In some sense, z is invariant under local contrast changes. To be more precise, we observe that if $u = \sum_{i=1}^p c_i \chi_{B_i}$ where B_i are sets of finite perimeter such that $\mathcal{H}^{N-1}((B_i \cup \partial^* B_i) \cap (B_j \cup \partial^* B_j)) = 0$ for $i \neq j$, $c_i \in \mathbb{R}$, and

$$-\operatorname{div}\left(\frac{Du}{|Du|}\right) = f \in L^2(\mathbb{R}^N), \tag{3.14}$$

then also $-\operatorname{div}\left(\frac{Dv}{|Dv|}\right) = f$ for any $v = \sum_{i=1}^p d_i \chi_{B_i}$ where $d_i \in \mathbb{R}$ and $\operatorname{sign}(d_i) = \operatorname{sign}(c_i)$. Indeed, there is a vector field $z \in L^\infty(\mathbb{R}^N; \mathbb{R}^N)$ such that $\|z\|_\infty \leq 1$, $-\operatorname{div}(z) = f$ and (3.9) holds. Then one can check that

$$\|D\chi_{B_i}\| = \operatorname{sign}(c_i)(z, D\chi_{B_i})$$

as measures in $\mathbb{R}^N$ and, as a consequence, $(z, Dv) = \|Dv\|$ as measures in $\mathbb{R}^N$.

Let us also observe that the solutions of (3.14) are not unique. Indeed, if $u \in L^2(\mathbb{R}^N) \cap BV(\mathbb{R}^N)$ is a solution of (3.14) and $g \in C^1(\mathbb{R})$ with $g'(r) > 0$ for all $r \in \mathbb{R}$, then $w = g(u)$ is also a solution of (3.14). In other words, a global contrast change of u produces a new solution of (3.14). In an informal way, the previous remark can be rephrased by saying that also local contrast changes of a given solution of (3.14) produce new solutions of it. To express this nonuniqueness in a more general way we suppose that $(u_1, v), (u_2, v) \in \mathcal{B}$, i.e., there are vector fields $z_i \in X(\mathbb{R}^N)_2$ with $\|z_i\|_\infty \le 1$, such that

$$-\mathrm{div}(z_i) = v, \qquad \int_{\mathbb{R}^N} (z_i, Du_i) = \int_{\mathbb{R}^N} \|Du_i\|, \qquad i = 1, 2.$$

Then

$$\begin{aligned} 0 &= -\int_{R^N} (\mathrm{div}(z_1) - \mathrm{div}(z_2))(u_1 - u_2)\, dx = \int_{R^N} (z_1 - z_2, Du_1 - Du_2) \\ &= \int_{R^N} \|Du_1\| - (z_2, Du_1) + \int_{\mathbb{R}^N} \|Du_2\| - (z_1, Du_2). \end{aligned}$$

Hence

$$\int_{R^N} \|Du_1\| = \int_{\mathbb{R}^N} (z_2, Du_1) \quad \text{and} \quad \int_{\mathbb{R}^N} \|Du_2\| = \int_{\mathbb{R}^N} (z_1, Du_2).$$

In other words, z_1 is in some sense a unit vector field of normals to the level sets of u_2 and a similar thing can be said of z_2 with respect to u_1. Any two solutions of (3.14) should be related in this way.

Definition 3.4. Let $u \in C([0,T]; L^2(\mathbb{R}^N)) \cap W^{1,1}_{\mathrm{loc}}(]0,T[; L^2(\mathbb{R}^N))$. We say that u is a *supersolution* of (3.1) and (3.2) if $u \in L^1_w(]0,T[; BV(\mathbb{R}^N))$ and there exists a vector field $z \in L^\infty(]0,T[\times\mathbb{R}^N; \mathbb{R}^N)$ such that $\|z\|_\infty \le 1$, $\mathrm{div}(z(t)) \in L^2_{\mathrm{loc}}(\mathbb{R}^N)$ for almost all $t \in \,]0,T[$,

$$\int_{\mathbb{R}^N} (z(t), Du(t)) = \int_{\mathbb{R}^N} \|Du(t)\| \quad \text{a.e. } t \in \,]0,T[, \tag{3.15}$$

$u_t \ge \mathrm{div}(z)$ and $u(0) \ge u_0$.

For convenience, the proof of the following proposition will be given at the end of Section 3.3.

Proposition 3.5. *Let $u_0 \in L^2(\mathbb{R}^N)$. Let u be a supersolution of* (3.1), (3.2) *and let v be the strong solution of* (3.1), (3.2). *Then $u \ge v$.*

3.2 The Notion of Entropy Solution

Recall that by $L^1_w(0,T;BV(\mathbb{R}^N))$ we denote the space of functions $w : [0,T] \to BV(\mathbb{R}^N)$ such that $w \in L^1(]0,T[\times\mathbb{R}^N)$, the maps

$$t \in [0,T] \to \int_{\mathbb{R}^N} \phi \, dDw(t)$$

are measurable for every $\phi \in C_0^1(\mathbb{R}^N;\mathbb{R}^N)$ and $\int_0^T \|Dw(t)\|(\mathbb{R}^N)\, dt < \infty$. By $L^1_w(0,T;BV_{\rm loc}(\mathbb{R}^N))$ we denote the space of functions $w : [0,T] \to BV_{\rm loc}(\mathbb{R}^N)$ such that $w\varphi \in L^1_w(]0,T[;BV(\mathbb{R}^N))$ for all $\varphi \in C_0^\infty(\mathbb{R}^N)$. We need to consider the set of truncation functions

$$\mathcal{P} := \left\{p \in W^{1,\infty}(\mathbb{R}) : p' \geq 0, \ \operatorname{supp}(p') \text{ compact}\right\}. \tag{3.16}$$

Definition 3.6. A function $u \in C([0,T];L^1_{\rm loc}(\mathbb{R}^N))$ is called an *entropy solution* of (3.1), (3.2) if $u(t)$ converges to u_0 in $L^1_{\rm loc}(\mathbb{R}^N)$ as $t \to 0^+$,

$$p(u) \in L^1_w(0,T;BV_{\rm loc}(\mathbb{R}^N)) \qquad \forall p \in \mathcal{P},$$

and there exists $z \in L^\infty\left(]0,T[\times\mathbb{R}^N;\mathbb{R}^N\right)$ with $\|z\|_\infty \leq 1$ such that

$$u_t = \operatorname{div}(z) \qquad \text{in } \mathcal{D}'\left(]0,T[\times\mathbb{R}^N\right) \tag{3.17}$$

and

$$-\int_0^T\int_{\mathbb{R}^N} j(u-l)\eta_t + \int_0^T\int_{\mathbb{R}^N} \eta\, d\left\|D\left(p(u-l)\right)\right\| + \int_0^T\int_{\mathbb{R}^N} z\cdot\nabla\eta\, p(u-l) \leq 0 \tag{3.18}$$

for all $l \in \mathbb{R}$, all $\eta \in C^\infty\left(]0,T[\times\mathbb{R}^N\right)$, with $\eta \geq 0$, $\eta(t,x) = \phi(t)\psi(x)$, being $\phi \in C_0^\infty\left(]0,T[\right)$, $\psi \in C_0^\infty(\mathbb{R}^N)$, and all $p \in \mathcal{P}$, where $j(r) := \int_0^r p(s)\, ds$.

Inequality (3.18) is a weak way to impose equality (3.3); indeed if we integrate by parts, we formally substitute (3.17), using $\|z\|_\infty \leq 1$ and the fact that η is nonnegative, we get

$$\begin{aligned} \int_{\mathbb{R}^N} z\cdot\nabla\eta\, p(u-l) &= -\int_{\mathbb{R}^N} j(u-l)_t\eta - \int_{\mathbb{R}^N}\eta\, d(z, D\left(p(u-l)\right)) \\ &\geq -\int_{\mathbb{R}^N} j(u-l)_t\eta - \int_{\mathbb{R}^N}\eta\, d|(z, D\left(p(u-l)\right)|, \end{aligned}$$

which, after integration in time, shows that the opposite inequality in (3.18) is satisfied.

Remark 3.7. As we shall prove in Section 3.4, if $u_0 \in L^2(\mathbb{R}^N)$, then the strong solution of (3.1), (3.2) coincides with the entropy solution.

The aim of Sections 3.3 and 3.4 is to prove the following result.

Theorem 3.8. *Let $u_0 \in L^1_{\rm loc}(\mathbb{R}^N)$. Then there exists a unique entropy solution of (3.1) and (3.2) in $[0,T] \times \mathbb{R}^N$ for all $T > 0$. Moreover, if $u_0, u_{0k} \in L^1_{\rm loc}(\mathbb{R}^N)$ are such that $u_{0k} \to u_0$ in $L^1_{\rm loc}(\mathbb{R}^N)$ and u, u_k denote the corresponding entropy solutions, then $u_k \to u$ in $C([0,T]; L^1_{\rm loc}(\mathbb{R}^N))$ as $k \to +\infty$.*

3.3 Uniqueness in $L^1_{\rm loc}(\mathbb{R}^N)$

To prove uniqueness we use the doubling variables technique introduced by Kruzhkov ([143]). The same method will be used throughout this monograph in order to prove uniqueness of solutions.

Let $\alpha > N$, $T_k^+(r) = \max(T_k(r), 0)$ $(k \geq 0)$ and let j_α be the primitive of $\alpha T_k^+(r)^{\alpha-1}$ vanishing at $r = 0$. If $N = 1$, we take $\alpha \geq 2$, so that $j'_\alpha \in W^{1,\infty}(\mathbb{R})$.

Proposition 3.9. *Let $u_0, \overline{u_0} \in L^1_{\rm loc}(\mathbb{R}^N)$. Let $u, \overline{u}$ be two entropy solutions of (3.1) with initial conditions $u_0, \overline{u_0}$, respectively. Then*

$$\int_{\mathbb{R}^N} j_\alpha(u(t) - \overline{u}(t)) \leq \int_{\mathbb{R}^N} j_\alpha(u_0 - \overline{u}_0) \qquad \forall t > 0. \tag{3.19}$$

Proof. Let $T > 0$ and $Q_T := \,]0,T[\times\mathbb{R}^N$. Write $j = j_\alpha$, $j^*(r) := j(-r)$, $p(r) := \alpha T_k^+(r)^{\alpha-1}$, $p^*(r) := j^{*\prime}(r) = -p(-r)$. Let $z, \overline{z} \in L^\infty(Q_T; \mathbb{R}^N)$ with $\|z\|_\infty \leq 1$, $\|\overline{z}\|_\infty \leq 1$ and such that, if $r, \overline{r} \in \mathbb{R}^N$, with $\|r\| \leq 1$, $\|\overline{r}\| \leq 1$ and $l_1, l_2 \in \mathbb{R}$, then

$$\begin{aligned} &-\int_0^T \int_{\mathbb{R}^N} j(u - l_1)\eta_t + \int_0^T \int_{\mathbb{R}^N} \eta \, d\|D\left(p(u - l_1)\right)\| \\ &+ \int_0^T \int_{\mathbb{R}^N} (z - r) \cdot \nabla\eta \, p(u - l_1) + \int_0^T \int_{\mathbb{R}^N} r \cdot \nabla\eta \, p(u - l_1) \leq 0, \end{aligned} \tag{3.20}$$

and

$$\begin{aligned} &-\int_0^T \int_{\mathbb{R}^N} j^*(\overline{u} - l_2)\eta_t + \int_0^T \int_{\mathbb{R}^N} \eta \, d\|D\left(p^*(\overline{u} - l_2)\right)\| \\ &+ \int_0^T \int_{\mathbb{R}^N} (\overline{z} - \overline{r}) \cdot \nabla\eta \, p^*(\overline{u} - l_2) + \int_0^T \int_{\mathbb{R}^N} \overline{r} \cdot \nabla\eta \, p^*(\overline{u} - l_2) \leq 0, \end{aligned} \tag{3.21}$$

for all $\eta \in C^\infty(Q_T)$, with $\eta \geq 0$, $\eta(t,x) = \phi(t)\psi(x)$, being $\phi \in C_0^\infty(]0,T[)$, $\psi \in C_0^\infty(\mathbb{R}^N)$.

We choose two different pairs of variables (t,x), (s,y) and consider u, z as functions of (t,x) and $\overline{u}$, $\overline{z}$ as functions of (s,y). Let $0 \leq \phi \in C_0^\infty(]0,T[)$,

$0 \leq \psi \in C_0^\infty(\mathbb{R}^N)$, (ρ_n) a standard sequence of mollifiers in $\mathbb{R}^N$ and $(\tilde{\rho}_n)$ a sequence of mollifiers in $\mathbb{R}$. Define

$$\eta_n(t,x,s,y) := \tilde{\rho}_n(t-s)\rho_n(x-y)\phi\left(\frac{t+s}{2}\right)\psi\left(\frac{x+y}{2}\right) \geq 0.$$

Note that for n sufficiently large,

$$(t,x) \mapsto \eta_n(t,x,s,y) \in C_0^\infty\left(]0,T[\times\mathbb{R}^N\right) \quad \forall\, (s,y) \in Q_T,$$
$$(s,y) \mapsto \eta_n(t,x,s,y) \in C_0^\infty\left(]0,T[\times\mathbb{R}^N\right) \quad \forall\, (t,x) \in Q_T.$$

Hence, for (s,y) fixed, if we take $l_1 = \overline{u}(s,y)$ and $r = \overline{z}(s,y)$ in (3.20), we get

$$\begin{aligned} &-\int_0^T\int_{\mathbb{R}^N} j(u-\overline{u}(s,y))(\eta_n)_t + \int_0^T\int_{\mathbb{R}^N} \eta_n \, d\|D_x\left(p(u-\overline{u}(s,y))\right)\| \\ &+\int_0^T\int_{\mathbb{R}^N} (z-\overline{z}(s,y))\cdot\nabla_x\eta_n \; p(u-\overline{u}(s,y)) \\ &+\int_0^T\int_{\mathbb{R}^N} \overline{z}(s,y)\cdot\nabla_x\eta_n \; p(u-\overline{u}(s,y)) \leq 0. \end{aligned} \tag{3.22}$$

Similarly, for (t,x) fixed, if we take $l_2 = u(t,x)$ and $\overline{r} = z(t,x)$ in (3.21), we get

$$\begin{aligned} &-\int_0^T\int_{\mathbb{R}^N} j^*(\overline{u}-u(t,x))(\eta_n)_s + \int_0^T\int_{\mathbb{R}^N} \eta_n \, d\|D_y\left(p^*(\overline{u}-u(t,x))\right)\| \\ &+\int_0^T\int_{\mathbb{R}^N} (\overline{z}-z(t,x))\cdot\nabla_y\eta_n \; p^*(\overline{u}-u(t,x)) \\ &+\int_0^T\int_{\mathbb{R}^N} z(t,x)\cdot\nabla_y\eta_n \; p^*(\overline{u}-u(t,x)) \leq 0. \end{aligned} \tag{3.23}$$

Now, since $p^*(r) = -p(-r)$ and $j^*(r) = j(-r)$, we can rewrite (3.23) as

$$\begin{aligned} &-\int_0^T\int_{\mathbb{R}^N} j(u(t,x)-\overline{u})(\eta_n)_s + \int_0^T\int_{\mathbb{R}^N} \eta_n \, d\|D_y\left(p(u(t,x)-\overline{u})\right)\| \\ &+\int_0^T\int_{\mathbb{R}^N} (z(t,x)-\overline{z})\cdot\nabla_y\eta_n \; p(u(t,x)-\overline{u}) \\ &-\int_0^T\int_{\mathbb{R}^N} z(t,x)\cdot\nabla_y\eta_n \, p(u(t,x)-\overline{u}) \leq 0. \end{aligned} \tag{3.24}$$

Integrating (3.22) with respect to (s,y) and (3.24) with respect to (t,x) and taking the sum yields

$$\begin{aligned}
&-\int_{Q_T\times Q_T} j(u(t,x)-\overline{u}(s,y))\big((\eta_n)_t+(\eta_n)_s\big)\\
&+\int_{Q_T\times Q_T} \eta_n\, d\|D_x\,(p(u-\overline{u}(s,y)))\,\| + \int_{Q_T\times Q_T} \eta_n\, d\|D_y\,(p(u(t,x)-\overline{u}(s))\|\\
&+\int_{Q_T\times Q_T} \big(z(t,x)-\overline{z}(s,y)\big)\cdot\big(\nabla_x\eta_n+\nabla_y\eta_n\big)\, p(u(t,x)-\overline{u}(s,y))\\
&+\int_{Q_T\times Q_T} \overline{z}(s,y)\cdot\nabla_x\eta_n\, p(u(t,x)-\overline{u}(s,y))\\
&-\int_{Q_T\times Q_T} z(t,x)\cdot\nabla_y\eta_n\, p(u(t,x)-\overline{u}(s,y))\le 0.
\end{aligned} \tag{3.25}$$

Now, by Green's formula we have

$$\begin{aligned}
&\int_{Q_T\times Q_T} \overline{z}(s,y)\cdot\nabla_x\eta_n\ p(u(t,x)-\overline{u}(s,y))\\
&+\int_{Q_T\times Q_T} \eta_n\, d\|D_x\,(p(u(t,x)-\overline{u}(s,y)))\,\|\\
&=-\int_{Q_T\times Q_T} \eta_n\,(\overline{z}(s,y),D_xp(u(t,x)-\overline{u}(s,y)))\\
&\quad+\int_{Q_T\times Q_T} \eta_n\, d\|D_x\,(p(u(t,x)-\overline{u}(s,y)))\,\|\ge 0,
\end{aligned}$$

and

$$\begin{aligned}
&-\int_{Q_T\times Q_T} z(t,x)\cdot\nabla_y\eta_n\ p(u(t,x)-\overline{u}(s,y))\\
&+\int_{Q_T\times Q_T} \eta_n\, d\|D_y\,(p(u(t,x)-\overline{u}(s,y)))\,\|\\
&=\int_{Q_T\times Q_T} \eta_n\,(z(t,x),D_yp(u(t,x)-\overline{u}(s,y)))\\
&\quad+\int_{Q_T\times Q_T} \eta_n\, d\|D_y\,(p(u(t,x)-\overline{u}(s,y)))\,\|\ge 0.
\end{aligned}$$

Hence, from (3.25), it follows that

$$\begin{aligned}
&-\int_{Q_T\times Q_T} j(u(t,x)-\overline{u}(s,y))\big((\eta_n)_t+(\eta_n)_s\big)\\
&+\int_{Q_T\times Q_T} \big(z(t,x)-\overline{z}(s,y)\big)\cdot\big(\nabla_x\eta_n+\nabla_y\eta_n\big)\, p(u(t,x)-\overline{u}(s,y))\le 0.
\end{aligned} \tag{3.26}$$

Since

$$(\eta_n)_t + (\eta_n)_s = \tilde{\rho}_n(t-s)\rho_n(x-y)\phi'\left(\frac{t+s}{2}\right)\psi\left(\frac{x+y}{2}\right)$$

and

$$\nabla_x\eta_n + \nabla_y\eta_n = \tilde{\rho}_n(t-s)\rho_n(x-y)\phi\left(\frac{t+s}{2}\right)\nabla\psi\left(\frac{x+y}{2}\right),$$

passing to the limit in (3.26) as $n \to +\infty$ yields

$$\begin{aligned} &-\int_{Q_T} j(u(t,x)-\overline{u}(t,x))\phi'(t)\psi(x) \\ &+\int_{Q_T} \big(z(t,x)-\overline{z}(t,x)\big)\cdot\nabla\psi(x)\;\phi(t)p(u(t,x)-\overline{u}(t,x)) \leq 0. \end{aligned} \tag{3.27}$$

Let us choose $\psi = \varphi^\alpha$, $\varphi \in C_0^\infty(\mathbb{R}^N)$, $\varphi \geq 0$. Since (3.27) holds for any $\phi \in C_0^\infty(]0,T[)$, it follows that

$$\begin{aligned} &\frac{d}{dt}\int_{\mathbb{R}^N} j(u(t,x)-\overline{u}(t,x))\varphi(x)^\alpha \\ &\leq \int_{\mathbb{R}^N} \big(\overline{z}(t,x)-z(t,x)\big)\cdot\nabla\varphi(x)^\alpha\; p(u(t,x)-\overline{u}(t,x)). \end{aligned}$$

Therefore

$$\begin{aligned} &\frac{d}{dt}\int_{\mathbb{R}^N} j(u(t,x)-\overline{u}(t,x))\varphi(x)^\alpha \\ &\leq 2\alpha\int_{\mathbb{R}^N} |p(u(t,x)-\overline{u}(t,x))|\varphi^{\alpha-1}|\nabla\varphi| \\ &\leq 2\alpha\left(\int_{\mathbb{R}^N}\left(|p(u(t,x)-\overline{u}(t,x))|\varphi^{\alpha-1}\right)^{\frac{\alpha}{\alpha-1}}\right)^{\frac{\alpha-1}{\alpha}}\left(\int_{\mathbb{R}^N}|\nabla\varphi|^\alpha\right)^{\frac{1}{\alpha}} \\ &\leq 2\alpha^2\left(\int_{\mathbb{R}^N}|T_k^+(u(t,x)-\overline{u}(t,x))^\alpha|\varphi^\alpha\right)^{\frac{\alpha-1}{\alpha}}\left(\int_{\mathbb{R}^N}|\nabla\varphi|^\alpha\right)^{\frac{1}{\alpha}}. \end{aligned} \tag{3.28}$$

Now, we observe that $T_k^+(r)^\alpha \leq j_\alpha(r)$ for all $r \in \mathbb{R}$. Hence

$$\begin{aligned} &\frac{d}{dt}\int_{\mathbb{R}^N} j(u(t,x)-\overline{u}(t,x))\varphi^\alpha \\ &\leq 2\alpha^2\left(\int_{\mathbb{R}^N} j(u(t,x)-\overline{u}(t,x))\varphi^\alpha\right)^{\frac{\alpha-1}{\alpha}}\left(\int_{\mathbb{R}^N}|\nabla\varphi|^\alpha\right)^{\frac{1}{\alpha}}, \end{aligned}$$

and, therefore,

$$\frac{d}{dt}\left(\int_{\mathbb{R}^N} j(u(t,x)-\overline{u}(t,x))\varphi^\alpha\right)^{\frac{1}{\alpha}} \leq 2\alpha\left(\int_{\mathbb{R}^N}|\nabla\varphi|^\alpha\right)^{\frac{1}{\alpha}}.$$

Setting $\varphi_n(x) := \varphi(\frac{x}{n})$ instead of $\varphi(x)$ we get

$$\begin{aligned}&\frac{d}{dt}\left(\int_{\mathbb{R}^N} j(u(t,x)-\overline{u}(t,x))\varphi_n^\alpha\right)^{\frac{1}{\alpha}}\\&\le 2\alpha\left(\int_{\mathbb{R}^N}|\nabla\varphi_n|^\alpha\right)^{\frac{1}{\alpha}} = 2\alpha n^{\frac{N-\alpha}{\alpha}}\left(\int_{\mathbb{R}^N}|\nabla\varphi|^\alpha\right)^{\frac{1}{\alpha}}.\end{aligned}$$

Integrating from 0 to T and using the facts that $u(t)\to u_0$, $\overline{u}(t)\to\overline{u}_0$ in $L^1_{\text{loc}}(\mathbb{R}^N)$ as $t\to 0^+$, we have

$$\begin{aligned}&\left(\int_{\mathbb{R}^N} j(u(T,x)-\overline{u}(T,x))\varphi_n^\alpha\right)^{\frac{1}{\alpha}}\\&\le\left(\int_{\mathbb{R}^N} j(u_0-\overline{u}_0)\varphi_n^\alpha\right)^{\frac{1}{\alpha}} + 2\alpha T n^{\frac{N-\alpha}{\alpha}}\left(\int_{\mathbb{R}^N}|\nabla\varphi|^\alpha\right)^{\frac{1}{\alpha}}.\end{aligned}\tag{3.29}$$

Letting $n\to\infty$ and recalling that $\alpha > N$, we obtain that

$$\int_{\mathbb{R}^N} j(u(T,x)-\overline{u}(T,x)) \le \int_{\mathbb{R}^N} j(u_0-\overline{u}_0).\tag{3.30}$$

□

Corollary 3.10. *Let $u_0,\overline{u_0}\in L^1_{\text{loc}}(\mathbb{R}^N)$. Let $u,\overline{u}$ be two entropy solutions of* (3.1) *with initial conditions $u_0,\overline{u_0}$, respectively. If $u_0\le\overline{u_0}$, then $u\le\overline{u}$. In particular, the entropy solution of* (3.1) *is unique.*

Corollary 3.11. *Let $S(t)$ be the semigroup in $L^1_{loc}(\mathbb{R}^N)$ constructed from the entropy solutions. Then $S(t)$ acts as an order preserving contraction semigroup in $L^p(\mathbb{R}^N)$.*

Proof. The result is a consequence of estimate (3.30). □

Proof of the last assertion of Theorem 3.8. Writing (3.29) for $u(t,x)$ and $u_k(t,x)$, we have

$$\begin{aligned}&\left(\int_{\mathbb{R}^N} j(u(t,x)-u_k(t,x))\varphi_n^\alpha\right)^{\frac{1}{\alpha}}\\&\le\left(\int_{\mathbb{R}^N} j(u_0-u_{0k})\varphi_n^\alpha\right)^{\frac{1}{\alpha}} + 2\alpha t n^{\frac{N-\alpha}{\alpha}}\left(\int_{\mathbb{R}^N}|\nabla\varphi|^\alpha\right)^{\frac{1}{\alpha}},\end{aligned}$$

for any $t\in[0,T]$ and any $n,k\ge 1$. Given $p\in\mathbb{N}$, let $n_p\in\mathbb{N}$ be such that

$$2\alpha T n_p^{\frac{N-\alpha}{\alpha}}\left(\int_{\mathbb{R}^N}|\nabla\varphi|^\alpha\right)^{\frac{1}{\alpha}} \le \frac{1}{p}.$$

Choose now $\varphi \in C_0^\infty(\mathbb{R}^N)$ of the form $\varphi(x) = \phi(\|x\|)$ where ϕ is a decreasing function. By our choice of φ we have that

$$\left(\int_{\mathbb{R}^N} j(u(t,x) - u_k(t,x))\varphi_n^\alpha\right)^{\frac{1}{\alpha}} \leq \left(\int_{\mathbb{R}^N} j(u(t,x) - u_k(t,x))\varphi_{n_p}^\alpha\right)^{\frac{1}{\alpha}}$$
$$\leq \left(\int_{\mathbb{R}^N} j(u_0 - u_{0k})\varphi_{n_p}^\alpha\right)^{\frac{1}{\alpha}} + \frac{1}{p}$$

for any $t \in [0,T]$ and any $k \geq 1$. Now, let $k_p \in \mathbb{N}$ be such that

$$\left(\int_{\mathbb{R}^N} j(u_0 - u_{0k})\varphi_{n_p}^\alpha\right)^{\frac{1}{\alpha}} \leq \frac{1}{p}$$

for any $k \geq k_p$. Then

$$\left(\int_{\mathbb{R}^N} j(u(t,x) - u_k(t,x))\varphi^\alpha\right)^{\frac{1}{\alpha}} \leq \frac{2}{p}$$

for any $t \in [0,T]$ and any $k \geq k_p$. We conclude that

$$u_k \to u \quad \text{in } C([0,T]; L^1_{\text{loc}}(\mathbb{R}^N)). \qquad \square$$

Remark 3.12. The same proof above yields that $\{u_k\}$ is a Cauchy sequence in $C([0,T]; L^1_{\text{loc}}(\mathbb{R}^N))$ when $\{u_{0k}\}$ is a Cauchy sequence in $L^1_{\text{loc}}(\mathbb{R}^N)$.

We observe that Proposition 3.5 can be proved along the same lines as Proposition 3.9. But, in this case the proof is simpler and we sketch it.

Proof of Proposition 3.5. Let z_u, z_v be the vector fields associated to u, v, respectively. Using the same notations of the proof of Proposition 3.9, we have

$$\begin{aligned}
\frac{d}{dt}\int_{\mathbb{R}^N} j(v(t) - u(t))\varphi^\alpha\,dx &= \int_{\mathbb{R}^N} p(v(t) - u(t))(v_t(t) - u_t(t))\varphi^\alpha\,dx \\
&\leq \int_{\mathbb{R}^N} p(v(t) - u(t))(\operatorname{div}(z_v(t)) - \operatorname{div}(z_u(t)))\varphi^\alpha\,dx \\
&= -\int_{\mathbb{R}^N} \varphi^\alpha\, d\,(D(p(v(t) - u(t))), z_v(t) - z_u(t))\,dx \\
&\quad - \int_{\mathbb{R}^N} p(v(t) - u(t))(z_v(t) - z_u(t))\cdot\nabla\varphi^\alpha\,dx \\
&\leq -\int_{\mathbb{R}^N} p(v(t) - u(t))(z_v(t) - z_u(t))\cdot\nabla\varphi^\alpha\,dx.
\end{aligned}$$

To conclude the proof, we proceed as in the proof of Proposition 3.9 after formula (3.28). $\square$

The following estimate, which is a consequence of the homogeneity of $\mathcal{B}$, will be useful to prove the regularity in time of the solution when the initial condition is in $L^1_{\text{loc}}(\mathbb{R}^N)$ (see Proposition 3.20 of Section 3.6 below). The proof is similar to the proof of (2.35).

Proposition 3.13. *Let $u_0 \in L^p(\mathbb{R}^N)$, $1 \leq p \leq \infty$, and let $u(t)$ be the entropy solution of (3.1) and (3.2). Then*

$$\| u'(t) \|_p \leq \frac{\| u(t) \|_p}{t} \quad \text{for a.e. } t > 0. \tag{3.31}$$

If $u_0 \geq 0$, then we have that

$$u'(t) \leq \frac{u(t)}{t} \quad \text{for a.e. } t > 0.$$

Similarly, if $u_0 \leq 0$, then $u'(t) \geq \frac{u(t)}{t}$ for almost every $t > 0$.

3.4 Existence in $L^1_{\text{loc}}(\mathbb{R}^N)$

Lemma 3.14. *Let $u_0 \in L^2(\mathbb{R}^N)$ and let u be the strong solution of (3.1) and (3.2). Let $T > 0$, $p \in \mathcal{P}$, set $j(r) := \int_0^r p(s)\, ds$, and let $\varphi \in C^\infty([0,T] \times \mathbb{R}^N)$ with compact support in x. Then*

$$\begin{aligned}\int_{\mathbb{R}^N} j(u(T))\varphi(T) - \int_0^T \int_{\mathbb{R}^N} j(u)\varphi_t + \int_0^T \int_{\mathbb{R}^N} \varphi \, d\|D\,(p(u))\,\| \\ = -\int_0^T \int_{\mathbb{R}^N} z \cdot \nabla\varphi \; p(u) + \int_{\mathbb{R}^N} j(u_0)\varphi(0).\end{aligned} \tag{3.32}$$

In particular, u is an entropy solution of (3.1).

Proof. Since

$$\begin{aligned}\frac{d}{dt}\int_{\mathbb{R}^N} j(u)\varphi &= \int_{\mathbb{R}^N} p(u)u_t\varphi + \int_{\mathbb{R}^N} j(u)\varphi_t \\ &= -\int_{\mathbb{R}^N} \varphi \; d(z, D(p(u))) - \int_{\mathbb{R}^N} z \cdot \nabla\varphi \; p(u) + \int_{\mathbb{R}^N} j(u)\varphi_t,\end{aligned}$$

integrating both terms of the above equality in $]0,T[$, and using the fact that

$$\int_{\mathbb{R}^N} \varphi \; d(z(t), D(p(u(t)))) = \int_{\mathbb{R}^N} \varphi \; d\|D(p(u(t)))\| \qquad \text{for a.e. } t \in \;]0,T[,$$

which is a consequence of Corollary C.16 and the equality

$$\int_{\mathbb{R}^N} \varphi \; d(z(t), Du(t)) = \int_{\mathbb{R}^N} \varphi \; d\|Du(t)\| \qquad \text{for a.e. } t \in \;]0,T[,$$

we obtain

$$\int_{\mathbb{R}^N} j(u(T))\varphi(T) - \int_0^T \int_{\mathbb{R}^N} j(u)\varphi_t + \int_0^T \int_{\mathbb{R}^N} \varphi \, d\|D\,(p(u))\,\| \qquad (3.33)$$
$$= - \int_0^T \int_{\mathbb{R}^N} z \cdot \nabla\varphi \, p(u) + \int_{\mathbb{R}^N} j(u_0)\varphi(0),$$

and (3.32) holds. □

Proof of existence. Let $u_0 \in L^1_{\mathrm{loc}}(\mathbb{R}^N)$. Let $u_{0n} \in L^2(\mathbb{R}^N)$ be such that $u_{0n} \to u_0$ in $L^1_{\mathrm{loc}}(\mathbb{R}^N)$. Let u_n be the strong solutions of (3.1) corresponding to the initial conditions u_{0n}. By Remark 3.12, $\{u_n\}$ is a Cauchy sequence in $C([0,T]; L^1_{\mathrm{loc}}(\mathbb{R}^N))$. Thus we may assume that $u_n \to u$ in $C([0,T]; L^1_{\mathrm{loc}}(\mathbb{R}^N))$ for some $u \in C([0,T]; L^1_{\mathrm{loc}}(\mathbb{R}^N))$. In particular, we have that $u(t) \to u_0$ in $L^1_{\mathrm{loc}}(\mathbb{R}^N)$ as $t \to 0+$.

Now, let $p \in \mathcal{P}$ and let $\varphi \in C_0^\infty(]0,T[\times\mathbb{R}^N)$. Inserting $u = u_n$ into (3.32) gives

$$-\int_0^T \int_{\mathbb{R}^N} j(u_n)\varphi_t + \int_0^T \int_{\mathbb{R}^N} \varphi \, d\|D\,(p(u_n))\,\| = - \int_0^T \int_{\mathbb{R}^N} z_n \cdot \nabla\varphi p(u_n). \quad (3.34)$$

In particular, the choice of $j(r) = r$, i.e., $p(r) = 1$, gives

$$\int_0^T \int_{\mathbb{R}^N} u_n \varphi_t = \int_0^T \int_{\mathbb{R}^N} z_n \cdot \nabla\varphi. \qquad (3.35)$$

Possibly passing to a subsequence, we may assume that

$$z_n \to z \quad \text{weakly}^* \text{ in } (L^\infty(]0,T[\times\mathbb{R}^N))^N.$$

Letting $n \to \infty$ in (3.35) we have

$$\int_0^T \int_{\mathbb{R}^N} u\varphi_t = \int_0^T \int_{\mathbb{R}^N} z \cdot \nabla\varphi. \qquad (3.36)$$

We conclude $u_t = \operatorname{div}(z)$ in $\mathcal{D}'(]0,T[\times\mathbb{R}^N)$. As $j(u_n) \to j(u)$ and $p(u_n) \to p(u)$ in $C([0,T]; L^1_{\mathrm{loc}}(\mathbb{R}^N))$, letting $n \to \infty$ in (3.34) we obtain

$$-\int_0^T \int_{\mathbb{R}^N} j(u)\varphi_t + \int_0^T \int_{\mathbb{R}^N} \varphi \, d\|D\,(p(u))\,\| \leq - \int_0^T \int_{\mathbb{R}^N} z \cdot \nabla\varphi \, p(u)$$

provided $\varphi \geq 0$. In particular, since $j(u), p(u) \in C([0,T]; L^1_{\mathrm{loc}}(\mathbb{R}^N))$ we have

$$p(u) \in L^1_w(0,T; BV_{\mathrm{loc}}(\mathbb{R}^N)) \qquad \forall p \in \mathcal{P},$$

and we conclude that u is an entropy solution of (3.1). □

3.5 Initial Conditions in $L^1(\mathbb{R}^N)$

Definition 3.15. A function $u \in C([0,T];L^1(\mathbb{R}^N))$ is called a *strong solution* of (3.1) if

$$u \in W^{1,1}_{\mathrm{loc}}(0,T;L^1(\mathbb{R}^N)), \quad T_k(u) \in L^1_w(0,T;BV(\mathbb{R}^N)) \quad \forall k > 0$$

and there exists $z \in L^\infty(]0,T[\times\mathbb{R}^N;\mathbb{R}^N)$ with $\|z\|_\infty \le 1$ such that

$$u_t = \mathrm{div}(z) \qquad \text{in } \mathcal{D}'(]0,T[\times\mathbb{R}^N)$$

and

$$\int_{\mathbb{R}^N} (T_k(u(t)) - w)u_t(t) \le \int_{\mathbb{R}^N} (z(t), Dw) - \int_{\mathbb{R}^N} \|DT_k(u(t))\| \tag{3.37}$$

for all $w \in L^\infty(\mathbb{R}^N) \cap BV(\mathbb{R}^N)$, a.e. $t \in [0,T]$.

As it happens in the Neumann problem (see Chapter 2), we note that the equality sign also holds in (3.37).

Theorem 3.16. *If $u_0 \in L^1(\mathbb{R}^N)$, then the entropy solution $u(t)$ is a strong solution, and conversely.*

Proof. Assume first that u is an entropy solution of (3.1) and (3.2). Then we know that $T_k(u(t)) \in L^1_w(0,T;BV_{loc}(\mathbb{R}^N))$, and, by Proposition 3.13, $u \in W^{1,1}_{\mathrm{loc}}(0,T;L^1(\mathbb{R}^N))$. By taking $p = T_k$, approximating u_0 by functions in $L^2(\mathbb{R}^N)$, and letting $\varphi \uparrow 1$ in (3.32) we obtain that $T_k(u(t)) \in L^1_w(0,T;BV(\mathbb{R}^N))$ for all $k > 0$. Now, approximating u_0 in $L^1(\mathbb{R}^N)$ by initial conditions in $L^2(\mathbb{R}^N)$ it follows that

$$u_t = \mathrm{div}(z) \quad \text{in } \mathcal{D}'(]0,T[\times\mathbb{R}^N). \tag{3.38}$$

Observe that using $\varphi \to \varphi(t) \in \mathcal{D}'(]0,T[)$ in (3.32) we obtain

$$\int_{\mathbb{R}^N} T_k(u(t))u_t(t) + \int_{\mathbb{R}^N} \|DT_k(u(t))\| \le 0. \tag{3.39}$$

By approximation, using (3.38) it follows that

$$-\int_{\mathbb{R}^N} wu_t(t) = \int_{\mathbb{R}^N} (z(t), Dw) \tag{3.40}$$

for all $w \in L^\infty(\mathbb{R}^N) \cap W^{1,1}(\mathbb{R}^N)$, a.e. $t \in [0,T]$. A further approximation proves that (3.40) holds for any $w \in L^\infty(\mathbb{R}^N) \cap BV(\mathbb{R}^N)$. Adding (3.39) and (3.40) we obtain that $u(t)$ is a strong solution of (3.1). Conversely, if $u(t)$ is a strong solution of (3.1), since $u_t \in L^1_{loc}(0,T;L^1(\mathbb{R}^N))$ it is immediate to check that $u(t)$ is an entropy solution of (3.1). □

3.6 Time Regularity

Let us recall the basic estimates of semigroups generated by subdifferentials. According to *Step* 2 of Theorem 3.2 and Theorem A.33 we have that

$$\operatorname*{ess\ sup}_{s\in\,]t,\infty[} \int_{\mathbb{R}^N} |u_t(s,x)|^2 dx \leq \frac{1}{t}\int_{\mathbb{R}^N} |u_0|^2 dx \qquad \forall t>0, \tag{3.41}$$

$$\int_0^T \int_{\mathbb{R}^N} |u_t(t,x)|^2 t dx dt \leq \frac{1}{2}\int_{\mathbb{R}^N} |u_0|^2 dx \tag{3.42}$$

and if $u_0 \in BV(\mathbb{R}^N)$,

$$\int_0^T \int_{\mathbb{R}^N} |u_t(t,x)|^2 dx dt \leq \int_{\mathbb{R}^N} \|Du_0\|. \tag{3.43}$$

Our purpose is to localize estimates (3.42), (3.43). To cover the case of initial conditions in $L^1_{\rm loc}(\mathbb{R}^N)$, we need to consider the family $\mathcal{T} \subseteq \mathcal{P}$ of truncations $T_{a,b}$, with $a<b$, defined by

$$T_{a,b}(r) := \begin{cases} a & \text{if } r<a, \\ r & \text{if } a\leq r\leq b, \\ b & \text{if } r>b. \end{cases}$$

Proposition 3.17. *Let $u_0 \in L^2(\mathbb{R}^N)$ and let u be the strong solution of* (3.1) *and* (3.2). *Then*

$$p(u)_t \in L^2_{\rm loc}(0,T;L^2(\mathbb{R}^N)), \qquad t^{\frac{1}{2}}p(u)_t \in L^2(0,T;L^2(\mathbb{R}^N)), \qquad \forall p\in\mathcal{T}.$$

Moreover, for any $\varphi \in C_0^\infty(\mathbb{R}^N)$ and any $s<t$ such that $p(u(s))\in BV_{\rm loc}(\mathbb{R}^N)$ we have the estimate

$$\begin{aligned} &\frac{1}{2}\int_s^t \int_{\mathbb{R}^N} |p(u)_t|^2\varphi^2 + \int_{\mathbb{R}^N} \varphi^2 \ d\|D\left(p(u(t))\right)\| \\ &\leq \int_{\mathbb{R}^N} \varphi^2 d\|D\left(p(u(s))\right)\| + 2(t-s)\int_{\mathbb{R}^N} |\nabla\varphi|^2, \end{aligned} \tag{3.44}$$

and, if T is such that $u(T)\in BV_{\rm loc}(\mathbb{R}^N)$, also

$$\begin{aligned} &\frac{1}{2}\int_0^T \int_{\mathbb{R}^N} t|p(u)_t|^2\varphi^2 + T\int_{\mathbb{R}^N} \varphi^2 d\|Dp(u(T))\| \\ &\leq \int_0^T \int_{\mathbb{R}^N} \varphi^2 d\|Dp(u(t))\| + T^2\int_{\mathbb{R}^N} |\nabla\varphi|^2. \end{aligned} \tag{3.45}$$

Proof. Let $\varphi \in C_0^\infty(\mathbb{R}^N)$ and set

$$I := \left\{ s \in \,]0,T[\, :\ u(s) \in BV_{\rm loc}(\mathbb{R}^N) \int_{\mathbb{R}^N} |u_t(s,x)|^2 \, dx \le \frac{1}{s} \int_{\mathbb{R}^N} |u_0|^2 \, dx \right\}.$$

We recall that $]0,T[\setminus I$ has measure zero. Let $s,t \in I$. Multiply the equation $u_t(t) = \operatorname{div}(z(t))$ by $(p(u(t)) - p(u(s)))\varphi^2$ and integrate over $\mathbb{R}^N$. After integrating by parts, we obtain

$$\begin{aligned} &\int_{\mathbb{R}^N} \varphi^2 d\left(\|D\left(p(u(t))\right)\| - \|D\left(p(u(s))\right)\|\right) \\ &\le -\int_{\mathbb{R}^N} u_t(t)\left[p(u(s)) - p(u(t))\right]\varphi^2 \int_{\mathbb{R}^N} z(t)\cdot\nabla\varphi^2\left[p(u(t)) - p(u(s))\right]. \end{aligned} \tag{3.46}$$

Let $\delta > 0$ and let $s,t \in I$, $s,t \ge \delta$. Using (3.41), we have

$$\begin{aligned} &\int_{\mathbb{R}^N} \varphi^2 d\Big(\|D\left(p(u(t))\right)\| - \|D\left(p(u(s))\right)\|\Big) \\ &\le \frac{1}{\delta}\|u_0\|_2 \|\left[p(u(s)) - p(u(t))\right]\varphi^2\|_2 + \int_{\mathbb{R}^N} |\nabla\varphi^2||p(u(t)) - p(u(s))|. \end{aligned}$$

Since a similar inequality holds with s and t interchanged, we have

$$\begin{aligned} &\left|\int_{\mathbb{R}^N} \varphi^2 d\Big(\|D\left(p(u(t))\right)\| - \|D(p(u(s)))\|\Big)\right| \\ &\le \frac{1}{\delta}\|u_0\|_2 \|(p(u(s)) - p(u(t)))\varphi^2\|_2 + \int_{\mathbb{R}^N} |\nabla\varphi^2||p(u(t)) - p(u(s))|. \end{aligned} \tag{3.47}$$

As $u \in W^{1,2}_{\rm loc}(0,T;L^2(\mathbb{R}^N))$, i.e., is a locally absolutely continuous function of time, then also $p(u)$ is and, from (3.47), we deduce that $\int_{\mathbb{R}^N} \varphi^2 d\|D\left(p(u)\right)\|$ is absolutely continuous in $]\delta,T[$ for any $\delta > 0$ sufficiently small. Put $s = t - h \in I$ in (3.46), divide by $h > 0$, and let $h \to 0^+$. We obtain, at any differentiability point t of u and $\int_{\mathbb{R}^N} \varphi^2 d\|D\left(p(u)\right)\|$,

$$\begin{aligned} \int_{\mathbb{R}^N} p'(u)u_t^2\varphi^2 \ &+\ \frac{d}{dt}\int_{\mathbb{R}^N} \varphi^2 d\|D\left(p(u)\right)\| \le 2\int_{\mathbb{R}^N} |p(u)_t||\varphi||\nabla\varphi| \\ &\le\ 2\left(\int_{\mathbb{R}^N} |p(u)_t|^2\varphi^2\right)^{1/2}\left(\int_{\mathbb{R}^N} |\nabla\varphi|^2\right)^{1/2} \\ &\le\ \frac{1}{2}\int_{\mathbb{R}^N} |p(u)_t|^2\varphi^2 + 2\|\nabla\varphi\|_2^2. \end{aligned}$$

Since $p'(r) \in \{0,1\}$ for almost every r, we have

$$\frac{1}{2}\int_{\mathbb{R}^N} |p(u)_t|^2\varphi^2 + \frac{d}{dt}\int_{\mathbb{R}^N} \varphi^2 d\|D\left(p(u)\right)\| \le 2\|\nabla\varphi\|_2^2. \tag{3.48}$$

Observe that inequality (3.48) holds almost everywhere in $]0,T[$. Choosing $s \in I$ and integrating (3.48) in $]s,t[$ we obtain (3.44). Since φ does not depend on time, from (3.32) it follows that

$$\int_{\mathbb{R}^N} j(u(T))\varphi^2 + \int_0^T \int_{\mathbb{R}^N} \varphi^2 \, d\|D(p(u))\| \leq \int_0^T \int_{\mathbb{R}^N} |\nabla\varphi^2||p(u)| + \int_0^T \int_{\mathbb{R}^N} j(u_0)\varphi^2. \tag{3.49}$$

Inequality (3.49) proves that

$$\int_{\mathbb{R}^N} \varphi^2 d\|D\,(p(u))\,\| \in L^1(0,T).$$

Hence

$$t_n \int_{\mathbb{R}^N} \varphi^2 \|D\,(p(u(t_n)))\,\| \to 0$$

for a subsequence $t_n \to 0^+$, $t_n \in I$. Multiplying (3.48) by t and integrating on $]t_n, T[$ we obtain

$$\frac{1}{2}\int_{t_n}^T \int_{\mathbb{R}^N} t|p(u)_t|^2\varphi^2 + \int_{t_n}^T t\frac{d}{dt}\int_{\mathbb{R}^N} \varphi^2 \, d\|D(p(u))\| \leq (T^2 - t_n^2)\int_{\mathbb{R}^N} |\nabla\varphi^2|.$$

Integrating by parts with respect to time we get

$$\begin{aligned}
&\frac{1}{2}\int_{t_n}^T \int_{\mathbb{R}^N} t|p(u)_t|^2\varphi^2 + T\int_{\mathbb{R}^N} \varphi^2 \, d\|D(p(u(T)))\| \\
&\leq \int_{t_n}^T \int_{\mathbb{R}^N} \varphi^2 \, d\|D(p(u))\| + t_n \int_{\mathbb{R}^N} \varphi^2 \, d\|D(p(u(t_n)))\| + (T^2 - t_n^2)\int_{\mathbb{R}^N} |\nabla\varphi^2|.
\end{aligned}$$

Letting $n \to \infty$, we obtain (3.45). □

Corollary 3.18. *Let $u_0 \in L^1_{\text{loc}}(\mathbb{R}^N)$. Let u be the entropy solution of* (3.1) *and* (3.2). *Then*

$$p(u)_t \in L^2_{\text{loc}}(0,\infty; L^2_{\text{loc}}(\mathbb{R}^N)), \qquad t^{\frac{1}{2}}p(u)_t \in L^2_{\text{loc}}([0,\infty[; L^2_{\text{loc}}(\mathbb{R}^N)), \quad \forall p \in \mathcal{T}.$$

Proof. Let $(u_{0n}) \subset L^2(\mathbb{R}^N)$ be a sequence such that $u_{0n} \to u_0$ in $L^1_{\text{loc}}(\mathbb{R}^N)$. Let u_n be the strong solution of (3.1) corresponding to the initial condition u_{0n}. Inserting $u = u_n$ into (3.32) and using the fact that the corresponding vector fields z_n satisfy $\| z_n \|_\infty \leq 1$, we obtain

$$\begin{aligned}
&\int_{\mathbb{R}^N} j(u_n(T))\varphi^2 + \int_0^T \int_{\mathbb{R}^N} \varphi^2 \, d\|D\,(p(u_n))\,\| \\
&\leq \int_0^T \int_{\mathbb{R}^N} |\nabla\varphi^2||p(u_n)| + \int_{\mathbb{R}^N} j(u_n(0))\varphi^2
\end{aligned} \tag{3.50}$$

for any $p \in \mathcal{P}$, $T > 0$, $\varphi \in C_0^\infty(\mathbb{R}^N)$ and $n \in \mathbb{N}$. Since the right-hand side of (3.50) is bounded by

$$C := \| p \|_\infty \, T \int_{\mathbb{R}^N} |\nabla \varphi^2| \, dx + \sup_n \int_{\mathbb{R}^N} j(u_n(0))\varphi^2,$$

we have

$$\int_0^T \int_{\mathbb{R}^N} \varphi^2 \, d\| D\,(p(u_n)) \| \leq C. \tag{3.51}$$

Choose now $T > 0$ such that $u_n(T) \in BV_{\rm loc}(\mathbb{R}^N))$ for all $n \in \mathbb{N}$. Using (3.45) and (3.51) we have

$$\frac{1}{2}\int_0^T \int_{\mathbb{R}^N} t|p(u_n)_t|^2\varphi^2 \leq C + T^2 \int_{\mathbb{R}^N} |\nabla\varphi|^2. \tag{3.52}$$

Since $p(u_n) \to p(u)$ in $C([0,T]; L^1_{\rm loc}(\mathbb{R}^N))$, letting $n \to \infty$ in (3.52) we obtain

$$\frac{1}{2}\int_0^T \int_{\mathbb{R}^N} tp(u)_t^2\varphi^2 \leq C + T^2 \int_{\mathbb{R}^N} |\nabla\varphi|^2.$$

Since this holds for almost every $T > 0$, the conclusion follows. □

Remark 3.19. If $p(u_0) \in BV_{\rm loc}(\mathbb{R}^N)$ we have

$$p(u) \in L^1_w(0,T; BV_{\rm loc}(\mathbb{R}^N)),$$
$$p(u) \in W^{1,2}(0,T; L^2_{\rm loc}(\mathbb{R}^N)) \subseteq C([0,T]; L^2_{\rm loc}(\mathbb{R}^N))$$

for any $p \in \mathcal{T}$. Indeed, this follows from (3.44) instead of using (3.45) in the above argument.

If u is the entropy solution of (3.1) and (3.2) for $u_0 \in L^1_{\rm loc}(\mathbb{R}^N)$ and $K \in \mathbb{R}$, then $v(t) := u(t) + K$ is the entropy solution of (3.1) whose initial condition is $v(0) = u_0 + K$. If we denote by $S(t)$ the semigroup in $L^1_{\rm loc}(\mathbb{R}^N)$ constructed from the entropy solutions, we may write $S(t)(u_0 + K) = S(t)u_0 + K$ for any $u(0) = u_0 \in L^1_{\rm loc}(\mathbb{R}^N)$ and $K \in \mathbb{R}$.

Proposition 3.20. *Let $u_0 \in L^1_{\rm loc}(\mathbb{R}^N)$ with $u_0 \geq -M$ for some $M > 0$. If u is the entropy solution of* (3.1) *and* (3.2) *we have*

$$u'(t) \leq \frac{u(t) + M}{t} \qquad \text{for a.e. } t > 0.$$

Moreover, $u_t \in L^1_{\rm loc}(0,T; L^1_{\rm loc}(\mathbb{R}^N))$ for any $T > 0$. A similar statement holds if $u_0 \leq M$ for some $M > 0$.

Proof. Let $0 \leq v_{0n} \in L^2(\mathbb{R}^N)$ be such that $v_{0n} \to u_0 + M$ in $L^1_{\rm loc}(\mathbb{R}^N)$. Let $v_n(t) := S(t)(v_{0n})$. By Proposition 3.13 we have

$$v'_{nt} \leq \frac{v_n}{t} \qquad \text{for a.e. } t > 0.$$

Since

$$v_n(t) \to S(t)(u_0 + M) = S(t)(u_0) + M = u(t) + M \quad \text{in } L^1(0,T; L^1_{\rm loc}(\mathbb{R}^N)),$$

it follows that

$$u_t \leq \frac{u+M}{t} \qquad \text{in } \mathcal{D}'(]0,T[\times\mathbb{R}^N). \tag{3.53}$$

By estimate (3.53), u_t is a Radon measure in $]s,t[\times B_R(0)$, for all $0 < s < t$ and $R > 0$. Thus

$$\int_s^t \int_{B_R(0)} |u_t| < \infty \tag{3.54}$$

in any ball $B_R(0)$, $R > 0$. Now, taking $p = T_{ab}$, the estimate in Corollary 3.18 says that u_t is a function in $L^2(Q_{a,b} \cap B_R(0))$, for all $a < b$, where $Q_{a,b} := \{(t,x) \in Q \, : \, a < u(t,x) < b\}$, and all $R > 0$. This observation together with (3.54) proves that $u_t \in L^1_{\rm loc}(0,T; L^1_{\rm loc}(\mathbb{R}^N))$. $\square$

We conclude this section with the following observation. The existence and uniqueness results for (3.1) and (3.2) may be used to prove an estimate for the time derivative of the solution of

$$\frac{\partial v}{\partial t} = \operatorname{div}\left(\frac{Dv}{\sqrt{1+|Dv|^2}}\right) \qquad \text{in }]0,\infty[\times\mathbb{R}^N, \tag{3.55}$$

when the initial datum $v(0,x) = v_0(x) \in L^1(\mathbb{R}^N)$. First, we observe that existence and uniqueness results for (3.55) when $v_0 \in L^1(\mathbb{R}^N)$ can be obtained as a consequence of the results in Chapter 7. Next we notice that if v is the solution of (3.55) corresponding to the initial condition $v_0 \in L^1(\mathbb{R}^N)$, then $u(t,x,x_{N+1}) = v(t,x) - x_{N+1}$ is the entropy solution of (3.1) in $\mathbb{R}^{N+1}$ such that $u(0,x,x_{N+1}) = v_0(x) - x_{N+1}$. In other words, the semigroups $T(t)$ and $S(t)$ associated with (3.55) and (3.1) satisfy

$$S(t)(v_0 - x_{N+1}) = T(t)v_0 - x_{N+1} \quad \text{for any } v_0 \in L^1(\mathbb{R}^N).$$

Now, proceeding as in the proof of Proposition 3.13 with $\lambda = \frac{t+h}{t}$ we obtain

$$\begin{aligned} v(t+h) - v(t) &= u(t+h) - u(t) = \frac{h}{t+h} u(t+h) + S(t)\left(\lambda^{-1}(v_0 - x_{N+1})\right) - u(t) \\ &= \frac{h}{t+h} u(t+h) + T(t)\left(\lambda^{-1} v_0\right) - T(t)v_0 + \frac{h}{t+h} x_{N+1} \\ &= \frac{h}{t+h} v(t+h) + T(t)\left(\lambda^{-1} v_0\right) - T(t)v_0. \end{aligned}$$

This implies that

$$\left\| \frac{v(t+h)-v(t)}{h} \right\|_1 \leq \frac{2}{t+h} \|v_0\|_1.$$

From this, and using the techniques of completely accretive operators (see Section A.7) as in the proof of Theorem 2.8 it can be proved that $\|v_t\|_1 \leq \frac{2}{t}\|v_0\|_1$.

3.7 An L^N-L^∞ Regularizing Effect

As in the Neumann problem (see Section 2.5), there is no L^1-L^∞ or L^1-L^2 regularizing effect for the Cauchy problem. To see this, it is enough to observe that the function

$$v(t,x) = \left(\frac{1}{\|x\|^{N/2}} - \frac{t}{\|x\|} \right) \chi_{B_1(0)}(x),$$

with associated vector field

$$z(t)(x) = \begin{cases} \frac{1}{1-N} \frac{x}{\|x\|} & x \in B_1(0), \\ \\ \frac{1}{1-N} \frac{x}{\|x\|^N} & x \in \mathbb{R}^N \setminus \overline{B_1(0)}, \end{cases}$$

solves (3.1) in $(0,1) \times \mathbb{R}^N$ with initial datum $v_0(x) = \frac{1}{\|x\|^{N/2}} \chi_{B_1(0)}(x)$. Obviously, $v(t) \in L^1(\mathbb{R}^N) \setminus L^2(\mathbb{R}^N)$, $0 \leq t < 1$.

Using the same technique as in Section 2.5, if the initial condition is in $L^N(\mathbb{R}^N)$ a regularizing effect can be obtained.

Theorem 3.21. *Let $u(t)$ be the strong solution of (3.1) such that $u(0) = u_0$. If $u_0 \in L^N(\mathbb{R}^N)$, then $u(t) \in L^p(\mathbb{R}^N) \cap L^\infty(\mathbb{R}^N)$ for any $p \geq 1$ and $t > 0$.*

The result will be a consequence of the homogeneity estimate (3.31) and the next result whose proof is similar to the one of Theorem 2.17 .

Theorem 3.22. *Let $u \in BV(\mathbb{R}^N)$. Assume that there is $z \in X(\mathbb{R}^N)_N$ with $\|z\|_\infty \leq 1$ such that $(z, Du) = \|Du\|$. Then $u \in L^r(\mathbb{R}^N)$ for all $r < \infty$. If $z \in X(\mathbb{R}^N)_q$ with $q > N$, then $u \in L^\infty(\mathbb{R}^N)$.*

3.8 Measure Initial Conditions

In [19], we have studied the Cauchy problem (3.1) coupled with $u(0) = \mu$, μ being a bounded Radon measure in $\mathbb{R}^N$. Since (3.1) is well posed in $L^1(\mathbb{R}^N)$ we can approximate μ by functions in $u_{0,n} \in L^1(\mathbb{R}^N)$, compute the corresponding solutions $u_n(t)$ and pass to the limit to obtain a function $u(t)$ taking values in the space of Radon measures. For a later purpose let us denote $u(t) = u(t)_{ac} + u(t)_s$, where $u(t)_{ac}$ and $u(t)_s$ denote the absolutely continuous and singular parts of $u(t)$

with respect to Lebesgue measure in $\mathbb{R}^N$. We did not consider general measures, instead we restricted ourselves to the case of measures

$$\mu = h + \alpha \mathcal{H}^k \llcorner S, \tag{3.56}$$

where $h \in L^1(\mathbb{R}^N) \cap L^\infty(\mathbb{R}^N)$, $\alpha \geq 0$, $\mathcal{H}^k$ is the k-dimensional Hausdorff measure in $\mathbb{R}^N$ and S is an orientable k-manifold in $\mathbb{R}^N$ without boundary of class $W^{3,\infty}$. Note that one can use many different approximations $u_{0,n}$ to the measure μ. To start with, we approximate the singular part of μ, i.e., the measure $\alpha \mathcal{H}^k \llcorner S$ by constant functions on a band around S. Indeed, using essentially the ideas of Minkowski's content (see [10]) we know that

$$\alpha \frac{\mathcal{H}^k(S)}{|I_n(S)|} \chi_{I_n(S)} \rightharpoonup \alpha \mathcal{H}^k \llcorner S \quad \text{weakly}^* \text{ as measures as } n \to \infty, \tag{3.57}$$

where $I_n(S) = \{x \in \mathbb{R}^N : d(x,S) \leq \frac{1}{n}\}$. Then, given a measure μ of the form (3.56), if $u_{0,n}(\mu)$ is the L^1-function defined by

$$u_{0,n}(\mu) := \mu_{ac} + \frac{\alpha \mathcal{H}^k(S)}{|I_n(S)|} \chi_{I_n(S)}, \tag{3.58}$$

we have that $u_{0,n} \rightharpoonup \mu$ weakly* as measures. Since $u_{0,n}(\mu) \in L^1(\mathbb{R}^N)$, we know (Theorem 3.16) that there exists a unique strong solution u_n of the problem (3.1) with initial datum $u_{0,n}(\mu)$. Then we have that

$$u_n \rightharpoonup u \qquad \text{in } C_w([0,T], \mathcal{M}_b(\mathbb{R}^N)),$$

and we say that $u(t)$ is a *limit solution* of (3.1) corresponding to the initial condition μ ([19]). We compute some explicit limit solutions for initial measures which have some radial symmetry, in particular for sums of Dirac measures concentrated at points or circles. These explicit solutions exhibit some curious behaviour, namely, Dirac measures concentrated at a finite number of points do not move, while the measure $\mu := \alpha \mathcal{H}^{N-1} \llcorner \partial B(0,R)$ has a more complex evolution described by the Radon measures in $\mathbb{R}^N$

$$u(t) = \begin{cases} \left(1 - \dfrac{2}{\alpha} t\right) \mu + \dfrac{N}{R} t \chi_{B_R(0)}, & 0 < t \leq \dfrac{\alpha}{2}, \\[2ex] \left(\alpha \dfrac{N}{R} - \dfrac{N}{R} t\right)^+ \chi_{B_R(0)}, & t \geq \dfrac{\alpha}{2}. \end{cases} \tag{3.59}$$

In particular, we deduce that there is no regularizing effect for (3.1) when the initial condition is a measure. On the other hand, this makes explicit that solutions have a very different behaviour according to the Hausdorff dimension of the support of the measure. If this dimension is $k < N - 1$ it seems that the singular part of the measure does not move, while it moves when $k = N-1$. This behaviour is explored in [19] where it is proved for measures of the form (3.56) and the behaviour of limit solutions is characterized. Let us consider first the case $k = N-1$. Let C_2 denote the unbounded connected component of $\mathbb{R}^N \setminus S$ and C_1 its complement in $\mathbb{R}^N \setminus S$. In the time interval $[0, \frac{\alpha}{2}]$ we have that $u(t) = u(t)_{ac} + u(t)_s$ with $u(t)_s = (1 - \frac{2}{\alpha}t)\mu_s$ and $u(t)_{ac}|_{C_i}$, $i = 1, 2$, is the strong solution of the Dirichlet problem

$$\begin{cases} v_t = \operatorname{div}\left(\dfrac{Dv}{|Dv|}\right) & \text{in } (0,T) \times C_i, \\ v = C_T & \text{on } (0,T) \times S, \\ v(0) = \mu_{ac} & \text{in } C_i. \end{cases} \tag{3.60}$$

Note that $u(\frac{\alpha}{2})_s = 0$. In the time interval $[\frac{\alpha}{2}, \infty)$, $u(t) = u(t)_{ac}$ is the strong solution of (3.1) with initial condition $u(\frac{\alpha}{2})_{ac}$. In case $k < N-1$, we have that $u(t)_s = \mu_s$ for all $t \geq 0$, and $u(t)_{ac}$ is the strong solution of (3.1) with initial condition μ_{ac}. Furthermore, $u(t)$ satisfies an entropy condition which characterizes in some way the solution of (3.1). The paper [19] contains also the description of limit solutions when the initial measure μ is approximated by functions

$$u_{n0} = \mu_{ac} + \rho_n * \mu_s$$

where $\rho_n(x) = n^N \rho(nx)$ and ρ is a radial, smooth, positive convolution kernel with compact support, and $\mu_s = \alpha \mathcal{H}^k \llcorner S$ with $k < N-1$. The limit solution obtained in this case coincides with the limit solution obtained using the approximation (3.57). At this moment we do not know if the analogous result holds when $k = N-1$.

Chapter 4

Asymptotic Behaviour and Qualitative Properties of Solutions

The purpose of this chapter is to give some qualitative properties of the flow

$$\frac{\partial u}{\partial t} = \operatorname{div}\left(\frac{Du}{|Du|}\right) \qquad \text{in }]0,\infty[\times\mathbb{R}^N. \tag{4.1}$$

First, we prove that the flow decreases the $\mathcal{H}^{N-1}$ Hausdorff measure of the level sets, and that local maxima (resp. minima) strictly decrease (resp. increase) with time. For that, we shall need to compute some explicit radial solutions of (4.1). Then, we shall prove that, for bounded initial conditions, the solution vanishes in finite time, and the extinction profile is a solution of the eigenvalue problem

$$-\operatorname{div}\left(\frac{Du}{|Du|}\right) = u. \tag{4.2}$$

Next we compute some explicit solutions of this eigenvalue problem, which give also explicit solutions of (4.1). We shall use them to compute explicit solutions of the total variation denoising model in image processing.

4.1 Radially Symmetric Explicit Solutions

We shall construct explicit radial solutions of (4.1) in $\mathbb{R}^N$ or in a ball of $\mathbb{R}^N$.

Let $g \in C^2(\mathbb{R}^+)$, $g'(r) < 0$, for $r > 0$, $g(0+) > 0$. Let $M > 0$, $M < g(0+)$. Let $U(x) = \min(g(\|x\|), M)$, $x \in \mathbb{R}^N$. Let $r_0 > 0$ be such that $g(r_0) = M$. Observe that $U(x) = M$ for $\|x\| \leq r_0$ and $U(x) = g(\|x\|)$ for $\|x\| > r_0$, i.e.,

$U(x)$ has a flat zone in the ball $B_{r_0}(0)$ and a radial decreasing profile outside it. Let $R > r_0$, $B_R = B_R(0)$. Let us construct a function $U(t,x)$ such that $U \in C([0,T], B_R) \cap W^{1,1}(0,T; L^1(B_R))$, $U \in L^1_w(0,T; BV(B_R))$, $U(0,x) = U(x)$ and there exists $Z(t) \in X(B_R)_1$ with $\|Z(t)\|_\infty \leq 1$, such that for $T > 0$ small enough and $t \in [0,T]$ we have

$$U_t = \operatorname{div}(Z(t)) \qquad \text{in } \mathcal{D}'(B_R) \tag{4.3}$$

and

$$\int_{B_R} Z(t) \cdot DU(t)\, dx = \int_{B_R} \|DU(t)\|. \tag{4.4}$$

Observe that if $u(x) = h(\|x\|)$, where $h \in C^2(\mathbb{R}^+)$, $h(r) > 0$, $h'(r) < 0$ for $r > 0$, then $\operatorname{div}\left(\frac{Du}{|Du|}\right) = -\frac{N-1}{\|x\|}$. We expect the solution $U(t,x)$ to be a radial function. If the flat zone of $U(x)$ has to remain flat, then we should have

$$U_t = \begin{cases} -\lambda & \text{if } \|x\| < r(t), \\ -\frac{N-1}{\|x\|} & \text{if } \|x\| > r(t), \end{cases} \tag{4.5}$$

for some $\lambda > 0$ and some $r(t) > 0$ such that $r(0) = r_0$. The choice $Z(t) = -\frac{x}{\|x\|}$ for $\|x\| > r(t)$ is consistent with (4.3). Now, in $B_{r(t)}(0)$, $Z(t)$ must be such that

$$\operatorname{div}(Z(t)) = -\lambda \qquad \text{with } \; Z(t)\,|_{\partial B_{r(t)}(0)} = -\frac{x}{\|x\|},\; \|Z(t)\|_\infty \leq 1. \tag{4.6}$$

Integrating the previous equation in $B_{r(t)}(0)$ we have

$$\begin{aligned} -\lambda\, \mathcal{L}^N(B_{r(t)}(0)) &= \int_{B_{r(t)}(0)} \operatorname{div}(Z(t))\, dx \\ &= \int_{\partial B_{r(t)}(0)} Z(t) \cdot \nu\, d\mathcal{H}^{N-1} = -\mathcal{H}^{N-1}(\partial B_{r(t)}(0)) \end{aligned}$$

and, therefore,

$$\lambda = \frac{H^{N-1}(\partial B_{r(t)}(0))}{\mathcal{L}^N(B_{r(t)}(0))} = \frac{N}{r(t)}.$$

Then, we may take $Z(t) = -\frac{x}{r(t)}$ when $\|x\| < r(t)$. Such a choice of $Z(t)$ satisfies (4.6). We have to choose $r(t)$ such that $U(t,x)$ remains a Lipschitz function, in particular continuous for $\|x\| = r(t)$. Assume for the time being that $r'(t) > 0$. Let us observe that

$$U(t,0) = U(0) - N \int_0^t \frac{ds}{r(s)}.$$

Now, if $x \in B_R$ is a point such that $\|x\| > r(t)$, then

$$U(t,x) = U(x) + \int_0^t U_t(s,x)\, ds = U(x) - \frac{t(N-1)}{\|x\|}.$$

Since the value $U(t, r(t)-)$ must coincide with the value of $U(t, 0)$ and $U(t, r(t)-) = U(t, r(t)+)$, then

$$U(0) - N \int_0^t \frac{ds}{r(s)} = U(r(t)+) - \frac{t(N-1)}{|r(t)|}.$$

Thus, differentiating the above expression we see that $r(t)$ must satisfy the differential equation

$$r'(t) = -\frac{r(t)}{(N-1)t + g'(r(t))r(t)^2}. \tag{4.7}$$

We take $r(t)$ to be the solution of (4.7) such that $r(0) = r_0$. Then, the function

$$U(t,x) = \begin{cases} U(0) - N \displaystyle\int_0^t \frac{ds}{r(s)} & \text{if } \|x\| < r(t), \\ U(x) - (N-1)\frac{t}{\|x\|} & \text{if } \|x\| > r(t) \end{cases} \tag{4.8}$$

satisfies (4.3) for $t > 0$ small, $Z(t, x)$ being given by

$$Z(t,x) = \begin{cases} -\frac{x}{r(t)} & \text{if } \|x\| < r(t), \\ -\frac{x}{\|x\|} & \text{if } \|x\| > r(t). \end{cases} \tag{4.9}$$

Observe that $\|Z(t)\|_\infty \leq 1$ and satisfies (4.4) for $t > 0$ small.

If $g(r) = a - kr^p$, $a, k, p > 0$, the radius $r(t)$ can be determined explicitly. Indeed, $r(t)$ satisfies the differential equation

$$r'(t) = \frac{r(t)}{kpr(t)^{p+1} - (N-1)t}. \tag{4.10}$$

Looking for a solution of (4.10) of the form $r(t) = \lambda t^\alpha$ we find that

$$r(t) = \left(\frac{N+p}{kp}\right)^{\frac{1}{p+1}} t^{\frac{1}{p+1}}$$

solves (4.10).

Thus, the following result is established.

Proposition 4.1. *There is some $T > 0$ such that the function U defined in (4.8) is a solution of (4.3) in $\mathcal{D}'((0,T) \times B_R)$ satisfying (4.4) for all $t \in [0,T]$ and such that $U(0,x) = U(x)$. Moreover, $U \in C([0,T], B_R) \cap W^{1,1}(0,T; L^1(B_R))$ and $U \in L^1(0,T; BV(B_R))$.*

Lemma 4.2. *Let $u_0 = k\chi_{B_r(0)}$. Then the unique solution $u(t,x)$ of problem (4.1) with initial datum u_0 is given by*

$$u(t,x) = \text{sign}(k)\frac{N}{r}\left(\frac{|k|r}{N} - t\right)^+ \chi_{B_r(0)}(x).$$

Observe that we may write

$$u(t,x) = \mathrm{sign}(k)\left(|k| - \frac{\mathcal{H}^{N-1}(\partial B_r(0))}{\mathcal{L}^N(B_r(0))}t\right)^+ \chi_{B_r(0)}(x).$$

Proof. Suppose that $k > 0$, the solution for $k < 0$ being constructed in a similar way. We look for a solution of (4.1) of the form $u(t,x) = \alpha(t)\chi_{B_r(0)}(x)$ on some time interval $(0,T)$. Then, we shall look for some $z(t) \in X(\mathbb{R}^N)_1$ with $\|z\|_\infty \leq 1$, such that

$$u'(t) = \mathrm{div}(z(t)) \qquad \text{in } \mathcal{D}'(\mathbb{R}^N), \tag{4.11}$$

$$\int_{\mathbb{R}^N} (z(t), Du(t)) = \int_{\mathbb{R}^N} \|Du(t)\|. \tag{4.12}$$

If we take $z(t)(x) = -\frac{x}{r}$ for $x \in \partial B_r(0)$, integrating equation (4.11) in $B_r(0)$ we obtain

$$\alpha'(t)\,\mathcal{L}^N(B_r(0)) = \int_{B_r(0)} \mathrm{div}(z(t))\,dx = \int_{\partial B_r(0)} z(t)\cdot\nu\,d\mathcal{H}^{N-1} = -\mathcal{H}^{N-1}(\partial B_r(0)).$$

Thus

$$\alpha'(t) = -\frac{N}{r},$$

and, therefore,

$$\alpha(t) = k - \frac{N}{r}t.$$

In that case, T must be given by $T = \frac{kr}{N}$.

To construct z in $(0,T) \times (\mathbb{R}^N \setminus B_r(0))$ we shall look for z of the form $z = \rho(\|x\|)\frac{x}{\|x\|}$ such that $\mathrm{div}(z(t)) = 0$, $\rho(r) = -1$. Since

$$\mathrm{div}(z(t)) = \nabla\rho(\|x\|)\cdot\frac{x}{\|x\|} + \rho(\|x\|)\mathrm{div}\left(\frac{x}{\|x\|}\right) = \rho'(\|x\|) + \rho(\|x\|)\frac{N-1}{\|x\|},$$

we must have

$$\rho'(s) + \rho(s)\frac{N-1}{s} = 0 \qquad \text{for} \quad s > r. \tag{4.13}$$

The solution of (4.13) such that $\rho(r) = -1$ is

$$\rho(s) = -r^{N-1}s^{1-N}.$$

Thus, in $\mathbb{R}^N \setminus B_r(0)$,

$$z(t) = -r^{N-1}\frac{x}{\|x\|^N}.$$

Consequently, the candidate for $z(t)$ is the vector field

$$z(t) := \begin{cases} -\dfrac{x}{r} & \text{if } x \in B_r(0) \ \text{ and } \ 0 \le t \le T, \\ -r^{N-1}\dfrac{x}{\|x\|^N} & \text{if } x \in \mathbb{R}^N \setminus \overline{B_r(0)}, \ \text{ and } \ 0 \le t \le T, \\ 0 & \text{if } x \in \mathbb{R}^N \ \text{ and } \ t > T, \end{cases}$$

and the corresponding function $u(t,x)$ is

$$u(t,x) = \left(k - \frac{N}{r}t\right)\chi_{B_r(0)}(x)\chi_{[0,T]}(t),$$

where $T = \dfrac{kr}{N}$. Let us check that $u(t,x)$ satisfies (4.11), (4.12). If $\varphi \in \mathcal{D}(\mathbb{R}^N)$ and $0 \le t \le T$, we have

$$\begin{aligned} \int_{\mathbb{R}^N} \frac{\partial z_i(t)}{\partial x_i}\varphi\, dx &= -\frac{1}{r}\int_{B_r(0)} \varphi\, dx + \int_{\partial B_r(0)} \frac{x_i}{r}\frac{x_i}{r}\varphi\, d\mathcal{H}^{N-1} \\ &\quad -\int_{\mathbb{R}^N \setminus B_r(0)} \frac{\partial}{\partial x_i}\left(\frac{r^{N-1}x_i}{\|x\|^N}\right)\varphi\, dx - \int_{\partial B_r(0)} \frac{r^{N-1}}{r^N}x_i\frac{x_i}{r}\varphi\, d\mathcal{H}^{N-1}. \end{aligned}$$

Hence

$$\int_{\mathbb{R}^N} \operatorname{div}(z(t))\varphi\, dx = -\frac{N}{r}\int_{B_r(0)} \varphi\, dx,$$

and consequently, (4.11) holds. Finally, if $0 \le t \le T$, by Green's formula, we have

$$\begin{aligned} \int_{\mathbb{R}^N} (z(t), Du(t)) &= -\int_{\mathbb{R}^N} \operatorname{div}(z(t))u(t)\, dx = -\int_{B_r(0)} \left(k - \frac{N}{r}t\right)\operatorname{div}(z(t))\, dx \\ &= \int_{B_r(0)} \left(k - \frac{N}{r}t\right)\frac{N}{r}\, dx = \left(k - \frac{N}{r}t\right)\frac{N}{r}\mathcal{L}^N(B_r(0)) \\ &= \left(k - \frac{N}{r}t\right)\mathcal{H}^{N-1}(\partial B_r(0)) = \int_{\mathbb{R}^N} \|Du(t)\|. \end{aligned}$$

Therefore (4.12) holds, and consequently $u(t,x)$ is the solution of (4.1) with initial datum $u_0 = k\chi_{B_r(0)}$. □

Lemma 4.3. *Let $\Omega = B_R(0) \setminus \overline{B_r(0)}$, $0 < r < R$ and $u_0 = k\chi_\Omega$. Then the unique solution $u(t,x)$ of problem* (4.1) *with initial datum u_0 is*

$$u(t,x) = \operatorname{sign(k)}\left(|k| - \frac{Per(\Omega)}{\mathcal{L}^N(\Omega)}t\right)\chi_\Omega(x) + \frac{\operatorname{Per}(B_r(0))}{\mathcal{L}^N(B_r(0))}t\chi_{B_r(0)}(x), \tag{4.14}$$

$t \in [0, T_1]$, $x \in \mathbb{R}^N$, where T_1 is such that

$$T_1 \cdot \left(\frac{Per(\Omega)}{\mathcal{L}^N(\Omega)} + \frac{Per(B_r(0))}{\mathcal{L}^N(B_r(0))} \right) = |k|$$

and $u(t,x)$ evolves as the solution given in Lemma 4.2 until its extinction.

Proof. Let $\xi : \mathbb{R}^N \to \mathbb{R}^N$ be the vector field defined as

$$\xi(x) := \begin{cases} \dfrac{x}{r} & \text{for } x \in B_r(0), \\ \left((Rr)^{N-1} \dfrac{R+r}{\|x\|^N} - (R^{N-1} + r^{N-1}) \right) \dfrac{x}{R^N - r^N}, & x \in B_R(0) \setminus \overline{B_r(0)}, \\ -\dfrac{R^{N-1}}{\|x\|^N} x & \text{for } x \in \mathbb{R}^N \setminus \overline{B_R(0)}. \end{cases}$$

Then $\|\xi\|_\infty \leq 1$, $\mathrm{div}(\xi) = \frac{N}{r} = \frac{\mathrm{Per}(B_r(0)}{\mathcal{L}^N(B_r(0))}$ on $B_r(0)$, $\mathrm{div}(\xi) = -\frac{\mathrm{Per}(\Omega)}{\mathcal{L}^N(\Omega)}$ on $B_R(0) \setminus \overline{B_r(0)}$, $\mathrm{div}(\xi) = 0$ on $\mathbb{R}^N \setminus \overline{B_R(0)}$, and $\xi \cdot \nu^{B_r(0)} = 1$ on $\partial B_r(0)$, $\xi \cdot \nu^{B_R(0)} = -1$ on $\partial B_R(0)$. Therefore, one can check that the solution u of (4.1) with initial condition $u_0 = \chi_\Omega$ in $[0, T_1]$ is given by (4.14). At $t = T_1$, the two evolving sets reach the same height and $u(T_1, x) = \alpha \chi_{B_R(0)}$ for some $\alpha > 0$. For $t > T_1$ the solution u is equal to the solution starting from $\alpha \chi_{B_R(0)}$ (at time T_1) as it is described in Lemma 4.2. □

Remark 4.4. The above results show that there is no spatial smoothing effect, for $t > 0$, similar to the case of the linear heat equation and many other quasilinear parabolic equations. In our case, the solution is discontinuous and has the minimal required spatial regularity: $u(t,.) \in BV(\mathbb{R}^N) \setminus W^{1,1}(\mathbb{R}^N)$.

Remark 4.5. The solution given in Lemma 4.2 also gives the explicit solution of the Dirichlet problem in any domain Ω such that $B_r(0) \subseteq \Omega$ (see [15]).

Remark 4.6. For the Neumann problem, we can also compute explicitly the evolution of the characteristic function of a ball $B_r(p)$ when Ω is a ball centered at p. To fix ideas, let $p = 0$, $\Omega = B_R(0)$ and $u_0(x) = k\chi_{B_r(0)}$, where $0 < r < R$ and $k > 0$. Then we look for a solution of (2.1) of the form $u(t) = \alpha(t)\chi_{B_r(0)} + \beta(t)\chi_{B_R(0)\setminus B_r(0)}$ on some time interval $(0, T)$ defined by the inequalities $\alpha(t) > \beta(t)$ for all $t \in (0, T)$, and $\alpha(0) = k$, $\beta(0) = 0$. As above, we look for some $z \in L^\infty((0,T) \times B_R(0))$ with $\|z\|_\infty \leq 1$, such that

$$\alpha'(t) = \mathrm{div}(z) \qquad \text{in } (0,T) \times B_r(0), \tag{4.15}$$

$$\beta'(t) = \mathrm{div}(z) \quad \text{in } (0,T) \times (B_R(0) \setminus B_r(0)),$$

$$z = -\frac{x}{\|x\|} \qquad \text{on } (0,T) \times \partial B_r(0), \tag{4.16}$$

$$z \cdot \nu = 0 \qquad \text{on } (0,T) \times \partial B_R(0),$$

and

$$\int_{B_R(0)} (z, Du) = \int_{B_R(0)} \|Du\| \qquad \text{for all } t \in (0,T). \tag{4.17}$$

Proceeding as we did in Lemma 4.2 we compute

$$z(t,x) := \begin{cases} -\dfrac{x}{r} & \text{for } x \in B_r(0), \\ \left(1 - \dfrac{R^N}{\|x\|^N}\right) \dfrac{r^{N-1}}{R^N - r^N} x & \text{for } x \in B_R(0) \setminus \overline{B_r(0)}, \end{cases}$$

and

$$u(t,x) = \left(k - \frac{N}{r} t\right) \chi_{B_r(0)}(x) + \frac{N r^{N-1}}{R^N - r^N} t \chi_{B_R(0) \setminus B_r(0)}(x)$$

in $(0,T) \times B_R(0)$, where T is given by

$$T\left(\frac{N}{r} + N \frac{r^{N-1}}{R^N - r^N}\right) = k. \tag{4.18}$$

After time T the solution is

$$u(t,x) = \left(k - \frac{N}{r} T\right) \chi_{B_R(0)}(x) = \frac{N r^{N-1}}{R^N - r^N} T \chi_{B_R(0)}(x)$$

and we may take as the corresponding vector field $z(t,x) = 0$. We leave as an exercise to check that $u(t,x)$ is the solution of (2.1) in $(0,\infty) \times B_R(0)$ with initial datum $u_0(x)$. Exact solutions for the minimization problem with constraints (1.12) in $\mathbb{R}^N$ with $N = 1,2,3$ have been given in ([186]).

4.2 Some Qualitative Properties

We shall prove that the length of the level curves of the solution is a decreasing function of time, as should be expected. We shall also prove that flat zones which are local maxima (minima) immediately decrease (respectively, increase) with time.

Recall that $T_{a,b}(r) = \min(\max(a,r),b)$, $a,b,r \in \mathbb{R}$, $a < b$. Let $\psi : BV(\mathbb{R}^N) \to [0,+\infty)$ be defined by

$$\psi(u) = \int_{\mathbb{R}^N} \|Du\|, \quad u \in BV(\mathbb{R}^N).$$

Proposition 4.7. *Let $u_0 \in L^1(\mathbb{R}^N)$. Let $u(t,x)$ be the strong solution of problem (4.1) with initial datum u_0. Then*

$$\psi(T_{a,b}(u(t))) \le \psi(T_{a,b}(u(s))) \tag{4.19}$$

a.e. in $s,t \in (0,\infty)$, $t > s$ and all $a < b$.

Proof. Assume first that $u_0 \in L^2(\mathbb{R}^N)$. Let $\delta > 0$ and $t, s \geq \delta$ such that $(u(t), -u_t(t))$, $(u(s), -u_t(s)) \in \mathcal{B}$, where $\mathcal{B}$ is the operator introduced in the proof of Theorem 3.2. Let $a, b \in \mathbb{R}$. Assume that $0 \leq a < b$. We have

$$\int_{\mathbb{R}^N} (T_b(u(t)) - w)u_t(t)\,dx \leq \int_{\mathbb{R}^N} (z(t), Dw) - \int_{\mathbb{R}^N} \|DT_b(u(t))\| \tag{4.20}$$

for all $w \in BV(\mathbb{R}^N) \cap L^\infty(\mathbb{R}^N)$. Using as test function $w - a + T_{-b,a}(u(t))$, we get

$$\begin{aligned} &\int_{\mathbb{R}^N} (T_{a,b}(u(t)) - w)u_t(t)\,dx \\ &\leq \int_{\mathbb{R}^N} (z(t), Dw) + \int_{\mathbb{R}^N} (z(t), DT_{-b,a}(u(t))) - \int_{\mathbb{R}^N} \|DT_b(u(t))\| \\ &\leq \int_{\mathbb{R}^N} (z(t), Dw) - \int_{\mathbb{R}^N} \|DT_{a,b}(u(t))\| \end{aligned}$$

for all $w \in BV(\mathbb{R}^N) \cap L^\infty(\mathbb{R}^N)$. In a similar way, we have

$$\int_{\mathbb{R}^N} (T_{a,b}(u(t)) - w)u_t(t)\,dx \leq \int_{\mathbb{R}^N} (z(t), Dw) - \int_{\mathbb{R}^N} \|DT_{a,b}(u(t))\| \tag{4.21}$$

for all $w \in BV(\mathbb{R}^N) \cap L^\infty(\mathbb{R}^N)$ and all $a, b \in \mathbb{R}$, $a < b$. Setting $w = T_{a,b}(u(s))$ in (4.21) we have

$$\int_{\mathbb{R}^N} \|DT_{a,b}(u(t))\| - \int_{\mathbb{R}^N} \|DT_{a,b}(u(s))\| \leq \int_{\mathbb{R}^N} u_t(t)(T_{a,b}(u(s)) - T_{a,b}(u(t)))\,dx.$$

Using estimate (3.41) we may write

$$\int_{\mathbb{R}^N} \|DT_{a,b}(u(t))\| - \int_{\mathbb{R}^N} \|DT_{a,b}(u(s))\| \leq \frac{2}{\delta}\|u_0\|_2 \|T_{a,b}(u(s)) - T_{a,b}(u(t))\|_2.$$

Since a similar estimate holds with s and t interchanged, we have

$$\left|\int_{\mathbb{R}^N} \|DT_{a,b}(u(t))\| - \int_{\mathbb{R}^N} \|DT_{a,b}(u(s))\|\right| \leq \frac{2}{\delta}\|u_0\|_2 \|T_{a,b}(u(s)) - T_{a,b}(u(t))\|_2. \tag{4.22}$$

Since $u \in W^{1,1}_{loc}(0, T, L^2(\mathbb{R}^N))$, i.e, is a locally absolutely continuous function of time, then also $T_{a,b}(u)$ is, and, from (4.22), we deduce that $\psi(T_{a,b}(u))$ is absolutely continuous in $[0, T]$ for all $T > 0$. Let $t \in [0, \infty)$ be such that u, $T_{a,b}(u)$, $\psi(T_{a,b}(u))$ are differentiable at t and $(u(t), -u_t(t)) \in \mathcal{B}$. Set $w = T_{a,b}(u(t+\epsilon))$, $w = T_{a,b}(u(t-\epsilon))$ in (4.21) to obtain

$$\int_{\mathbb{R}^N} (T_{a,b}(u(t)) - T_{a,b}(u(t \pm \epsilon)))u_t(t) \leq \psi(T_{a,b}(u(t \pm \epsilon))) - \psi(T_{a,b}(u(t))).$$

Letting $\epsilon \to 0^+$ we have

$$\frac{d}{dt}\psi(T_{a,b}(u(t))) = -\int_{\mathbb{R}^N} T'_{a,b}(u(t))u_t(t)^2 \leq 0.$$

Hence, $\psi(T_{a,b}(u(t)))$ is a decreasing function of time. In particular, for all $s < t$ such that $T_{a,b}(u(s)), T_{a,b}(u(t)) \in BV(\mathbb{R}^N)$, hence a.e. in s, t, $s < t$, we have

$$\psi(T_{a,b}(u(t))) \leq \psi(T_{a,b}(u(s))). \tag{4.23}$$

In particular, if $u(0) \in L^2(\mathbb{R}^N)$ and $T_{a,b}(u(0)) \in BV(\mathbb{R}^N)$ for all $a < b$, then

$$\psi(T_{a,b}(u(t))) \leq \psi(T_{a,b}(u(0))) \tag{4.24}$$

almost everywhere in $t \in (0, \infty)$, for all $a < b$.

Now, let $u_0 \in L^1(\mathbb{R}^N)$ and $u(t)$ be the corresponding strong solution of (4.1). Let $s > 0$ be such that $T_{a,b}(u(s)) \in BV(\mathbb{R}^N)$ for all $a < b$ (which is possible by the proof of Theorem B.15). Take s as the origin of time. Let $u^n(s) \in L^2(\mathbb{R}^N) \cap BV(\mathbb{R}^N)$ such that $u^n(s) \to u(s)$ in $L^1(\mathbb{R}^N)$ and $\|DT_{a,b}(u^n(s))\| \to \|DT_{a,b}(u(s))\|$ as $n \to \infty$, for all $a < b$. Let $u^n(t, x)$ be the strong solution of (4.1) with initial condition at $t = s$, $u^n(s, x) = u^n(s)(x)$. Then $u^n(t) \to u(t)$ in $C([s, T], L^1(\mathbb{R}^N))$ for all $T > s$. Using (4.24) we have

$$\psi(T_{a,b}(u^n(t))) \leq \psi(T_{a,b}(u^n(s)))$$

for almost all $t \in (s, \infty)$ and all $a < b$. Letting $n \to \infty$, we get

$$\psi(T_{a,b}(u(t))) \leq \psi(T_{a,b}(u(s))) \tag{4.25}$$

for almost all $t \in (s, \infty)$ and all $a < b$. Hence (4.25) holds for almost all $s, t \in (0, \infty)$, $s < t$, and all $a < b$. □

Lemma 4.8. *Let $g(t, \lambda) \in L^1_{loc}((0, \infty) \times \mathbb{R})$. Let $\Delta = \{(s, t) \in (0, \infty) \times (0, \infty) : s < t\}$. Suppose that*

$$\int_a^b g(t, \lambda)d\lambda \leq \int_a^b g(s, \lambda)d\lambda \tag{4.26}$$

a.e. in $(s, t) \in \Delta$ and all $a < b$. Then

$$g(t, \lambda) \leq g(s, \lambda)$$

a.e. in $(s, t, \lambda) \in \Delta \times \mathbb{R}$.

Proof. Consider the function $G(s, t, \lambda)$ defined in $\Delta \times \mathbb{R}$ by

$$G(s, t, \lambda) = g(t, \lambda) - g(s, \lambda).$$

Let (s_0, t_0, λ_0) be a Lebesgue point of G. By assumption we know that for $r > 0$ and almost all $(s,t) \in \Delta$,

$$\int_{\lambda_0 - r}^{\lambda_0 + r} G(s,t,\lambda)\, d\lambda \leq 0.$$

Hence,

$$G(s_0, t_0, \lambda_0) = \lim_{r \to 0+} \frac{1}{(2r)^3} \int_{s_0 - r}^{s_0 + r} \int_{t_0 - r}^{t_0 + r} \int_{\lambda_0 - r}^{\lambda_0 + r} G(s,t,\lambda)\, dsdtd\lambda \leq 0.$$

Since almost all points of $\Delta \times \mathbb{R}$ are Lebesgue points of G, the lemma follows. □

Corollary 4.9. *Let $u_0 \in L^1(\mathbb{R}^N)$. Let $u(t,x)$ be the strong solution of (4.1) with initial datum u_0. Then, for almost all $\lambda \in \mathbb{R}$,*

$$\int_{\mathbb{R}^N} \|D\chi_{\{u(t)>\lambda\}}\| \leq \int_{\mathbb{R}^N} \|D\chi_{\{u(s)>\lambda\}}\| \tag{4.27}$$

a.e. in $s, t \in (0,\infty)$, $t > s > 0$.

Proof. Let $g(t,\lambda) = \int_{\mathbb{R}^N} \|D\chi_{\{u(t)>\lambda\}}\|$. By Proposition 4.7 and the coarea formula, g satisfies the assumptions of Lemma 4.8. Then the conclusion follows from that lemma. □

Note that

$$\int_{\mathbb{R}^N} \|D\chi_{\{u(t)>\lambda\}}\| = \mathcal{H}^{N-1}(\partial^*\{u(t) > \lambda\}),$$

where $\partial^*\{u(t) > \lambda\}$ is the reduced boundary of the set $\{x \in \mathbb{R}^N : u(t) > \lambda\}$ (see Theorem B.26).

Next, we prove that flat zones which are local maxima (minima) immediately decrease (respectively, increase) with time.

Proposition 4.10. *Let $u_0 \in BUC(\mathbb{R}^N) \cap L^1(\mathbb{R}^N)$, $\lambda \in \mathbb{R}$ be such that $\{x \in \mathbb{R}^N : u_0(x) = \lambda\} = K \subseteq B$ for some ball B and $u_0(x) < \lambda$ on ∂B. Let u be the entropy solution of problem (4.1) with initial datum u_0. Then $u(t,x) < \lambda$, for all $t > 0$ small enough, $x \in B$.*

Remark. If $\lambda = \max_{x \in \mathbb{R}^N} u_0(x)$ and $u_0(x) < \lambda$ for all $x \notin B$, then we have that $u(t,x) < \lambda$, for all $t > 0$, and all $x \in B$.

We shall use a comparison principle for the Dirichlet problem together with explicit supersolutions to prove Proposition 4.10.

Proof. Without loss of generality we may assume that $B = B_R(0)$. Let $U(x) = \min(g(\|x\|), \lambda)$ where $g \in C^2(\mathbb{R}^+)$, $g(r) > 0$, $g'(r) < 0$, for $r > 0$. Assume that $g(0+) > 1$, $g(R) < \lambda$. Assume also that $\sup_{x \in \partial B} u_0 < g(R)$ and $u_0(x) \leq U(x)$ for $x \in B$. We observe that $u \in C([0,T], BUC(\mathbb{R}^N))$, for all $T > 0$, when $u_0 \in BUC(\mathbb{R}^N) \cap L^1(\mathbb{R}^N)$. For that, assume that $u_0 \in BUC(\mathbb{R}^N) \cap W^{1,\infty}(\mathbb{R}^N)$. Then $u(t) \in BUC(\mathbb{R}^N) \cap W^{1,\infty}(\mathbb{R}^N)$ for all $t > 0$. Indeed, we know that the solution u_ϵ of

$$\frac{\partial u}{\partial t} = \operatorname{div}\left(\frac{Du}{(\epsilon^2 + |Du|^2)^{1/2}}\right) + \epsilon \Delta u \qquad \text{in } Q = (0, \infty) \times \mathbb{R}^N, \quad (4.28)$$

$$u(0,x) = u_0(x) \qquad \text{in } x \in \mathbb{R}^N \quad (4.29)$$

converges in $L^2(\mathbb{R}^N)$ to the strong solution of (4.1) (see [177]). By the maximum principle, we have

$$\| u_\epsilon(t) \|_\infty \leq \| u_0 \|_\infty .$$

Now, using the Bernstein method as in [7], it is easy to see that

$$\|\nabla u_\epsilon(t)\|_\infty \leq \|\nabla u_0\|_\infty, \quad \text{for all } t > 0.$$

Letting $\epsilon \to 0$ we obtain that the previous estimates also hold for $u(t)$. Then it follows that $u(t) \in BUC(\mathbb{R}^N) \cap W^{1,\infty}(\mathbb{R}^N)$ for all $t > 0$ with Lipschitz constant $\|\nabla u_0\|_\infty$. On the other hand, since $\mathcal{B}$ is m-completely accretive in $L^1(\mathbb{R}^N)$, $\mathcal{B} \cap \left(L^\infty(\mathbb{R}^N) \cap L^1(\mathbb{R}^N) \times L^\infty(\mathbb{R}^N) \cap L^1(\mathbb{R}^N)\right)$ generates a strongly continuous semigroup in $L^\infty(\mathbb{R}^N) \cap L^1(\mathbb{R}^N)$. Consequently, u is a continuous function in (t,x). Now, let $u_0 \in BUC(\mathbb{R}^N)$, and $u(t,x)$ be the corresponding strong solution of (4.1). Let $u_{0n} \in BUC(\mathbb{R}^N) \cap W^{1,\infty}(\mathbb{R}^N)$ be such that $u_{0n} \to u_0$ in $BUC(\mathbb{R}^N) \cap W^{1,\infty}(\mathbb{R}^N)$ as $n \to \infty$. If u_n is the entropy solution of (4.1) with initial datum u_{0n}, then $u_n \in BUC(\mathbb{R}^N) \cap W^{1,\infty}(\mathbb{R}^N)$ converges to u in $C([0,T], BUC(\mathbb{R}^N))$ as $n \to \infty$. Therefore, $u \in C([0,T], BUC(\mathbb{R}^N))$.

Since $u(t,x)$ is a continuous function of (t,x), for some $T > 0$, we have that $u(t,x) \leq U(t,x)$ for all $t \in [0,T]$, $x \in \partial B$. Let us prove that $u(t,x) \leq U(t,x)$, for all $t \in [0,T]$, $x \in B$, where $U(t,x)$ is the solution constructed in Proposition 4.1. Since u is a strong solution, there exists $z(t) \in X(\mathbb{R}^N)_1$ with $\| z \|_\infty \leq 1$ such that

$$u_t(t) = \operatorname{div}(z(t)), \quad \text{a.e.} \quad t \in [0,T] \quad (4.30)$$

and

$$\int_{\mathbb{R}^N} (z(t), Du(t)) = \int_{\mathbb{R}^N} \|Du\|, \quad \text{a.e. } t \in [0,T]. \quad (4.31)$$

On the other hand, using the notation of Section 4.1 we have

$$U_t = \operatorname{div}(Z(t)) \text{ in } B, \text{ a.e. } t \in [0,T]. \quad (4.32)$$

Let $p \in \mathcal{P}$, $p \in C^1(\mathbb{R})$. Multiplying (4.30) and (4.32) by $p(u(t) - U(t))$, taking its difference, and integrating in B, we obtain, after integration by parts,

$$\int_B p(u(t) - U(t))(u_t(t) - U_t(t))\,dx$$

$$= -\int_B (z(t) - Z(t), Dp(u(t) - U(t))) + \int_{\partial B} [z(t) - Z(t), \nu]\, p((u(t) - U(t)).$$

Working as in the proof of Theorem 2.8, we have that

$$-\int_B (z(t) - Z(t), Dp(u(t) - U(t))) \leq 0.$$

Since $u(t) \leq U(t)$ on $\partial B, t \in [0,T]$, if we take p converging to sign^+, it follows that

$$\frac{d}{dt}\int_B (u(t) - U(t))^+ \leq 0 \quad \text{for } t \in [0,T].$$

Since $u_0(x) \leq U(x)$, it follows that $u(t,x) \leq U(t,x) < \lambda$ for $t \in (0,T]$, $x \in B$. □

4.3 Asymptotic Behaviour

We shall study the asymptotic behaviour of the solutions of

$$(P)\begin{cases} \dfrac{\partial u}{\partial t} = \operatorname{div}\left(\dfrac{Du}{|Du|}\right) & \text{in} \quad Q = (0,\infty) \times \mathbb{R}^N, \\ u(0,x) = u_0(x) & \text{in} \quad x \in \mathbb{R}^N \end{cases}$$

for bounded initial data with compact support. The main goal will be to describe the behaviour of solutions of (P) near the extinction time (we shall prove that it is finite). We shall prove that this behaviour is described by a function which is a solution of an eigenvalue problem for the operator $-\operatorname{div}\left(\frac{Du}{|Du|}\right)$ and we shall describe the solutions of this eigenvalue problem in the radial case. Moreover, we point out other qualitative properties which are peculiar to this special class of quasilinear equations. For instance, there is an infinite "waiting time", i.e., there is no propagation of the support of the initial datum.

The main result in this section is the following.

Theorem 4.11. *Let $u_0 \in L^\infty(\mathbb{R}^N)$ with support contained in a ball B of radius $R > 0$ and let $u(t,x)$ be the unique solution of problem (P). Then* $\text{supp}(u) \subseteq B$. *If $T^*(u_0) = \inf\{t > 0 : \ u(t) = 0\}$, then*

$$T^*(u_0) \leq \frac{R\|u_0\|_\infty}{N}. \tag{4.33}$$

Let

$$w(t,x) := \begin{cases} \dfrac{u(t,x)}{T^*(u_0)-t} & \text{if } 0 \le t < T^*(u_0), \\ 0 & \text{if } t \ge T^*(u_0). \end{cases}$$

Then, there exists an increasing sequence $t_n \to T^*(u_0)$ *and a solution* $v^* \neq 0$ *of the eigenvalue problem*

$$-\mathrm{div}\left(\frac{Dv}{|Dv|}\right) = v \quad \text{in } \mathbb{R}^N \tag{4.34}$$

such that

$$\lim_{n\to\infty} w(t_n) = v^* \quad \text{in} \quad L^p(\mathbb{R}^N)$$

for all $1 \le p < \infty$. *Moreover* v^* *is a minimizer of* $\Psi(\cdot) - \langle \cdot, v^* \rangle$ *in* $BV(\mathbb{R}^N) \cap L^2(\mathbb{R}^N)$, *where* Ψ *is the functional defined by* 3.7.

Notice that Theorem 4.11 improves a previous result proved in [129] showing that the solutions of the Dirichlet problem stabilize as $t \to \infty$ by converging in the L^1-norm to zero.

Lemma 4.12. *Assume that* $u_0 \in L^\infty(\mathbb{R}^N)$ *has its support contained in a ball* B *of radius* $R > 0$ *and let* $u(t,x)$ *be the unique solution of problem (P). Then,* $supp(u(t)) \subseteq B$ *for all* $t \ge 0$ *and we have*

$$\|u(t)\|_\infty \le \frac{N}{R}\left(\frac{R\|u_0\|_\infty}{N} - t\right)^+. \tag{4.35}$$

Thus, if $T^*(u_0) = \inf\{t > 0 : \ u(t) = 0\}$, *then*

$$T^*(u_0) \le \frac{R\|u_0\|_\infty}{N}. \tag{4.36}$$

Proof. Take

$$\overline{u}(t,x) := \frac{N}{R}\left(\frac{R\|u_0\|_\infty}{N} - t\right)^+ \chi_B(x),$$

and use the comparison principle (2.8) to conclude that

$$-\overline{u}(t) \le u(t) \le \overline{u}(t),$$

and (4.35) follows. □

Remark 4.13. The above result could be compared with what happens in the study of the parabolic problem associated to the p-Laplacian operator. Consider the Cauchy problem for the p-Laplacian:

$$(P_p)\begin{cases} \dfrac{\partial u}{\partial t} = \mathrm{div}\left(|Du|^{p-2}Du\right) & \text{in} \quad Q = (0,\infty)\times\mathbb{R}^N, \\ u(0,x) = u_0(x) & \text{in} \quad x \in \mathbb{R}^N, \end{cases}$$

with $1 < p < \infty$. It is well known (see [93], [94], [132]) that if $p > 2$ then there is *finite speed of propagation* (i.e., if $\mathrm{supp}(u_0) \subset B_r(0,)$, then the solution of problem (P_p) satisfies that $\mathrm{supp}(u(t))$ is a compact set for any $t > 0$, but, if $1 < p \leq 2$ and $u_0 \geq 0$, $u_0 \neq 0$, then $u(t) > 0$ or $u(t) = 0$ in $\mathbb{R}^N$ for all $t > 0$ ([94], [132]). Observe that (P) can be considered as the limit case $p = 1$ of problem (P_p) and the above result shows that there is no propagation of the support of the initial datum (or equivalently, there is an infinite waiting time). Finite time extinction of the solutions of (P_p) when $\frac{2N}{N+2} \leq p < 2$, $N \geq 2$, was proved in [31], and, for $1 < p < \frac{2N}{N+1}$, in [131] (see also [197], [24]). The same approach also proves the finite time extinction of solutions of (P) (see inequality (4.44) in the proof of Lemma 4.15).

Lemma 4.14. *Assume that $u_0 \in L^\infty(\mathbb{R}^N)$ has compact support contained in a ball B of radius $R > 0$ and denote by $u(t)$ the solution of problem (P)(at time t) with initial datum u_0. Then we have that*

$$\|u(t)\|_\infty \geq \frac{N}{R}(T^*(u_0) - t) \qquad \text{for } \ 0 \leq t \leq T^*(u_0). \tag{4.37}$$

Proof. Take $k > 0$, such that $\frac{kR}{N} = T^*(u_0)$. By Lemma 4.2, we know that

$$v(t,x) = \frac{N}{R}\left(\frac{kR}{N} - t\right)^+ \chi_B(x)$$

is the solution of problem (P) with initial datum $v_0 = k\chi_B$. The proof of (4.37) follows from the inequality

$$\|u(t)\|_{L^\infty(\mathbb{R}^N)} \geq \|v(t)\|_{L^\infty(\mathbb{R}^N)}.$$

By contradiction, suppose there exists $0 < t_0 < T^*(u_0)$ such that

$$\|u(t_0)\|_{L^\infty(\mathbb{R}^N)} < \|v(t_0)\|_{L^\infty(\mathbb{R}^N)}$$

and let $\epsilon > 0$ be such that

$$\|u(t_0)\|_{L^\infty(\mathbb{R}^N)} < \|v(t_0)\|_{L^\infty(\mathbb{R}^N)} - \epsilon = k - \frac{t_0 N}{R} - \epsilon = k_1. \tag{4.38}$$

Consider now the functions

$$v_1(t,x) := \frac{N}{R}\left(\frac{k_1 R}{N} - t\right)^+ \chi_B(x), \quad v_2(t,x) := -\frac{N}{R}\left(\frac{k_1 R}{N} - t\right)^+ \chi_B(x).$$

By (4.38), we have that $v_2(0) \leq u(t_0) \leq v_1(0)$. Hence, by Proposition 3.5, it follows that $v_2(t) \leq u(t_0 + t) \leq v_1(t)$. Hence,

$$T^*(u_0) - t_0 = T^*(u(t_0)) \leq \frac{k_1 R}{N} = \frac{R}{N}\left(k - \frac{t_0 N}{R} - \epsilon\right) = T^*(u_0) - t_0 - \frac{\epsilon R}{N},$$

which is a contradiction, and the proof concludes. □

To study the behaviour of $u(t)$ near the *finite extinction time* $T^*(u_0)$, we follow the method introduced in [47] (see also [95]) . Before giving the proof of Theorem 4.11, we establish lower and upper bounds on the rate of decay of $\|u(t)\|_N$ and $\|u(t)\|_\infty$, respectively. In order to get the upper bound, observe firstly we use the homogeneity estimate proved in Proposition 3.13.

Lemma 4.15. *Assume that $u_0 \in L^\infty(\mathbb{R}^N)$ has support contained in a ball B of radius $R > 0$ and let $u(t,x)$ be the unique solution of problem (P). Then we have:*

(i) *There exists a constant C independent of the initial datum, such that*

$$\|u(t)\|_N \geq C(T^*(u_0) - t) \qquad \text{for} \quad 0 \leq t \leq T^*(u_0). \tag{4.39}$$

(ii) *Given $0 < \tau < T^*(u_0)$, we have*

$$\|u(t)\|_\infty \leq \frac{2\|u_0\|_\infty}{\tau}(T^*(u_0) - t) \qquad \text{for} \quad \tau \leq t \leq T^*(u_0). \tag{4.40}$$

Proof. (i) By Theorem 3.2 there exists $z(t) \in X(\mathbb{R}^N)_1$, $\|z(t)\|_\infty \leq 1$, satisfying

$$\int_{\mathbb{R}^N} (z(t), Du(t)) = \int_{\mathbb{R}^N} \|Du(t)\|, \tag{4.41}$$

$$-\int_{\mathbb{R}^N} (w - u(t))u'(t)\, dx \leq \int_{\mathbb{R}^N} (z(t), Dw) - \int_{\mathbb{R}^N} \|Du(t)\| \tag{4.42}$$

for every $w \in BV(\mathbb{R}^N) \cap L^2(\mathbb{R}^N)$. Let $q \geq 1$, and $\varphi(r) := |r|^{q-1}r$. Then, taking $w = u(t) - \varphi(u(t))$ as test function in (4.42), it yields

$$\int_{\mathbb{R}^N} \varphi(u(t))u'(t)\, dx \leq -\int_{\mathbb{R}^N} (z(t), D\varphi(u(t))).$$

Now, by Corollary C.16 and having in mind (4.41), we have

$$\int_{\mathbb{R}^N} (z(t), D\varphi(u(t))) = \int_{\mathbb{R}^N} \theta(z(t), D\varphi(u(t)), x)\|D\varphi(u(t))\| = \int_{\mathbb{R}^N} \|D\varphi(u(t))\|.$$

Consequently, we get

$$\frac{1}{q+1}\frac{d}{dt}\int_{\mathbb{R}^N} |u(t)|^{q+1}\, dx + \int_{\mathbb{R}^N} \|D\varphi(u(t))\| \leq 0. \tag{4.43}$$

We denote $v(t)(x) := \varphi(u(t))(x)$. By Sobolev's inequality for BV functions (Theorem B.18) we obtain that

$$\||u(t)|^q\|_{L^{N/N-1}(\mathbb{R}^N)} = \|v(t)\|_{L^{N/N-1}(\mathbb{R}^N)} \leq C\int_{\mathbb{R}^N} \|Dv(t)\|.$$

Therefore, from (4.43), we obtain that

$$\frac{1}{q+1}\frac{d}{dt}\int_{\mathbb{R}^N}|u(t)|^{q+1}+\frac{1}{C}\||u(t)|^q\|_{L^{N/N-1}(\mathbb{R}^N)}\leq 0.$$

Then, taking $q = N-1$, we get

$$\frac{d}{dt}\int_{\mathbb{R}^N}|u(t)|^N+\frac{N}{C}\left(\int_{\mathbb{R}^N}|u(t)|^N\right)^{\frac{N-1}{N}}\leq 0. \tag{4.44}$$

Hence

$$\frac{d}{dt}\left[\left(\int_{\mathbb{R}^N}|u(t)|^N\right)^{\frac{1}{N}}\right]+\frac{1}{C}\leq 0. \tag{4.45}$$

Then, given $0\leq t\leq T^*(u_0)$, integrating (4.45) from t to $T^*(u_0)$ we obtain (4.39).

(ii) Since, $u(T^*(u_0)) = 0$, from Proposition 3.13 , if $t\geq\tau>0$, we get

$$\begin{aligned}\left|\frac{u(t,x)}{T^*(u_0)-t}\right| &= \frac{|u(T^*(u_0),x)-u(t,x)|}{T^*(u_0)-t} = \frac{1}{T^*(u_0)-t}\left|\int_t^{T^*(u_0)} u'(s)\,ds\right| \\ &\leq \frac{1}{T^*(u_0)-t}\int_t^{T^*(u_0)}\frac{2}{s}\|u_0\|_\infty\,ds\leq\frac{2}{\tau}\|u_0\|_\infty,\end{aligned}$$

and (4.40) follows. □

Proof of Theorem 4.11. Since $u(t)\in BV(\mathbb{R}^N)$ for almost any $t>0$, without loss of generality, we may assume that $u_0\in BV(\mathbb{R}^N)$. We make a change of scale in time $t=\varphi(\tau)$ so that $\varphi(+\infty)=T^*(u_0)$. Let $\varphi(\tau):=T^*(u_0)(1-e^{-\tau})$. Hence, if we define

$$v(\tau):=\frac{u(\varphi(\tau))}{T^*(u_0)}e^\tau,$$

we have

$$v'(\tau)=u'(\varphi(\tau))+v(\tau).$$

Let Ψ be the functional defined in (3.7). Since the operator $\partial\Psi$ is positively homogeneous of degree zero, we have that

$$(v(\tau),-v'(\tau)+v(\tau))\in\partial\Psi \quad \text{for almost all } \tau>0. \tag{4.46}$$

Therefore, $v(\tau)$ is a strong solution of the problem

$$v'(\tau)+\partial\Psi(v(\tau))\ni v(\tau).$$

Let us see that there exists an increasing sequence $\tau_n\to+\infty$ and a function $v^*\in BV(\mathbb{R}^N)$, such that $\lim_{n\to\infty}v(\tau_n)=v^*$ in $L^p(\mathbb{R}^N)$, which implies the existence of an increasing sequence $t_n\to T^*(u_0)$ such that $\lim_{n\to\infty}w(t_n)=v^*$ in $L^p(\mathbb{R}^N)$.

First, observe that, using (4.40), we have

$$\|v(\tau)\|_\infty = \frac{e^\tau}{T^*(u_0)}\,\|u(\varphi(\tau))\|_\infty \leq \frac{2\|u_0\|_\infty}{\tau_0} \quad \text{for all } \tau \geq \tau_0 > 0. \tag{4.47}$$

On the other hand, by (A.35), we have

$$\frac{d}{d\tau}\Psi\left(v(\tau)\right) = (-v'(\tau)+v(\tau), v'(\tau)) = -\int_{\mathbb{R}^N} v'(\tau)^2\,dx + \int_{\mathbb{R}^N} v(\tau)v'(\tau)\,dx,$$

i.e.,

$$\frac{d}{d\tau}\left(\int_{\mathbb{R}^N} \|Dv(\tau)\| - \frac{1}{2}\int_{\mathbb{R}^N} v(\tau)^2\,dx\right) = -\int_{\mathbb{R}^N} v'(\tau)^2\,dx \leq 0. \tag{4.48}$$

Integrating from 0 to τ we obtain

$$\int_{\mathbb{R}^N} \|Dv(\tau)\| - \frac{1}{2}\int_{\mathbb{R}^N} v(\tau)^2\,dx \leq \int_{\mathbb{R}^N} \|Dv(0)\| - \frac{1}{2}\int_{\mathbb{R}^N} v(0)^2\,dx \quad \forall\, \tau \geq 0. \tag{4.49}$$

Since the support of v is contained in B, estimates (4.47) and (4.49) prove that $\{v(\tau) \,:\, \tau \geq 0\}$ is bounded in $BV(\mathbb{R}^N)$. Having compact support in B, by Theorem B.21, $\{v(\tau) \,:\, \tau \geq 0\}$ is relatively compact in $L^p(\mathbb{R}^N)$ for $1 \leq p < \frac{N}{N-1}$, and consequently, there exists $\tau_n \to \infty$ and $v^* \in L^p(\mathbb{R}^N) \cap BV(\mathbb{R}^N)$ such that $v(\tau_n) \to v^*$ in $L^p(\mathbb{R}^N)$. Moreover, by (4.47) we can assume that $v(\tau_n) \to v^*$ in $L^q(\mathbb{R}^N)$ for all $1 \leq q < \infty$. On the other hand, by (4.39), we have that

$$\|v(\tau)\|_N \geq C \qquad \forall\, \tau \geq 0.$$

Then, we get $v^* \neq 0$.

Finally, let us prove that v^* is a solution of the stationary problem (4.34) which minimizes $\Psi(\cdot) - \langle \cdot, v^* \rangle$ in $BV(\mathbb{R}^N) \cap L^2(\mathbb{R}^N)$. Let $(T(t))_{t\geq 0}$ be the semigroup in $L^1(\mathbb{R}^N)$ generated by $\partial\Psi - I$. Then, we prove that $T(t)v^* = v^*$ for all $t \geq 0$. In fact, by (4.48), we have

$$\int_t^s \int_{\mathbb{R}^N} v'(\tau)^2\;d\tau dx \leq \int_{\mathbb{R}^N} \|Dv(t)\| + \frac{1}{2}\int_{\mathbb{R}^N} v(s)^2\,dx \leq M \tag{4.50}$$

for all $0 < t \leq s$. Now,

$$\|v(t+\tau_n) - v(\tau_n)\|_2^2 = \int_{\mathbb{R}^N} \left| \int_{\tau_n}^{t+\tau_n} v'(s)\;ds \right|^2 dx \leq t \int_{\mathbb{R}^N} \int_{\tau_n}^{t+\tau_n} \left|v'(s)\right|^2\,ds\,dx,$$

hence by (4.50), it follows that there exists $\epsilon_n \to 0$ such that

$$\|v(t+\tau_n) - v(\tau_n)\|_2^2 \leq t\epsilon_n \qquad \forall\, n \in \mathbb{N}. \tag{4.51}$$

Fix $t > 0$. Then, since $v(t) = T(t)\left(\frac{u_0}{T^*(u_0)}\right)$, we have

$$\begin{aligned}\|T(t)v^* - v^*\|_2 &\le \|T(t)v^* - v(t+\tau_n)\|_2 + \|v(t+\tau_n) - v(\tau_n)\|_2 + \|v(\tau_n) - v^*\|_2 \\ &\le e^t\, \|v(\tau_n) - v^*\|_2 + \|v(t+\tau_n) - v(\tau_n)\|_2 + \|v(\tau_n) - v^*\|_2,\end{aligned}$$

and, having in mind (4.51), it follows that $T(t)v^* = v^*$. Thus $0 \in \partial\Psi(v^*) - v^*$, in other words, v^* minimizes $\Psi(\cdot) - \langle\cdot, v^*\rangle$ in $BV(\mathbb{R}^N) \cap L^2(\mathbb{R}^N)$. □

Remark 4.16. Using the same techniques as before, similar results can be proved for the Dirichlet problem for the total variation flow in a bounded domain (i.e., for the problem (5.1) with $\varphi = 0$), see [15].

4.3.1 Solutions of Problem (4.34) in the Radial Case

In Theorem 4.11 we have shown that the asymptotic profile of the solutions of problem (4.1) are solutions of problem (4.34). In this section we are going to study this class of solutions of problem (4.34) in the radial case. To do that one of our tools is the decomposition of any set of finite perimeter into M-connected components given in [9] (see Section B.7).

Proposition 4.17. *Let v be a solution of problem* (4.34) *which is a minimizer of $\Psi(\cdot) - \langle\cdot, v\rangle$ in $BV(\mathbb{R}^2)$.*

(i) *Assume that $v \ge 0$ has compact support contained in a ball $B \subseteq \mathbb{R}^2$. Then, for almost all $k \ge 0$, the M-connected components of $[v \ge k] := \{x \in \mathbb{R}^2 : v(x) \ge k\}$ are convex.*

(ii) *Assume that $v \ge 0$ is a radially symmetric function and it has compact support contained in $\Omega = B_R(0)$, $R > 0$. Then, for almost all $k \in \mathbb{R}$, the M-connected components of $[v \ge k]$ are convex and consequently, $v(x) = g(\|x\|)$, where g is a decreasing function of $r > 0$.*

Proof. (i) Let k be such that $[v \ge k]$ is a set of finite perimeter in $\mathbb{R}^2$. Let $X_i(k)$, $i \in I$, be the M-connected components of $[v \ge k]$ ([9], Section B.7). Let $co(X_i(k))$ be the convex envelope of $X_i(k)$, $i \in I$. Let $A(k) = \cup_{i \in I} co(X_i(k))$. Now, observe that if $k \ge k'$ are such that $[v \ge k]$, $[v \ge k']$ are sets of finite perimeter in $\mathbb{R}^2$, then $A(k) \subseteq A(k')$ (modulo a null set). Indeed, since $k \ge k'$, we have that $X_i(k) \subseteq X_i(k')$ (modulo a null set), and, hence, also $co(X_i(k)) \subseteq co(X_i(k'))$. Thus, $A(k) \subseteq A(k')$. Let w be the L^∞ function such that $[w \ge k] = A(k)$ a.e. for almost all $k \in \mathbb{R}$ ([9], Section B.7). Since $[v \ge k] \subseteq A(k)$ for almost all $k \in \mathbb{R}$, we have that $v \le w$. Now, since $\mathcal{H}^1(\partial^M co(X_i(k))) \le \mathcal{H}^1(\partial^M X_i(k))$, using the coarea formula, we have that

$$\int_{\mathbb{R}^2} \|Dw\| \le \int_{\mathbb{R}^2} \|Dv\|.$$

Hence, $w \in BV(\mathbb{R}^2)$. Now, if for a nonnull set K of $k \in \mathbb{R}$, $X_i(k)$ is not convex, we have that $\mathcal{H}^1(\partial^M co(X_i(k))) < \mathcal{H}^1(\partial^M X_i(k))$, then

$$\int_{\mathbb{R}^2} \|Dw\| < \int_{\mathbb{R}^2} \|Dv\|.$$

Therefore

$$\int_{\mathbb{R}^2} \|Dw\| - \int_{\mathbb{R}^2} wv\, dx < \int_{\mathbb{R}^2} \|Dv\| - \int_{\mathbb{R}^2} v^2\, dx,$$

and v cannot be a minimizer of $\Psi(\cdot) - \langle \cdot, v\rangle$ in $BV(\mathbb{R}^2)$.

(ii) In this case the proof is similar to the one of (i). Since almost all upper level sets of v have convex M-connected components and v is radially symmetric, this implies that, for almost all $k \in \mathbb{R}$, $[v \geq k]$ is a ball centered at 0. Thus, we have that $v(x) = g(\|x\|)$ where g is a decreasing function of $r > 0$. □

By Proposition 4.17, we know that the positive radial solutions v of (4.34) with compact support contained in $\Omega = B_R(0)$, $R > 0$, are of the form $v(x) = g(\|x\|)$ for some decreasing function $g(r)$. By modifying, if necessary, v in a set of measure zero, we may assume that g is upper semi-continuous in $[0, R]$. Consequently, the set $[v \geq k] = \{x \in \mathbb{R}^2 : \|x\| \leq f(k)\}$, where f is the decreasing function $f(k) := \sup\{r \in [0, +\infty[\,:\, g(r) \geq k\}$, $k \in [g(R), g(0)]$. Moreover, since

$$\mathrm{Per}([v \geq k]) = \mathrm{Per}(\{x \in B(0,R) : \|x\| \leq f(k)\}) = 2\pi f(k),$$

$f(k)$ can be identified as

$$f(k) = \frac{1}{2\pi}\mathrm{Per}([v \geq k]).$$

Let us prove that

$$\mathrm{Per}([v \geq k]) = \int_{[v \geq k]} v(x)\, dx \qquad \forall\, k \in]g(R), g(0)]. \tag{4.52}$$

Indeed, since v is a solution of (4.34) there exists $z \in X(\mathbb{R}^2)_1$ satisfying: $v = -\mathrm{div}(z)$ in $\mathcal{D}'(\mathbb{R}^2)$ and $\int_{\mathbb{R}^2} (z, Dv) = \|Dv\|(\mathbb{R}^2)$. Hence, if $k > g(R)$, using Green's formula we have

$$\int_{[v\geq k]} v\, dx = \int_{\mathbb{R}^2} v\chi_{[v\geq k]}\, dx = -\int_{\mathbb{R}^2} \mathrm{div}(z)\chi_{[v\geq k]}\, dx = \int_{\mathbb{R}^2} (z, D\chi_{[v\geq k]}).$$

Now, by the coarea formula, we have

$$\begin{aligned} \int_0^\infty \|D\chi_{[v\geq t]}\|(\mathbb{R}^2)\, dt &= \int_{\mathbb{R}^2} \|Dv\| = \int_{\mathbb{R}^2} (z, Dv) \\ &= -\int_{\mathbb{R}^2} \mathrm{div}(z)\, v\, dx = \int_0^\infty \int_{\mathbb{R}^2} -\mathrm{div}(z)\, \chi_{[v\geq t]}\, dx\, dt \\ &= \int_0^\infty \int_{\mathbb{R}^2} (z, D\chi_{[v\geq t]})\, dt \leq \int_0^\infty \|D\chi_{[v\geq t]}\|(\mathbb{R}^2)\, dt. \end{aligned}$$

It follows that

$$\int_{\mathbb{R}^2} (z, D\chi_{[v\geq t]}) = \int_{\mathbb{R}^2} \|D\chi_{[v\geq t]}\|,$$

and, consequently, (4.52) holds.

On the other hand, since $0 \leq v(x) = g(\|x\|)$ and g is decreasing, we have that

$$\begin{aligned}\int_{[v\geq k]} v(x)\,dx &= \int_{[v\geq k]} \int_0^{+\infty} \chi_{[v\geq t]}(x)\,dt\,dx \\ &= \int_{[v\geq k]} (k + \int_k^{g(0)} \chi_{[v\geq t]}(x)\,dt)\,dx = k|[v\geq k]| + \int_k^{g(0)} |[v\geq t]|\,dt.\end{aligned}$$

Then, a.e. in $k \in [g(R), g(0)]$, we have that

$$\frac{d}{dk}\mathrm{Per}([v\geq k]) = k\frac{d}{dk}|[v\geq k]|,$$

which, written in terms of $f(k)$ is

$$\frac{d}{dk}2\pi f(k) = k\frac{d}{dk}\pi f(k)^2,$$

i.e.,

$$\frac{d}{dk}f(k) = kf(k)\frac{d}{dk}f(k).$$

Consequently, we have that either $f(k) = \frac{1}{k}$ or $f'(k) = 0$ for almost all $k \in [g(R), g(0)]$. Since f is a (pseudo)inverse of g, in terms of g this gives that either $g(r) = \frac{1}{r}$ or $g'(r) = 0$, a.e. in $r \in (0, R)$. Summarizing, we have proved the following result.

Corollary 4.18. *Let $u_0 \geq 0$ be a radial function with compact support in $B_R(0)$, $R > 0$. If v^* is the asymptotic profile of the solution of (P) with initial datum u_0, then there exists a decreasing function $g : [0, R] \to [0, \|u_0\|_\infty]$ satisfying $g(r) = \frac{1}{r}$ or $g'(r) = 0$, a.e. in $r \in (0, R)$, such that $v^*(x) = g(\|x\|)$.*

The computations leading to Corollary 4.18 also hold in $\mathbb{R}^N$, $N \geq 3$, and a result similar to this could be stated.

Proof. The result follows as a consequence of the above computations having in mind that, since u_0 is a radially symmetric function, we have that v^* is also a radially symmetric function. □

Let us give some examples of radial explicit solutions.

Proposition 4.19. *The function*

$$u(x) = \frac{\mathrm{Per}(B_r(0))}{\mathcal{L}^N(B_r(0))}\chi_{B_r(0)}(x)$$

is a solution of (4.34).

Proof. Working as in the proof of Lemma 4.2 it is easy to see that u is a solution of (4.34) whose associated vector field is

$$z(x) = \begin{cases} \dfrac{-x}{r} & \text{if} \quad x \in B_r(0), \\ -r^{N-1}\dfrac{x}{\|x\|^N} & \text{if} \quad x \in \mathbb{R}^N \setminus B_r(0). \end{cases}$$

□

Examples of oscillating solutions. Let $0 = R_0 < R_1 < \cdots < R_p < R_{p+1} = +\infty$, so that $B_{R_0}(0) = \emptyset$, $B_{R_{p+1}}(0) = \mathbb{R}^2$. Set for simplicity $B_i := B_{R_i}(0)$, for $i = 0, \ldots, p+1$. Let $\Omega_i := B_i \setminus \overline{B}_{i-1}$, $i = 1, \ldots, p+1$. Let $a_1, \ldots, a_{p+1}$ be real numbers such that $a_i \neq a_{i-1}$, $a_i \neq a_{i+1}$, $i = 2, \ldots, p$, and $a_{p+1} = 0$. Let $\overline{u} := \sum_{i=1}^p a_i \chi_{\Omega_i}$. We claim that choosing a_i appropriately we have that $\overline{u}$ is a solution of (4.34). To be more precise, we say that we have specified a qualitative ordering of $a_1, \ldots, a_{p+1}$ if a_1 is above a_2 (i.e., $a_1 > a_2$) or below a_2 (i.e., $a_1 < a_2$), a_2 is above or below $a_3, \ldots, a_p$ is above or below a_{p+1}. Then, for each qualitative ordering of $a_1, \ldots, a_{p+1}$, the values of $a_1, \ldots, a_{p+1}$ can be uniquely specified so that u is a solution of (4.34). This will be a consequence of the following observations.

If $(\overline{u}, z)$, with $\overline{u} = \sum_{i=1}^p a_i \chi_{\Omega_i}$, is a solution of (4.34), then integrating $\mathrm{div}(z)$ in B_i we get

$$\int_{\partial B_i} z \cdot \nu^{B_i}\, d\mathcal{H}^1 = \epsilon_i \mathrm{Per}(B_i) \tag{4.53}$$

where $\epsilon_i := \mathrm{sign}(a_{i+1} - a_i)$. Now, integrating (4.34) in Ω_i and using (4.53) we obtain

$$a_i = \frac{\epsilon_{i-1}\mathrm{Per}(B_{i-1}) - \epsilon_i \mathrm{Per}(B_i)}{\mathcal{L}^N(B_i) - \mathcal{L}^N(B_{i-1})} \tag{4.54}$$

where $Per(B_0) = 0$ and $\mathcal{L}^N(B_0) = 0$.

If $B_R := B_R(0)$, we recall that the vector fields $\xi(x) := \frac{x}{R}$ and $z(x) := R\frac{x}{\|x\|^2}$ satisfy

$$-\mathrm{div}(\xi) = \frac{\mathrm{Per}(B_R)}{\mathcal{L}^N(B_R)} \quad \text{in } B_R, \quad \xi|_{\partial B_R} = \frac{x}{\|x\|},$$

respectively,

$$-\mathrm{div}(z) = 0 \quad \text{in } \mathbb{R}^2 \setminus \overline{B}_R, \quad z|_{\partial B_R} = \frac{x}{\|x\|}.$$

The following lemma follows by a simple computation and we shall omit its proof.

Lemma 4.20. *Let $0 < r < R$. The vector field*

$$\xi^{-,-}(x) = -\left(R^{N-1} - r^{N-1} + (R-r)\frac{r^{N-1}R^{N-1}}{\|x\|^N}\right)\frac{x}{R^N - r^N}$$

satisfies

$$-\operatorname{div}(\xi^{-,-}) = \frac{\operatorname{Per}(B_R) - \operatorname{Per}(B_r)}{\mathcal{L}^N(B_R) - \mathcal{L}^N(B_r)} \quad \text{in } B_R \setminus \overline{B}_r,$$

$$\xi^{-,-}|_{\partial B_R} = -\frac{x}{\|x\|}, \quad \xi^{-,-}|_{\partial B_r} = -\frac{x}{\|x\|}.$$

The vector field

$$\xi^{-,+}(x) := \left((R+r)\frac{r^{N-1}R^{N-1}}{\|x\|^N} - (R^{N-1} + r^{N-1})\right)\frac{x}{R^N - r^N}$$

satisfies

$$-\operatorname{div}(\xi^{-,+}) = \frac{\operatorname{Per}(B_R) + \operatorname{Per}(B_r)}{\mathcal{L}^N(B_R) - \mathcal{L}^N(B_r)} \quad \text{in } B_R \setminus \overline{B}_r,$$

$$\xi^{-,+}|_{\partial B_R} = -\frac{x}{\|x\|}, \quad \xi^{-,+}|_{\partial B_r} = \frac{x}{\|x\|}.$$

The vector field

$$\xi^{+,-}(x) := \left((R^{N-1} + r^{N-1}) - (R+r)\frac{r^{N-1}R^{N-1}}{\|x\|^N}\right)\frac{x}{R^N - r^N}$$

satisfies

$$-\operatorname{div}(\xi^{+,-}) = -\frac{\operatorname{Per}(B_R) + \operatorname{Per}(B_r)}{\mathcal{L}^N(B_R) - \mathcal{L}^N(B_r)} \quad \text{in } B_R \setminus \overline{B}_r,$$

$$\xi^{+,-}|_{\partial B_R} = \frac{x}{\|x\|}, \quad \xi^{+,-}|_{\partial B_r} = -\frac{x}{\|x\|}.$$

The vector field

$$\xi^{+,+}(x) = \left(R^{N-1} - r^{N-1} + (R-r)\frac{r^{N-1}R^{N-1}}{\|x\|^N}\right)\frac{x}{R^N - r^N}$$

satisfies

$$-\operatorname{div}(\xi^{+,+}) = -\frac{\operatorname{Per}(B_R) - \operatorname{Per}(B_r)}{\mathcal{L}^N(B_R) - \mathcal{L}^N(B_r)} \quad \text{in } B_R \setminus \overline{B}_r,$$

$$\xi^{+,+}|_{\partial B_R} = \frac{x}{\|x\|}, \quad \xi^{+,+}|_{\partial B_r} = \frac{x}{\|x\|}.$$

In all cases $\|\xi^{\pm,\pm}\|_\infty \leq 1$.

Finally, let us check that given a qualitative ordering of $a_1, \ldots, a_{p+1}$ there is a corresponding solution of (4.34) of the form $\overline{u} = \sum_{i=1}^{p} a_i \chi_{\Omega_i}$. First we observe that once we have specified ϵ_1, the value of a_1 is given by $a_1 = -\epsilon_1 \frac{\mathrm{Per}(B_1)}{\mathcal{L}^N(B_1)}$. Thus, it will be sufficient to check that given three consecutive values a_{i-1}, a_i, a_{i+1} with their qualitative ordering, we can uniquely determine the value of a_i. For simplicity let us call these values a_1, a_2, a_3. Let us prove the compatibility of the values of a_1, a_2, a_3 given by (4.54) with its qualitative ordering, if this is specified in advance. There are four cases to be considered: (i) $a_3 < a_2$, $a_1 < a_2$, (ii) $a_3 < a_2$, $a_1 > a_2$, (iii) $a_3 > a_2$, $a_1 > a_2$, (iv) $a_3 > a_2$, $a_1 < a_2$.

Assume that we are in case (i). Then $\epsilon_1 = 1$ and $\epsilon_2 = -1$. Then, by Lemma 4.20, we have

$$a_1 = \frac{\epsilon_0 \mathrm{Per}(B_0) - \mathrm{Per}(B_1)}{\mathcal{L}^N(B_1) - \mathcal{L}^N(B_0)}, \qquad a_2 = \frac{\mathrm{Per}(B_2) + \mathrm{Per}(B_1)}{\mathcal{L}^N(B_2) - \mathcal{L}^N(B_1)},$$

$$a_3 = \frac{-\mathrm{Per}(B_2) - \epsilon_3 \mathrm{Per}(B_3)}{\mathcal{L}^N(B_3) - \mathcal{L}^N(B_2)}.$$

Independently of the values of $\epsilon_0, \epsilon_3 \in \{+1, -1\}$ we have

$$a_1 \le \frac{\mathrm{Per}(B_0) - \mathrm{Per}(B_1)}{\mathcal{L}^N(B_1) - \mathcal{L}^N(B_0)} < a_2, \qquad a_3 \le \frac{-\mathrm{Per}(B_2) + \mathrm{Per}(B_3)}{\mathcal{L}^N(B_3) - \mathcal{L}^N(B_2)} < a_2.$$

Thus, the value of a_2 is consistent with the qualitative ordering specified in advance. The other three cases can be checked in a similar way. Thus, having specified the qualitative ordering of $a_1, \ldots, a_{p+1}$, the values of ϵ_i are given, and formula (4.54) gives the corresponding value of a_i. We have checked the consistency of this choice. In that case, $\overline{u} = \sum_{i=1}^{p} a_i \chi_{\Omega_i}$ is a solution of (4.34).

4.4 Evolution of Sets in $\mathbb{R}^2$: The Connected Case

Throughout this section, as well as in Sections 4.5, 4.6, 4.7, we take $N = 2$. Let $B \subset \mathbb{R}^2$ be an open set; we say that ∂B is of class $C^{1,1}$ if ∂B can be written, locally around each point, as the graph (with respect to a suitable orthogonal coordinate system) of a function f of class C^1 with Lipschitz continuous gradient, and B can be written (locally) as the epigraph of f. If ∂B is of class $C^{1,1}$, we denote by $\kappa_{\partial B}$ the ($\mathcal{H}^1$-almost everywhere defined) curvature of ∂B.

Let $\Omega \subset \mathbb{R}^2$ be a bounded set of finite perimeter. We set

$$\lambda_\Omega := \frac{\mathrm{Per}(\Omega)}{|\Omega|}, \quad \text{where } |\Omega| = \mathcal{L}^2(\Omega).$$

We want to study when the function

$$u(t,x) := (1 - \lambda_\Omega t)^+ \chi_\Omega(x) \tag{4.55}$$

is the entropy solution of (4.1) and (3.2) when we choose $u_0 = \chi_\Omega$.

Remark 4.21. The function u defined in (4.55) is the solution of (4.1) with $u(0,x) = \chi_\Omega(x)$ if and only if the function $v := \chi_\Omega$ satisfies the equation

$$-\mathrm{div}\left(\frac{Dv}{|Dv|}\right) = \lambda_\Omega v, \tag{4.56}$$

i.e., if and only if there exists a vector field $\xi \in L^\infty(\mathbb{R}^2;\mathbb{R}^2)$ such that $\|\xi\|_\infty \le 1$,

$$-\mathrm{div}(\xi) = \lambda_\Omega v \tag{4.57}$$

and

$$\int_{\mathbb{R}^2} (\xi, Dv) = \int_{\mathbb{R}^2} \|Dv\|. \tag{4.58}$$

With a little abuse of notation, we also write that the pair (v,ξ) is a solution of (4.56).

It is clear that if v is a solution of (4.56), then $\lambda_\Omega v$ is a solution of (4.34). If χ_Ω is a solution of (4.56) and C is a connected component of Ω, using (4.57) and (4.58) it follows that

$$\lambda_C = \lambda_\Omega. \tag{4.59}$$

Definition 4.22. Let $\Omega \subseteq \mathbb{R}^2$ be a set of finite perimeter. We say that Ω is $-$*calibrable* if there exists a vector field $\xi_\Omega^- : \mathbb{R}^2 \to \mathbb{R}^2$ with the following properties:

(i) $\xi_\Omega^- \in L^2_{\mathrm{loc}}(\mathbb{R}^2;\mathbb{R}^2)$ and $\mathrm{div}(\xi_\Omega^-) \in L^2_{\mathrm{loc}}(\mathbb{R}^2)$;

(ii) $|\xi_\Omega^-| \le 1$ almost everywhere in Ω;

(iii) $\mathrm{div}(\xi_\Omega^-)$ is constant on Ω;

(iv) $\theta(\xi_\Omega^-, -D\chi_\Omega, x) = -1$ for $\mathcal{H}^1$-almost every $x \in \partial^*\Omega$.

We say that Ω is $+$*calibrable* if there exists a vector field $\xi_\Omega^+ : \mathbb{R}^2 \to \mathbb{R}^2$ satisfying properties (i), (ii), (iii), and such that $\theta(\xi_\Omega^+, -D\chi_\Omega, x) = 1$ for $\mathcal{H}^1$-almost every $x \in \partial^*\Omega$.

Heuristically, condition (iv) says that the inner (resp. outer) normal trace of ξ_Ω^- (resp. of ξ_Ω^+) is 1.

It is clear that Ω is $-$calibrable if and only if Ω is $+$ calibrable (it is sufficient to define $\xi_\Omega^+ := -\xi_\Omega^-$). Moreover, if Ω is bounded and $-$calibrable, the constant in (iii) equals $-\lambda_\Omega$, i.e., $-\mathrm{div}(\xi_\Omega^-) \equiv \lambda_\Omega$ on Ω.

The following remark should be compared with (a) of Proposition 4.31.

Remark 4.23. Let $\Omega \subset \mathbb{R}^2$ be a bounded set of finite perimeter which is $-$calibrable. Then

$$\frac{\mathrm{Per}(\Omega)}{|\Omega|} \le \frac{\mathrm{Per}(D)}{|D|} \qquad \forall D \subseteq \Omega,\ D \text{ of finite perimeter.} \tag{4.60}$$

Indeed,

$$\lambda_\Omega = \frac{1}{|D|}\int_D -\mathrm{div}(\xi_\Omega^-)\, dx \le \frac{1}{|D|}\mathrm{Per}(D).$$

Remark 4.24. Let $\Omega \subset \mathbb{R}^2$ be a bounded set of finite perimeter. Assume that Ω is $-$calibrable and that $\mathbb{R}^2 \setminus \Omega$ is $+$calibrable. Define

$$\xi := \begin{cases} \xi_\Omega^- & \text{on } \Omega, \\ \xi_{\mathbb{R}^2\setminus\Omega}^+ & \text{on } \mathbb{R}^2 \setminus \Omega. \end{cases}$$

Then $\xi \in L^\infty(\mathbb{R}^2;\mathbb{R}^2)$ and $\mathrm{div}(\xi) \in L^\infty(\mathbb{R}^2)$.

Lemma 4.25. *Let $\Omega \subset \mathbb{R}^2$ be a bounded set of finite perimeter. Then $v := \chi_\Omega$ is a solution of* (4.56) *if and only if Ω is $-$calibrable with $-\mathrm{div}(\xi_\Omega^-) = \lambda_\Omega$ in Ω and $\mathbb{R}^2 \setminus \Omega$ is $+$calibrable, with $\mathrm{div}(\xi_{\mathbb{R}^2\setminus\Omega}^+) = 0$ in $\mathbb{R}^2 \setminus \Omega$.*

Proof. If (χ_Ω, ξ) is a solution of (4.56), then $\xi_\Omega^- := \xi$, $\xi_{\mathbb{R}^2\setminus\Omega}^+ := \xi$ satisfy (i)–(iii) of Definition 4.22. Moreover, by (4.58) and (C.12),

$$\int_{\partial^*\Omega} \theta(\xi_\Omega^-, D\chi_\Omega, x)d\mathcal{H}^1(x) = \mathrm{Per}(\Omega) = \int_{\partial^*\Omega} \theta(\xi_{\mathbb{R}^2\setminus\Omega}^+, -D\chi_{\mathbb{R}^2\setminus\Omega}, x)d\mathcal{H}^1(x),$$

so that (iv) of Definition 4.22 is satisfied. Conversely, it is enough to define $\xi := \xi_\Omega^-\chi_\Omega + \xi_{\mathbb{R}^2\setminus\Omega}^+\chi_{\mathbb{R}^2\setminus\Omega}$, and to use Remark 4.24 to check that (χ_Ω, ξ) solves (4.56). □

We are precisely interested in characterizing the sets of Lemma 4.25. The following theorem gives an answer to this question, under the additional assumption that Ω is connected; thanks to Remark 4.21, we can characterize those sets Ω such that the function u in (4.55) is the solution of (4.1) and (3.2) with $u_0 = \chi_\Omega$. In Theorems 4.40 and 4.42 of Section 4.5 we consider the general situation.

Theorem 4.26. *Let $C \subset \mathbb{R}^2$ be a bounded set of finite perimeter, and assume that C is connected. The function $v := \chi_C$ is a solution of* (4.56) *if and only if the following three conditions hold:*

(i) *C is convex;*

(ii) *∂C is of class $C^{1,1}$;*

(iii) *the following inequality holds:*

$$\operatorname*{ess\,sup}_{p\in\partial C} \kappa_{\partial C}(p) \le \frac{\mathrm{Per}(C)}{|C|}. \tag{4.61}$$

To prove Theorem 4.26, we need several intermediate steps. We start with the proof of the implication

$$\chi_C \text{ solution of (4.56)} \Rightarrow \text{(i)–(iii) hold,} \tag{4.62}$$

which will be given after Lemma 4.35.

Given any set $D \subseteq \mathbb{R}^2$, we define

$$D_\rho := \bigcup \{B_\rho : B_\rho \text{ open ball of radius } \rho \text{ contained in } C\},$$

where $\rho > 0$ is small enough such that D_ρ is nonempty.

The result of the next lemma, without an estimate on the curvature, is proved in Proposition 2.4.3 of [134]. Since in the following the estimate on the curvature plays a crucial role, we need to include the proof.

Lemma 4.27. *Let $C \subset \mathbb{R}^2$ be a bounded open convex set. The following conditions are equivalent:*

(a) *there exists $\rho > 0$ such that $C = C_\rho$;*

(b) *∂C is of class $C^{1,1}$ and* $\operatorname*{ess\,sup}_{p \in \partial C} \kappa_{\partial C}(p) \leq \dfrac{1}{\rho}$.

Proof. *(a)* $\Rightarrow$ *(b)*. Assume that $C = C_\rho$ for some $\rho > 0$ and fix a point $z \in \partial C$. Up to a translation and rotation of coordinates, we can suppose that $z = 0$, that ∂C can be written, in a neighborhood of 0, as the graph Γ_f, with respect to the x-variable, of a nonnegative convex function f vanishing at 0 (therefore the open epigraph of f coincides with C in a neighborhood of z). Since $C = C_\rho$, the open ball of radius ρ contained in the epigraph of f and tangent to Γ_f at $(0,0)$ lies locally above f. Therefore we can choose a parabola tangent to Γ_f at $(0,0)$, lying locally inside the epigraph of f and above the ball, whose graph has curvature at zero equals to $\frac{1}{\rho} + \epsilon$. Precisely, for any $\epsilon > 0$ sufficiently small there exists $\delta > 0$ such that $f(x) \leq \left(\frac{1}{2\rho} + \epsilon\right) x^2$ for any $|x| \leq \delta$. It follows that f is differentiable at $x = 0$ with $f'(0) = 0$, i.e., ∂C is differentiable at z. Therefore ∂C is differentiable at any point. Since ∂C is convex and differentiable at any point, it follows that ∂C is of class C^1.

Let us now prove that ∂C is of class $C^{1,1}$. The idea is the same as before, but now we need a family of parabolas locally above f, passing to an arbitrary point $(t, f(t))$ for $|t| \leq \delta$ and tangent (at the same point) to Γ_f. It will follow that ∂C is locally an infimum of parabolas with second derivative larger than $\frac{1}{\rho}$ (up to ϵ). Precisely, as $C = C_\rho$, given $\epsilon > 0$ sufficiently small and possibly reducing δ, we have

$$f(x) \leq \phi_t(x) := \left(\frac{1}{2\rho} + \epsilon\right) (x - a(t))^2 + b(t) \qquad \forall |x|, |t| \leq \delta,$$

where $a(t) := t - \frac{f'(t)}{(1/\rho)+2\epsilon}$ and $b(t) := f(t) - \frac{f'(t)^2}{(2/\rho)+4\epsilon}$ (note that $f \in C^1$, so that a and b are well defined). Since

$$f = \inf_{|t|\leq\delta} \phi_t \qquad \text{on } |x| \leq \delta,$$

and since ϕ_t are semiconcave with semiconcavity constant equal to $\frac{1}{2\rho} + \epsilon$ for any $|t| \leq \delta$, it follows that f is semiconcave on $[-\delta, \delta]$ with semiconcavity constant equal to $\frac{1}{2\rho} + \epsilon$. Hence f is of class $C^{1,1}$ in $[-\delta, \delta]$ and $f'' \leq \frac{1}{\rho} + \frac{\epsilon}{2}$ almost everywhere in $[-\delta, \delta]$. Therefore ∂C is of class $C^{1,1}$ and, since ϵ is arbitrary, $\operatorname{ess\,sup}_{p\in\partial C} \kappa_{\partial C}(p) \leq \frac{1}{\rho}$.

The implication $(b) \Rightarrow (a)$ is a particular case of [38, Lemma 9.2], with the choices $P = C$, $\widetilde{\phi}(\xi_1, \xi_2) = \sqrt{\xi_1^2 + \xi_2^2}$ and $\lambda = \rho$. □

Remark 4.28. If condition (a) of Lemma 4.27 holds, then $C = C_\sigma$ for any $\sigma \in [0, \rho]$, since any ball B_ρ of radius ρ is the union of all balls B_σ of radius $\sigma \in [0, \rho]$ contained in B_ρ.

Lemma 4.29. *Let $a, b \in \mathbb{R}$, $a < b$, $\lambda > 0$ and $G_\lambda : H_0^1([a, b]) \to \mathbb{R}$ be defined as*

$$G_\lambda(u) := \int_{[a,b]} \left[\sqrt{1 + (u'(s))^2} - \lambda u(s)\right] \, d\mathcal{H}^1(s). \tag{4.63}$$

Assume that there exists a function $u_\lambda \in H_0^1([a, b])$ whose graph is contained in a translation of $\partial B_{\frac{1}{\lambda}}$. Then u_λ is the unique minimizer of G_λ in $H_0^1([a, b])$.

Proof. It is a particular case of [38, Lemma 8.4] with the choice $\widetilde{\phi}(\xi_1, \xi_2) = \sqrt{\xi_1^2 + \xi_2^2}$. □

Lemma 4.30. *Let $\Omega \subset \mathbb{R}^2$ be a bounded set of finite perimeter. Assume that $\mathbb{R}^2 \setminus \Omega$ is $+$calibrable. Then $\operatorname{div}(\xi^+_{\mathbb{R}^2\setminus\Omega}) = 0$ on $\mathbb{R}^2 \setminus \Omega$.*

Proof. Let for simplicity $\xi := \xi^+_{\mathbb{R}^2\setminus\Omega}$. Let $R > 0$ be such that $B_R \supset \Omega$ and let U be the unbounded component of $\mathbb{R}^2 \setminus \Omega$. By assumption we have that $\operatorname{div}(\xi) = \alpha$ on $U \cap B_R$ for some real constant α. Using (C.12) and the properties of ξ (see (ii) and (iv) of Definition 4.22) we have

$$-2\pi R + \operatorname{Per}(U) \leq \int_{U\cap B_R} \operatorname{div}(\xi)\, dx \leq 2\pi R + \operatorname{Per}(U).$$

If we denote by λ the (finite) measure of the union of all connected components of $\mathbb{R}^2 \setminus \Omega$ contained in B_R, it follows that

$$\frac{-2\pi R + \operatorname{Per}(U)}{\pi R^2 - |\Omega| - \lambda} \leq \alpha = \frac{\int_{U\cap B_R} \operatorname{div}(\xi)\, dx}{|U \cap B_R|} \leq \frac{2\pi R + \operatorname{Per}(U)}{\pi R^2 - |\Omega| - \lambda} v.$$

Letting $R \to +\infty$ we deduce $\alpha = 0$. □

Proposition 4.31. *Let $\Omega \subset \mathbb{R}^2$ be a bounded set of finite perimeter which is $-$calibrable and such that $\mathbb{R}^2 \setminus \Omega$ is $+$calibrable. Then*

(a) *the following relations hold:*

$$\frac{\mathrm{Per}(\Omega)}{|\Omega|} \le \frac{\mathrm{Per}(D)}{|\Omega \cap D|} \qquad \forall D \subseteq \mathbb{R}^2,\ D \text{ of finite perimeter;} \tag{4.64}$$

(b) *each connected component of Ω is convex.*

Proof. Let $\xi \in L^\infty(\mathbb{R}^2;\mathbb{R}^2)$, $\|\xi\|_\infty \le 1$ be the vector field defined by $\xi := \xi_\Omega^- \chi_\Omega + \xi^+_{\mathbb{R}^2\setminus\Omega}\chi_{\mathbb{R}^2\setminus\Omega}$. By Remark 4.24 we have that $\mathrm{div}(\xi) \in L^\infty(\mathbb{R}^2)$. Let $D \subseteq \mathbb{R}^2$ be a set of finite perimeter. Using Lemma 4.30 and the fact that $-\mathrm{div}(\xi_\Omega^-) \equiv \lambda_\Omega$ on Ω, we have

$$-\int_{\mathbb{R}^2} \chi_D \mathrm{div}(\xi)\, dx = -\int_{\mathbb{R}^2} \chi_\Omega \chi_D \mathrm{div}(\xi)\, dx = \lambda_\Omega \int_{\mathbb{R}^2} \chi_{\Omega\cap D}\, dx = \lambda_\Omega |\Omega \cap D|.$$

Hence

$$\lambda_\Omega |\Omega \cap D| \le \mathrm{Per}(D), \tag{4.65}$$

and (4.64) follows.

Moreover from (4.65) it follows that

$$\mathrm{Per}(\Omega) \le \mathrm{Per}(D) \qquad \forall D \supseteq \Omega,\ D \text{ of finite perimeter.}$$

We conclude that each connected component of Ω must be convex. □

Definition 4.32. Given $\lambda \in \mathbb{R}$ we define the functional $\mathcal{G}_\lambda$ as

$$\mathcal{G}_\lambda(D) := \mathrm{Per}(D) - \lambda|D|, \qquad D \subseteq \mathbb{R}^2,\ D \text{ of finite perimeter.}$$

Proposition 4.33. *Let C be a bounded open convex set, and assume that C is $-$calibrable. Then ∂C is of class $C^{1,1}$.*

Proof. Set for simplicity $\xi := -\xi_C^-$ and recall that $\mathrm{div}(\xi) = \lambda_C$ on C. For any $\lambda > \lambda_C$ and any finite perimeter set B strictly contained in C we then have

$$\mathcal{G}_\lambda(B) \ge \int_B (\mathrm{div}(\xi) - \lambda)\, dx > \int_C (\mathrm{div}(\xi) - \lambda)\, dx = \mathcal{G}_\lambda(C). \tag{4.66}$$

Assume now by contradiction that ∂C is not of class $C^{1,1}$. By Lemma 4.27 it follows that C_ρ is strictly contained in C for some $\rho > 0$. Fix $\sigma < \rho$ such that $\sigma\lambda_C < 1$. By Remark 4.28 we have that C_σ is strictly contained in C. Applying Lemma 4.29 to the connected components of $\partial C_\sigma \setminus \partial C$, we get

$$\mathcal{G}_{\frac{1}{\sigma}}(C_\sigma) \le \mathcal{G}_{\frac{1}{\sigma}}(C),$$

which contradicts (4.66). □

Remark 4.34. (i) If $\Omega \subset \mathbb{R}^2$ is a bounded set of finite perimeter satisfying (4.60) it follows that $\mathcal{G}_{\lambda_\Omega}(D) \geq 0$ for any $D \subseteq \Omega$ of finite perimeter, while obviously $\mathcal{G}_{\lambda_\Omega}(\Omega) = 0$. Therefore Ω minimizes $\mathcal{G}_{\lambda_\Omega}$ among all finite perimeter sets $D \subseteq \Omega$.

(ii) By the proof of Proposition 4.33, it follows that if C is a bounded open convex set which is $-$calibrable, then C minimizes $\mathcal{G}_\lambda$ among all finite perimeter sets $B \subseteq C$ and where $\lambda > \lambda_C$.

In order to prove the implication (4.62) of Theorem 4.26 we need one more lemma.

Lemma 4.35. *Let $C \subset \mathbb{R}^2$ be a bounded open convex set with $C^{1,1}$ boundary satisfying* (4.60) *with C in place of Ω. Then* (4.61) *holds.*

Proof. Let U be a neighborhood of ∂C and let $h \in C_0^1(U)$. Let $\alpha \in \mathbb{R}$ be sufficiently small, and let $\Psi_\alpha(x,y) := (x,y) + \alpha h(x,y)\nu(x,y)$, where $\nu \in C^1(U;\mathbb{R}^2)$ is a vector field satisfying $|\nu| = 1$ on U, and $\nu = \nu^C$ on ∂C. Extend Ψ_α as $\Psi_\alpha(x,y) = (x,y)$ outside U. Let $C_\alpha := \Psi_\alpha(C)$. By Remark 4.34 it follows that C minimizes $\mathcal{G}_{\lambda_C}$ among all finite perimeter sets contained in C. Therefore, if h is nonpositive,

$$0 \leq \lim_{\alpha \to 0^+} \frac{\mathcal{G}_{\lambda_C}(C_\alpha) - \mathcal{G}_{\lambda_C}(C)}{\alpha} = \int_{\partial C} \left[\kappa_{\partial C} - \lambda_C\right] h \, d\mathcal{H}^1.$$

It follows that $\kappa_{\partial C}(x) \leq \lambda_C$ for $\mathcal{H}^1$-almost every $x \in \partial C$. □

We are now in the position to prove the implication (4.62) of Theorem 4.26. If χ_C is a solution of (4.56), by Lemma 4.25 (applied with $\Omega = C$) it follows that C is $-$calibrable with $-\mathrm{div}(\xi_C^-) = \lambda_C$ in C and $\mathbb{R}^2 \setminus C$ is $+$calibrable with $\mathrm{div}(\xi^+_{\mathbb{R}^2 \setminus C}) = 0$ in $\mathbb{R}^2 \setminus C$. Therefore by (b) of Proposition 4.31 (applied with $\Omega = C$) and the assumption that C is connected it follows that C is convex. Hence by Proposition 4.33 we have that ∂C is of class $C^{1,1}$. Moreover, inequality (4.60) holds. Therefore we can apply Lemma 4.35 to conclude that (4.61) holds.

The converse result is also true and was proved by Weinberger and Giusti in [123].

Theorem 4.36. *Let Ω be a convex domain in $\mathbb{R}^2$, with boundary of class C^1, and let $k(x,y)$ be the curvature of $\partial\Omega$ at (x,y). Then for every set of finite perimeter $E \subset \Omega$, $E \neq \emptyset, \Omega$, we have*

$$\mathcal{G}_{\lambda_\Omega}(E) > 0 \tag{4.67}$$

if and only if

$$\sup_{(x,y) \in \partial\Omega} k(x,y) \leq \lambda_\Omega. \tag{4.68}$$

Proof. The necessity of (4.68) has been proved in Lemma 4.35. To prove the sufficiency, let $L = \inf \mathcal{G}_{\lambda_\Omega}(E)$ where the infimum is taken with respect to all sets of finite perimeter $E \subset \Omega$. Note that $L \leq 0$, since $\mathcal{G}_{\lambda_\Omega}(\Omega) = 0$. From the lower semi-continuity of the perimeter it follows that $\mathcal{G}_{\lambda_\Omega}$ has a minimum, hence there exists a set of finite perimeter $E_0 \subset \Omega$ such that $\mathcal{G}_{\lambda_\Omega}(E_0) = L$. We shall prove that $E_0 = \Omega$ and that for every set of finite perimeter $E \neq \emptyset, \Omega$ we have

$$\mathcal{G}_{\lambda_\Omega}(E) > L.$$

Let E_{0i} be the M-connected components of E_0 (see Section B.7). Suppose that there are countably many of them. By the isoperimetric inequality, we have that

$$|E_{0i}| \leq \frac{1}{4\pi}\operatorname{Per}(E_{0i})^2,$$

hence

$$\operatorname{Per}(E_{0i}) - k_0|E_{0i}| \geq \operatorname{Per}(E_{0i})\left(1 - \frac{k_0}{4\pi}\operatorname{Per}(E_{0i})\right) > 0$$

for $i > m$ for some $m \in \mathbb{N}$. Thus replacing E_0 by $\cup_{i=1}^m E_{0i}$ we may decrease $\mathcal{G}(E)$. We conclude that E_0 has a finite number of M-connected components.

Note that since $\mathcal{G}_{\lambda_\Omega}(E_0) = \sum_{i=1}^m \mathcal{G}_{\lambda_\Omega}(E_{0i})$ we have that

$$\mathcal{G}_{\lambda_\Omega}(E_{0i}) \leq 0 \quad \forall i = 1, \dots, m. \tag{4.69}$$

Otherwise we would decrease the energy by taking out the sets E_{0i} for which the above quantity is positive.

We observe that each M-connected component of E_0 must be convex. In fact, let E be a nonconvex M-connected set and let $\hat{E}$ be its convex envelope. Since

$$\operatorname{Per}(\hat{E}) \leq \operatorname{Per}(E) \tag{4.70}$$

and

$$|\hat{E}| > |E|, \tag{4.71}$$

we have

$$\mathcal{G}(E) > \mathcal{G}(\hat{E}) \geq L.$$

Thus by replacing an M-connected component of E_0 by its convex envelope, we decrease the energy. This proves that each M-connected component of E_0 is convex. Moreover, a similar argument proves that they are at positive distance from each other.

Now we prove that each M-connected component of E_0 must touch $\partial\Omega$. Suppose that for some $i \in \{1, \dots, m\}$, $E_{0i} \subset\subset \Omega$. Take $X_0 \in E_{0i}$ and for each $t > 0$ define

$$E_{0it} = \left\{Y \in \mathbb{R}^2 \,:\, Y = X_0 + t(X - X_0),\, X \in E_{0i}\right\}.$$

Then we have

$$\operatorname{Per}(E_{0it}) = t\operatorname{Per}(E_{0i}), \quad \text{and} \quad |E_{0it}| = t^2|E_{0i}|.$$

From $E_{0i} \subset\subset \Omega$ it follows that $E_{0it} \subset \Omega$ for some $t > 1$. Let

$$E_0' = (E_0 \setminus E_{0i}) \cup E_{0it}.$$

Then observe that

$$\begin{aligned} \mathcal{G}_{\lambda_\Omega}(E_0) - \mathcal{G}_{\lambda_\Omega}(E_0') &= \mathcal{G}_{\lambda_\Omega}(E_{0i}) - \mathcal{G}_{\lambda_\Omega}(E_{0it}) \\ &= (t-1)\operatorname{Per}(E_{0i}) - \lambda_\Omega(1-t^2)|E_{0i}| > 0 \end{aligned}$$

the last quantity being positive because of (4.69). It follows that E_{0i} must touch $\partial\Omega$.

Let E one of the M-connected components of E_0. Let us consider the set $\partial\Omega \setminus \overline{E}$, and let Γ be one of its connected components. Let P_1 and P_2 be the endpoints of Γ and let ν_1 and ν_2 be the unit normal vectors to $\partial\Omega$ at P_1,P_2, respectively. Finally, let θ be the angle between ν_1 and ν_2.

If $\theta > \pi$, in a neighborhood of $\partial\Omega \cap \overline{E} = \partial\Omega \setminus \Gamma$, both $\partial\Omega$ and ∂E are representable as graphs: $y = \omega(x)$ and $y = e(x)$, respectively, with $e \geq \omega$. In this case, a small translation of E in the y direction takes $\overline{E}$ in the interior of Ω, leaving $\mathcal{G}_{\lambda_\Omega}(E)$ unchanged. We would obtain a minimum of $\mathcal{G}_{\lambda_\Omega}$ with an M-connected component not touching $\partial\Omega$, a contradiction. We are then left with the case $\theta \leq \pi$ for every connected component Γ of $\partial\Omega \setminus \overline{E}$.

Let us observe that there cannot be any M-connected component of E_0 in $\Omega \setminus \overline{E}$. Indeed if F would be one of them, then the arc of $\partial\Omega \setminus \overline{F}$ would span an angle $> \pi$, but, as we have seen, this is impossible. Thus E coincides with E_0.

To conclude the argument, suppose first that $\theta < \pi$. Then in a neighborhood of Γ, $\partial\Omega$ is representable as the graph of the function

$$y = \omega(x), \quad a \leq x \leq b$$

such that $y < \omega(x)$ in Ω. Similarly, the connected component of $\partial E \cap \Omega$ corresponding to Γ can be represented as

$$y = e(x), \quad a \leq x \leq b$$

with $y < e(x)$ in E and

$$e(x) \leq \omega(x), \tag{4.72}$$

$$e(a) = \omega(a), \quad \mathbf{e}(b) = \omega(b). \tag{4.73}$$

From (4.68) we have

$$k(x, \omega(x)) = -\frac{d}{dx}\left(\frac{\omega'}{\sqrt{1+\omega'^2}}\right) \leq \lambda_\Omega$$

and hence $\omega(x)$ is a subsolution for the functional

$$G_{\lambda_\Omega}(f) = \int_a^b \sqrt{1+f'^2}\,dx - \lambda_\Omega \int_a^b f\,dx.$$

From (4.72) it follows that

$$G_{\lambda_\Omega}(\omega) \le G_{\lambda_\Omega}(e) \tag{4.74}$$

and, from the strict convexity of G_{λ_Ω}, the inequality is strict unless $e = \omega$.

Let now

$$F = \{(x,y) \in \mathbb{R}^2 : a < x < b,\ e(x) < y < \omega(x)\}$$

and let

$$E' = E \cup F.$$

We have

$$\mathrm{Per}(E') = \mathrm{Per}(E) - \int_a^b \sqrt{1+e'^2}\,dx + \int_a^b \sqrt{1+\omega'^2}\,dx,$$

$$|E'| = |E| + \int_a^b (\omega - e)\,dx$$

and, hence, if $E' \neq E$, we have

$$\mathrm{Per}(E') - \lambda_\Omega |E'| < \mathrm{Per}(E) - \lambda_\Omega |E|,$$

and therefore $E = E_0$ could not be a minimum of $\mathcal{G}_{\lambda_\Omega}$.

When $\theta = \pi$ the preceding argument is not directly applicable as it is, since condition (4.73) is not satisfied in general. We have instead

$$\lim_{x\to a} T\omega(x) = 1, \quad \lim_{x\to b} T\omega(x) = -1, \tag{4.75}$$

where

$$T\omega(x) = \frac{\omega'(x)}{\sqrt{1+\omega'^2(x)}}.$$

If we set

$$G_{\lambda_\Omega,1}(f) = G_{\lambda_\Omega}(f) + f(b) + f(a),$$

we have

$$\delta G_{\lambda_\Omega,1}(f;\eta) = \frac{d}{dt} G_{\lambda_\Omega,1}(f+t\eta)|_{t=0} = \int_a^b \eta' Tf\,dx - \lambda_\Omega \int_a^b \eta\,dx + \eta(b) + \eta(a) \tag{4.76}$$

and from (4.75)

$$\delta G_{\lambda_\Omega,1}(\omega;\eta) = \int_a^b \eta[k(x,\omega(x)) - \lambda_\Omega]\,dx \ge 0$$

for every $\eta \leq 0$. In conclusion, from the convexity of $G_{\lambda_\Omega,1}$ we get

$$G_{\lambda_\Omega,1}(\omega) < G_{\lambda_\Omega,1}(e)$$

unless $e = \omega$. Arguing as before and observing that

$$\mathrm{Per}(E') = \mathrm{Per}(E) + \int_a^b \sqrt{1+\omega'^2}\,dx - \int_a^b \sqrt{1+e'^2}\,dx + \omega(a) - e(a) + \omega(b) - e(b)$$

we get again $\mathcal{G}_{\lambda_\Omega}(E) > L$ provided that $E \neq E'$.

We have thus proved that for every set of finite perimeter $B \subset \Omega$, $B \neq \emptyset, \Omega$, we have

$$\mathcal{G}_{\lambda_\Omega}(B) > L = \min \mathcal{G}_{\lambda_\Omega}.$$

We can therefore conclude that $L = \mathcal{G}_{\lambda_\Omega}(\Omega) = 0 = \mathcal{G}_{\lambda_\Omega}(\emptyset)$, and that

$$\mathcal{G}_{\lambda_\Omega}(B) > 0$$

for every set of finite perimeter $B \subseteq \Omega$, $B \neq \emptyset, \Omega$. □

Remark 4.37. We note that the assumption that $\partial\Omega$ is C^1 is redundant, since the convexity of Ω and the bound (4.68) imply that $\partial\Omega$ is $C^{1+\alpha}$ for every $\alpha < 1$.

Let us now prove the opposite implication of Theorem 4.26, that is

$$\text{(i)–(iii)} \Rightarrow \chi_C \text{ is a solution of (4.56).} \tag{4.77}$$

Assume that C is a bounded open $C^{1,1}$ convex set satisfying (4.61). According to Theorem 4.36, (4.61) is a necessary and sufficient condition for C to be a minimizer of the functional $\mathcal{G}_{\lambda_C}$ among all sets of finite perimeter $D \subseteq C$. In this case the function $f := \lambda_C \chi_C$ satisfies $\|f\|_* \leq 1$. Indeed, if $w \in L^2(\mathbb{R}^2) \cap BV(\mathbb{R}^2)$ is nonnegative, we have

$$\begin{aligned}\int_{\mathbb{R}^2} f(x)w(x)\,dx &= \int_0^\infty \int_{\mathbb{R}^2} \lambda_C \chi_C \chi_{\{w\geq t\}}\,dx\,dt = \int_0^\infty \lambda_C |C \cap \{w \geq t\}|\,dt \\ &\leq \int_0^\infty \mathrm{Per}(C \cap \{w \geq t\})\,dt \leq \int_0^\infty \mathrm{Per}(\{w \geq t\})\,dt = \int_{\mathbb{R}^2} \|Dw\|,\end{aligned}$$

where we have used that for all $t \geq 0$ for which $\{w \geq t\}$ is a set of finite perimeter we have that

$$\mathrm{Per}(C \cap \{w \geq t\}) \leq \mathrm{Per}(\{w \geq t\})$$

which is a consequence of the convexity of C. Splitting any function $\omega \in L^2(\mathbb{R}^2) \cap BV(\mathbb{R}^2)$ into its positive and negative part, using the above inequality one can prove that

$$\left| \int_{\mathbb{R}^2} f(x)\omega(x)\,dx \right| \leq \int_{\mathbb{R}^2} \|D\omega\|.$$

It follows that $\|f\|_* \leq 1$. Then, by Lemma 3.3, there is a vector field $\xi \in L^\infty(\mathbb{R}^2;\mathbb{R}^2)$ with $\| \xi \|_\infty \leq 1$ such that

$$-\mathrm{div}(\xi) = f = \lambda_C \chi_C. \tag{4.78}$$

Now, multiplying (4.78) by χ_C and integrating by parts, we obtain

$$\int_{\mathbb{R}^2} (\xi, D\chi_C) = \lambda_C \int_{\mathbb{R}^2} \chi_C \, dx = \mathrm{Per}(C) = \int_{\mathbb{R}^2} \|D\chi_C\|,$$

hence χ_C is a solution of (4.56). The proof of Theorem 4.26 is concluded.

Remark 4.38. We conclude this section by recalling that in the paper [123], condition (4.61) was used as a necessary and sufficient condition for the existence of a solution u with $\nabla u \in L^\infty_{\mathrm{loc}}(C;\mathbb{R}^2)$ of the equation

$$-\mathrm{div}\left(\frac{\nabla u}{\sqrt{1+|\nabla u|^2}}\right) = \lambda_C \quad \text{in } C \tag{4.79}$$

with boundary condition

$$\lim_{C \ni y \to x} \frac{\nabla u(y)}{\sqrt{1+|\nabla u(y)|^2}} = -\nu^C(x) \quad \text{for any } \ x \in \partial C.$$

Remark 4.39. The functional

$$\int_{\mathbb{R}^2} \|Du\|$$

can be regarded, up to a constant, as the anisotropic perimeter [39] of the set $\{(x,y) \in \mathbb{R}^N \times \mathbb{R} : y < u(x)\}$, corresponding to the anisotropy given by the cylindrical norm $\phi(z,\zeta) := \max\{\|z\|, |\zeta|\}$, for $(z,\zeta) \in \mathbb{R}^N \times \mathbb{R}$. Therefore, equation (4.1) is similar (even if not exactly the same) to the equation defining the anisotropic mean curvature flow corresponding to ϕ. Interestingly enough, it turns out that, when $N = 2$, the problem of determining those bounded connected sets Ω whose characteristic function evolve by decreasing its height is close to the problem of determining which planar horizontal facets of a given solid subset of $\mathbb{R}^2 \times \mathbb{R}$ do not break or bend under the ϕ-anisotropic mean curvature flow. This problem has been considered in [37], [38] and the techniques developed there can be adapted, to some extent, to the present situation (see in particular Theorem 4.26).

4.5 Evolution of Sets in $\mathbb{R}^2$: The Nonconnected Case

The aim of this section is to generalize Theorem 4.26 to nonconnected sets (see Theorems 4.40 and 4.42). Theorem 4.42 is basically a further generalization of Theorem 4.40.

Theorem 4.40. *Let $\Omega \subset \mathbb{R}^2$ be a bounded set of finite perimeter. If $v := \chi_\Omega$ is a solution of* (4.56), *then Ω has a finite number of connected components $C_1, \ldots, C_m$, and*

(i) *C_i is convex for any $i = 1, \ldots, m$;*

(ii) *∂C_i is of class $C^{1,1}$ for any $i = 1, \ldots, m$;*

(iii) *the following inequalities hold:*

$$\operatorname*{ess\ sup}_{p \in \partial C_i} \kappa_{\partial C_i}(p) \leq \frac{\mathrm{Per}(C_i)}{|C_i|} \qquad \forall i = 1, \ldots, m;$$

(iv) $\frac{\mathrm{Per}(C_i)}{|C_i|} = \frac{\mathrm{Per}(C_j)}{|C_j|}$ *for any $i, j \in \{1, \ldots, m\}$;*

(v) *let $0 \leq k \leq m$ and let $\{i_1, \ldots, i_k\} \subseteq \{1, \ldots, m\}$ be any k-tuple of indices; if we denote by $E_{i_1,\ldots,i_k}$ a solution of the variational problem*

$$\min \left\{ \mathrm{Per}(E) : \quad E \text{ of finite perimeter }, \bigcup_{j=1}^{k} C_{i_j} \subseteq E \subseteq \mathbb{R}^2 \setminus \bigcup_{j=k+1}^{m} C_{i_j} \right\}, \tag{4.80}$$

we have

$$\mathrm{Per}(E_{i_1,\ldots,i_k}) \geq \sum_{j=1}^{k} \mathrm{Per}(C_{i_j}). \tag{4.81}$$

Conversely, assume that $\Omega \subset \mathbb{R}^2$ is a bounded open set which is the union of a finite number $C_1, \ldots, C_m$ of connected components satisfying (i)–(v). *Then $v := \chi_\Omega$ is a solution of* (4.56).

This theorem can be proved along the lines described in [34]. We shall prove it as a consequence of Theorem 4.42. We start with the following observation.

Lemma 4.41. *Let $\alpha_i > 0$ and $B_i \subseteq \mathbb{R}^2$ be bounded measurable sets, for $i = 1, \ldots, m$. Let $g := \sum_{i=1}^{m} \alpha_i \chi_{B_i}$. Then $\|g\|_* \leq 1$ if and only if*

$$\sum_{i=1}^{m} \alpha_i |B_i \cap D| \leq \mathrm{Per}(D) \tag{4.82}$$

for all $D \subset \mathbb{R}^2$, D bounded of finite perimeter.

Proof. Assume that $\|g\|_* \leq 1$. Let $D \subseteq \mathbb{R}^2$ be a bounded set of finite perimeter. Then

$$\sum_{i=1}^{m} \alpha_i |B_i \cap D| = \int_{\mathbb{R}^2} g \chi_D \, dx \leq \int_{\mathbb{R}^2} \|D\chi_D\| = \mathrm{Per}(D).$$

Conversely, assume that (4.82) holds. Let $v \in L^2(\mathbb{R}^2) \cap BV(\mathbb{R}^2)$ be nonnegative. We have

$$\begin{aligned}\int_{\mathbb{R}^2} gv\,dx &= \sum_{i=1}^m \alpha_i \int_0^\infty \int_{\mathbb{R}^2} \chi_{B_i}\chi_{\{v\geq t\}}\,dx\,dt = \sum_{i=1}^m \alpha_i \int_0^\infty |B_i \cap \{v \geq t\}|\,dt \\ &\leq \int_0^\infty \mathrm{Per}(\{v \geq t\})\,dt = \int_{\mathbb{R}^2} \|Dv\|.\end{aligned}$$

Splitting into the positive and negative parts, the above inequality holds for a generic $v \in L^2(\mathbb{R}^2) \cap BV(\mathbb{R}^2)$. Therefore $\|g\|_* \leq 1$. □

The following result is essentially a generalization of Theorem 4.40.

Theorem 4.42. *Let $\Omega \subset \mathbb{R}^2$ be a bounded set of finite perimeter and assume that Ω consists of a finite number of connected components $C_1, \ldots, C_m$. Let $b_i > 0$ for $i = 1, \ldots, m$. The function $u := \sum_{i=1}^m b_i\chi_{C_i}$ is a solution of* (4.34) *if and only if*

(a) $b_i = \frac{\mathrm{Per}(C_i)}{|C_i|}$ *for all* $i = 1, \ldots, m$;

(b) *conditions* (i)–(iii) *and* (v) *of Theorem* 4.40 *hold.*

Proof. Assume that (u, ξ) is a solution of (4.34), where $u = \sum_{i=1}^m b_i\chi_{C_i}$. The identity $(\xi, Du) = \|Du\|$ implies that $(\xi, D\chi_{C_i}) = \|D\chi_{C_i}\|$ as measures in $\mathbb{R}^2$, for all $i = 1, \ldots, m$. Using this observation and integrating the equality $-\mathrm{div}(\xi) = u$ in C_i it follows that $b_i = \lambda_{C_i}$. Now, let $D \subseteq \mathbb{R}^2$ be a set of finite perimeter. Multiplying the equation $-\mathrm{div}(\xi) = u$ by χ_D and integrating in $\mathbb{R}^2$ we obtain

$$\mathrm{Per}(D) \geq -\int_{\mathbb{R}^2} \chi_D \mathrm{div}(\xi)\,dx = \sum_{i=1}^m b_i|C_i \cap D| \geq b_j|C_j \cap D|, \tag{4.83}$$

i.e., $\lambda_{C_j} \leq \frac{\mathrm{Per}(D)}{|C_j \cap D|}$ for each $j = 1, \ldots, m$. As in the proof of Theorem 4.40, it follows that (i)–(iii) hold. Finally, let us prove that condition (v) holds. If we write (4.83) for $D = E_{i_1,\ldots,i_k}$ we have

$$\sum_{i=1}^m \lambda_{C_i}|C_i \cap E_{i_1,\ldots,i_k}| \leq \mathrm{Per}(E_{i_1,\ldots,i_k}),$$

which gives (4.81) since $C_{i_j} \cap E_{i_1,\ldots,i_k} = C_{i_j}$ for $j = 1, \ldots, k$, while $C_i \cap E_{i_1,\ldots,i_k} = \emptyset$ for $i \notin \{i_1, \ldots, i_k\}$.

Conversely, assume that conditions (a) and (b) hold. Reasoning as in the proof of (4.77) it follows that each C_i is $-$calibrable. We shall prove that $g := \sum_{i=1}^m \lambda_{C_i}\chi_{C_i}$ satisfies $\|g\|_* \leq 1$. According to Lemma 4.41, it will be sufficient to prove that

$$\sum_{i=1}^m \lambda_{C_i}|C_i \cap D| \leq \mathrm{Per}(D) \qquad \forall\, D \text{ bounded of finite perimeter.} \tag{4.84}$$

By additivity of the area and the perimeter, it is sufficient to prove (4.84) when D is also indecomposable. Let $D \subseteq \mathbb{R}^2$ be such a set. Since C_i are $-$calibrable sets, by Remark 4.23 (applied with $\Omega := C_i$ and $D := D \cap C_i$), we have that

$$\lambda_{C_i}|C_i \cap D| \leq \mathrm{Per}(C_i \cap D).$$

Then, to prove (4.84), it will be sufficient to prove that

$$\sum_{i=1}^{m} \mathrm{Per}(C_i \cap D) \leq \mathrm{Per}(D) \quad \forall\, D \text{ bounded indecomposable of finite perimeter.} \tag{4.85}$$

Let us identify D with its representative $\mathring{D}^{\mathrm{M}}$ (see Appendix B). Denote by $C_{i_1}, \ldots$, C_{i_k} the connected components of Ω such that $D \cup \bigcup_{j=1}^{k} C_{i_j}$ is connected. Those components intersect either D or $\partial^* D$. Let $E_{i_1,\ldots,i_k}$ be a minimizer of problem (4.80). Using (4.81) and the minimality of $E_{i_1,\ldots,i_k}$ we then have

$$\sum_{j=1}^{k} \mathrm{Per}(C_{i_j}) \leq \mathrm{Per}(E_{i_1,\ldots,i_k}) \leq \mathrm{Per}\left(D \cup \bigcup_{j=1}^{k} C_{i_j}\right). \tag{4.86}$$

We claim that

$$\mathrm{Per}\left(D \cup \bigcup_{j=1}^{k} C_{i_j}\right) \leq \mathrm{Per}(D, \mathbb{R}^2 \setminus \overline{\Omega}) + \sum_{j=1}^{k} \mathrm{Per}(C_{i_j}) - \mathcal{H}^1\left(D \cap \left(\bigcup_{j=1}^{k} \partial C_{i_j}\right)\right). \tag{4.87}$$

Indeed, since $\partial^*(D \cup X) \subseteq (\partial^* D \setminus X) \cup (\partial X \setminus D)$ where $X := \bigcup_{j=1}^{k} C_{i_j}$, we have

$$\mathrm{Per}(D \cup X) \leq \mathcal{H}^1(\partial^* D \setminus X) + \mathcal{H}^1(\partial X \setminus D) - \mathcal{H}^1(\partial^* D \cap \partial X)$$

since the term with a minus sign was counted twice by the first two terms at the right-hand side. Thus

$$\begin{aligned} \mathrm{Per}(D \cup X) &\leq \mathcal{H}^1(\partial^* D \setminus \overline{X}) + \mathcal{H}^1(\partial X \setminus D) \\ &= \mathrm{Per}(D, \mathbb{R}^2 \setminus \overline{X}) + Per(X) - \mathcal{H}^1(\partial X \cap D) \\ &= \mathrm{Per}(D, \mathbb{R}^2 \setminus \overline{\Omega}) + \mathrm{Per}(X) - \mathcal{H}^1(\partial X \cap D) \end{aligned}$$

which proves claim (4.87).

Inserting (4.87) into (4.86), we obtain

$$\mathcal{H}^1\left(D \cap \left(\bigcup_{j=1}^{k} \partial C_{i_j}\right)\right) \leq \mathrm{Per}(D, \mathbb{R}^2 \setminus \overline{\Omega}). \tag{4.88}$$

On the other hand, since $\partial^*(C_i \cap D) \subseteq (\partial^* D \cap C_i) \cup (\partial C_i \cap D) \cup (\partial^* D \cap \partial C_i)$, we have, using (4.88),

$$\begin{aligned}\sum_{i=1}^{N} \operatorname{Per}(C_i \cap D) &= \sum_{j=1}^{k} \operatorname{Per}(C_{i_j} \cap D)\\ &\leq \operatorname{Per}(D,\Omega) + \mathcal{H}^1\left(D \cap \left(\bigcup_{j=1}^{k} \partial C_{i_j}\right)\right) + \mathcal{H}^1\left(\partial^* D \cap \left(\bigcup_{j=1}^{k} \partial C_{i_j}\right)\right)\\ &\leq \operatorname{Per}(D,\Omega) + \operatorname{Per}(D, \mathbb{R}^2 \setminus \overline{\Omega}) + \mathcal{H}^1\left(\partial^* D \cap \left(\bigcup_{j=1}^{k} \partial C_{i_j}\right)\right) = \operatorname{Per}(D).\end{aligned}$$

We have proved that $\|g\|_* \leq 1$. According to Lemma 3.3 there is a vector field $\xi \in L^\infty(\mathbb{R}^2;\mathbb{R}^2)$ with $\|\xi\|_\infty \leq 1$ such that $-\operatorname{div}(\xi) = u$. Multiplying this equation by u and integrating in $\mathbb{R}^2$ we obtain

$$\int_{\mathbb{R}^2} (\xi, Du) = \int_{\mathbb{R}^2} u^2\, dx = \sum_{i=1}^{m} \frac{\operatorname{Per}(C_i)^2}{|C_i|} = \int_{\mathbb{R}^2} \|Du\|.$$

Therefore, u is a solution of (4.34). □

Proof of Theorem 4.40. Assume that $v := \chi_\Omega$ is a solution of (4.56) and let ξ be the vector field of Remark 4.21. By Lemma 4.25 we have that Ω is $-$calibrable and $\mathbb{R}^2 \setminus \Omega$ is $+$calibrable. By (b) of Proposition 4.31 we have that each connected component C of Ω is convex, and by Proposition 4.33 we have that ∂C is of class $C^{1,1}$. By Remark 4.23 and Lemma 4.35 we also know that (4.61) holds. Therefore, as Ω is bounded and recalling (4.59), it follows that Ω consists of a finite number of connected components $C_1, \ldots, C_m$. Integrating $-\operatorname{div}(\xi)$ on each C_i we obtain

$$\frac{\operatorname{Per}(\Omega)}{|\Omega|} = \frac{\operatorname{Per}(C_i)}{|C_i|} = \frac{\operatorname{Per}(C_j)}{|C_j|} \qquad \forall i,j \in \{1,\ldots,m\}.$$

The theorem is now a consequence of Theorem 4.42. □

Remark 4.43. Theorem 4.40 has been extended in two different directions. In [35] the authors constructed tower solutions of the eigenvalue problem (4.34). In [4] the authors studied the evolution of general convex sets or finite unions of them (satisfying assumption (v) of Theorem 4.40) and give further solutions of (4.34) and the denoising problem to be studied in Section 4.7.

4.6 Some Examples

In order to clarify the conditions given in Sections 4.4 and 4.5, we shall discuss some explicit examples.

Example 1. Let $\Omega = co(B_r((-\frac{L}{2},0)) \cup B_r((\frac{L}{2},0)) \subset \mathbb{R}^2$, $L, r > 0$, be the set of Figure 4.1. It is easy to check that Ω satisfies the assumptions of Theorem 4.26, since Ω is a convex set with $C^{1,1}$ boundary and there holds

$$\operatorname*{ess\,sup}_{p\in\partial\Omega} \kappa_{\partial\Omega}(p) = \frac{1}{r} < \frac{2\pi r + 2L}{\pi r^2 + 2rL} = \frac{\mathrm{Per}(\Omega)}{|\Omega|}. \tag{4.89}$$

Moreover, since the inequality in (4.89) is always strict, the solution of (4.1) starting from $\chi_{\Omega'}$ remains a characteristic function for any convex set Ω' of class $C^{1,1}$ close enough to Ω in the $C^{1,1}$-norm.

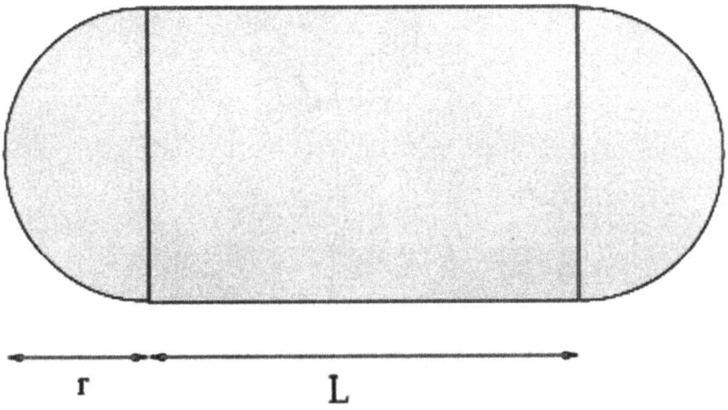

Figure 4.1: a bean–shaped set as initial datum for the solution

Example 2. Let $\Omega \subset \mathbb{R}^2$ be the union of two disjoint balls of radius r, whose centers are at distance L (see Figure 4.2). Then condition (4.81) of Theorem 4.40 reads as

$$L \geq \pi r.$$

Under this condition the solution of (4.1) and (3.2) with $u_0 = \chi_\Omega$ remains a characteristic function.

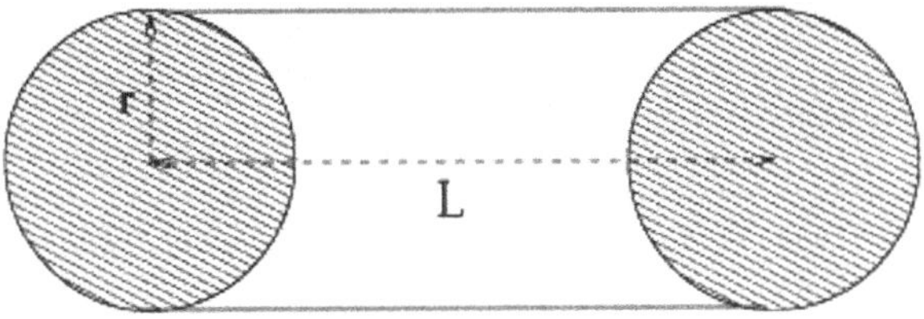

Figure 4.2: two balls as initial datum for the solution

Example 3. Consider now three disjoint balls of radius r, whose centers are on the vertices of an equilateral triangle with edges of length 1 (see Figure 4.3). In this case, condition (4.81) reads as

$$r \leq \frac{3}{4\pi}.$$

Notice that this condition is more restrictive than the condition holding for two balls, which has been discussed in Example 1 and gives $r \leq \frac{1}{\pi}$. This implies that it is not enough to consider only pairs of sets in condition (v) of Theorem 4.40.

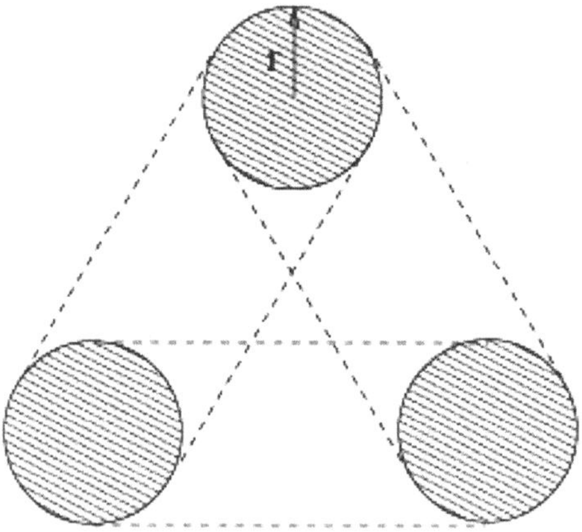

Figure 4.3: three balls as initial datum for the solution

4.7 Explicit Solutions for the Denoising Problem

The previous results allow us to explicitly compute the minimum of the denoising problem

$$\min_{u \in L^2(\mathbb{R}^2) \cap BV(\mathbb{R}^2)} \left\{ \int_{\mathbb{R}^2} \|Du\| + \frac{1}{2\lambda} \int_{\mathbb{R}^2} (u - f)^2 \, dx \right\}, \tag{4.90}$$

where $\lambda > 0$ and some given $f \in L^2(\mathbb{R}^2)$.

Proposition 4.44. *Let $\lambda > 0$, $b \in \mathbb{R}$ and $a := \operatorname{sign}(b)(|b| - \lambda)^+$. If $\overline{u} \in BV(\mathbb{R}^2)$ is a solution of (4.34), then the function $a\overline{u}$ is the solution of the variational problem (4.90) with $f := b\overline{u}$. Conversely, if $a\overline{u}$ is the solution of (4.90) with $f = b\overline{u}$ and $b - a = \pm\lambda$, then $\overline{u} \in BV(\mathbb{R}^2)$ is a solution of (4.34).*

In particular, if Ω satisfies the conditions listed in Theorem 4.40, then $a\lambda_\Omega \chi_\Omega$ is a solution of (4.90) with $f = b\lambda_\Omega \chi_\Omega$. The converse statement holds if $b - a = \pm\lambda$.

Proof. Recall (see Lemma 3.3) that a function $u \in BV(\mathbb{R}^2)$ is the solution of (4.90) if and only if u is the solution of

$$u - \lambda \operatorname{div}\left(\frac{Du}{|Du|}\right) = f. \tag{4.91}$$

Let $f := b\overline{u}$ where $\overline{u}$ satisfies (4.34). Without loss of generality we may assume that $b \geq 0$ (the case $b < 0$ can be obtained by changing $b \to -b$ and $u \to -u$). Suppose first that $b > \lambda$, so that $a = b - \lambda$. Since

$$-\lambda \operatorname{div}\left(\frac{D\overline{u}}{|D\overline{u}|}\right) = \lambda\overline{u} = (b-a)\overline{u},$$

it follows that $u := a\overline{u}$ satisfies (4.91). Now, assume that $0 \leq b \leq \lambda$, so that $a = 0$. Let $\xi \in L^\infty(\mathbb{R}^2;\mathbb{R}^2)$ be such that $\|\xi\|_\infty \leq 1$ and $-\operatorname{div}(\xi) = \overline{u}$. Obviously, if $z := \frac{b}{\lambda}\xi$, then $\|z\|_\infty \leq 1$, and $-\operatorname{div}(z) = -\frac{b}{\lambda}\operatorname{div}(\xi) = \frac{b}{\lambda}\overline{u}$, that is, $-\lambda\operatorname{div}(z) = b\overline{u} = f$. Since

$$\int_{\mathbb{R}^N} (z, D0) = 0 = \int_{\mathbb{R}^N} \|D0\|,$$

it follows that $u = 0$ solves (4.91). The converse statement follows by substituting $f = b\overline{u}$ and $u = a\overline{u}$ into (4.91).

The last assertion follows from Theorem 4.40 and the first part of the proof. □

Let us prove an extension of the above result.

Proposition 4.45. *Let Ω be a bounded set of finite perimeter which consists of a finite number $C_1, \ldots, C_m$ of connected components. Let $b_i \in \mathbb{R}$ for $i = 1, \ldots, m$. Assume that the function $\overline{u} := \sum_{i=1}^m \lambda_{C_i}\chi_{C_i}$ solves (4.34). Let $\lambda > 0$ and $a_i := \operatorname{sign}(b_i)(|b_i| - \lambda)^+$. Then the function $u := \sum_{i=1}^m a_i\lambda_{C_i}\chi_{C_i}$ is the solution of the variational problem (4.90) with $f = \sum_{i=1}^m b_i\lambda_{C_i}\chi_{C_i}$. The converse statement holds if a_i, b_i are such that $b_i - a_i = \lambda$, or $b_i - a_i = -\lambda$, for all $i = 1, \ldots, m$.*

Proof. As in the proof of Proposition 4.44, we have to prove that u is the solution of (4.91). We observe that this is obviously true if $b_i \geq \lambda$, or $b_i \leq -\lambda$, for all $i = 1, \ldots, m$. In the general case, let $I_\lambda := \{i \in \{1, \ldots, m\} : |b_i| \geq \lambda\}$, $J_\lambda := \{i \in \{1, \ldots, m\} : |b_i| < \lambda\}$. Since, in this case,

$$f - u = \lambda \sum_{i \in I_\lambda} \operatorname{sign}(b_i)\lambda_{C_i}\chi_{C_i} + \sum_{i \in J_\lambda} b_i\lambda_{C_i}\chi_{C_i},$$

to prove that u is a solution of (4.91) we have to construct a vector field $\xi \in L^\infty(\mathbb{R}^2;\mathbb{R}^2)$ with $\|\xi\|_\infty \leq 1$, such that

$$-\operatorname{div}(\xi) = \sum_{i \in I_\lambda} \operatorname{sign}(b_i)\lambda_{C_i}\chi_{C_i} + \sum_{i \in J_\lambda} \frac{b_i}{\lambda}\lambda_{C_i}\chi_{C_i} \tag{4.92}$$

and $(\xi, Du) = \|Du\|$. Let $F \in L^2(\mathbb{R}^2)$ denote the right-hand side of (4.92), and let $F^+ = \sup(F, 0)$, $F^- = \sup(-F, 0)$. By Lemma 3.3, a solution $\xi \in L^\infty(\mathbb{R}^2, \mathbb{R}^2)$ of (4.92) with $\|\xi\|_\infty \leq 1$ exists if and only if $\|F\|_* \leq 1$ where

$$\|F\|_* := \sup\left\{\left|\int_{\mathbb{R}^2} F(x)v(x)\,dx\right| : v \in L^2(\mathbb{R}^2) \cap BV(\mathbb{R}^2),\ \int_{\mathbb{R}^2} |Du| \leq 1\right\}.$$

Let us prove that $\|F\|_* \leq 1$. For that let $v \in BV(\mathbb{R}^2)$. Since

$$\int_{\mathbb{R}^2} F(x)v(x)\,dx \leq \int_{\mathbb{R}^2} (F^+v^+ + F^-v^-)\,dx$$

and

$$\int_{\mathbb{R}^2} \|Dv\| = \int_{\mathbb{R}^2} \|Dv^+\| + \int_{\mathbb{R}^2} \|Dv^-\|,$$

the inequality

$$\int_{\mathbb{R}^2} F(x)v(x)\,dx \leq \int_{\mathbb{R}^2} \|Dv\|$$

follows if we prove that

$$\int_{\mathbb{R}^2} F^+v^+\,dx \leq \int_{\mathbb{R}^2} \|Dv^+\| \quad \text{and} \quad \int_{\mathbb{R}^2} F^-v^-\,dx \leq \int_{\mathbb{R}^2} \|Dv^-\|.$$

Thus, without loss of generality, we may assume that $F \geq 0$ (i.e. all b_i appearing in the definition of F are ≥ 0) and $v \in BV(\mathbb{R}^2)$, $v \geq 0$. Then, using that $\frac{b_i}{\lambda} \leq 1$ for any $i \in J_\lambda$, we have that

$$\begin{aligned}
&\int_{\mathbb{R}^2} F(x)v(x)\,dx = \int_0^\infty \int_{\mathbb{R}^2} F\chi_{\{v\geq t\}}\,dx\,dt \\
&= \sum_{i\in I_\lambda} \lambda_{C_i} \int_0^\infty \int_{\mathbb{R}^2} \chi_{C_i}\chi_{\{v\geq t\}}\,dx\,dt + \sum_{i\in J_\lambda} \frac{b_i}{\lambda}\lambda_{C_i} \int_0^\infty \int_{\mathbb{R}^2} \chi_{C_i}\chi_{\{v\geq t\}}\,dx\,dt \\
&\leq \sum_{i=1}^m \lambda_{C_i} \int_0^\infty |C_i \cap \{v \geq t\}|\,dx\,dt \leq \int_0^\infty \mathrm{Per}([v \geq t])\,dt = \int_{\mathbb{R}^2} \|Dv\|.
\end{aligned}$$

Therefore $\|F\|_* \leq 1$. By Lemma 3.3, there is a vector field $\xi \in L^\infty(\mathbb{R}^2; \mathbb{R}^2)$ such that $\|\xi\|_\infty \leq 1$, satisfying (4.92). Since $a_i = 0$ for all $i \in J_\lambda$, it follows that

$$\begin{aligned}
\int_{\mathbb{R}^2} \|Du\| &= \sum_{i\in I_\lambda} |a_i|\lambda_{C_i}\mathrm{Per}(C_i) = \sum_{i\in I_\lambda} a_i\lambda_{C_i} \int_{\mathbb{R}^2} (-\mathrm{div}(\xi))\chi_{C_i}\,dx \\
&= \sum_{i=1}^m a_i\lambda_{C_i} \int_{\mathbb{R}^2} (\xi, D\chi_{C_i}) = \int_{\mathbb{R}^2} (\xi, Du)
\end{aligned}$$

which, in turn implies that $(\xi, Du) = \|Du\|$, since $\|\xi\|_\infty \leq 1$.

The converse statement is obvious. □

Proposition 4.45 proves that a_i is a soft thresholding of b_i with threshold λ. This is in coincidence with the soft thresholding rule used in the wavelet shrinkage method for denoising (see [92], [103], [102], [105], [153]). As proved by Meyer in [153], a soft thresholding applied to the wavelet coefficients of the function $f \in L^2(\mathbb{R}^2)$ gives a quasi-optimal solution of the denoising problem (4.90) (Theorem 1.11). Let us also mention that it has been proved recently that the wavelet coefficients of a BV function are somewhere between ℓ^1 and weak ℓ^1 (see [82], [81], [163], [153]).

Finally, that a solution of (4.90) when Ω is a ball was given by the above formula was already observed by Meyer in [153] and Strong–Chan in [186].

Chapter 5

The Dirichlet Problem for the Total Variation Flow

5.1 Introduction

Suppose that Ω is an open bounded domain with a Lipschitz boundary. The purpose of this chapter is to study the Dirichlet problem

$$\begin{cases} \dfrac{\partial u}{\partial t} = \operatorname{div}\left(\dfrac{Du}{|Du|}\right) & \text{in} \quad Q = (0,\infty) \times \Omega, \\ u(t,x) = \varphi(x) & \text{on} \quad S = (0,\infty) \times \partial\Omega, \\ u(0,x) = u_0(x) & \text{in} \quad x \in \Omega, \end{cases} \tag{5.1}$$

where $u_0 \in L^1(\Omega)$ and $\varphi \in L^1(\partial\Omega)$. This evolution equation is related to the gradient descent method used to solve the problem

$$\underset{u \in BV(\Omega)}{\text{Minimize}} \int_\Omega \|Du\| + \int_\Omega f u \, dx + \int_{\partial\Omega} |u - \varphi| \, d\mathcal{H}^{N-1} \tag{5.2}$$

where $f \in L^1(\Omega)$, $\varphi \in L^\infty(\partial\Omega)$ (existence for this variational problem was proved in [118], Theorem 1.4).

One of the motivations for studying this problem comes from a numerical approach introduced in [29] to extend a function u defined in $\mathbb{R}^2 \setminus \Omega$ inside Ω along the integral curves of a vector field $\theta^\perp$ which is the counterclockwise rotation of a vector field $\theta : \mathbb{R}^2 \to \mathbb{R}^2$ satisfying $|\theta| \leq 1$ and $\operatorname{div}(\theta) \in L^p(\Omega)$, $p \geq 1$. The proposal was to compute a function $u \in BV(\Omega)$ such that $(\theta, Du) = \|Du\|$. In practice, the proposal of [29] was to minimize the functional

$$F(u) = \int_\Omega \|Du\| - \int_\Omega \theta \cdot Du$$

defined in the set of functions of bounded variation $BV(\Omega)$ whose trace at the boundary is given by φ. Formally, if we integrate by parts in the second term of $F(u)$ we obtain

$$F(u) = \int_{\Omega} \|Du\| + \int_{\Omega} \operatorname{div}(\theta) u \, dx - \int_{\partial\Omega} \theta \cdot \vec{n} u \; d\mathcal{H}^{N-1}.$$

Since u, θ are known at the boundary, after incorporating the Dirichlet boundary condition in a weak sense, minimizing F amounts to minimizing functional (5.2) with $f = \operatorname{div}(\theta)$.

The other motivation for the study of (5.1) comes from [20] and [41]. The general purpose of these works being the study of elliptic and parabolic problems in divergence form with initial data in L^1. Existence and uniqueness results of entropy solutions when the associated variational energy has a growth at infinity of order p with $p > 1$ are proved in [41] (see also [21], [49] and [196]).

The case of equation

$$u_t = \operatorname{div}\left(\frac{Du}{|Du|}\right) \tag{5.3}$$

with Neumann boundary conditions was considered in Chapter 2, where we proved existence and uniqueness of weak solutions. In that case, this equation generates a nonlinear contraction semigroup in $L^1(\Omega)$ which is homogeneous of degree 0, and this fact implies the regularity in time of the solutions of (5.3). Indeed, the homogeneity of the operator permits the conclusion that $u_t(t) \in L^1(\Omega)$ a.e. for $t > 0$. This was used to prove uniqueness of solutions of (5.3) in case of Neumann boundary conditions. For Dirichlet boundary conditions this property is lost and a different approach is needed. Our aim in this chapter is to introduce a new concept of solution for problem (5.1) for which existence and uniqueness is proved for initial data in $L^1(\Omega)$ and boundary data in $L^1(\partial\Omega)$.

5.2 Definitions and Preliminary Facts

To make precise our notion of solution we need to introduce a weak trace on $\partial\Omega$ of the normal component of certain vector fields in Ω. We define the space

$$Z(\Omega) := \left\{(z,\xi) \in L^\infty(\Omega, \mathbb{R}^N) \times BV(\Omega)^* \ : \ \operatorname{div}(z) = \xi \ \ in \ \ \mathcal{D}'(\Omega)\right\}.$$

We denote $R(\Omega) := W^{1,1}(\Omega) \cap L^\infty(\Omega) \cap C(\Omega)$. For $(z,\xi) \in Z(\Omega)$ and $w \in R(\Omega)$ we define

$$\langle (z,\xi), w\rangle_{\partial\Omega} := \langle \xi, w\rangle_{BV(\Omega)^*, BV(\Omega)} + \int_{\Omega} z \cdot \nabla w \, dx.$$

Then, working as in the proof of Theorem C.2, we obtain that if $w, v \in R(\Omega)$ and $w = v$ on $\partial\Omega$ one has

$$\langle (z,\xi), w\rangle_{\partial\Omega} = \langle (z,\xi), v\rangle_{\partial\Omega} \qquad \forall \ (z,\xi) \in Z(\Omega). \tag{5.4}$$

As a consequence of (5.4), we can give the following definition: Given $u \in BV(\Omega) \cap L^\infty(\Omega)$ and $(z,\xi) \in Z(\Omega)$, we define $\langle (z,\xi), u\rangle_{\partial\Omega}$ by setting

$$\langle (z,\xi), u\rangle_{\partial\Omega} := \langle (z,\xi), w\rangle_{\partial\Omega}$$

where w is any function in $R(\Omega)$ such that $w = u$ on $\partial\Omega$. Again, working as in the proof of Theorem C.2, we can prove that for every $(z,\xi) \in Z(\Omega)$ there exists $M_{z,\xi} > 0$ such that

$$|\langle (z,\xi), u\rangle_{\partial\Omega}| \leq M_{z,\xi} \|u\|_{L^1(\partial\Omega)} \qquad \forall\, u \in BV(\Omega) \cap L^\infty(\Omega). \tag{5.5}$$

Now, taking a fixed $(z,\xi) \in Z(\Omega)$, we consider the linear functional $F : L^\infty(\partial\Omega) \to \mathbb{R}$ defined by

$$F(v) := \langle (z,\xi), w\rangle_{\partial\Omega}$$

where $v \in L^\infty(\partial\Omega)$ and $w \in BV(\Omega) \cap L^\infty(\Omega)$ is such that $w_{|\partial\Omega} = v$. By estimate (5.5), there exists $\gamma_{z,\xi} \in L^\infty(\partial\Omega)$ such that

$$F(v) = \int_{\partial\Omega} \gamma_{z,\xi}(x) v(x)\, d\mathcal{H}^{N-1}.$$

Consequently there exists a linear operator $\gamma : Z(\Omega) \to L^\infty(\partial\Omega)$, with $\gamma(z,\xi) := \gamma_{z,\xi}$, satisfying

$$\langle (z,\xi), w\rangle_{\partial\Omega} = \int_{\partial\Omega} \gamma_{z,\xi}(x) w(x)\, d\mathcal{H}^{N-1} \qquad \forall\, w \in BV(\Omega) \cap L^\infty(\Omega).$$

In case $z \in C^1(\overline{\Omega}, \mathbb{R}^N)$, we have $\gamma_z(x) = z(x) \cdot \nu(x)$ for all $x \in \partial\Omega$. Hence, the function $\gamma_{z,\xi}(x)$ is the weak trace of the normal component of (z,ξ). For simplicity of the notation, we shall denote $\gamma_{z,\xi}(x)$ by $[z,\nu](x)$.

We need to consider the space $BV(\Omega)_2$, defined as $BV(\Omega) \cap L^2(\Omega)$ endowed with the norm

$$\|w\|_{BV(\Omega)_2} := \|w\|_{L^2(\Omega)} + \|Dw\|(\Omega).$$

It is easy to see that $L^2(\Omega) \subset BV(\Omega)_2^*$ and

$$\|w\|_{BV(\Omega)_2^*} \leq \|w\|_{L^2(\Omega)} \qquad \forall\, w \in L^2(\Omega). \tag{5.6}$$

Now, it is well known (see [179]) that the dual $\left(L^1(0,T; BV(\Omega)_2)\right)^*$ is isometric to the space $L^\infty(0,T; BV(\Omega)_2^*, BV(\Omega)_2)$ of all weakly* measurable functions $f : [0,T] \to BV(\Omega)_2^*$, such that $v(f) \in L^\infty([0,T])$, where $v(f)$ denotes the supremum of the set $\{|\langle w, f\rangle| \ : \ \|w\|_{BV(\Omega)_2} \leq 1\}$ in the vector lattice of measurable real functions. Moreover, the dual pair of the isometry is defined by

$$\langle w, f\rangle = \int_0^T \langle w(t), f(t)\rangle\, dt,$$

for $w \in L^1(0,T; BV(\Omega)_2)$ and $f \in L^\infty(0,T; BV(\Omega)_2^*, BV(\Omega)_2)$.

To make precise our notion of solution we need the following definitions (see [98]).

Definition 5.1. Let (A, Σ, μ) be a finite measure space and let X be a Banach space. If $f : A \to X$ is weakly μ-measurable (i.e., $\langle x^*, f\rangle$ are measurable functions for any $x^* \in X^*$) such that $\langle x^*, f\rangle \in L^1(\mu)$ for all $x^* \in X^*$, then f is called *Dunford integrable*. The *Dunford integral* of f over $E \in \Sigma$ is defined to be the element $x_E^{**} \in X^{**}$ such that

$$\langle x_E^{**}, x^*\rangle = \int_E \langle x^*, f(s)\rangle \, d\mu(s)$$

for all $x^* \in X^*$, and we write $x_E^{**} = (D)\int_E f\, d\mu$. In case that $(D)\int_E f\, d\mu \in X$ for each $E \in \Sigma$, then f is called Pettis integrable .

Definition 5.2. Let $\Psi \in L^1(0,T; BV(\Omega))$. We say Ψ admits a *weak derivative* in $L^1_w(0,T; BV(\Omega)) \cap L^\infty(Q_T)$ if there is a function $\Theta \in L^1_w(0,T; BV(\Omega)) \cap L^\infty(Q_T)$ such that $\Psi(t) = \displaystyle\int_0^t \Theta(s)ds$, the integral being taken as a Pettis integral.

Definition 5.3. Let $\xi \in \left(L^1(0,T; BV(\Omega)_2)\right)^*$. We say that ξ is *the time derivative* in the space $\left(L^1(0,T; BV(\Omega)_2)\right)^*$ of a function $u \in L^1((0,T) \times \Omega)$ if

$$\int_0^T \langle \xi(t), \Psi(t)\rangle dt = -\int_0^T \int_\Omega u(t,x)\Theta(t,x)dxdt$$

for all test functions $\Psi \in L^1(0,T; BV(\Omega))$ which admit a weak derivative $\Theta \in L^1_w(0,T; BV(\Omega)) \cap L^\infty(Q_T)$ and have compact support in time.

Observe that if $w \in L^1(0,T; BV(\Omega)) \cap L^\infty(Q_T)$ and $z \in L^\infty(Q_T, \mathbb{R}^N)$ is such that there exists $\xi \in \left(L^1(0,T; BV(\Omega))\right)^*$ with $\mathrm{div}(z) = \xi$ in $\mathcal{D}'(Q_T)$, we can define, associated to the pair (z, ξ), the distribution (z, Dw) in Q_T by

$$\langle (z, Dw), \phi\rangle := -\int_0^T \langle \xi(t), w(t)\phi(t)\rangle \, dt - \int_0^T \int_\Omega z(t,x)w(t,x)\nabla_x\phi(t,x)\, dxdt \tag{5.7}$$

for all $\phi \in \mathcal{D}(Q_T)$.

Definition 5.4. Let $\xi \in \left(L^1(0,T; BV(\Omega)_2)\right)^*$, $z \in L^\infty(Q_T, \mathbb{R}^N)$. We say that $\xi = \mathrm{div}(z)$ in $\left(L^1(0,T; BV(\Omega)_2)\right)^*$ if (z, Dw) is a Radon measure in Q_T with normal boundary values $[z,\nu] \in L^\infty((0,T) \times \partial\Omega)$, such that

$$\int_{Q_T} (z, Dw) + \int_0^T \langle \xi(t), w(t)\rangle \, dt = \int_0^T \int_{\partial\Omega} [z(t,x), \nu] w(t,x) d\mathcal{H}^{N-1}\, dt,$$

for all $w \in L^1(0,T; BV(\Omega)) \cap L^\infty(Q_T)$.

Let $T_k(r) = [k - (k - |r|)^+]\mathrm{sign}_0(r)$, $k \geq 0$, $r \in \mathbb{R}$. We consider the set $\mathcal{T} = \{T_k, T_k^+, T_k^- \, : \, k > 0\}$. We need to consider a more general set of truncation functions, concretely, the set $\mathcal{P}$ defined in (3.16). Obviously, $\mathcal{T} \subset \mathcal{P}$.

5.3 The Main Result

In this section we give the concept of solution for the Dirichlet problem (5.1) and we state the existence and uniqueness result for this type of solutions.

Definition 5.5. A measurable function $u : (0,T) \times \Omega \to \mathbb{R}$ is an *entropy solution* of (5.1) in $Q_T = (0,T) \times \Omega$ if $u \in C([0,T]; L^1(\Omega))$, $p(u(\cdot)) \in L^1_w(0,T; BV(\Omega))$ $\forall\ p \in \mathcal{T}$ and there exist $(z(t), \xi(t)) \in Z(\Omega)$ with $\|z(t)\|_\infty \leq 1$, and $\xi \in \left(L^1(0,T; BV(\Omega)_2)\right)^*$ such that ξ is the time derivative of u in $\left(L^1(0,T; BV(\Omega)_2)\right)^*$, $\xi = \mathrm{div}(z)$ in $\left(L^1(0,T; BV(\Omega)\right)^*$ and $[z(t), \nu] \in \mathrm{sign}\big(p(\varphi) - p(u(t))\big)$ a.e. in $t \in [0,T]$, satisfying

$$-\int_0^T \int_\Omega j(u(t) - l)\eta_t + \int_0^T \int_\Omega \eta(t)\|Dp(u(t) - l)\| + z(t) \cdot D\eta(t) p(u(t) - l)$$
$$\leq \int_0^T \int_{\partial\Omega} [z(t), \nu]\eta(t) p(u(t) - l),$$

for all $l \in \mathbb{R}$, for all $\eta \in C^\infty(\overline{Q_T})$, with $\eta \geq 0$, $\eta(t,x) = \phi(t)\psi(x)$, being $\phi \in \mathcal{D}(]0,T[)$, $\psi \in C^\infty(\overline{\Omega})$ and $p \in \mathcal{T}$, where $j(r) = \displaystyle\int_0^r p(s)\, ds$.

The main result of this chapter is the following.

Theorem 5.6. *Let $u_0 \in L^1(\Omega)$, and $\varphi \in L^1(\partial\Omega)$. Then there exists a unique entropy solution of* (5.1) *in $(0,T) \times \Omega$ for every $T > 0$ such that $u(0) = u_0$. Moreover, if $u(t), \hat{u}(t)$ are the entropy solutions corresponding to initial data $u_0, \hat{u}_0$, respectively, then*

$$\|\big(u(t) - \hat{u}(t)\big)^+\|_1 \leq \|\big(u_0 - \hat{u}_0\big)^+\|_1 \quad \text{and} \quad \|u(t) - \hat{u}(t)\|_1 \leq \|u_0 - \hat{u}_0\|_1 \tag{5.8}$$

for all $t \geq 0$.

5.4 The Semigroup Solution

To prove Theorem 5.6 we shall use the techniques of completely accretive operators and the Crandall–Liggett semigroup generation theorem. So we shall associate a completely accretive operator $\mathcal{A}_\varphi$ to the formal differential expression $-\mathrm{div}(\frac{Du}{|Du|})$ together with the Dirichlet boundary condition.

Let us introduce the following operator $\mathcal{A}_\varphi$ in $L^1(\Omega)$.

Definition 5.7. $(u,v) \in \mathcal{A}_\varphi$ if and only if $u, v \in L^1(\Omega)$, $p(u) \in BV(\Omega)$ for all $p \in \mathcal{P}$ and there exists $z \in X(\Omega)_1$ with $\|z\|_\infty \leq 1$, $v = -\mathrm{div}(z)$ in $\mathcal{D}'(\Omega)$ such that

$$\int_\Omega (w - p(u))v \leq \int_\Omega z \cdot \nabla w - \int_\Omega \|Dp(u)\| + \int_{\partial\Omega} |w - p(\varphi)| - \int_{\partial\Omega} |p(u) - p(\varphi)|,$$

$\forall w \in W^{1,1}(\Omega) \cap L^\infty(\Omega)$ and $\forall p \in \mathcal{P}$.

Theorem 5.8. *Let $\varphi \in L^1(\partial\Omega)$. The operator $\mathcal{A}_\varphi$ is m-completely accretive in $L^1(\Omega)$ with dense domain.*

To prove this theorem, we need first to consider the following operator, which is related with the p-Laplacian operator with Dirichlet boundary condition. For $p > 1$, let $\varphi \in W^{1-1/p,p}(\partial\Omega)$, and

$$W^{1,p}_\varphi(\Omega) := \left\{ u \in W^{1,p}(\Omega) \ : \ u|_{\partial\Omega} = \varphi \quad \mathcal{H}^{N-1} - \text{a.e. on} \ \ \partial\Omega \right\}.$$

We define the operator $A_{\varphi,p}$ in $L^1(\Omega)$ as:

$$(u,v) \in A_{\varphi,p} \quad \text{if and only if } u \in W^{1,p}_\varphi(\Omega) \cap L^\infty(\Omega),\ v \in L^1(\Omega) \text{ and}$$

$$\int_\Omega (w-u)v \le \int_\Omega |\nabla u|^{p-2}\nabla u \cdot \nabla(w-u)$$

for every $w \in W^{1,p}_\varphi(\Omega) \cap L^\infty(\Omega)$.

Proposition 5.9. *Let $\varphi \in L^\infty(\partial\Omega) \cap W^{1-1/p,p}(\partial\Omega)$. The operator $A_{\varphi,p}$ is completely accretive and $L^\infty(\Omega) \subseteq R(I + A_{\varphi,p})$.*

Proof. Let $p \in P_0$ and $(u,v),(\hat{u},\hat{v}) \in A_{\varphi,p}$. Since $(u,v), \in A_{\varphi,p}$, taking $w = u - p(u-\hat{u})$ as test function in the definition of the operator $A_{\varphi,p}$ we get

$$\int_\Omega p(u-\hat{u})v \ge \int_\Omega |\nabla u|^{p-2}\nabla u \cdot \nabla p(u-\hat{u}).$$

Similarly, since $(\hat{u},\hat{v}), \in A_{\varphi,p}$, taking $w = \hat{u} + p(u-\hat{u})$ as test function in the definition of the operator $A_{\varphi,p}$ we get

$$\int_\Omega p(u-\hat{u})\hat{v} \le \int_\Omega |\nabla \hat{u}|^{p-2}\nabla \hat{u} \cdot \nabla p(u-\hat{u}).$$

Hence

$$\int_\Omega (v-\hat{v})p(u-\hat{u}) \ge \int_\Omega \left(|\nabla u|^{p-2}\nabla u - |\nabla \hat{u}|^{p-2}\nabla \hat{u}\right) \cdot \nabla p(u-\hat{u}) \ge 0.$$

Therefore, by Corollary A.38, $A_{\varphi,p}$ is completely accretive.

Let us see now that $L^\infty(\Omega) \subseteq R(I + A_{\varphi,p})$. Let $v \in L^\infty(\Omega)$; we need to prove that there exists $u \in W^{1,p}_\varphi(\Omega) \cap L^\infty(\Omega)$ such that $(u, v-u) \in A_{\varphi,p}$, i.e.,

$$\int_\Omega (w-u)(v-u) \le \int_\Omega |\nabla u|^{p-2}\nabla u \cdot \nabla(w-u) \quad \forall\, w \in W^{1,p}_\varphi(\Omega) \cap L^\infty(\Omega). \tag{5.9}$$

For $n \in \mathbb{N}$, let $\gamma_n(s) := T_n(s) + \frac{1}{n}|s|^{p-2}s$, and consider the operators $A_n : W^{1,p}_\varphi(\Omega) \to \left(W^{1,p}(\Omega)\right)^*$, defined by

$$\langle A_n u, w \rangle := \int_\Omega |\nabla u|^{p-2}\nabla u \cdot \nabla w + \int_\Omega \gamma_n(u) w.$$

It is easy to see that A_n is monotone, coercive and continuous on finite-dimensional subspaces. Then, by classical results (see for instance [138]), given v there exists $u_n \in W^{1,p}_{\varphi}(\Omega)$, such that

$$\langle A_n u_n, u_n - w\rangle \leq \int_{\Omega} v(u_n - w) \qquad \forall\, w \in W^{1,p}_{\varphi}(\Omega).$$

That is

$$\int_{\Omega} (w - u_n)(v - \gamma_n(u_n)) \leq \int_{\Omega} |\nabla u_n|^{p-2} \nabla u_n \cdot \nabla(w - u_n) \tag{5.10}$$

for all $w \in W^{1,p}_{\varphi}(\Omega)$. Let $k > 0$ be such that $\|\varphi\|_\infty \leq k$. If we take $w = T_k(u_n)$ in (5.10), we get

$$\int_{\Omega} (T_k(u_n) - u_n)(v - \gamma_n(u_n)) \leq \int_{\Omega} |\nabla u_n|^{p-2} \nabla u_n \cdot \nabla(T_k(u_n) - u_n).$$

Hence, if

$$A_n(k) := \{x \in \Omega \;:\; |u_n(x)| > k\},$$

we have that

$$\begin{aligned}
\int_{\Omega} |\nabla(u_n - T_k(u_n))|^p &= \int_{A_n(k)} |\nabla u_n|^{p-2} \nabla u_n \cdot \nabla u_n \\
&= \int_{\Omega} |\nabla u_n|^{p-2} \nabla u_n \cdot \nabla(u_n - T_k(u_n)) \\
&\leq \int_{\Omega} v(u_n - T_k(u_n)) - \int_{\Omega} \gamma_n(u_n)(u_n - T_k(u_n)) \\
&\leq \int_{\Omega} v(u_n - T_k(u_n)).
\end{aligned}$$

Now, by Young's inequality

$$\int_{\Omega} v(u_n - T_k(u_n)) \leq C_\epsilon \|v\|_\infty^{p'} \mathcal{L}^N(A_n(k)) + \epsilon C \int_{\Omega} |u_n - T_k(u_n)|^p.$$

Since $u_n - T_k(u_n) \in W^{1,p}_0(\Omega)$, using the Poincaré inequality, we obtain that

$$\|u_n - T_k(u_n)\|_{1,p} \leq R\mathcal{L}^N(A_n(k))^{1/p},$$

hence, by applying the classical Stampacchia methods (see for instance, Appendix B in [138]), it follows that there exists a constant M_1 depending on $\|v\|_\infty$ and $\|\varphi\|_\infty$, such that

$$\|u_n\|_\infty \leq M_1 \qquad \forall\, n \in \mathbb{N}. \tag{5.11}$$

On the other hand, taking w_0 as test function in (5.10), and applying Young's inequality, we obtain

$$\begin{aligned}\int_\Omega |\nabla u_n|^p &\leq \int_\Omega |\nabla u_n|^{p-2}\nabla u_n \cdot \nabla w_0 + \int_\Omega v(u_n - w_0) + \int_\Omega \gamma_n(u_n)(w_0 - u_n) \\ &\leq \epsilon C \int_\Omega |\nabla u_n|^p + C_\epsilon \int_\Omega |\nabla w_0|^p + \int_\Omega v u_n + \int_\Omega w_0(\gamma_n(u_n) - v).\end{aligned}$$

It follows from this that there exists a constant M_2 depending on $\mathcal{L}^N(\Omega)$, $\|v\|_\infty$, $\|\varphi\|_\infty$ and $\|w_0\|_{1,p}$, such that

$$\int_\Omega |\nabla u_n|^p \leq M_2 \qquad \forall\, n \in \mathbb{N}. \tag{5.12}$$

As a consequence of (5.11) and (5.12), $\{u_n\}_{n\in\mathbb{N}}$ is bounded in $W^{1,p}(\Omega)$. Hence, there exists a subsequence, still denoted u_n, such that $u_n \to u \in W^{1,p}(\Omega)$ weakly in $W^{1,p}(\Omega)$. Moreover, by the Rellich–Kondrachov theorem, $u_n \to u$ in $L^p(\Omega)$, and by Theorem 3.4.5 in [157], $u_n \to u$ in $L^p(\partial\Omega)$. After passing to a suitable subsequence, we can assume that $u_n \to u$ a.e. in Ω. Thus, by (5.11), $\|u\|_\infty \leq M_1$. Therefore we have that $u \in W^{1,p}_\varphi(\Omega) \cap L^\infty(\Omega)$.

Proceeding as in the proof of step 3 of Theorem 2.1 in [20], we obtain that

$$|\nabla u_n|^{p-2}\nabla u_n \to |\nabla u|^{p-2}\nabla u \qquad \text{in measure, and a.e.}$$

Now, by (5.12), we have that $\{|\nabla u_n|^{p-2}\nabla u_n\}_{n\in\mathbb{N}}$ is bounded in $\left(L^{p'}(\Omega)\right)^N$. Hence

$$|\nabla u_n|^{p-2}\nabla u_n \to |\nabla u|^{p-2}\nabla u \quad \text{weakly in } \left(L^{p'}(\Omega)\right)^N. \tag{5.13}$$

Given $w \in W^{1,p}_\varphi(\Omega) \cap L^\infty(\Omega)$, by (5.13), we get

$$\int_\Omega |\nabla u_n|^{p-2}\nabla u_n \cdot \nabla w \to \int_\Omega |\nabla u|^{p-2}\nabla u \cdot \nabla w, \tag{5.14}$$

and, using Fatou's lemma, we have

$$\int_\Omega |\nabla u|^{p-2}\nabla u \cdot \nabla u \leq \liminf_{n\to\infty} \int_\Omega |\nabla u_n|^{p-2}\nabla u_n \cdot \nabla u_n. \tag{5.15}$$

On the other hand, since $u_n \to u$ in $L^p(\Omega)$ we have

$$\lim_{n\to\infty} \int_\Omega (w - u_n)(v - \gamma_n(u_n)) = \int_\Omega (w - u)(v - u). \tag{5.16}$$

From (5.14), (5.15) and (5.16), passing to the limit in (5.10) we get (5.9), and the proof concludes. □

To prove Theorem 5.8, we need to give the following characterization of the operator $\mathcal{A}_\varphi$.

Proposition 5.10. *The following assertions are equivalent:*

(a) $(u,v) \in \mathcal{A}_\varphi$.

(b) $u,v \in L^1(\Omega)$, $p(u) \in BV(\Omega)$ *for all* $p \in \mathcal{P}$, *and there exists* $z \in X(\Omega)_1$, *with* $\|z\|_\infty \leq 1$, $v = -\mathrm{div}(z)$ *in* $\mathcal{D}'(\Omega)$ *such that*

$$\int_\Omega (w - p(u))v\,dx \leq \int_\Omega (z, Dw) - \int_\Omega \|Dp(u)\| + \int_{\partial\Omega} |w - p(\varphi))|\,d\mathcal{H}^{N-1} - \int_{\partial\Omega} |p(u) - p(\varphi)|\,d\mathcal{H}^{N-1} \tag{5.17}$$

for every $w \in BV(\Omega) \cap L^\infty(\Omega)$ *and* $p \in \mathcal{P}$.

(c) $u,v \in L^1(\Omega)$, $p(u) \in BV(\Omega)$ *for all* $p \in \mathcal{P}$, *and there exists* $z \in X(\Omega)_1$, *with* $\|z\|_\infty \leq 1$, $v = -\mathrm{div}(z)$ *in* $\mathcal{D}'(\Omega)$ *such that*

$$\int_\Omega (w - p(u))v\,dx \leq \int_\Omega (z, Dw) - \int_\Omega \|Dp(u)\| - \int_{\partial\Omega} [z,\nu](w - p(\varphi))\,d\mathcal{H}^{N-1} - \int_{\partial\Omega} |p(u) - p(\varphi)|\,d\mathcal{H}^{N-1} \tag{5.18}$$

for every $w \in BV(\Omega) \cap L^\infty(\Omega)$ *and* $p \in \mathcal{P}$.

(d) $u,v \in L^1(\Omega)$, $p(u) \in BV(\Omega)$ *for all* $p \in \mathcal{P}$, *and there exists* $z \in X(\Omega)_1$, *with* $\|z\|_\infty \leq 1$, $v = -\mathrm{div}(z)$ *in* $\mathcal{D}'(\Omega)$ *such that*

$$\int_\Omega (z, Dp(u)) = \int_\Omega \|Dp(u)\| \qquad \forall\, p \in \mathcal{P}, \tag{5.19}$$

$$[z,\nu] \in \mathrm{sign}(p(\varphi) - p(u)) \quad \mathcal{H}^{N-1}-\text{a.e. on } \partial\Omega, \ \forall\, p \in \mathcal{P}. \tag{5.20}$$

Proof. Let $(u,v) \in \mathcal{A}_\varphi$. Then, there exists $z \in X(\Omega)_1$ with $\|z\|_\infty \leq 1, v = -\mathrm{div}(z)$ in $\mathcal{D}'(\Omega)$, such that

$$\int_\Omega (w - p(u))v\,dx \leq \int_\Omega z \cdot \nabla w\,dx - \int_\Omega \|Dp(u)\| + \int_{\partial\Omega} |w - p(\varphi)|\,d\mathcal{H}^{N-1} - \int_{\partial\Omega} |p(u) - p(\varphi)|\,d\mathcal{H}^{N-1} \tag{5.21}$$

for every $w \in W^{1,1}(\Omega) \cap L^\infty(\Omega)$ and every $p \in \mathcal{P}$. Let $w \in BV(\Omega) \cap L^\infty(\Omega)$, $p \in \mathcal{P}$. Using Theorem B.3 and Lemma C.8 we know that there exists a sequence

$w_n \in W^{1,1}(\Omega) \cap L^\infty(\Omega)$ such that

$$
\begin{aligned}
& w_n \to w \qquad \text{in } L^1(\Omega), \\
& \int_\Omega |\nabla w_n| \, dx \to \int_\Omega \|Dw\|, \\
& \int_\Omega z \cdot \nabla w_n \, dx = \int_\Omega (z, Dw_n) \to \int_\Omega (z, Dw)
\end{aligned}
\tag{5.22}
$$

and $w_n|_{\partial\Omega} = w|_{\partial\Omega}$, $\|w_n\|_\infty \le \|w\|_\infty$, $\forall\, n \in \mathbb{N}$. Then taking w_n as test function in (5.21) and letting $n \to \infty$ we get that (5.21) holds for all $w \in BV(\Omega) \cap L^\infty(\Omega)$ and all $p \in \mathcal{P}$. Thus (a) and (b) are equivalent.

Since

$$
-\int_{\partial\Omega} [z, \nu](w - p(\varphi)) \, d\mathcal{H}^{N-1} \le \int_{\partial\Omega} |w - p(\varphi)| \, d\mathcal{H}^{N-1},
$$

to prove the equivalence between (b) and (c), it is enough to show that if $(u, v) \in \mathcal{A}_\varphi$, then (5.18) is satisfied. In fact, since $(u, v) \in \mathcal{A}_\varphi$, there exists $z \in X(\Omega)_1$ with $\|z\|_\infty \le 1$, $v = -\mathrm{div}(z)$ in $\mathcal{D}'(\Omega)$, such that

$$
\begin{aligned}
\int_\Omega (w - p(u)) v \, dx \le & \int_\Omega (z, Dw) - \int_\Omega \|Dp(u)\| \\
& + \int_{\partial\Omega} |w - p(\varphi)| \, d\mathcal{H}^{N-1} - \int_{\partial\Omega} |p(u) - p(\varphi)| \, d\mathcal{H}^{N-1}
\end{aligned}
\tag{5.23}
$$

for every $w \in BV(\Omega) \cap L^\infty(\Omega)$ and every $p \in \mathcal{P}$. Now, given $w \in BV(\Omega) \cap L^\infty(\Omega)$ and $p \in \mathcal{P}$, by Theorem B.3 and Lemma C.1, there exists $w_n \in W^{1,1}(\Omega) \cap L^\infty(\Omega)$ such that $w_n \to w$ in $L^1(\Omega)$, $w_n|_{\partial\Omega} = p(\varphi)$ and $\|w_n\|_\infty \le \|w\|_\infty + \|p(\varphi)\|_\infty$, for all $n \in \mathbb{N}$. Then taking w_n as test function in (5.21) and using Green's formula (C.10), we get

$$
\begin{aligned}
& \int_\Omega (w_n - p(u)) v \, dx \le \int_\Omega (z, Dw_n) - \int_\Omega \|Dp(u)\| - \int_{\partial\Omega} |p(u) - p(\varphi)| \, d\mathcal{H}^{N-1} \\
& = -\int_\Omega \mathrm{div}(z) w_n \, dx + \int_{\partial\Omega} [z, \nu] p(\varphi) \, d\mathcal{H}^{N-1} - \int_\Omega \|Dp(u)\| - \int_{\partial\Omega} |p(u) - p(\varphi)| \, d\mathcal{H}^{N-1}.
\end{aligned}
$$

Letting $n \to \infty$, it follows that

$$
\begin{aligned}
\int_\Omega (w - p(u)) v \, dx \le & -\int_\Omega \mathrm{div}(z) w \, dx + \int_{\partial\Omega} [z, \nu] p(\varphi) \, d\mathcal{H}^{N-1} \\
& - \int_\Omega \|Dp(u)\| - \int_{\partial\Omega} |p(u) - p(\varphi)| \, d\mathcal{H}^{N-1}.
\end{aligned}
$$

Therefore, applying again Green's formula, we obtain (5.18).

Suppose now that (b), or, equivalently, (c) is satisfied. Taking $w = p(u)$ in (5.21) we obtain

$$0 \leq \int_\Omega (z, Dp(u)) - \int_\Omega \|Dp(u)\|.$$

Thus,

$$\int_\Omega (z, Dp(u)) \leq \|z\|_\infty \int_\Omega \|Dp(u)\| \leq \int_\Omega \|Dp(u)\| \leq \int_\Omega (z, Dp(u)),$$

and (5.19) holds. Let us prove (5.20). Since $p(\varphi) \in L^\infty(\partial\Omega)$, by Lemma C.1, there exist $w_n \in W^{1,1}(\Omega) \cap L^\infty(\Omega)$ satisfying:

$$w_n|_{\partial\Omega} = p(\varphi) \qquad \forall\, n \in \mathbb{N},$$

$$\int_\Omega |\nabla w_n|\, dx \leq \int_{\partial\Omega} |p(\varphi)|\, d\mathcal{H}^{N-1} + \frac{1}{n} \qquad \forall\, n \in \mathbb{N},$$

$$\|w_n\|_1 \leq \frac{1}{n}, \quad \|w_n\|_\infty \leq \|p(\varphi)\|_\infty \qquad \forall\, n \in \mathbb{N}.$$

Taking $w = w_n$ in (5.21) and using Green's formula, we get

$$\begin{aligned}\int_\Omega (w_n - p(u))v\, dx \leq\ & -\int_\Omega \operatorname{div}(z) w_n\, dx + \int_{\partial\Omega} [z,\nu] p(\varphi)\, d\mathcal{H}^{N-1} \\ & - \int_\Omega \|Dp(u)\| - \int_{\partial\Omega} |p(u) - p(\varphi)|\, d\mathcal{H}^{N-1}.\end{aligned} \tag{5.24}$$

Then, letting $n \to \infty$ in (5.24), we obtain

$$-\int_\Omega p(u)v\, dx \leq \int_{\partial\Omega} [z,\nu] p(\varphi)\, d\mathcal{H}^{N-1} - \int_\Omega \|Dp(u)\| - \int_{\partial\Omega} |p(u) - p(\varphi)|\, d\mathcal{H}^{N-1}.$$

Now, by (5.19), and applying Green's formula, we have that

$$\int_\Omega \|Dp(u)\| = \int_\Omega (z, Dp(u)) = \int_\Omega v p(u)\, dx + \int_{\partial\Omega} [z,\nu] p(u)\, d\mathcal{H}^{N-1}.$$

Hence,

$$0 \leq \int_{\partial\Omega} \Big([z,\nu]\big(p(\varphi) - p(u)\big) - |p(u) - p(\varphi)| \Big)\, d\mathcal{H}^{N-1}.$$

Since

$$[z,\nu]\big(p(\varphi) - p(u)\big) - |p(u) - p(\varphi)| \leq 0,$$

we have that

$$[z,\nu]\big(p(\varphi) - p(u)\big) = |p(u) - p(\varphi)| \quad \mathcal{H}^{N-1} - \text{a.e. on } \partial\Omega,$$

and we obtain (5.20). Finally, to prove that (d) implies (c), we only need to apply Green's formula. □

Remark 5.11. (1) As a consequence of the proof of the above proposition we can put equality in the definition of the operator, that is, the following characterization of the operator $\mathcal{A}_\varphi$ holds.

$(u,v) \in \mathcal{A}_\varphi$ if and only if $u, v \in L^1(\Omega)$, $p(u) \in BV(\Omega)$ for all $p \in \mathcal{P}$ and there exists $z \in X(\Omega)_1$ with $\|z\|_\infty \leq 1$, $v = -\mathrm{div}(z)$ in $\mathcal{D}'(\Omega)$ such that

$$\int_\Omega (w - p(u))v\, dx = \int_\Omega (z, Dw) - \int_\Omega \|Dp(u)\|$$
$$+ \int_{\partial\Omega} |w - p(\varphi)|\, d\mathcal{H}^{N-1} - \int_{\partial\Omega} |p(u) - p(\varphi)|\, d\mathcal{H}^{N-1},$$

$\forall w \in BV(\Omega) \cap L^\infty(\Omega)$ and $\forall p \in \mathcal{P}$.

(2) As a consequence of the above proposition, if $(u,v) \in \mathcal{A}_\varphi$, we have that $\theta(z, DT_k(u), x) = 1$ a.e. with respect to the measure $\|DT_k(u)\|$. In case that $z \in C(\Omega, \mathbb{R}^N)$, this implies that

$$z(x) \cdot \frac{DT_k(u)}{\|DT_k(u)\|}(x) = 1, \qquad \|DT_k(u)\|\text{-a.e.}$$

where $\frac{DT_k(u)}{\|DT_k(u)\|}$ denotes the density of $DT_k(u)$ with respect to $\|DT_k(u)\|$ (see Theorem C.14). Heuristically, this amounts to saying that $z = \frac{Du}{\|Du\|}$. When z is not continuous we have that

$$z(x) \cdot \frac{DT_k(u)}{\|DT_k(u)\|}(x) = 1, \qquad \|\nabla T_k(u)\|\text{-a.e.}$$

where $\|\nabla T_k(u)\|$ denotes the absolutely continuous part of $\|DT_k(u)\|$ with respect to the Lebesgue measure in $\mathbb{R}^N$ (see Theorem C.14). In particular, if $u \in W^{1,1}(\Omega) \cap L^\infty(\Omega)$ we have that

$$z(x) \cdot \frac{\nabla u}{\|\nabla u\|}(x) = 1, \qquad \|\nabla u\|\text{-a.e.}$$

(3) Observe that by (d) in the above proposition, if $u \in L^\infty(\Omega)$, then the truncations are redundant in the definition of $\mathcal{A}_\varphi$.

To prove the following result, we need to introduce the functional $\Phi : L^1(\Omega) \to (-\infty, +\infty]$ defined by

$$\Phi(u) = \begin{cases} \displaystyle\int_\Omega \|Du\| + \int_{\partial\Omega} |u - \varphi|\, d\mathcal{H}^{N-1} & \text{if } u \in BV(\Omega), \\ +\infty & \text{if } u \in L^1(\Omega) \setminus BV(\Omega). \end{cases} \tag{5.25}$$

The functional Φ is convex and lower semi-continuous in $L^1(\Omega)$ (see Appendix B).

Proposition 5.12. *Let $\varphi \in L^1(\partial\Omega)$. Then $L^\infty(\Omega) \subset R(I + \mathcal{A}_\varphi)$ and $D(\mathcal{A}_\varphi)$ is dense in $L^1(\Omega)$.*

Proof. Suppose first that $\varphi \in W^{1/2,2}(\partial\Omega) \cap L^\infty(\partial\Omega)$. Let $v \in L^\infty(\Omega)$. We shall find $u \in BV(\Omega) \cap L^\infty(\Omega)$ such that $(u, v-u) \in \mathcal{A}_\varphi$, i.e., there is $z \in X(\Omega)_1$ with $\|z\|_\infty \leq 1$ such that $v - u = -\mathrm{div}(z)$ and

$$\int_\Omega (w-u)(v-u)\,dx \leq \int_\Omega z \cdot \nabla w\,dx - \int_\Omega \|Du\| + \int_{\partial\Omega} |w-\varphi|\,d\mathcal{H}^{N-1} - \int_{\partial\Omega} |u-\varphi|\,d\mathcal{H}^{N-1} \tag{5.26}$$

for every $w \in W^{1,1}(\Omega) \cap L^\infty(\Omega)$.

Since $\varphi \in W^{1-1/p,p}(\partial\Omega)$ for all $p > 1$, by Proposition 5.9, we know that for any $1 < p \leq 2$ there is $u_p \in W^{1,p}_\varphi(\Omega) \cap L^\infty(\Omega)$ such that $(u_p, v-u_p) \in A_{\varphi,p}$. Hence

$$\int_\Omega (w-u_p)(v-u_p)\,dx \leq \int_\Omega |\nabla u_p|^{p-2}\nabla u_p \cdot \nabla(w-u_p)\,dx, \tag{5.27}$$

for every $w \in W^{1,p}_\varphi(\Omega) \cap L^\infty(\Omega)$.

Let $M := \sup\{\|\varphi\|_\infty, \|v\|_\infty\}$. Then, taking $w = u_p - (u_p - M)^+$ as test function in (5.27), we obtain

$$\int_\Omega (u_p - M)^+(u_p - v)\,dx \leq 0.$$

Hence,

$$\int_{\{u_p>M\}} (u_p-M)^2\,dx \leq \int_{\{u_p>M\}} (u_p-M)(u_p-v)\,dx = \int_\Omega (u_p-M)^+(u_p-v)\,dx \leq 0.$$

Consequently, $u_p \leq M$ a.e. in Ω. Analogously, taking $w = u_p + (u_p+M)^-$ as test function, we get $-M \leq u_p$ a.e. in Ω. Therefore,

$$\|u_p\|_\infty \leq M \qquad \text{for all } 1 < p \leq 2. \tag{5.28}$$

Taking $w = w_0 \in W^{1,p}_\varphi(\Omega) \cap L^\infty(\Omega)$ in (5.27) and applying Young's inequality we obtain

$$\begin{aligned}\int_\Omega |\nabla u_p|^p\,dx &\leq \int_\Omega |\nabla u_p|^{p-2}\nabla u_p \cdot \nabla w_0\,dx - \int_\Omega (w_0-u_p)(v-u_p)\,dx \\ &\leq \epsilon \int_\Omega |\nabla u_p|^p\,dx + C_\epsilon \int_\Omega |\nabla w_0|^p\,dx + C(\|v\|_\infty, \|w_0\|_\infty).\end{aligned}$$

Thus

$$\int_\Omega |\nabla u_p|^p\,dx \leq M_1 \qquad \forall\, 1 < p \leq 2, \tag{5.29}$$

where M_1 depends on $\|v\|_\infty$, $\|w_0\|_\infty$ and $\|w_0\|_{1,2}$. Using Hölder's inequality we also have that

$$\int_\Omega |\nabla u_p|\, dx \leq M_2 \quad \forall\, 1 < p \leq 2, \tag{5.30}$$

where M_2 does not depend on p. Thus, $\{u_p\}_{p>1}$ is bounded in $W^{1,1}(\Omega)$ and we may extract a subsequence such that u_p converges in $L^1(\Omega)$ and almost everywhere to some $u \in L^1(\Omega)$ as $p \to 1+$. Now, by (5.28) and (5.30), we have that $u \in BV(\Omega) \cap L^\infty(\Omega)$.

Let us prove that $\{|\nabla u_p|^{p-2}\nabla u_p\}_{p>1}$ is weakly relatively compact in $L^1(\Omega, \mathbb{R}^N)$. For that, using (5.29), we observe that

$$\int_\Omega |\nabla u_p|^{p-1}\, dx \leq \left(\int_\Omega |\nabla u_p|^p\, dx\right)^{\frac{p-1}{p}} \mathcal{L}^N(\Omega)^{\frac{1}{p}} \leq M_3,$$

where M_3 does not depend on p. On the other hand, for any measurable subset $E \subseteq \Omega$ such that $\mathcal{L}^N(E) < 1$, we have

$$\left|\int_E |\nabla u_p|^{p-2}\nabla u_p\, dx\right| \leq \int_E |\nabla u_p|^{p-1}\, dx \leq M_1^{\frac{p-1}{p}} \mathcal{L}^N(E)^{\frac{1}{p}} \leq M_4 \mathcal{L}^N(E)^{\frac{1}{2}}.$$

Thus, $\{|\nabla u_p|^{p-2}\nabla u_p\}_{p>1}$, being bounded and equi-integrable in $L^1(\Omega, \mathbb{R}^N)$, is weakly relatively compact in $L^1(\Omega, \mathbb{R}^N)$. We may assume that

$$|\nabla u_p|^{p-2}\nabla u_p \rightharpoonup z \quad \text{as } p \to 1^+, \text{ weakly in } L^1(\Omega, \mathbb{R}^N). \tag{5.31}$$

Given $\psi \in C_0^\infty(\Omega)$, taking $w = u_p \pm \psi$ in (5.27) and letting $p \to 1^+$, we obtain

$$\int_\Omega (v - u)\psi\, dx = \int_\Omega z \cdot \nabla\psi\, dx,$$

that is, $v - u = -\mathrm{div}(z)$ in $\mathcal{D}'(\Omega)$. Let us prove that $\|z\|_\infty \leq 1$. For any $k > 0$, let $B_{p,k} = \{x \in \Omega : |\nabla u_p(x)| > k\}$. As a consequence of (5.29) we have that

$$\mathcal{L}^N(B_{p,k}) \leq \frac{M_1}{k^p} \qquad \text{for every } p > 1,\ k > 0. \tag{5.32}$$

As above, there is some $g_k \in L^1(\Omega, \mathbb{R}^N)$ such that

$$|\nabla u_p|^{p-2}\nabla u_p \chi_{B_{p,k}} \rightharpoonup g_k$$

weakly in $L^1(\Omega, \mathbb{R}^N)$ as $p \to 1$. Now for any $\phi \in L^\infty(\Omega, \mathbb{R}^N)$ with $\|\phi\|_\infty \leq 1$, we easily prove that

$$\left|\int_\Omega |\nabla u_p|^{p-2}\nabla u_p \cdot \phi\chi_{B_{p,k}}\, dx\right| \leq \frac{M_1}{k}.$$

Letting $p \to 1$, we get that

$$\int_\Omega |g_k|\, dx \le \frac{M_1}{k} \qquad \text{for every } k > 0. \tag{5.33}$$

Since we have that

$$\left| |\nabla u_p|^{p-2} \nabla u_p \chi_{\Omega \setminus B_{p,k}} \right| \le k^{p-1} \quad \text{for any } p > 1,$$

letting $p \to 1$, we obtain that

$$|\nabla u_p|^{p-2} \nabla u_p \chi_{\Omega \setminus B_{p,k}}$$

weakly converges in $L^1(\Omega, \mathbb{R}^N)$ to some function $f_k \in L^1(\Omega, \mathbb{R}^N)$ such that $\|f_k\|_\infty \le 1$. Hence, for any $k > 0$, we may write $z = f_k + g_k$ with $\|f_k\|_\infty \le 1$ and g_k satisfying (5.33). It follows that $\|z\|_\infty \le 1$.

For every $w \in W^{1,2}_\varphi(\Omega) \cap L^\infty(\Omega)$, by (5.27) and Young's inequality, we get

$$\int_\Omega |\nabla u_p|\, dx + \int_{\partial\Omega} |u_p - \varphi|\, d\mathcal{H}^{N-1}$$
$$\le \frac{p-1}{p} \mathcal{L}^N(\Omega) - \frac{1}{p} \int_\Omega (w - u_p)(v - u_p)\, dx + \frac{1}{p} \int_\Omega |\nabla u_p|^{p-2} \nabla u_p \cdot \nabla w\, dx.$$

Then, using the lower semi-continuity of the functional Φ defined by (5.25), letting $p \to 1^+$, we obtain

$$\int_\Omega \|Du\| + \int_{\partial\Omega} |u - \varphi|\, d\mathcal{H}^{N-1} \le - \int_\Omega (w - u)(v - u)\, dx + \int_\Omega z \cdot \nabla w\, dx, \tag{5.34}$$

for every $w \in W^{1,2}_\varphi(\Omega) \cap L^\infty(\Omega)$.

Now, to prove (5.26), we assume first that there exists $w_0 \in W^{1,2}(\Omega) \cap L^\infty(\Omega)$, such that $\varphi = w_0|_{\partial\Omega}$ (i.e., φ is the trace of w_0). Let $w \in W^{1,1}(\Omega) \cap L^\infty(\Omega)$ and let $w_n \in W^{1,2}_\varphi(\Omega) \cap L^\infty(\Omega)$ be such that $w_n \to w$ in $L^1(\Omega)$ as $n \to \infty$ and $\|w_n\|_\infty \le \|w\|_\infty$. Using w_n as test function in (5.34) and applying Green's formula (C.10), we may write

$$\int_\Omega (w_n - u)(v - u)\, dx \le \int_\Omega z \cdot \nabla w_n\, dx - \int_\Omega \|Du\| - \int_{\partial\Omega} |u - \varphi|\, d\mathcal{H}^{N-1}$$
$$= - \int_\Omega \operatorname{div}(z) w_n\, dx + \int_{\partial\Omega} [z, \nu] \varphi\, d\mathcal{H}^{N-1} - \int_\Omega \|Du\| - \int_{\partial\Omega} |u - \varphi|\, d\mathcal{H}^{N-1}.$$

From here, letting $n \to \infty$ and applying again Green's formula, we get

$$\begin{aligned}
&\int_\Omega (w-u)(v-u)\,dx \\
&\leq -\int_\Omega \operatorname{div}(z) w\,dx + \int_{\partial\Omega} [z,\nu]\varphi\,d\mathcal{H}^{N-1} - \int_\Omega \|Du\| - \int_{\partial\Omega} |u-\varphi|\,d\mathcal{H}^{N-1} \\
&= \int_\Omega z\cdot\nabla w\,dx - \int_{\partial\Omega} [z,\nu] w\,d\mathcal{H}^{N-1} + \int_{\partial\Omega} [z,\nu]\varphi\,d\mathcal{H}^{N-1} - \int_\Omega \|Du\| - \int_{\partial\Omega} |u-\varphi|\,d\mathcal{H}^{N-1} \\
&\leq \int_\Omega z\cdot\nabla w\,dx - \int_\Omega \|Du\| + \int_{\partial\Omega} |w-\varphi|\,d\mathcal{H}^{N-1} - \int_{\partial\Omega} |u-\varphi|\,d\mathcal{H}^{N-1},
\end{aligned}$$

and the proof of (5.26), in this particular case, concludes.

Suppose now we are in the general case, that is, $\varphi \in L^1(\partial\Omega)$. Take $v_n \in W^{1,2}(\Omega)\cap L^\infty(\Omega)$, such that $\varphi_n := v_n|_{\partial\Omega} \to \varphi$ in $L^1(\partial\Omega)$. From the above, there exists $u_n \in BV(\Omega)\cap L^\infty(\Omega)$ and $z_n \in X(\Omega)_1$ with $\|z_n\|_\infty \leq 1$ such that $v - u_n = -\operatorname{div}(z_n)$ and

$$\begin{aligned}
\int_\Omega (w-u_n)(v-u_n)\,dx \leq & \int_\Omega z_n\cdot\nabla w\,dx - \int_\Omega \|Du_n\| \\
& + \int_{\partial\Omega} |w-\varphi_n|\,d\mathcal{H}^{N-1} - \int_{\partial\Omega} |u_n-\varphi_n|\,d\mathcal{H}^{N-1}
\end{aligned} \tag{5.35}$$

for every $w \in W^{1,1}(\Omega)\cap L^\infty(\Omega)$. Moreover, by (5.28), we have $\|u_n\|_\infty \leq \max\{\|v\|_\infty, \|\varphi_n\|_\infty\}$. We can assume that $z_n \to z$ weakly* in $L^\infty(\Omega)$. Now, taking $w=0$ in (5.35), we get

$$-\int_\Omega u_n v\,dx + \int_\Omega (u_n)^2\,dx + \int_\Omega \|Du_n\| + \int_{\partial\Omega} |u_n-\varphi_n|\,d\mathcal{H}^{N-1} \leq \int_{\partial\Omega} |\varphi_n|\,d\mathcal{H}^{N-1}.$$

Hence

$$\begin{aligned}
\|u_n\|_2^2 + \int_\Omega \|Du_n\| &\leq \int_\Omega u_n v\,dx + \int_{\partial\Omega} |\varphi_n|\,d\mathcal{H}^{N-1} \\
&\leq \frac{1}{2}\|u_n\|_2^2 + \frac{1}{2}\|v\|_2^2 + \int_{\partial\Omega} |\varphi_n|\,d\mathcal{H}^{N-1}.
\end{aligned}$$

Thus, $\{u_n\}$ is a bounded sequence in $BV(\Omega)\cap L^2(\Omega)$. Then, since $BV(\Omega)$ is compactly embedded in $L^1(\Omega)$ (see Theorem B.21), there is a subsequence, still denoted by $\{u_n\}$, such that $u_n \to u$ in $L^1(\Omega)$. Finally, taking limits in (5.35), we obtain that $(u, v-u) \in \mathcal{A}_\varphi$.

To prove the density of $D(\mathcal{A}_\varphi)$ in $L^1(\Omega)$, we prove that $C_0^\infty(\Omega) \subseteq \overline{D(\mathcal{A}_\varphi)}^{L^1(\Omega)}$. Let $v \in C_0^\infty(\Omega)$. By the above, $v \in R(I+\frac{1}{n}\mathcal{A}_\varphi)$ for all $n \in \mathbb{N}$. Thus, for each $n \in \mathbb{N}$ there exists $u_n \in D(\mathcal{A}_\varphi)$ such that $(u_n, n(v-u_n)) \in \mathcal{A}_\varphi$ and, therefore there exists some $z_n \in X(\Omega)_1$ with $\|z_n\|_\infty \leq 1$, $n(v - u_n) = -\mathrm{div}(z_n)$ in $\mathcal{D}'(\Omega)$ such that

$$\int_\Omega (w-T_k(u_n))n(v-u_n)\,dx \leq \int_\Omega z_n \cdot \nabla w\,dx - \int_\Omega \|DT_k(u_n)\| \\ + \int_{\partial\Omega} |w-T_k(\varphi)|\,d\mathcal{H}^{N-1} - \int_{\partial\Omega} |T_k(u_n)-T_k(\varphi)|\,d\mathcal{H}^{N-1}$$

for every $w \in W^{1,1}(\Omega) \cap L^\infty(\Omega)$. Taking $w = T_k(v)$ and applying Fatou's Lemma we have that

$$\int_\Omega (v-u_n)^2\,dx \leq \frac{1}{n}\left(\int_\Omega |\nabla v| + \int_{\partial\Omega} |\varphi|\,d\mathcal{H}^{N-1}\right).$$

Letting $n \to \infty$, it follows that $u_n \to v$ in $L^2(\Omega)$. Therefore $v \in \overline{D(\mathcal{A}_\varphi)}^{L^1(\Omega)}$. $\square$

Proof of Theorem 5.8. Let $(u,v), (\hat{u},\hat{v}) \in \mathcal{A}_\varphi$, $p \in P_0$. by Corollary A.38, we have to prove that

$$\int_\Omega p(u-\hat{u})(v-\hat{v})\,dx \geq 0. \tag{5.36}$$

Let $z, \hat{z} \in X(\Omega)_1$, $\|z\|_\infty \leq 1, \|\hat{z}\|_\infty \leq 1$, be such that $v = -\mathrm{div}(z)$, $\hat{v} = -\mathrm{div}(\hat{z})$,

$$\begin{aligned} \int_\Omega (w-T_k(u))v\,dx &\leq \int_\Omega (z, Dw) - \int_\Omega \|DT_k(u)\| \\ &- \int_{\partial\Omega} [z,\nu](w-T_k(\varphi))\,d\mathcal{H}^{N-1} - \int_{\partial\Omega} |T_k(u)-T_k(\varphi)|\,d\mathcal{H}^{N-1}, \end{aligned} \tag{5.37}$$

and

$$\begin{aligned} \int_\Omega (w-T_k(\hat{u}))\hat{v}\,dx &\leq \int_\Omega (\hat{z}, Dw) - \int_\Omega \|DT_k(\hat{u})\| \\ &- \int_{\partial\Omega} [\hat{z},\nu](w-T_k(\varphi))\,d\mathcal{H}^{N-1} - \int_{\partial\Omega} |T_k(\hat{u})-T_k(\varphi)|\,d\mathcal{H}^{N-1}, \end{aligned} \tag{5.38}$$

for any $w \in BV(\Omega) \cap L^\infty(\Omega)$ and any $k > 0$. As observed in the previous remark, $\theta(z, DT_k(u), x) = 1$ $\|DT_k(u)\|$ – a.e., and, using Corollary C.7, we obtain that

$$\int_B (z, DT_k(u)) = \int_B \theta(z, DT_k(u), x)\|DT_k(u)\| = \int_B \|DT_k(u)\|,$$

$$\left|\int_B (\hat{z}, DT_k(u))\right| \leq \int_B \|DT_k(u)\|$$

for any Borel set $B \subseteq \Omega$. Similarly,

$$\int_B (\hat{z}, DT_k(\hat{u})) = \int_B \|DT_k(\hat{u})\|,$$

$$\left| \int_B (z, DT_k(\hat{u})) \right| \leq \int_B \|DT_k(\hat{u})\|$$

for any Borel set $B \subseteq \Omega$. It follows that

$$\int_B (z - \hat{z}, D(T_k(u) - T_k(\hat{u}))) \geq 0$$

for any Borel set $B \subseteq \Omega$. This implies that

$$\theta(z - \hat{z}, D(T_k(u) - T_k(\hat{u})), x) \geq 0 \quad \|D(T_k(u) - T_k(\hat{u}))\|\text{-a.e..}$$

Since, according to Corollary C.16, we have that

$$\theta(z - \hat{z}, Dp(T_k(u) - T_k(\hat{u})), x) = \theta(z - \hat{z}, D(T_k(u) - T_k(\hat{u})), x)$$

a.e. with respect to the measures $\|D(T_k(u) - T_k(\hat{u}))\|$ and $\|Dp(T_k(u) - T_k(\hat{u}))\|$. We conclude that

$$\theta(z - \hat{z}, Dp(T_k(u) - T_k(\hat{u})), x) \geq 0, \quad \|Dp(T_k(u) - T_k(\hat{u}))\|\text{-a.e.} \tag{5.39}$$

Taking $w = T_k(u) - p(T_k(u) - T_k(\hat{u}))$ in (5.37) and $w = T_k(\hat{u}) + p(T_k(u) - T_k(\hat{u}))$ in (5.38), adding both terms, and using (5.20) and (5.39), we obtain

$$\begin{aligned}\int_\Omega p(T_k(u) - T_k(\hat{u}))(\hat{v} - v)\, dx &\leq \int_\Omega (\hat{z} - z, Dp(T_k(u) - T_k(\hat{u}))) \\ &\quad + \int_{\partial\Omega} ([z, \nu] - [\hat{z}, \nu]) p(T_k(u) - T_k(\hat{u}))\, d\mathcal{H}^{N-1} \\ &= - \int_\Omega \theta(z - \hat{z}, Dp(T_k(u) - T_k(\hat{u})), x) \|Dp(T_k(u) - T_k(\hat{u}))\| \\ &\quad + \int_{\partial\Omega} ([z, \nu] - [\hat{z}, \nu]) p(T_k(u) - T_k(\hat{u}))\, d\mathcal{H}^{N-1} \leq 0.\end{aligned}$$

The inequality (5.36) follows by letting $k \to \infty$. Therefore $\mathcal{A}_\varphi$ is completely accretive.

In view of Proposition 5.12, to prove that $\mathcal{A}_\varphi$ satisfies the range condition, it is enough to prove that $\mathcal{A}_\varphi$ is closed. Let $(u_n, v_n) \in \mathcal{A}_\varphi$ such that $(u_n, v_n) \to (u, v)$ in $L^1(\Omega) \times L^1(\Omega)$. Let us see that $(u, v) \in \mathcal{A}_\varphi$. Since $(u_n, v_n) \in \mathcal{A}_\varphi$, there exists $z_n \in X(\Omega)_1$, $\|z_n\|_\infty \leq 1$ with $v_n = -\mathrm{div}(z_n)$ in $\mathcal{D}'(\Omega)$ such that

$$\begin{aligned}\int_\Omega (w - p(u_n)) v_n\, dx &\leq \int_\Omega (z_n, Dw) - \int_\Omega \|Dp(u_n))\| \\ &\quad + \int_{\partial\Omega} |w - p(\varphi)|\, d\mathcal{H}^{N-1} - \int_{\partial\Omega} |p(u_n) - p(\varphi)|\, d\mathcal{H}^{N-1}\end{aligned} \tag{5.40}$$

for every $w \in BV(\Omega) \cap L^\infty(\Omega)$ and all $p \in \mathcal{P}$. Since $\|z_n\|_\infty \leq 1$ we may assume that $z_n \rightharpoonup z$ in the weak* topology of $L^\infty(\Omega, \mathbb{R}^N)$ with $\|z\|_\infty \leq 1$. Moreover, since $v_n \to v$ in $L^1(\Omega)$, we have $v = -\mathrm{div}(z)$ in $\mathcal{D}'(\Omega)$, and

$$\lim_{n\to\infty} \int_\Omega (z_n, Dw) = \int_\Omega (z, Dw).$$

Now, letting $n \to \infty$ in (5.40), and having in mind the lower semi-continuity of the functional Φ, defined in (5.25), we obtain that

$$\begin{aligned}&\int_\Omega (w - p(u))v\,dx \\ &\leq \int_\Omega (z, Dw) - \int_\Omega \|Dp(u)\| + \int_{\partial\Omega} |w - p(\varphi)|\,d\mathcal{H}^{N-1} - \int_{\partial\Omega} |p(u) - p(\varphi)|\,d\mathcal{H}^{N-1}.\end{aligned}$$

Consequently, $(u, v) \in \mathcal{A}_\varphi$. □

5.5 Strong Solutions for Data in $L^2(\Omega)$

In this section we are going to see that when the initial datum is in $L^2(\Omega)$, then the semigroup solution is a strong solution.

Let $\{S(t)\}_{t\geq 0}$ be the contraction semigroup in $L^1(\Omega)$ generated by the operator $\mathcal{A}_\varphi$ via the Crandall–Liggett exponential formula. Since $\mathcal{A}_\varphi$ is an m-completely accretive operator, $S(t)(L^2(\Omega)) \subset L^2(\Omega)$. Let $\Psi_\varphi : L^2(\Omega) \to]-\infty, +\infty]$, the restriction to $L^2(\Omega)$ of the functional Φ defined by (5.25), i.e.,

$$\Psi_\varphi(u) = \begin{cases} \displaystyle\int_\Omega \|Du\| + \int_{\partial\Omega} |u - \varphi|\,d\mathcal{H}^{N-1} & \text{if } u \in BV(\Omega) \cap L^2(\Omega), \\ +\infty & \text{if } u \in L^2(\Omega) \setminus BV(\Omega) \cap L^2(\Omega). \end{cases} \tag{5.41}$$

Since the functional Ψ_φ is convex and lower semi-continuous in $L^2(\Omega)$, we have that $\partial\Psi_\varphi$ is a maximal monotone operator in $L^2(\Omega)$, and consequently (see Theorem A.33), if $\{T(t)\}_{t\geq 0}$ is the semigroup in $L^2(\Omega)$ generated by $\partial\Psi_\varphi$, for every $u_0 \in L^2(\Omega)$, $u(t) := T(t)u_0$ is a strong solution of the problem

$$\begin{cases} \dfrac{du}{dt} + \partial\Psi_\varphi u(t) \ni 0, \\ u(0) = u_0. \end{cases} \tag{5.42}$$

Lemma 5.13. *Let* $B_\varphi := \mathcal{A}_\varphi \cap (L^2(\Omega) \times L^2(\Omega))$. *Then* $B_\varphi = \partial\Psi_\varphi$.

Proof. Let $(u,v) \in B_\varphi$. Then, $u, v \in L^2(\Omega)$, $p(u) \in BV(\Omega)$ for all $p \in \mathcal{P}$ and there exists $z \in X(\Omega)_1$ with $\|z\|_\infty \leq 1$, $v = -\mathrm{div}(z)$ in $\mathcal{D}'(\Omega)$ such that

$$\int_\Omega (w - p(u))v\,dx \leq \int_\Omega (z, Dw) - \int_\Omega \|Dp(u)\| + \int_{\partial\Omega} |w - p(\varphi)|\,d\mathcal{H}^{N-1} - \int_{\partial\Omega} |p(u) - p(\varphi)|\,d\mathcal{H}^{N-1},$$

$\forall w \in BV(\Omega) \cap L^\infty(\Omega)$ and $\forall p \in \mathcal{P}$. Letting $p = T_k$ and $k \to \infty$ we obtain that

$$\int_\Omega (w-u)v\,dx \leq \int_\Omega (z, Dw) - \int_\Omega \|Du\| + \int_{\partial\Omega} |w - \varphi|\,d\mathcal{H}^{N-1} - \int_{\partial\Omega} |u - \varphi|\,d\mathcal{H}^{N-1},$$

$\forall w \in BV(\Omega) \cap L^\infty(\Omega)$. To prove that $(u,v) \in \partial\Psi_\varphi$, we have to prove that

$$\int_\Omega (w-u)v\,dx \leq \int_\Omega \|Dw\| - \int_\Omega \|Du\| + \int_{\partial\Omega} |w - \varphi|\,d\mathcal{H}^{N-1} - \int_{\partial\Omega} |u - \varphi|\,d\mathcal{H}^{N-1} \tag{5.43}$$

for every $w \in L^2(\Omega) \cap BV(\Omega)$. Now, given $w \in L^2(\Omega) \cap BV(\Omega)$, since $(u,v) \in B_\varphi$, by the first observation of the lemma, there exists $z \in X(\Omega)_1$, with $\|z\|_\infty \leq 1$, $v = -\mathrm{div}(z)$ in $\mathcal{D}'(\Omega)$ such that

$$\int_\Omega (T_k(w) - u)v\,dx \leq \int_\Omega (z, DT_k(w)) - \int_\Omega \|Du\| + \int_{\partial\Omega} |T_k(w)\,d\mathcal{H}^{N-1} - \varphi| - \int_{\partial\Omega} |u - \varphi|\,d\mathcal{H}^{N-1},$$

for every $k > 0$. From this, it follows that

$$\int_\Omega (T_k(w) - u)v\,dx \leq \int_\Omega \|DT_k(w)\| - \int_\Omega \|Du\| + \int_{\partial\Omega} |T_k(w) - \varphi|\,d\mathcal{H}^{N-1} - \int_{\partial\Omega} |u - \varphi|\,d\mathcal{H}^{N-1}. \tag{5.44}$$

Now, since $\lim_{k\to\infty} T_k(w) = w$ in $L^2(\Omega)$, and $\displaystyle\int_\Omega \|DT_k(w)\| \leq \int_\Omega \|Dw\|$, we have

$$\int_\Omega \|Dw\| \leq \liminf_{k\to\infty} \int_\Omega \|DT_k(w)\| \leq \limsup_{k\to\infty} \int_\Omega \|DT_k(w)\| \leq \int_\Omega \|Dw\|.$$

Therefore, letting $k \to \infty$ in (5.44), we obtain (5.43). We have proved that $B_\varphi \subset \partial\Psi_\varphi$. By Proposition 5.12, we have that $L^\infty(\Omega) \subset R(I + B_\varphi)$. Hence, $\partial\Psi_\varphi = \overline{B_\varphi}^{L^2(\Omega)}$. It follows that $\partial\Psi_\varphi = \mathcal{A}_\varphi \cap (L^2(\Omega) \times L^2(\Omega))$. $\square$

Using the above lemma and having in mind Proposition 5.10, we have the following result.

Theorem 5.14. *Let $\varphi \in L^1(\partial\Omega)$. Given $u_0 \in L^2(\Omega)$, $u(t) = S(t)u_0$ is a strong solution of (5.42). Moreover, $u'(t) \in L^2(\Omega)$, $p(u(t)) \in BV(\Omega)$ for all $p \in \mathcal{P}$, and there exists $z(t) \in X(\Omega)_1$, $\|z(t)\|_\infty \leq 1$ and $u'(t) = \operatorname{div}(z(t))$ in $\mathcal{D}'(\Omega)$ a.e. $t \in [0, +\infty[$, satisfying*

$$\begin{aligned} -\int_\Omega (w - p(u(t)))u'(t)\,dx \leq \int_\Omega (z(t), Dw) - \int_\Omega \|Dp(u(t))\| \\ - \int_{\partial\Omega} [z(t),\nu](w - p(\varphi))\,d\mathcal{H}^{N-1} - \int_{\partial\Omega} |p(u(t)) - p(\varphi)|\,d\mathcal{H}^{N-1} \end{aligned} \tag{5.45}$$

for every $w \in BV(\Omega) \cap L^\infty(\Omega)$ and $p \in \mathcal{P}$.

Moreover, $u(t)$ is also characterized as follows: there exists $z(t) \in X(\Omega)_1$, $\|z(t)\|_\infty \leq 1$ and $u'(t) = \operatorname{div}(z(t))$ in $\mathcal{D}'(\Omega)$ a.e. $t \in [0, +\infty[$, satisfying

$$\int_\Omega (z(t), Dp(u(t))) = \int_\Omega \|Dp(u(t))\| \qquad \forall\, p \in \mathcal{P}, \tag{5.46}$$

$$[z(t),\nu] \in \operatorname{sign}(p(\varphi) - p(u(t))) \quad \mathcal{H}^{N-1} - \text{a.e. on } \partial\Omega, \ \forall\, p \in \mathcal{P}. \tag{5.47}$$

Remark 5.15. Note that under the assumptions of Theorem 5.14, since $u(t) \in BV(\Omega)$, applying the lower semi-continuity of Ψ_φ, if we set $p = T_k$ and take limits when $k \to \infty$, we obtain that (5.45), (5.46) and (5.47) are true when p is the identity map.

We have the following weak form of the maximum principle.

Theorem 5.16. *Let u_1 and u_2 be two strong solutions of*

$$\begin{cases} \dfrac{du_i}{dt} + \partial\Psi_{\varphi_i} u_i(t) \ni 0, \\ u_i(0) = u_{i,0}, \quad i = 1, 2, \end{cases} \tag{5.48}$$

where $u_{i,0} \in L^2(\Omega)$ and $\varphi_i \in L^1(\partial\Omega)$. Suppose that $u_{1,0} \geq u_{2,0}$ and $\varphi_1 \geq \varphi_2$. Then we have $u_1 \geq u_2$.

Proof. By Theorem 5.14 and the above remark, we have that $u_i(t), u_i'(t) \in L^2(\Omega)$, and there exist $z_i(t) \in X(\Omega)_1$, $\|z_i(t)\|_\infty \leq 1$ and $u_i'(t) = \operatorname{div}(z_i(t))$ in $\mathcal{D}'(\Omega)$, satisfying:

$$\int_\Omega (z_i(t), D(u_i(t))) = \int_\Omega \|D(u_i(t))\|, \tag{5.49}$$

$$[z_i(t),\nu] \in \operatorname{sign}(\varphi_i - u_i(t)) \quad \mathcal{H}^{N-1} - \text{a.e. on } \partial\Omega. \tag{5.50}$$

Since $\frac{d}{dt}\big(u_2(t)-u_1(t)\big) = \operatorname{div}\big(z_2(t)-z_1(t)\big)$ in $L^2(\Omega)$, multiplying by $\big(u_2(t)-u_1(t)\big)^+$, integrating, and using Green's formula (C.10), we get

$$\begin{aligned}\frac{1}{2}\int_\Omega \frac{d}{dt}\Big[\big(u_2(t)-u_1(t)\big)^+\Big]^2 dx &= \int_\Omega \operatorname{div}\left(z_2(t)-z_1(t)\right)\left(u_2(t)-u_1(t)\right)^+ dx \\ &= -\int_\Omega \Big(z_2(t)-z_1(t), D\big(\big(u_2(t)-u_1(t)\big)^+\big)\Big) \\ &\quad + \int_{\partial\Omega} [z_2(t)-z_1(t),\nu]\left(u_2(t)-u_1(t)\right)^+ d\mathcal{H}^{N-1}.\end{aligned} \tag{5.51}$$

Now, by (5.49) it follows that

$$\theta(z_2(t)-z_1(t), D(u_2(t)-u_1(t)), x) \geq 0 \qquad \|D(u_2(t)-u_1(t))\| - \text{a.e.}$$

According to Corollary C.16, we have

$$\theta(z_2(t)-z_1(t), D(u_2(t)-u_1(t)), x) = \theta(z_2(t)-z_1(t), D\big(u_2(t)-u_1(t)\big)^+, x)$$

a.e. with respect to $\|D(u_2(t)-u_1(t))\|$ and $\|D(u_2(t)-u_1(t))^+\|$. Hence we can conclude that

$$\theta(z_2(t)-z_1(t), D\big(u_2(t)-u_1(t)\big)^+, x) \geq 0, \quad \|D(u_2(t)-u_1(t))^+\| - \text{a.e.}$$

As a consequence, we have

$$\begin{aligned}&\int_\Omega \big(z_2(t)-z_1(t), D\big(\big(u_2(t)-u_1(t)\big)^+\big)\big) \\ &= \int_\Omega \theta(z_2(t)-z_1(t), D\big(u_2(t)-u_1(t)\big)^+, x)\|D(u_2(t)-u_1(t))^+\| \geq 0.\end{aligned} \tag{5.52}$$

On the other hand, since $\varphi_1 \geq \varphi_2$, from (5.50), it is easy to see that

$$\int_{\partial\Omega} [z_2(t)-z_1(t),\nu]\big(u_2(t)-u_1(t)\big)^+ d\mathcal{H}^{N-1} \leq 0. \tag{5.53}$$

From (5.51), (5.52) and (5.53), we obtain that

$$\frac{1}{2}\int_\Omega \frac{d}{dt}\Big[\big(u_2(t)-u_1(t)\big)^+\Big]^2 dx \leq 0.$$

Hence the initial condition $u_{1,0} \geq u_{2,0}$ gives $u_1 \geq u_2$, and the proof concludes. □

Proposition 5.17. *Let $0 \leq u_0 \in L^2(\Omega)$ and $0 \leq \varphi \in L^1(\partial\Omega)$. Then, if u is the strong solution of the problem (5.41), we have*

$$u'(t) \leq \frac{u(t)}{t} \qquad \text{for } t > 0.$$

The opposite inequality holds if $u_0, \varphi \leq 0$.

Proof. We shall prove the proposition only when $u_0, \varphi \geq 0$, the other case being similar. First, let us see that for $\lambda > 0$, we have

$$\lambda^{-1}u(\lambda t) = e^{-t\mathcal{A}_{\lambda^{-1}\varphi}}(\lambda^{-1}u_0). \tag{5.54}$$

By Crandall–Liggett's exponential formula, it is enough to prove that for all $\mu > 0$,

$$\left(I + \mu\mathcal{A}_{\lambda^{-1}\varphi}\right)^{-1}(\lambda^{-1}u_0) = \lambda^{-1}\left(I + \lambda\mu\mathcal{A}_{\varphi}\right)^{-1}(u_0). \tag{5.55}$$

In fact: $v_\mu := \left(I + \mu\mathcal{A}_{\lambda^{-1}\varphi}\right)^{-1}(\lambda^{-1}u_0)$ if and only if $\left(v_\mu, \dfrac{\lambda^{-1}u_0 - v_\lambda}{\mu}\right) \in \mathcal{A}_{\lambda^{-1}\varphi}$, which is equivalent to the existence of $z_\mu \in X(\Omega)_1$, such that

$$-\mathrm{div}(z_\mu) = \frac{\lambda^{-1}u_0 - v_\lambda}{\mu},$$

$$\int_\Omega (z_\mu, Dv_\mu) = \int_\Omega \|Dv_\mu\|,$$

$$[z_\mu, \nu] \in \mathrm{sign}\left(\lambda^{-1}\varphi - v_\mu\right).$$

Then, we have

$$-\mathrm{div}(z_\mu) = \frac{u_0 - \lambda v_\lambda}{\lambda\mu},$$

$$\int_\Omega (z_\mu, D\lambda v_\mu) = \int_\Omega \|D\lambda v_\mu\|,$$

$$[z_\mu, \nu] \in \mathrm{sign}\left(\varphi - \lambda v_\mu\right),$$

which is equivalent to saying that $\left(\lambda v_\mu, \dfrac{u_0 - \lambda v_\mu}{\lambda\mu}\right) \in \mathcal{A}_\varphi$, that is, we have $v_\mu = \lambda^{-1}\left(I + \lambda\mu\mathcal{A}_\varphi\right)^{-1}(\lambda^{-1}u_0)$, and (5.55) holds.

Fix $t > 0$. For $h > 0$, if $\lambda t = t + h$, applying (5.54), we obtain

$$\begin{aligned} u(t+h) - u(t) &= u(\lambda t) - u(t) = (1 - \lambda^{-1})u(\lambda t) + \lambda^{-1}u(\lambda t) - u(t) \\ &= \tfrac{h}{t+h}u(t+h) + e^{-t\mathcal{A}_{\lambda^{-1}\varphi}}(\lambda^{-1}u_0) - u(t). \end{aligned}$$

Now, since $\lambda^{-1}u_0 \leq u_0$ and $\lambda^{-1}\varphi \leq \varphi$, by Theorem 5.16, we get

$$e^{-t\mathcal{A}_{\lambda^{-1}\varphi}}(\lambda^{-1}u_0) \leq u(t).$$

Therefore,

$$u(t+h) - u(t) \leq \frac{h}{t+h}u(t+h),$$

and the result follows. □

5.6 Existence and Uniqueness for Data in $L^1(\Omega)$

In this section we are going to prove Theorem 5.6.

5.6.1 Proof of Theorem 5.6: Existence

Let $u_0 \in L^1(\Omega)$ and $\{S(t)\}_{t\geq 0}$ be the contraction semigroup in $L^1(\Omega)$ generated by $\mathcal{A}_\varphi$. We shall prove that $u(t) := S(t)u_0$ is an entropy solution of problem (5.1). We divide the proof in different steps.

Step 1. Since $\mathcal{D}(\mathcal{A}_\varphi) \cap L^\infty(\Omega)$ is dense in $L^1(\Omega)$, given $u_0 \in L^1(\Omega)$ there exists a sequence $u_{0,n} \in \mathcal{D}(\mathcal{A}_\varphi) \cap L^\infty(\Omega)$ such that $u_{0,n} \to u_0$ in $L^1(\Omega)$. Then, if $u_n(t) := S(t)u_{0,n}$, we have that $u_n \to u$ in $C([0,T]; L^1(\Omega))$ for every $T > 0$. As a consequence of Theorem 5.14, $u_n(t), u_n'(t) \in L^2(\Omega)$, $p(u_n(t)) \in BV(\Omega)$ for all $p \in \mathcal{P}$ and there exists $z_n(t) \in X(\Omega)_1$, $\|z_n(t)\|_\infty \leq 1$ and $u_n'(t) = \operatorname{div}(z_n(t))$ in $\mathcal{D}'(\Omega)$ a.e. $t \in [0, +\infty[$, satisfying

$$
\begin{aligned}
-\int_\Omega (w - p(u_n(t)))u_n'(t)\,dx &\leq \int_\Omega (z_n(t), Dw) - \int_\Omega \|Dp(u_n(t))\| \\
&- \int_{\partial\Omega} [z_n(t),\nu](w - p(\varphi))\,d\mathcal{H}^{N-1} - \int_{\partial\Omega} |p(u_n(t)) - p(\varphi)|\,d\mathcal{H}^{N-1}
\end{aligned} \tag{5.56}
$$

for every $w \in BV(\Omega) \cap L^\infty(\Omega)$ and $p \in \mathcal{P}$. Moreover

$$
\int_\Omega (z_n(t), Dp(u_n(t))) = \int_\Omega \|Dp(u_n(t))\| \qquad \forall\, p \in \mathcal{P} \tag{5.57}
$$

and

$$
[z_n(t),\nu] \in \operatorname{sign}(p(\varphi) - p(u_n(t))) \quad \mathcal{H}^{N-1} - \text{a.e. on } \partial\Omega, \ \forall\, p \in \mathcal{P}. \tag{5.58}
$$

Since $\|[z_n(t),\nu]\|_\infty \leq \|z_n(t)\|_\infty \leq 1$, we can suppose (up to extraction of a subsequence, if necessary) that

$$
[z_n(\cdot),\nu] \to \rho \quad \sigma(L^\infty(S_T), L^1(S_T)).
$$

Step 2. *Convergence of the derivatives and identification of the limit.* Since the map $t \mapsto u_n'(t)$ is strongly measurable from $[0,T]$ into $L^2(\Omega)$, and by (5.6),

$$
\|u_n'(t)\|_{BV(\Omega)_2^*} \leq \|u_n'(t)\|_{L^2(\Omega)},
$$

it follows that this map is strongly measurable from $[0,T]$ into $BV(\Omega)_2^*$. Moreover, for every $w \in BV(\Omega)_2$, by Green's formula (C.10) we have

$$
\int_\Omega u_n'(t)w\,dx = \int_\Omega \operatorname{div}(z_n(t))w\,dx = -\int_\Omega (z_n(t), Dw) + \int_{\partial\Omega} [z_n(t),\nu]w\,d\mathcal{H}^{N-1}.
$$

Hence

$$\left|\int_\Omega u_n'(t)w\,dx\right| \leq \int_\Omega \|Dw\| + \int_{\partial\Omega} |w|\,d\mathcal{H}^{N-1} \leq M\|w\|_{BV(\Omega)_2} \quad \forall\, n \in \mathbb{N}.$$

Thus,

$$\|u_n'(t)\|_{BV(\Omega)_2^*} \leq M \qquad \forall\, n \in \mathbb{N} \;\text{ and }\; t \in [0,T].$$

Thus, $\{u_n'\}_{n\in\mathbb{N}}$ is a bounded sequence in $L^\infty(0,T;BV(\Omega)_2^*)$. Since $L^\infty(0,T; BV(\Omega)_2^*)$ is a vector subspace of the dual space $\left(L^1(0,T;BV(\Omega)_2)\right)^*$, we can find a net $\{u_\alpha'\}$ such that

$$u_\alpha' \to \xi \in \left(L^1(0,T;BV(\Omega)_2)\right)^* \quad \text{weakly}^*. \tag{5.59}$$

Since $\|z_n(t)\|_\infty \leq 1$ for all $n \in \mathbb{N}$ and a.e. $t \in [0,T]$, we can suppose that

$$z_n \to z \in L^\infty(Q_T,\mathbb{R}^N) \qquad \text{weakly}^*. \tag{5.60}$$

Given $\eta \in \mathcal{D}(Q_T)$, since $\eta \in L^1(0,T;BV(\Omega)_2)$, we have

$$\begin{aligned}
\langle \xi,\eta\rangle &= \lim_\alpha \langle u_\alpha',\eta\rangle = \lim_\alpha \int_0^T \langle u_\alpha'(t),\eta(t)\rangle\,dt \\
&= \lim_\alpha \int_0^T\int_\Omega u_\alpha'(t)\eta(t)\,dx\,dt = \lim_\alpha \int_0^T\int_\Omega \operatorname{div}(z_\alpha(t))\eta(t)\,dx\,dt \\
&= -\lim_\alpha \int_0^T\int_\Omega z_\alpha(t)\cdot\nabla\eta(t)\,dx\,dt = -\int_{Q_T} z\cdot\nabla\eta = \langle \operatorname{div}_x(z),\eta\rangle.
\end{aligned}$$

Hence,

$$\xi = \operatorname{div}_x(z) \qquad \text{in } \mathcal{D}'(Q_T). \tag{5.61}$$

On the other hand, if we take $\eta(t,x) = \phi(t)\psi(x)$ with $\phi \in \mathcal{D}(]0,T[)$ and $\psi \in \mathcal{D}(\Omega)$, the same calculation as above shows that

$$\xi(t) = \operatorname{div}_x(z(t)) \qquad \text{in } \mathcal{D}'(\Omega) \;\text{ a.e. }\; t \in [0,T]. \tag{5.62}$$

We have proved that $(z(t),\xi(t)) \in Z(\Omega)$ for almost all $t \in [0,T]$, and therefore the normal trace $[z(t),\nu]$ defined as in Section 5.2 has sense.

Lemma 5.18. *ξ is the time derivative of u in the sense of Definition 5.3.*

Proof. Let $\Psi \in L^1(0,T;BV(\Omega))$ be the weak derivative of

$$\Theta \in L_w^1(0,T;BV(\Omega)) \cap L^\infty(Q_T),$$

i.e., $\Psi(t) = \displaystyle\int_0^t \Theta(s)ds$, the integral being taken as a Pettis integral. By (5.59) we have that

$$\int_0^T \langle \xi(t),\Psi(t)\rangle dt = \lim_\alpha \int_0^T \langle u_\alpha'(t),\Psi(t)\rangle\,dt.$$

Now,

$$\begin{aligned}\int_0^T \langle u'_\alpha(t), \Psi(t)\rangle \, dt &= \lim_h \int_0^T \int_\Omega \Psi(t) \frac{u_\alpha(t+h) - u(t)}{h} \, dx dt \\ &= \lim_h \int_0^T \int_\Omega \frac{\Psi(t-h) - \Psi(t)}{h} u_\alpha(t) \, dx dt \\ &= - \lim_h \int_0^T \int_\Omega \frac{1}{h} \int_{t-h}^t \Theta(s) ds \, u_\alpha(t) \, dx dt \\ &= - \int_0^T \int_\Omega \Theta(t,x) u_\alpha(t,x) \, dx dt.\end{aligned}$$

Passing to the limit in α in the above expression, we obtain

$$\int_0^T \langle \xi(t), \Psi(t)\rangle \, dt = - \int_0^T \int_\Omega \Theta(t,x) u(t,x) \, dx ds. \tag{5.63}$$

□

Step 3. *Convergence of the energy.* In this step we shall prove that for any $p \in \mathcal{P}$, we have

$$\begin{aligned}&\lim_{n\to\infty} \int_0^T \int_\Omega \|Dp(u_n(t))\| \, dt + \int_0^T \int_{\partial\Omega} |p(u_n(t)) - p(\varphi)| \, d\mathcal{H}^{N-1} dt \\ &= \int_0^T \int_\Omega \|Dp(u(t))\| \, dt + \int_0^T \int_{\partial\Omega} |p(u(t)) - p(\varphi)| \, d\mathcal{H}^{N-1} dt.\end{aligned} \tag{5.64}$$

Let us first prove the following result.

Lemma 5.19. *Let $w \in L^1(Q_T)$ be such that $w(t) \in BV(\Omega)$ for almost all $t \in [0,T]$. Then the map $t \mapsto \|w(t)\|_{BV(\Omega)}$ from $[0,T]$ into $\mathbb{R}$ is measurable, and the map $t \mapsto w(t)$ from $[0,T]$ into $BV(\Omega)$ is weakly measurable.*

Proof. Let $E := C_c(\Omega)^{N+1}$ and $S : BV(\Omega) \to E^*$ be the map defined by

$$S(w) := \left(w \, dx, \frac{\partial w}{\partial x_1}, \ldots, \frac{\partial w}{\partial x_N} \right).$$

Then, $\|w\|_{BV(\Omega)} \leq \|S(w)\|_{E^*} \leq N \|w\|_{BV(\Omega)}$. If we denote by F the closure in E of the set

$$\{(\phi_0, \phi_1, \ldots, \phi_N) \ : \ \phi_i \in \mathcal{D}(\Omega), \text{ and } \phi_0 = \operatorname{div}(\phi_1, \ldots, \phi_N)\},$$

then, it is proved in Remark B.7 that $S(BV(\Omega))$ is isomorphic to $(\frac{E}{F})^*$, that is, $G := \frac{E}{F}$ is the predual of the space $BV(\Omega)$. Now, if $\phi = (\phi_0, \phi_1, \ldots, \phi_N) \in \mathcal{D}(\Omega)^{N+1}$,

$$\langle S(w(t)), \phi\rangle = \int_\Omega w(t)\phi_0 \, dx - \sum_{i=1}^N \int_\Omega w(t) \frac{\partial \phi}{\partial x_i} \, dx.$$

Hence, the map $t \mapsto \langle w(t), \phi\rangle$ is measurable. Now, approximating the functions of $C_c(\Omega)^{N+1}$ by functions in $\mathcal{D}(\Omega)^{N+1}$, we get that for every $\phi \in G$, the function $t \mapsto \langle w(t), \phi\rangle$ is measurable. Thus, since G is separable, it follows that the map

$$t \mapsto \|w(t)\|_{BV(\Omega)} = \sup_{\phi\in G,\ \|\phi\|\leq 1} \langle S(w(t)), \phi\rangle$$

is measurable.

Given $v \in BV(\Omega)^*$, let $g(t) := \langle w(t), v\rangle$. To see that g is measurable, consider $v_\alpha \in G$, such that $v_\alpha \to v$ with respect to $\sigma(G^{**}, G^*) = \sigma(BV(\Omega)^*, BV(\Omega))$. From the above, we know that if $g_\alpha(t) := \langle S(w(t)), v_\alpha\rangle$, g_α is measurable, and $g_\alpha(t) \to g(t)$. Now, since

$$|g_\alpha(t)| \leq \|w(t)\|_{BV(\Omega)}\|v_\alpha\|_{BV(\Omega)^*} \leq R\|w(t)\|_{BV(\Omega)} = F(t) \in L^1(0,T),$$

and the order interval $[-F, F]$ in $L^1(0,T)$ is $\sigma(L^1(0,T), L^\infty(0,T))$-relatively compact, there exists a sequence g_{α_n}, such that

$$g_{\alpha_n} \to g \qquad \text{in } \sigma(L^1(0,T), L^\infty(0,T)).$$

Hence, g is measurable. □

Taking $w = 0$ in (5.56) we get

$$\begin{aligned}\int_\Omega \|Dp(u_n(t))\| + \int_{\partial\Omega} |p(u_n(t)) - p(\varphi)|\, d\mathcal{H}^{N-1} \\ \leq -\int_\Omega p(u_n(t))u_n'(t)\, dx + \int_{\partial\Omega} [z_n(t), \nu]p(\varphi)\, d\mathcal{H}^{N-1}.\end{aligned}$$

If we denote $J_p(r) := \int_0^r p(s)\, ds$, it follows that

$$\begin{aligned}&\int_0^T \int_\Omega \|Dp(u_n(t))\|\, dt + \int_0^T \int_{\partial\Omega} |p(u_n(t)) - p(\varphi)|\, d\mathcal{H}^{N-1}dt \\ &\leq -\int_0^T \frac{d}{dt}\int_\Omega J_p(u_n(t))\, dx + \int_0^T \int_{\partial\Omega} |p(\varphi)|\, d\mathcal{H}^{N-1}dt \\ &= \int_\Omega (J_p(u_{0,n}) - J_p(u_n(T))\, dx + \int_0^T \int_{\partial\Omega} |p(\varphi)|\, d\mathcal{H}^{N-1}dt \leq M_p.\end{aligned}$$

Since the functional $\Phi_p : L^1(\Omega) \to]-\infty, +\infty]$, defined by

$$\Phi_p(w) = \begin{cases} \int_\Omega \|Dw\| + \int_{\partial\Omega} |w - p(\varphi)|\, d\mathcal{H}^{N-1} & \text{if } w \in BV(\Omega), \\ +\infty & \text{if } w \in L^1(\Omega) \setminus BV(\Omega), \end{cases} \tag{5.65}$$

is lower semi-continuous in $L^1(\Omega)$, we have

$$\begin{aligned}\Phi_p(p(u(t)) &\leq \liminf_{n\to\infty} \Phi_p(p(u_n(t)) \\ &= \liminf_{n\to\infty} \left(\int_\Omega \|Dp(u_n(t))\| + \int_{\partial\Omega} |p(u_n(t)) - p(\varphi)| \, d\mathcal{H}^{N-1} \right).\end{aligned} \tag{5.66}$$

On the other hand, by Lemma 5.19, the map $t \mapsto \|p(u_n(t))\|_{BV(\Omega)}$ is measurable, then by Fatou's Lemma, it follows that

$$\begin{aligned}&\int_0^T \liminf_{n\to\infty} \left(\int_\Omega \|Dp(u_n(t))\| + \int_{\partial\Omega} |p(u_n(t)) - p(\varphi)| \, d\mathcal{H}^{N-1} \right) dt \\ &\leq \liminf_{n\to\infty} \int_0^T \left(\int_\Omega \|Dp(u_n(t))\| + \int_{\partial\Omega} |p(u_n(t)) - p(\varphi)| \, d\mathcal{H}^{N-1} \right) dt \leq M_p.\end{aligned} \tag{5.67}$$

As a consequence of (5.66) and (5.67), we obtain that $p(u(t)) \in BV(\Omega)$ for almost all $t \in [0, T]$.

From Lemma 5.19, if $0 \leq \eta \in \mathcal{D}(]0, T[)$, the map $t \mapsto p(u(t))\eta(t)$, from $[0, T]$ into $BV(\Omega)$ is weakly measurable.

Lemma 5.20. *For any $\tau > 0$, we define the function ψ^τ, as the Dunford integral (see Definition 5.1)*

$$\psi^\tau(t) := \frac{1}{\tau} \int_{t-\tau}^t \eta(s) p(u(s)) \, ds \in BV(\Omega)^{**},$$

that is,

$$\langle \psi^\tau(t), w \rangle = \frac{1}{\tau} \int_{t-\tau}^t \langle \eta(s) p(u(s)), w \rangle \, ds,$$

for any $w \in BV(\Omega)^*$. Then $\psi^\tau \in C([0, T]; BV(\Omega))$. Moreover, $\psi^\tau(t) \in L^2(\Omega)$, and, thus, $\psi^\tau(t) \in BV(\Omega)_2$.

Proof. Given $\phi \in \mathcal{D}(\Omega)$,

$$\begin{aligned}|\langle \psi^\tau(t), \phi \rangle| &\leq \frac{1}{\tau} \int_{t-\tau}^t |\eta(s)| |\langle p(u(s)), \phi \rangle| \, ds \\ &= \frac{1}{\tau} \int_{t-\tau}^t |\eta(s)| \left(\int_\Omega |p(u(s))| |\phi| \, dx \right) ds \leq C \|\phi\|_\infty.\end{aligned}$$

Thus, $\psi^\tau(t)$ is a finite Radon measure in Ω. Moreover, a similar calculation shows that for every $i = 1, 2, \ldots, N$, $\dfrac{\partial \psi^\tau(t)}{\partial x_i}$ is also a finite Radon measure in Ω. Hence, we have $\psi^\tau(t) \in BV(\Omega)$ (see, Exercise 3.2. in [10]), and the Dunford integral of the definition of $\psi^\tau(t)$ is a Pettis integral. Moreover, if $a_n \to 0$ (for simplicity suppose that $a_n > 0$), given $w \in BV(\Omega)^*$ with $\|w\| \leq 1$, we have

$$\begin{aligned}
&|\langle \psi^\tau(t+a_n) - \psi^\tau(t), w\rangle| \\
&= \left| \frac{1}{\tau} \int_{t+a_n-\tau}^{t+a_n} \eta(s)\langle p(u(s)), w\rangle \, ds - \frac{1}{\tau} \int_{t-\tau}^{t} \eta(s)\langle p(u(s)), w\rangle \, ds \right| \\
&\leq \left| \frac{1}{\tau} \int_{t}^{t+a_n} \eta(s)\langle p(u(s)), w\rangle \, ds - \frac{1}{\tau} \int_{t-\tau}^{t-\tau+a_n} \eta(s)\langle p(u(s)), w\rangle \, ds \right| \\
&\leq \frac{1}{\tau} \int_{t}^{t+a_n} |\eta(s)| \|p(u(s))\|_{BV(\Omega)} \, ds + \frac{1}{\tau} \int_{t-\tau}^{t-\tau+a_n} |\eta(s)| \|p(u(s))\|_{BV(\Omega)} \, ds.
\end{aligned}$$

Since the function $s \mapsto |\eta(s)| \|p(u(s))\|_{BV(\Omega)}$ is in $L^1([0,T])$,

$$\lim_{n\to\infty} \|\psi^\tau(t+a_n) - \psi^\tau(t)\|_{BV(\Omega)} = 0.$$

Thus, $\psi^\tau \in C([0,T]; BV(\Omega))$.

Moreover, $\psi^\tau(t) \in L^2(\Omega)$. In fact, given $g \in L^\infty(\Omega)$, with $\|g\|_2 \leq 1$, since $g \in BV(\Omega)^*$, we have

$$|\langle \psi^\tau(t), g\rangle| = \left| \frac{1}{\tau} \int_{t-\tau}^{t} \eta(s)\langle p(u(s)), g\rangle \, ds \right|$$

$$= \left| \frac{1}{\tau} \int_{t-\tau}^{t} \eta(s) \left(\int_\Omega p(u(s)) g \, dx \right) ds \right| \leq \frac{1}{\tau} \int_{t-\tau}^{t} |\eta(s)| \|p(u(s))\|_2 \|g\|_2 \leq M.$$

From the density of $L^\infty(\Omega)$ in $L^2(\Omega)$, we obtain that $\psi^\tau(t) \in L^2(\Omega)$. □

Lemma 5.21. *For $\tau > 0$ small enough, we have*

$$\int_0^T \langle \psi^\tau(t), \xi(t)\rangle \, dt \leq - \int_0^T \int_\Omega \frac{\eta(t-\tau) - \eta(t)}{-\tau} J_p(u(t)) \, dx dt. \tag{5.68}$$

Proof. Since $\psi^\tau \in C([0,T], BV(\Omega))$ admits a weak derivative in $L^1_w(0,T; BV(\Omega)) \cap L^\infty(Q_T)$, using (5.63) we have for $\tau > 0$ small enough that

$$\int_0^T \langle \psi^\tau(t), \xi(t)\rangle \, dt = \int_0^T \int_\Omega \frac{u(t+\tau) - u(t)}{\tau} \eta(t) p(u(t)) \, dx dt.$$

Now, since p is nondecreasing, we have

$$J_p(u(t)) - J_p(u(t+\tau)) \leq \big(u(t) - u(t+\tau)\big) p(u(t))$$

and, therefore, for $\tau > 0$ small enough, we obtain

$$\int_0^T \int_\Omega \frac{u(t+\tau)-u(t)}{\tau}\eta(t)p(u(t))\,dxdt \leq \int_0^T \int_\Omega \frac{J_p(u(t+\tau))-J_p(u(t))}{\tau}\eta(t)\,dxdt$$
$$= \int_0^T \int_\Omega \frac{\eta(s-\tau)-\eta(s)}{\tau} J_p(u(s))\,dxdt,$$

and this proves (5.68). □

Now, we can conclude the proof of Step 3. As a consequence of (5.68), using Green's formula, we have

$$\int_0^T \int_\Omega \frac{\eta(t-\tau)-\eta(t)}{-\tau} J_p(u(t))\,dxdt \leq -\int_0^T \langle \psi^\tau(t), \xi(t)\rangle\,dt$$
$$= -\lim_\alpha \int_0^T \langle \psi^\tau(t), u'_\alpha(t)\rangle\,dt = -\lim_\alpha \int_0^T \left(\frac{1}{\tau}\int_{t-\tau}^t \eta(s)\langle p(u(s)), u'_\alpha(t)\rangle\,ds\right)dt$$
$$= -\lim_\alpha \int_0^T \left(\frac{1}{\tau}\int_{t-\tau}^t \eta(s)\left(\int_\Omega p(u(s))\,\mathrm{div}\,z_\alpha(t)\,dx\right)ds\right)dt$$
$$= \lim_\alpha \int_0^T \left(\frac{1}{\tau}\int_{t-\tau}^t \eta(s)\int_\Omega (z_\alpha(t), Dp(u(s)))\,ds\right)dt$$
$$-\lim_\alpha \int_0^T \left(\frac{1}{\tau}\int_{t-\tau}^t \eta(s)\left(\int_{\partial\Omega}[z_\alpha(t),\nu]p(u(s))\,d\mathcal{H}^{N-1}\right)ds\right)dt$$
$$\leq \int_0^T \left(\frac{1}{\tau}\int_{t-\tau}^t \eta(s)\int_\Omega \|Dp(u(s))\|\,ds\right)dt$$
$$-\int_0^T \left(\frac{1}{\tau}\int_{t-\tau}^t \eta(s)\left(\int_{\partial\Omega}\rho(t)p(u(s))\,d\mathcal{H}^{N-1}\right)ds\right)dt.$$

Then, taking limit as $\tau \to 0^+$, we get

$$\int_0^T \int_\Omega \eta'(t)J_p(u(t))\,dxdt$$
$$\leq \int_0^T \eta(t)\int_\Omega \|Dp(u(t))\|\,dt - \int_0^T \eta(t)\int_{\partial\Omega}\rho(t)p(u(t))\,d\mathcal{H}^{N-1}dt.$$

Now, since this is true for all $0 \leq \eta \in \mathcal{D}(]0,T[)$, it follows that

$$-\frac{d}{dt}\int_\Omega J_p(u(t))\,dx \leq \int_\Omega \|Dp(u(t))\| - \int_{\partial\Omega}\rho(t)p(u(t))\,d\mathcal{H}^{N-1},$$

and, thus,

$$\begin{aligned}&\int_\Omega (J_p(u_0) - J_p(u(T)))\, dx \\ &\leq \int_0^T \int_\Omega \|Dp(u(t))\|\, dt - \int_0^T \int_{\partial\Omega} \rho(t) p(u(t))\, d\mathcal{H}^{N-1} dt.\end{aligned} \tag{5.69}$$

Finally, using (5.69), we obtain

$$\begin{aligned}&\int_0^T \int_\Omega \|Dp(u(t))\|\, dt + \int_0^T \int_{\partial\Omega} |p(u(t)) - p(\varphi)|\, d\mathcal{H}^{N-1} dt \\ &\leq \liminf_{n\to\infty} \int_0^T \int_\Omega \|Dp(u_n(t))\|\, dt + \int_0^T \int_{\partial\Omega} |p(u_n(t)) - p(\varphi)|\, d\mathcal{H}^{N-1} dt \\ &\leq \liminf_{n\to\infty} \left(\int_0^T \int_\Omega p(u_n(t)) u_n'(t)\, dx dt + \int_0^T \int_{\partial\Omega} [z_n(t), \nu] p(\varphi)\, d\mathcal{H}^{N-1} dt \right) \\ &= \int_\Omega J_p(u_0) - J_p(u(T))\, dx + \int_0^T \int_{\partial\Omega} \rho(t) p(\varphi)\, d\mathcal{H}^{N-1} dt \\ &\leq \int_0^T \int_\Omega \|Dp(u(t))\|\, dt + \int_0^T \int_{\partial\Omega} \rho(t) \big(p(\varphi) - p(u(t)\big)\, d\mathcal{H}^{N-1} dt \\ &\leq \int_0^T \int_\Omega \|Dp(u(t))\|\, dt + \int_0^T \int_{\partial\Omega} |p(u(t)) - p(\varphi)|\, d\mathcal{H}^{N-1} dt,\end{aligned}$$

which concludes the proof of (5.64). Moreover, we get that

$$\rho(t) \in \operatorname{sign}\big(p(\varphi) - p(u(t))\big) \quad \mathcal{H}^{N-1} - \text{a.e. on } \partial\Omega, \quad \text{a.e. } t \in [0,T]. \tag{5.70}$$

Step 4. *The boundary condition.* Let us now prove that

$$\rho(t) = [z(t), \nu] \quad \mathcal{H}^{N-1} - \text{a.e. on } \partial\Omega, \quad \text{a.e. } t \in [0,T]. \tag{5.71}$$

In fact, if $w \in BV(\Omega) \cap L^\infty(\Omega)$, and $v \in R(\Omega)$ is such that $v|_{\partial\Omega} = w|_{\partial\Omega}$, we have that

$$\int_0^t \langle z_\alpha(s), w \rangle_{\partial\Omega}\, ds = \int_0^t \langle \operatorname{div}(z_\alpha(s)), v \rangle\, ds + \int_0^t \int_\Omega z_\alpha(s) \cdot \nabla v\, dx ds.$$

Hence

$$\begin{aligned}\lim_\alpha \int_0^t \langle z_\alpha(s), w \rangle_{\partial\Omega}\, ds &= \int_0^t \langle \xi(s), v \rangle\, ds + \int_0^t \int_\Omega z(s) \cdot \nabla v\, dx ds \\ &= \int_0^t \langle z(s), w \rangle_{\partial\Omega}\, ds = \int_0^t \int_{\partial\Omega} [z(s), \nu] w\, d\mathcal{H}^{N-1} ds.\end{aligned} \tag{5.72}$$

On the other hand, since $z_\alpha(s) \in X(\Omega)_1$, if we apply Green's formula (C.10) we have that

$$\int_0^t \langle \mathrm{div}(z_\alpha(s)), v\rangle\, ds = -\int_0^t \int_\Omega z_\alpha(s) \cdot \nabla v\, dx ds + \int_0^t \int_{\partial\Omega} [z_\alpha(s), \nu] w\, d\mathcal{H}^{N-1} ds.$$

Hence

$$\int_0^t \langle z_\alpha(s), w\rangle_{\partial\Omega}\, ds = \int_0^t \int_{\partial\Omega} [z_\alpha(s), \nu] w\, d\mathcal{H}^{N-1} ds.$$

Taking limits in α, we get

$$\int_0^t \int_{\partial\Omega} \rho(s) w\, d\mathcal{H}^{N-1} ds = \int_0^t \int_{\partial\Omega} [z(s), \nu] w\, d\mathcal{H}^{N-1} ds \tag{5.73}$$

for all $w \in BV(\Omega) \cap L^\infty(\Omega)$, $t \in [0, T]$. Now, if $w \in L^1(\partial\Omega)$, we take $w_k \in BV(\Omega) \cap L^\infty(\Omega)$ such that $w_k|_{\partial\Omega} = T_k(w)$. By (5.73), we have

$$\int_0^t \int_{\partial\Omega} \rho(s) w_k\, d\mathcal{H}^{N-1} ds = \int_{\partial\Omega} [z(s), \nu] w_k\, d\mathcal{H}^{N-1} ds.$$

Letting $k \to \infty$, it follows that

$$\int_0^t \int_{\partial\Omega} \rho(s) w\, d\mathcal{H}^{N-1} ds = \int_0^t \int_{\partial\Omega} [z(s), \nu] w\, d\mathcal{H}^{N-1} ds$$

for all $w \in L^1(\partial\Omega)$, and $t \in [0, T]$, and (5.71) holds.

Step 5. Next, we prove that $\xi = \mathrm{div}(z)$ in $\left(L^1(0, T; BV(\Omega)_2)\right)^*$ in the sense of Definition 5.4. To do that let us first observe that (z, Dw), defined by (5.7), is a Radon measure in Q_T for all $w \in L^1_w(0, T; BV(\Omega)) \cap L^\infty(Q_T)$. Let $\phi \in \mathcal{D}(Q_T)$, then

$$\begin{aligned} \langle (z, Dw), \phi\rangle &= -\int_0^T \langle \xi(t) - u'_\alpha(t), w(t)\phi(t)\rangle\, dt \\ &\quad - \int_{Q_T} w(z - z_\alpha) \cdot \nabla_x \phi\, dx dt + \int_0^T \langle (z_\alpha(t), Dw(t)), \phi(t)\rangle\, dt. \end{aligned}$$

Then by (5.59), taking limits in α, we get

$$\langle (z, Dw), \phi\rangle = \lim_\alpha \int_0^T \langle (z_\alpha(t), Dw(t)), \phi(t)\rangle\, dt. \tag{5.74}$$

Therefore

$$|\langle (z, Dw), \phi\rangle| \leq \|\phi\|_\infty \int_0^T \int_\Omega \|Dw(t)\|\, dt,$$

that is, (z, Dw) is a Radon measure in Q_T. Moreover, from (5.74), applying Green's formula (C.10) we obtain that

$$\begin{aligned}\int_{Q_T}(z,Dw) &= \lim_\alpha \int_0^T (z_\alpha(t), Dw(t))\,dt\\ &= \lim_\alpha\left(-\int_0^T\int_\Omega \mathrm{div}(z_\alpha(t))w(t)\,dxdt + \int_0^T\int_{\partial\Omega}[z_\alpha(t),\nu]w(t)\,d\mathcal{H}^{N-1}dt\right)\\ &= -\int_0^T \langle \xi(t), w(t)\rangle\,dt + \int_0^T\int_{\partial\Omega}[z(t),\nu]w(t)\,d\mathcal{H}^{N-1}dt.\end{aligned}$$

Step 6. *Conclusion.* Finally, we are going to prove that u verifies

$$\begin{aligned}&-\int_0^T\int_\Omega j(u(t)-l)\eta_t + \int_0^T\int_\Omega \eta(t)\|Dp(u(t)-l)\|\\ &+\int_0^T\int_\Omega z(t)\cdot D\eta(t)p(u(t)-l) \le \int_0^T\int_{\partial\Omega}[z(t),\nu]\eta(t)p(u(t)-l),\end{aligned} \tag{5.75}$$

for all $\eta \in C^\infty(\overline{Q_T})$, with $\eta \ge 0$, $\eta(t,x) = \phi(t)\psi(x)$, being $\phi \in \mathcal{D}(]0,T[)$, $\psi \in C^\infty(\overline{\Omega})$, and $p \in \mathcal{P}$, where $j(r) = \int_0^r p(s)\,ds$.

Let $\eta \in C^\infty(\overline{Q_T})$, with $\eta \ge 0$, $\eta(t,x) = \phi(t)\psi(x)$, being $\phi \in \mathcal{D}(]0,T[)$, $\psi \in C^\infty(\overline{\Omega})$, and $p \in \mathcal{P}$, $a \in \mathbb{R}$. Let $H_p(r) := \int_a^r p(s)\,ds$. Since $u_n'(t) = \mathrm{div}(z_n(t))$, multiplying by $p(u_n(t))\eta(t)$ and integrating, we obtain that

$$\begin{aligned}&\int_0^T\int_\Omega \frac{d}{dt}H_p(u_n(t))\eta(t)\,dxdt = \int_0^T\int_\Omega p(u_n(t))u_n'(t)\eta(t)\,dxdt\\ &= \int_0^T\int_\Omega \mathrm{div}(z_n(t))p(u_n(t))\eta(t)\,dxdt\\ &= -\int_0^T\int_\Omega (z_n(t), D(p(u_n(t))\eta(t))\,dt + \int_0^T\int_{\partial\Omega}[z_n(t),\nu]p(u_n(t))\eta(t)\,d\mathcal{H}^{N-1}dt\\ &= -\int_0^T\int_\Omega \eta(t)\|Dp(u_n(t))\|\,dt - \int_0^T\int_\Omega z_n(t)\cdot\nabla\eta(t)\;p(u_n(t))dxdt\\ &\quad+\int_0^T\int_{\partial\Omega}[z_n(t),\nu]p(u_n(t))\eta(t)\,d\mathcal{H}^{N-1}dt\\ &= -\int_0^T\int_\Omega \eta(t)\|Dp(u_n(t))\|\,dt - \int_0^T\int_\Omega z_n(t)\cdot\nabla\eta(t)\;p(u_n(t))\,dxdt\\ &\quad-\int_0^T\int_{\partial\Omega}|p(u_n(t))-p(\varphi)|\eta(t)\,d\mathcal{H}^{N-1}dt + \int_0^T\int_{\partial\Omega}[z_n(t),\nu]p(\varphi)\eta(t)\,d\mathcal{H}^{N-1}dt.\end{aligned}$$

Hence, having in mind that $\eta(0) = \eta(T) = 0$, we get

$$\int_0^T \int_\Omega \eta(t)\|Dp(u_n(t))\| \, dt + \int_0^T \int_{\partial\Omega} |p(u_n(t)) - p(\varphi)|\eta(t) \, d\mathcal{H}^{N-1} dt$$

$$= -\int_0^T \int_\Omega z_n(t) \cdot \nabla\eta(t) \; p(u_n(t)) \, dxdt + \int_0^T \int_{\partial\Omega} [z_n(t), \nu] p(\varphi)\eta(t) \, d\mathcal{H}^{N-1} dt$$

$$- \int_0^T \int_\Omega \frac{d}{dt} H_p(u_n(t))\eta(t) \, dxdt$$

$$= -\int_0^T \int_\Omega z_n(t) \cdot \nabla\eta(t) \; p(u_n(t)) \, dxdt + \int_0^T \int_{\partial\Omega} [z_n(t), \nu] p(\varphi)\eta(t) \, d\mathcal{H}^{N-1} dt$$

$$- \int_0^T \int_\Omega \frac{d}{dt} \big(H_p(u_n(t))\eta(t)\big) \, dxdt + \int_0^T \int_\Omega H_p(u_n(t)) \; \eta_t \, dxdt$$

$$= -\int_0^T \int_\Omega z_n(t) \cdot \nabla\eta(t) \; p(u_n(t)) \, dxdt + \int_0^T \int_{\partial\Omega} [z_n(t), \nu] p(\varphi)\eta(t) \, d\mathcal{H}^{N-1} dt$$

$$+ \int_0^T \int_\Omega H_p(u_n(t)) \; \eta_t \, dxdt.$$

Letting $n \to \infty$, it follows that

$$\int_0^T \int_\Omega \eta(t)\|Dp(u(t))\| \, dt + \int_0^T \int_{\partial\Omega} |p(u(t)) - p(\varphi)|\eta(t) \, d\mathcal{H}^{N-1} dt$$

$$\leq \liminf_{n\to\infty} \left[\int_0^T \int_\Omega \eta(t)\|Dp(u_n(t))\| \, dt + \int_0^T \int_{\partial\Omega} |p(u_n(t)) - p(\varphi)|\eta(t) \, d\mathcal{H}^{N-1} dt \right]$$

$$= \liminf_{n\to\infty} \left[-\int_0^T \int_\Omega z_n(t) \cdot \nabla\eta(t) \; p(u_n(t)) \, dxdt \right.$$

$$\left. + \int_0^T \int_{\partial\Omega} [z_n(t), \nu] p(\varphi)\eta(t) \, d\mathcal{H}^{N-1} dt + \int_0^T \int_\Omega H_p(u_n(t)) \, \eta_t \, dxdt \right]$$

$$= -\int_0^T \int_\Omega z(t) \cdot \nabla\eta(t) \; p(u(t)) \, dxdt$$

$$+ \int_0^T \int_{\partial\Omega} [z(t), \nu] p(\varphi)\eta(t) \, d\mathcal{H}^{N-1} dt + \int_0^T \int_\Omega H_p(u(t)) \; \eta_t \, dxdt.$$

Now, using that $|p(u(t)) - p(\varphi)| = [z(t),\nu]\big(p(\varphi) - p(u(t))\big)$, we have

$$\begin{aligned}
&-\int_0^T \int_\Omega H_p(u(t))\eta_t \,dxdt + \int_0^T \int_\Omega \eta(t) \|Dp(u(t))\| \,dt \\
&+\int_0^T \int_\Omega z(t)\cdot\nabla\eta(t)\; p(u(t))\,dxdt \qquad (5.76)\\
&\le \int_0^T \int_{\partial\Omega} [z(t),\nu] p(u(t))\eta(t)\, d\mathcal{H}^{N-1} dt.
\end{aligned}$$

Finally, given $l \in \mathbb{R}$ and $p \in \mathcal{P}$, since $q(r) := p(r-l)$ is an element of $\mathcal{P}$, and taking $a = l$, we obtain (5.75) as a consequence of (5.76). The concludes the proof of existence.

5.6.2 Proof of Theorem 5.6: Uniqueness

To prove uniqueness we show that the entropy and the semigroup solutions coincide. As a consequence of semigroup theory, this will imply (5.8). The proof follows by using the doubling variables technique introduced by Kruzhkov ([143]) to prove the L^1-contraction property for entropy solutions of scalar conservation laws. Since the same method will be used in Chapter 7 for proving uniqueness of entropy solutions of a more general class of equations we shall not give the proof in detail here and we refer to Remark 7.17 in Subsection 7.4.2 (see also [14]) .

5.7 Regularity for Positive Initial Data

In this section we prove that when the initial data are nonnegative, semigroup solutions are strong solutions.

We need to consider truncations $T_{a,b}$, $a < b$ (see Section 3.6).

Proposition 5.22. *Let $u_0 \in L^1(\Omega)$, $\varphi \in L^1(\partial\Omega)$. Let $\big(S(t)\big)_{t\ge 0}$ be the semigroup generated by $\mathcal{A}_\varphi$. Then, if $u(t) = S(t)u_0$,*

(i) $\displaystyle\int_\Omega j(u(t))\,dx + \int_0^t \Phi(p(u(s)))\,ds \le \int_0^t\int_{\partial\Omega} |p(\varphi)|\,d\mathcal{H}^{N-1}ds + \int_\Omega j(u_0)\,dx$
where p is a truncation ($p = T_{a,b}$), j is the primitive of p and Φ is the functional defined by (5.25).

(ii) *$p(u)_t \in L^2_{loc}(0,\tau;L^2(\Omega))$, for every truncation p as above. Moreover, we have the estimate*

$$\Phi(p(u(t))) + \int_s^t \int_\Omega |p(u)_t|^2 \le C,$$

where $C > 0$ depends on s, $\|u_0\|_{L^1}$, $\|\varphi\|_{L^1}$ and p.

Proof. (i) Assume first that $u_0 \in L^2(\Omega)$. Then

$$\begin{aligned}\frac{d}{dt}\int_\Omega j(u) &= \int_\Omega p(u)u_t\,dx = \int_\Omega p(u)\mathrm{div}(z)\,dx \\ &= -\int_\Omega (z, Dp(u)) + \int_{\partial\Omega} [z,\nu]p(u)\,d\mathcal{H}^{N-1} \\ &= -\int_\Omega \|Dp(u)\| + \int_{\partial\Omega} [z,\nu](p(u)-p(\varphi)+p(\varphi))\,d\mathcal{H}^{N-1} \\ &= -\int_\Omega \|Dp(u)\| - \int_{\partial\Omega} |p(u)-p(\varphi)|\,d\mathcal{H}^{N-1} + \int_{\partial\Omega} [z,\nu]p(\varphi)\,d\mathcal{H}^{N-1}.\end{aligned}$$

Integrating this expression, we obtain

$$\int_\Omega j(u(t))\,dx + \int_0^t \Phi(p(u(s)))\,ds \le \int_0^t \int_{\partial\Omega} |p(\varphi)|\,d\mathcal{H}^{N-1}ds + \int_\Omega j(u_0)\,dx. \quad (5.77)$$

Since j has linear growth at infinity, if $u_0 \in L^1(\Omega)$, the estimate in (i) follows by approximating u_0 by functions $u_{0n} \in L^2(\Omega)$ and passing to the limit.

(ii) Assume first that $u_0 \in L^2(\Omega)$. Let $\delta > 0$ and $t, s \ge \delta$ such that $(u(t), -u_t(t))$, $(u(s), -u_t(s)) \in \mathcal{A}_\varphi$. We know that

$$\begin{aligned}\int_\Omega (p(u(t)) - w)u_t(t)\,dx &\le \int_\Omega (z(t), Dw) - \int_\Omega \|Dp(u(t))\| \\ &+ \int_{\partial\Omega} |w-p(\varphi)|\,d\mathcal{H}^{N-1} - \int_{\partial\Omega} |p(u(t))-p(\varphi)|\,d\mathcal{H}^{N-1}\end{aligned} \quad (5.78)$$

for all $w \in BV(\Omega)\cap L^\infty(\Omega)$. Setting $w = p(u(s))$ in the above expression we have

$$\Phi(p(u(t))) - \Phi(p(u(s))) \le \int_\Omega u_t(t)(p(u(s)) - p(u(t)))\,dx.$$

Using the estimate (A.36) for semigroups generated by subdifferentials in L^2 we have

$$\Phi(p(u(t))) - \Phi(p(u(s))) \le C(\delta)\|u_0\|_2\|p(u(s)) - p(u(t))\|_2.$$

Since a similar estimate holds with s and t interchanged, we have

$$|\Phi(p(u(t))) - \Phi(p(u(s)))| \le C(\delta)\|u_0\|_2\|p(u(s)) - p(u(t))\|_2. \quad (5.79)$$

Since $u \in W^{1,1}_{loc}(0,\tau;L^2(\Omega))$, i.e, u is a locally absolutely continuous function of time, then so also is $p(u)$, and, from (5.79), we deduce that $\Phi(p(u))$ is absolutely continuous in $[0,\tau]$ for all $\tau > 0$. Let $t \in [0,\infty)$ be such that u, $p(u)$, $\Phi(p(u))$ are differentiable at t and $(u(t), -u_t(t)) \in \mathcal{A}_\varphi$. Set $w = p(u(t+\epsilon))$, $w = p(u(t-\epsilon))$ in (5.78) to obtain

$$\int_\Omega (p(u(t)) - p(u(t\pm\epsilon)))u_t(t) \le \Phi(p(u(t\pm\epsilon))) - \Phi(p(u(t))).$$

Letting $\epsilon \to 0^+$ we have

$$\frac{d}{dt}\Phi(p(u(t))) + \int_\Omega p'(u(t))u_t(t)^2 = 0.$$

In particular, since p' is either 0 or 1, we have

$$\frac{d}{dt}\Phi(p(u(t))) + \int_\Omega |p(u)_t(t)|^2 \leq 0.$$

Hence $\Phi(p(u(t))$ is a decreasing function of t. If $u_0 \in BV(\Omega) \cap L^2(\Omega)$, integrating from 0 to t we get

$$\Phi(p(u(t))) + \int_0^t \int_\Omega |p(u)_t|^2 \leq \Phi(u_0).$$

Observe that by (i), if $u_0 \in L^2(\Omega)$, then $u(s) \in BV(\Omega) \cap L^2(\Omega)$ for almost all $s > 0$ and we have

$$\Phi(p(u)(t)) + \int_s^t \int_\Omega |p(u)_t|^2 \leq \Phi(p(u)(s)),$$

for almost all $0 < s < t$. Now, let $u_0 \in L^1(\Omega)$ and $u_{0n} \in L^2(\Omega)$ be such that $u_{0n} \to u_0$ in $L^1(\Omega)$. Then, if $u_n(t,x)$ denotes the solution corresponding to the initial datum u_{0n} we have

$$\Phi(p(u_n)(t)) + \int_s^t \int_\Omega |p(u_n)_t|^2 \leq \Phi(p(u_n)(s)), \tag{5.80}$$

for almost all $0 < s < t$ and all n. Now, observe that by the estimate in (i),

$$\int_0^t \Phi(p(u_n)(\tau))\, d\tau \leq C$$

for some constant $C > 0$. Let $\delta > 0$. Then

$$\int_0^t \Phi(p(u_n)(\tau))\, d\tau \geq \int_0^\delta \Phi(p(u_n)(\tau))\, d\tau \geq \Phi(p(u_n)(\delta))\delta,$$

and $\Phi(p(u_n)(s))$ is a bounded sequence for almost all $s > 0$. Thus, for a.e. $s > 0$, the left-hand side of (5.80) is bounded. Hence, we may assume that $p(u_n(t)) \to p(u(t))$ in $L^1(\Omega)$ for a.e. $t > 0$. Now, we may pass to the limit in (5.80) and use the lower semi-continuity of the left-hand side to obtain that

$$\Phi(p(u)(t)) + \int_s^t \int_\Omega |p(u)_t|^2 \leq C, \tag{5.81}$$

where C depends on $s, \|u_0\|_{L^1}, \|\varphi\|_{L^1}, p$. □

Theorem 5.23. *Let $u_0 \in L^1(\Omega)$, $\varphi \in L^1(\partial\Omega)$. Suppose that $u_0 + M \geq 0$, $\varphi + M \geq 0$ (or $u_0 - M \leq 0$, $\varphi - M \leq 0$) for some $M \geq 0$. Let $\left(S(t)\right)_{t\geq 0}$ be the semigroup generated by $\mathcal{A}_\varphi$. Then, if $u(t) = S(t)u_0$, $u_t \in L^1_{loc}(0,T; L^1(\Omega))$.*

Proof. It is easy to check via the resolvents that the semigroup solution corresponding to the data $u_0 \pm M, \varphi \pm M$ coincides with the semigroup solution corresponding to the data u_0, φ plus the constant M, i.e., $S(t)(u_0 \pm M, \varphi \pm M) = S(t)(u_0, \varphi) \pm M$. Thus, without loss of generality we may assume that $M = 0$. Let us prove the theorem in case $u_0, \varphi \geq 0$, the other case being analogous. We know, by the homogeneity estimate, Proposition 5.17, that u_t is a Radon measure in $(s,t) \times \Omega$, for all $0 < s < t$. Thus, its mass is bounded, i.e.,

$$\int_s^t \int_\Omega |u_t| < \infty.$$

Now, taking $p = T_{a,b}$, the estimate in (ii) of the previous proposition says that u_t is a function in $L^2(Q_{a,b})$, for all $a < b$, where $Q_{a,b} = \{(t,x) \in Q \ : \ a < u(t,x) < b\}$. Thus, this with the last integral bound, prove that $u_t \in L^1_{loc}(0,\tau; L^1(\Omega))$. □

Remark 5.24. Under the assumption of the above theorem, since u_t is an element of $L^1_{loc}(0,T; L^1(\Omega))$, working as in Chapter 2 we can prove that u is a strong solution. Consequently, existence and uniqueness can be obtained in an easier way than in the general case using the same technique as in Chapter 2.

Observe also that for $\varphi = 0$, the operator $\mathcal{A}_\varphi$ is homogeneous as in the Neumann problem. In that case, working as in Chapter 2, it can be proved that for every initial datum in $L^1(\Omega)$ the entropy solution of the homogeneous Dirichlet problem is a strong solution. Moreover, as it was shown for the Neumann and Cauchy problems, there is no regularizing effect in general, but the L^N-L^∞ regularizing effect holds.

Chapter 6

Parabolic Equations Minimizing Linear Growth Functionals: L^2-Theory

6.1 Introduction

Let Ω be a bounded set in $\mathbb{R}^N$ with boundary $\partial\Omega$ of class C^1. We are interested in the problem

$$\begin{cases} \dfrac{\partial u}{\partial t} = \operatorname{div} \mathbf{a}(x, Du) & \text{in} \quad Q = (0,\infty) \times \Omega, \\ u(t,x) = \varphi(x) & \text{on} \quad S = (0,\infty) \times \partial\Omega, \\ u(0,x) = u_0(x) & \text{in} \quad x \in \Omega, \end{cases} \tag{6.1}$$

where $\varphi \in L^1(\partial\Omega)$, $u_0 \in L^2(\Omega)$ and $\mathbf{a}(x,\xi) = \nabla_\xi f(x,\xi)$, f being a function with linear growth in $\|\xi\|$ as $\|\xi\| \to \infty$. One of the classical examples is the nonparametric area integrand for which $f(x,\xi) = \sqrt{1 + \|\xi\|^2}$. Problem (6.1) for this particular f is the time-dependent minimal surface equation, and has been studied in [145] and [90]. Other examples of problems of type (6.1) are the following: The evolution problem for plastic antiplanar shear, studied in [208], which corresponds to the plasticity functional f given by

$$f(\xi) = \begin{cases} \frac{1}{2}\|\xi\|^2 & \text{if} \quad \|\xi\| \leq 1, \\ \|\xi\| - \frac{1}{2} & \text{if} \quad \|\xi\| \geq 1; \end{cases} \tag{6.2}$$

evolution problems associated with Lagrangians

$$f(x,\xi) = \sqrt{1 + a_{ij}(x)\xi_i\xi_j},$$

where the functions a_{ij} are continuous and satisfy $a_{ij}(x) = a_{ji}(x)$, $\|\xi\|^2 \leq a_{ij}(x)\xi_i\xi_j \leq C\|\xi\|^2$ for all $\xi \in \mathbb{R}^N$; and the Lagrangian

$$g(x,\xi) = \sqrt{1 + x^2 + \|\xi\|^2},$$

which was considered by S. Bernstein ([46]). On the other hand, problem (6.1) was studied in [129] for some Lagrangians f, which do not include the nonparametric area integrand, but include instead the plasticity functional and the total variation flow for which $f(\xi) = \|\xi\|$. An application of this type of equations to faceted crystal growth is studied in [140].

The first results about existence and uniqueness of solutions for problems of type (6.1) were given for the time-dependent minimal surface equation by A. Lichnewsky and R. Temam in [145]. The corresponding steady-state problem is the *nonparametric Plateau problem*, which can be formulated in the following way: Find a real function u such that

$$\begin{cases} -\operatorname{div}\left(\dfrac{\nabla u}{\sqrt{1+|\nabla u|^2}}\right) = 0 & \text{in} \quad \Omega, \\ u = \varphi & \text{on} \quad \partial\Omega. \end{cases} \tag{6.3}$$

Jenkins and Serrin [128] showed that a necessary condition for the solvability of problem (6.3) is that the *mean curvature of $\partial\Omega$ with respect to the interior normal be everywhere nonnegative.* This condition turn out to be also sufficient (see [97], [119]). Moreover, from the a priori estimate for the gradient of solutions of the minimal surface equation given by Bombieri, De Giorgi and Miranda [56], follows the next existence result:

"Let Ω be a bounded open set in $\mathbb{R}^N$ with C^2-boundary of nonnegative mean curvature, and let φ be a continuous function on $\partial\Omega$. Then the Dirichlet problem for the minimal surface equation (6.3) *is solvable in $C^2(\Omega) \cap C(\overline{\Omega})$. Moreover, a function $u \in C^2(\Omega) \cap C(\overline{\Omega})$ is a solution of the Dirichlet problem* (6.3) *if and only if it minimizes the area integral among all functions taking boundary values φ on $\partial\Omega$".*

The *minimal surface equation* was studied, using a combination of techniques from geometric measure theory and PDEs, by many authors including Federer, Fleming, De Giorgi, Bombieri, Giusti, Miranda, Finn, Nitsche, Jenkins, Serrin and others. The interested reader may consult the books [124] and [97].

Temam in [191] (see also [109]) defines a "generalized solution" (also called pseudo-solution) of the problem

$$\min\left\{\int_\Omega \sqrt{1+|\nabla u|^2}\, dx \ : \ u \in W^{1,1}(\Omega), \ u = \varphi \ \text{ on } \ \partial\Omega\right\} \tag{6.4}$$

by making use of the relations between problem (6.4) and its dual in the sense of convex analysis. He proved that if $\varphi \in W^{1,1}(\Omega)$, then there exists a unique (modulo an additive constant) generalized solution of (6.4) (also called pseudo-solution) $u \in W^{1,1}(\Omega)$, which is a solution of the minimal surface equation. Moreover, every minimizing sequence $\{u_n\}$ of (6.4) converges to u in the following sense:

$$u_n \to u \quad \text{in } L^1(\Omega), \qquad \nabla u_n \to \nabla u \quad \text{in } (L^1_{loc}(\Omega))^N.$$

In [109] the authors proved that the pseudo-solution u of (6.4) is a solution of the variational problem

$$\min\left\{\int_\Omega \sqrt{1+|\nabla u|^2}\,dx + \int_{\partial\Omega} |\varphi - u|\,d\mathcal{H}^{N-1} \;:\; u \in W^{1,1}(\Omega)\right\}, \tag{6.5}$$

which makes the connection with De Giorgi's approach to the problem. Besides including the boundary datum φ in the functional, the Italian school (De Giorgi, Giusti, Miranda, ...) extended it to the class of functions of bounded variation.

In the same spirit of Lebesgue's definition of the area of graphs of continuous functions, the *relaxed area* of the graph of an L^1-function $u : \Omega \to \mathbb{R}$, where $\Omega \subset \mathbb{R}^N$ is an open bounded set, can be defined by ([182])

$$\mathcal{A}(u,\Omega) := \inf\left\{\liminf_{k\to\infty} \int_\Omega \sqrt{1+|\nabla u_k|^2}\,dx \;:\; u_k \in C^1(\Omega),\ u_k \to u \text{ in } L^1(\Omega)\right\}.$$

Now, it is well known (see for instance [119]) that

$$\mathcal{A}(u,\Omega) < \infty \iff u \in BV(\Omega)$$

and

$$\mathcal{A}(u,\Omega) =$$

$$\sup\left\{\int_\Omega (\theta_{N+1} + u\sum_{i=1}^N D_i\theta_i)\,dx \;:\; \theta = (\theta_1,\dots,\theta_{N+1}) \in C_0^1(\Omega,\mathbb{R}^{N+1}),\ |\theta| \le 1\right\},$$

i.e., for $u \in BV(\Omega)$, $\mathcal{A}(u,\Omega)$ is the total variation of the vector-valued measure $(D_1u,\dots,D_Nu,-\mathcal{L}^N)$, very often denoted by

$$\int_\Omega \sqrt{1+|Du|^2}.$$

In this framework, the *relaxed energy functional* is given by

$$\Phi_\varphi(u) := \sqrt{1+|Du|^2} + \int_{\partial\Omega} |\varphi - u|\,d\mathcal{H}^{N-1} \tag{6.6}$$

and for any $\varphi \in L^1(\partial\Omega)$ the functional Φ_φ attains its minimum in $BV(\Omega)$. Moreover, the following result is well known (see for instance [119]):

"Let Ω be a bounded open set in $\mathbb{R}^N$ with Lipschitz boundary of non-negative mean curvature, and let φ be a continuous function on $\partial\Omega$. Then the Dirichlet problem for the minimal surface equation (6.3) *is solvable in $C^2(\Omega) \cap C(\overline{\Omega})$. Moreover, the solution is unique and is the only minimizer of* (6.6) *in $BV(\Omega)$".*

Coming back to the time-dependent minimal surface equation, the approach given by A. Lichnewsky and R. Temam in [145] is closely related to the above one for the steady-state problem. They proved existence and uniqueness of a kind of solutions, named pseudo-solutions, for the problem

$$\begin{cases} \dfrac{\partial u}{\partial t} = \operatorname{div}\left(\dfrac{Du}{\sqrt{1+|Du|^2}}\right) & \text{in} \quad Q_T = (0,T) \times \Omega, \\ u(t,x) = \varphi(t,x) & \text{on} \quad S_T = (0,T) \times \partial\Omega, \\ u(0,x) = u_0(x) & \text{in} \quad x \in \Omega, \end{cases} \tag{6.7}$$

when the initial datum $u_0 \in L^2(\Omega) \cap H^1_{loc}(\Omega) \cap W^{1,1}(\Omega)$ and $\varphi \in H^1(Q_T)$. For simplicity, we assume that φ is independent of time. Then, the concept of pseudo-solution coincides with the one obtained by considering the abstract Cauchy problem in $L^2(\Omega)$ associated to the relaxed energy functional Φ_φ. Note that since Φ_φ is convex and lower semi-continuous in $L^2(\Omega)$ (with $\Phi_\varphi(u) = +\infty$ if $u \in L^2(\Omega) \setminus BV(\Omega)$), the existence and uniqueness of a solution of the abstract Cauchy problem

$$\begin{cases} u'(t) + \partial\Phi_\varphi(u(t)) \ni 0 & t \in]0,\infty[, \\ u(0) = u_0 & u_0 \in L^2(\Omega) \end{cases} \tag{6.8}$$

follows immediately from the nonlinear semigroup theory (see Appendix A). Now, to get the full strength of the abstract result derived from semigroup theory a characterization of $\partial\Phi_\varphi$ is needed. This was done by F. Demengel and R. Temam in [90] by means of the duality method of convex optimization introduced by R. T. Rockafellar in [170].

In [208], X. Zhou studies the evolution problem associated with the plasticity functional (6.2), more precisely the Dirichlet problem for the equation

$$\frac{\partial u}{\partial t} = \begin{cases} \operatorname{div}\left(\dfrac{\nabla u}{|\nabla u|}\right) & \text{if } |\nabla u| \geq 1, \\ \Delta u & \text{if } |\nabla u| < 1. \end{cases}$$

This problem arises from the study of plastic antiplanar shear deformation, where the scalar function u represents the vertical displacement of the homogeneous planar material in $\Omega \times \mathbb{R}$. In this problem, the portion of Ω where $|\nabla u| < 1$ is referred to as the elastic region while the complement is called the plastic region. The problem is studied in the same framework of Lichnewsky–Temam's paper and, with similar techniques, existence and uniqueness of solutions are proved when the initial datum is in $BV(\Omega)\cap L^\infty(\Omega)$ and the boundary datum is in $L^\infty(\partial\Omega)$. Using similar techniques, these results were generalized by R. Hardt and X. Zhou in [129] for some Lagrangians $f(\xi)$ which do not include the nonparametric area integrand, but include instead the plasticity functional and the total variation flow. Again, the concept of solution is the one obtained by considering the abstract Cauchy problem in $L^2(\Omega)$ associated to the relaxed energy, but the subdifferential of the energy functional is not characterized. We point out there is a viscosity approach to (6.1), given in [120] and [121], when the space dimension is 1.

In general, problem (6.1) does not have a classical solution. In this chapter we introduce a concept of solution of the Dirichlet problem (6.1), for which existence and uniqueness for initial data in $L^2(\Omega)$ is proved. To do that we use the nonlinear semigroup theory and we characterize the subdifferential of the energy associated with the problem. In the next chapter we study the same problem for initial conditions in $L^1(\Omega)$, as we did with the Dirichlet problem for the total variational flow in Chapter 5.

6.2 Preliminaries

In order to consider the relaxed energy we recall the definition of *function of a measure* (see for instance, [26] or [90]). Let $g:\Omega\times\mathbb{R}^N\to\mathbb{R}$ be a Carathéodory function such that

$$|g(x,\xi)|\le M(1+\|\xi\|)\qquad \forall\,(x,\xi)\in\Omega\times\mathbb{R}^N, \tag{6.9}$$

for some constant $M\ge 0$. Furthermore, we assume that g possesses an asymptotic function, i.e., for almost all $x\in\Omega$ there exists the finite limit

$$\lim_{t\to 0^+} tg\left(x,\frac{\xi}{t}\right)=g^0(x,\xi). \tag{6.10}$$

It is clear that the function $g^0(x,\xi)$ is positively homogeneous of degree 1 in ξ, i.e.,

$$g^0(x,s\xi)=sg^0(x,\xi)\qquad \text{for all } x,\xi \text{ and } s>0.$$

We denote by $\mathcal{M}(\Omega,\mathbb{R}^N)$ the set of all $\mathbb{R}^N$-valued bounded Radon measures on Ω. Given $\mu\in\mathcal{M}(\Omega,\mathbb{R}^N)$, we consider its Lebesgue decomposition

$$\mu=\mu^a+\mu^s,$$

where μ^a is the absolutely continuous part of μ with respect to the Lebesgue measure $\mathcal{L}^N$ of $\mathbb{R}^N$, and μ^s is singular with respect to $\mathcal{L}^N$. We denote by $\mu^a(x)$ the density of the measure μ^a with respect to $\mathcal{L}^N$ and by $(d\mu^s/d|\mu|^s)(x)$ the density of μ^s with respect to $|\mu|^s$.

Given $\mu \in \mathcal{M}(\Omega, \mathbb{R}^N)$, we define $\tilde{\mu} \in \mathcal{M}(\Omega, \mathbb{R}^{N+1})$ by

$$\tilde{\mu}(B) := \big(\mu(B), \mathcal{L}^N(B)\big),$$

for every Borel set $B \subset \mathbb{R}^N$. Then, we have

$$\tilde{\mu} = \tilde{\mu}^a + \tilde{\mu}^s = \tilde{\mu}^a(x)\mathcal{L}^N + \tilde{\mu}^s = (\mu^a(x), \chi_\Omega)\mathcal{L}^N + (\mu^s, 0).$$

Hence, we have

$$|\tilde{\mu}^s| = |\mu^s|, \qquad \frac{d\tilde{\mu}^s}{d|\tilde{\mu}^s|} = \left(\frac{d\mu^s}{d|\mu^s|}, 0\right) \quad |\mu^s| - \text{a.e.}$$

For $\mu \in \mathcal{M}(\Omega, \mathbb{R}^N)$ and g satisfying the above conditions, we define the measure $g(x, \mu)$ on Ω as

$$\int_B g(x,\mu) := \int_B g(x, \mu^a(x))\, dx + \int_B g^0\left(x, \frac{d\mu^s}{d|\mu|^s}(x)\right)\, d|\mu|^s \tag{6.11}$$

for all Borel sets $B \subset \Omega$. In formula (6.11) we may write $(d\mu/d|\mu|)(x)$ instead of $(d\mu^s/d|\mu|^s)(x)$, because the two functions coincide $|\mu|^s$-a.e.

Another way of writing the measure $g(x, \mu)$ is the following. Let us consider the function $\tilde{g} : \Omega \times \mathbb{R}^N \times [0, +\infty[\to \mathbb{R}$ defined as

$$\tilde{g}(x,\xi,t) := \begin{cases} g(x, \dfrac{\xi}{t})t & \text{if} \quad t > 0 \\ g^0(x,\xi) & \text{if} \quad t = 0. \end{cases} \tag{6.12}$$

As it is proved in [26], if g is a Carathéodory function satisfying (6.9), then one has

$$\int_B g(x,\mu) = \int_B \tilde{g}\left(x, \frac{d\mu}{d\alpha}(x), \frac{d\mathcal{L}^N}{d\alpha}(x)\right)\, d\alpha, \tag{6.13}$$

where α is any positive Borel measure such that $|\mu| + \mathcal{L}^N \ll \alpha$.

Due to the linear growth condition on the Lagrangian, the natural energy space to study (6.1) is the space of functions of bounded variation. For information concerning functions of bounded variation we refer to Appendix B.

Let g be a function satisfying (6.9). Then for every $u \in BV(\Omega)$ we have the measure $g(x, Du)$ defined by

$$\int_B g(x, Du) = \int_B g(x, \nabla u(x))\, dx + \int_B g^0(x, \overrightarrow{D^s u}(x))\, d|D^s u|$$

for all Borel sets $B \subset \Omega$. If we assume that Ω has Lipschitz boundary, and $g(x, \xi)$ is defined also for $x \in \partial\Omega$, we may consider the functional G in $BV(\Omega)$ defined by

$$G(u) := \int_\Omega g(x, Du) + \int_{\partial\Omega} g^0\big(x, \nu(x)[\varphi(x) - u(x)]\big)\, d\mathcal{H}^{N-1}, \tag{6.14}$$

where $\varphi \in L^1(\partial\Omega)$ is a given function and ν is the outer unit normal to $\partial\Omega$.

In [26], G. Anzellotti proves the lower semi-continuity of the functional G. In order to get this result let us give first some lemmas.

Lemma 6.1. *Assume $\tilde{g}(x,\xi,t)$ is continuous, and let $u, u_n \in BV(\Omega)$ be such that $u_n \to u$ in $L^1(\Omega)$. Then,*

(i) *if*

$$\int_\Omega \sqrt{1+\|Du_n\|^2} \to \int_\Omega \sqrt{1+\|Du\|^2} \quad \text{for } n \to \infty,$$

one also has

$$\int_\Omega g(x,Du_n) \to \int_\Omega g(x,Du) \quad \text{for } n \to \infty.$$

(ii) *If $\tilde{g}(x,\xi,t)$ is convex in (ξ,t) for all fix $x \in \Omega$ and*

$$\int_\Omega \|Du_n\| \leq C \qquad \forall\, n \in \mathbb{N},$$

one also has

$$\liminf_{n\to\infty} \int_\Omega g(x,Du_n) \geq \int_\Omega g(x,Du).$$

Proof. The proof is an immediate consequence of (6.13) and Reshetnyak's continuity and semi-continuity theorem (Theorem 3 and 2 of [169], see also [10]). □

Lemma 6.2. *Assume $\tilde{g}(x,\xi,t)$ is continuous. Then, for all $u \in BV(\Omega)$ and $\varphi \in L^1(\partial\Omega)$ there exists a sequence of functions $u_n \in C^1(\Omega) \cap BV(\Omega)$ such that $u_n|_{\partial\Omega} = \varphi$ and*

$$u_n \to u \quad \text{in } L^1(\Omega), \quad G(u_n) \to G(u).$$

Proof. By Theorem B.3, there exists a sequence of functions $v_n \in C^1(\Omega) \cap BV(\Omega)$ such that $v_n|_{\partial\Omega} = u|_{\partial\Omega}$, $v_n \to u$ in $L^1(\Omega)$ and

$$\int_\Omega \sqrt{1+|\nabla v_n|^2}\,dx \to \int_\Omega \sqrt{1+\|Du\|^2} \quad \text{for } n \to \infty.$$

On the other hand, by Lemma C.1 and having in mind Theorem B.3, we can find functions $w_n \in C^1(\Omega) \cap BV(\Omega)$ such that, for each $n \in \mathbb{N}$, we have

$$w_n|_{\partial\Omega} = \varphi - u|_{\partial\Omega}, \qquad w_n(x) = 0 \text{ if } \operatorname{dist}(x,\partial\Omega) > \frac{1}{n},$$

$$\int_\Omega |\nabla w_n|\,dx \leq \int_{\partial\Omega} |u-\varphi|\,d\mathcal{H}^{N-1} + \frac{1}{n}, \quad \int_\Omega |w_n|\,dx \leq \frac{1}{n}.$$

Set $u_n := v_n + w_n$. Obviously, we have that $u_{n|\partial\Omega} = \varphi$ for all $n \in \mathbb{N}$ and

$$u_n \to u \qquad \text{in } L^1(\Omega),$$

$$\lim_{n\to\infty}\int_\Omega \sqrt{1+|\nabla u_n|^2}\,dx = \int_\Omega \sqrt{1+\|Du\|^2} + \int_{\partial\Omega} |u-\varphi|\,d\mathcal{H}^{N-1}.$$

If we consider the $\mathbb{R}^N$-valued measures μ_n, μ on $\overline{\Omega}$ defined by

$$\mu_n(B) := \int_{B\cap\Omega} \nabla u_n\,dx, \quad \mu(B) := \int_{B\cap\Omega} \nabla u\,dx + \int_{B\cap\partial\Omega} (\varphi - u)\nu\,d\mathcal{H}^{N-1}$$

for all Borel sets $B \subset \overline{\Omega}$, and the $\mathbb{R}^{N+1}$-valued measures

$$\alpha_n(B) := \big(\mu_n(B), \mathcal{L}^N(B)\big), \qquad \alpha(B) := \big(\mu(B), \mathcal{L}^N(B)\big),$$

then we have

$$\alpha_n \rightharpoonup \alpha \quad \text{weakly as measures in } \overline{\Omega}, \qquad \lim_{n\to\infty} |\alpha_n|(\overline{\Omega}) = |\alpha|(\overline{\Omega}).$$

Therefore, since

$$G(u) = \int_{\overline{\Omega}} \tilde{g}(x,\alpha), \qquad G(u_n) = \int_{\overline{\Omega}} \tilde{g}(x,\alpha_n),$$

the proof concludes by using Reshetnyak's continuity theorem (Theorem 3 of [169], see also [10]). □

Lemma 6.3. *Assume that $\tilde{g}(x,\xi,t)$ is lower-semi-continuous on $\overline{\Omega}\times\mathbb{R}^N\times[0,+\infty[$, convex in (ξ,t) for each fixed $x\in\overline{\Omega}$, and (6.9) is satisfied. Then, for any fixed $\varphi\in L^1(\partial\Omega)$ and for any sequence $u_n\in BV(\Omega)$ such that $u_n\to u$ in $L^1(\Omega)$ one has*

$$\liminf_{n\to\infty} G(u_n) \geq G(u).$$

Proof. Let Ω_1 be some ball containing $\overline{\Omega}$ and consider the function $g^*:\Omega_1\times\mathbb{R}^N\to\mathbb{R}$ defined as

$$g^*(x,\xi) := \begin{cases} g(x,\xi) & \text{if } x\in\overline{\Omega}, \\ M(1+\|\xi\|) & \text{if } x\in\Omega_1\setminus\overline{\Omega}. \end{cases}$$

By (6.9) we know that g^* is lower-semi-continuous. Let $\phi\in W^{1,1}(\Omega_1)$ be such that $\phi|_{\partial\Omega}=\varphi$. For each function $u\in BV(\Omega)$, consider the function $u^*\in BV(\Omega_1)$ defined as

$$u^*(x) := \begin{cases} u(x) & \text{if } x\in\Omega \\ \phi(x) & \text{if } x\in\Omega_1\setminus\Omega. \end{cases}$$

Then,

$$\int_{\Omega_1} g^*(x,Du^*) = G(u) + M\int_{\Omega_1\setminus\Omega} (1+|\nabla\phi|)\,dx.$$

Since $u_n\to u$ in $L^1(\Omega)$ implies that $u_n^*\to u^*$ in $L^1(\Omega_1)$, to conclude the proof we only need to apply (ii) of Lemma 6.1 to g^*, having in mind that for this part of the lemma we only need the lower-semi-continuity of $\tilde{g}$. □

From Lemmas 6.2 and 6.3 we obtain the following result.

Theorem 6.4. *Assume that $\tilde{g}(x,\xi,t)$ satisfies the assumption of Lemmas 6.2 and 6.3, then G is the greatest functional on $BV(\Omega)$ which is lower-semi-continuous with respect to the $L^1(\Omega)$-convergence and satisfies $G(u) \le \int_\Omega g(x,\nabla u(x))\,dx$ for all functions $u \in C^1(\Omega) \cap W^{1,1}(\Omega)$ with $u = \varphi$ on $\partial\Omega$. Moreover one has*

$$\inf_{u\in BV(\Omega)} G(u) = \inf\left\{\int_\Omega g(x,\nabla u(x))\,dx \ : \ u \in BV(\Omega)\cap C^1(\Omega), \quad u = \varphi \text{ on } \partial\Omega\right\}.$$

6.3 The Existence and Uniqueness Result

In this section we define the concept of solution for the Dirichlet problem (6.1) and we state the existence and uniqueness result for this type of solutions when the initial data are in $L^2(\Omega)$.

Here we assume that Ω is an open bounded set in $\mathbb{R}^N$, $N \ge 2$, with boundary $\partial\Omega$ of class C^1, and the Lagrangian $f : \overline{\Omega} \times \mathbb{R}^N \to \mathbb{R}$ satisfies the following assumptions, which we shall refer to collectively as (H):

(H_1) f is continuous on $\overline{\Omega} \times \mathbb{R}^N$ and is a convex differentiable function of ξ with continuous gradient for each fixed $x \in \Omega$. Furthermore we require f to satisfy the linear growth condition

$$C_0\|\xi\| - C_1 \le f(x,\xi) \le M(\|\xi\| + C_2) \tag{6.15}$$

for some positive constants C_0, C_1, C_2. Moreover, f^0 exists and $f^0(x,-\xi) = f^0(x,\xi)$ for all $\xi \in \mathbb{R}^N$ and all $x \in \overline{\Omega}$.

(H_2) $\tilde{f}(x,\xi,t)$ is continuous on $\overline{\Omega} \times \mathbb{R}^N \times [0,+\infty[$ and convex in (ξ,t) for each fixed $x \in \overline{\Omega}$.

We consider the function $\mathbf{a}(x,\xi) = \nabla_\xi f(x,\xi)$ associated to the Lagrangian f. By the convexity of f,

$$\mathbf{a}(x,\xi)\cdot(\eta-\xi) \le f(x,\eta) - f(x,\xi), \tag{6.16}$$

and the following monotonicity condition is satisfied

$$(\mathbf{a}(x,\eta) - \mathbf{a}(x,\xi))\cdot(\eta-\xi) \ge 0. \tag{6.17}$$

Moreover, it is easy to see that

$$|\mathbf{a}(x,\xi)| \le M \qquad \forall\ (x,\xi) \in \Omega \times \mathbb{R}^N. \tag{6.18}$$

We consider the function $h : \Omega \times \mathbb{R}^N \to \mathbb{R}$ defined by

$$h(x,\xi) := \mathbf{a}(x,\xi)\cdot\xi.$$

From (6.16) and (6.15), it follows that

$$C_0\|\xi\| - D_1 \leq h(x,\xi) \leq M\|\xi\| \tag{6.19}$$

for some positive constant D_1.

We assume that

(H_3) $h(x,\xi) \geq 0$, h^0 exists and the function $\tilde{h}$ is continuous on $\overline{\Omega} \times \mathbb{R}^N \times [0,+\infty[$.

We need to consider the mapping $\mathbf{a}^\infty$ defined by

$$\mathbf{a}^\infty(x,\xi) := \lim_{t\to+\infty} \mathbf{a}(x,t\xi).$$

Observe that

$$h^0(x,\xi) = \mathbf{a}^\infty(x,\xi)\cdot\xi \quad \text{and} \quad C_0\|\xi\| \leq h^0(x,\xi) \leq M\|\xi\|.$$

(H_4) $\mathbf{a}^\infty(x,\xi) = \nabla_\xi f^0(x,\xi)$ for all $\xi \neq 0$ and all $x \in \overline{\Omega}$.

In particular, as a consequence of Euler's theorem, we have

$$f^0(x,\xi) = \mathbf{a}^\infty(x,\xi)\cdot\xi = h^0(x,\xi),$$

for all $\xi \in \mathbb{R}^N$ and all $x \in \overline{\Omega}$, and, therefore,

$$C_0\|\xi\| \leq f^0(x,\xi) \leq M\|\xi\| \qquad \forall \xi \in \mathbb{R}^N, \ \forall x \in \overline{\Omega}. \tag{6.20}$$

(H_5) $\mathbf{a}(x,\xi)\cdot\eta \leq h^0(x,\eta)$ for all $\xi,\eta \in \mathbb{R}^N$, and all $x \in \overline{\Omega}$.

Either from (H_4) or (H_5) it follows that $\mathbf{a}^\infty(x,\xi)\cdot\eta \leq h^0(x,\eta)$ for all $\xi,\eta \in \mathbb{R}^N$, $\xi \neq 0$, and all $x \in \overline{\Omega}$. Indeed, it suffices to replace ξ by $t\xi$ in (H_5) and let $t \to +\infty$.

Definition 6.5. Let $\varphi \in L^1(\partial\Omega)$ and $u_0 \in L^2(\Omega)$. A measurable function $u : (0,T)\times\Omega \to \mathbb{R}$ is a *solution* of (6.1) in $Q_T = (0,T)\times\Omega$ if $u \in C([0,T],L^2(\Omega))$, $u(0) = u_0$, $u'(t) \in L^2(\Omega)$, $u(t) \in BV(\Omega)\cap L^2(\Omega)$, $\mathbf{a}(x,\nabla u(t)) \in X(\Omega)_1$ a.e. $t \in [0,T]$, and for almost all $t \in [0,T]$ $u(t)$ satisfies:

$$u'(t) = \operatorname{div}(\mathbf{a}(x,\nabla u(t))) \quad \text{in } \mathcal{D}'(\Omega), \tag{6.21}$$

$$\mathbf{a}(x,\nabla u(t))\cdot D^s u(t) = f^0(x,D^s u(t)), \tag{6.22}$$

$$[\mathbf{a}(x,\nabla u(t)),\nu] \in \operatorname{sign}(\varphi - u(t)) f^0(x,\nu(x)) \quad \mathcal{H}^{N-1} - \text{a.e. on } \partial\Omega. \tag{6.23}$$

Our main result is the following:

Theorem 6.6. *Let $\varphi \in L^1(\partial\Omega)$ and assume we are under assumptions (H). Given $u_0 \in L^2(\Omega)$, there exists a unique solution u of (6.1) in Q_T for every $T > 0$ such that $u(0) = u_0$.*

6.4 Strong Solution for Data in $L^2(\Omega)$

To prove Theorem 6.6 we shall use the theory of nonlinear semigroups. Given $\varphi \in L^1(\partial\Omega)$ we define the energy functional associated with the problem (6.1) $\Phi_\varphi : L^2(\Omega) \to [0, +\infty]$ by

$$\Phi_\varphi(u) := \int_\Omega f(x, Du) + \int_{\partial\Omega} f^0(x, \nu(x)[\varphi - u]) \, d\mathcal{H}^{N-1},$$

if $u \in BV(\Omega) \cap L^2(\Omega)$ and

$$\Phi_\varphi(u) := +\infty \quad \text{if} \quad u \in L^2(\Omega) \setminus BV(\Omega).$$

Note that, on the boundary, the integrand can be written in the form

$$f^0(x, \nu(x)[\varphi - u]) = |\varphi - u| f^0(x, \nu(x)).$$

Functional Φ_φ is clearly convex and has the form given in (6.14). Then, as a consequence of Theorem 6.4, we have that Φ_φ is lower-semi-continuous. Therefore, the subdifferential $\partial\Phi_\varphi$ of Φ_φ, i.e., the operator in $L^2(\Omega)$ defined by

$$v \in \partial\Phi_\varphi(u) \iff \Phi_\varphi(w) - \Phi_\varphi(u) \geq \int_\Omega v(w - u) \, dx \quad \forall \, w \in L^2(\Omega),$$

is a maximal monotone operator in $L^2(\Omega)$ (see Appendix A). Hence, existence and uniqueness of a solution of the abstract Cauchy problem

$$\begin{cases} u'(t) + \partial\Phi_\varphi(u(t)) \ni 0, & t \in]0, \infty[\\ u(0) = u_0, & u_0 \in L^2(\Omega) \end{cases} \tag{6.24}$$

follows immediately from nonlinear semigroup theory (see Appendix A.1). Now, to get the full strength of the abstract result derived from semigroup theory we need to characterize $\partial\Phi_\varphi$. To get this characterization, we introduce the following operator $\mathcal{B}_\varphi$ in $L^2(\Omega)$.

$$(u, v) \in \mathcal{B}_\varphi \iff u \in BV(\Omega) \cap L^2(\Omega), v \in L^2(\Omega)$$

$$\text{and } \mathbf{a}(x, \nabla u) \in X(\Omega)_1 \text{ satisfies :}$$

$$-v = \text{div} \, \mathbf{a}(x, \nabla u) \quad \text{in } \mathcal{D}'(\Omega), \tag{6.25}$$

$$\mathbf{a}(x, \nabla u) \cdot D^s u = f^0(x, D^s u) = f^0(x, \overrightarrow{D^s u}) |D^s u|, \tag{6.26}$$

$$[\mathbf{a}(x, \nabla u), \nu] \in \text{sign } (\varphi - u) f^0(x, \nu(x)) \qquad \mathcal{H}^{N-1} - \text{a.e.} \tag{6.27}$$

Let $(u,v) \in \mathcal{B}_\varphi$, and $w \in BV(\Omega) \cap L^2(\Omega)$. Multiplying (6.25) by $w-u$, and using Green's formula (C.10), we obtain

$$\begin{aligned}
\int_\Omega (w-u)v dx &= -\int_\Omega (w-u) \operatorname{div} \mathbf{a}(x,\nabla u)\, dx \\
&= \int_\Omega (\mathbf{a}(x,\nabla u), Dw - Du) - \int_{\partial\Omega} [\mathbf{a}(x,\nabla u),\nu](w-u)\, d\mathcal{H}^{N-1} \\
&= \int_\Omega (\mathbf{a}(x,\nabla u), Dw) - \int_{\partial\Omega} [\mathbf{a}(x,\nabla u),\nu](w-\varphi)\, d\mathcal{H}^{N-1} \\
&\quad - \int_\Omega (\mathbf{a}(x,\nabla u), Du) - \int_{\partial\Omega} [\mathbf{a}(x,\nabla u),\nu](\varphi-u)\, d\mathcal{H}^{N-1} \\
&= \int_\Omega (\mathbf{a}(x,\nabla u), Dw) - \int_{\partial\Omega} [\mathbf{a}(x,\nabla u),\nu](w-\varphi)\, d\mathcal{H}^{N-1} \\
&\quad - \int_\Omega \mathbf{a}(x,\nabla u)\cdot \nabla u\, dx - \int_\Omega \mathbf{a}(x,\nabla u)\cdot D^s u \\
&\quad - \int_{\partial\Omega} |\varphi - u| f^0(x,\nu(x))\, d\mathcal{H}^{N-1} \\
&= \int_\Omega (\mathbf{a}(x,\nabla u), Dw) - \int_{\partial\Omega} [\mathbf{a}(x,\nabla u),\nu](w-\varphi)\, d\mathcal{H}^{N-1} \\
&\quad - \int_\Omega h(x,Du) - \int_{\partial\Omega} |\varphi - u| f^0(x,\nu(x))\, d\mathcal{H}^{N-1}.
\end{aligned}$$

Therefore, if $(u,v) \in \mathcal{B}_\varphi$, we have that

$$\begin{aligned}
\int_\Omega (w-u)v\, dx &= \int_\Omega (\mathbf{a}(x,\nabla u), Dw) - \int_{\partial\Omega} [\mathbf{a}(x,\nabla u),\nu](w-\varphi)\, d\mathcal{H}^{N-1} \\
&\quad - \int_\Omega h(x,Du) - \int_{\partial\Omega} |\varphi - u| f^0(x,\nu(x))\, d\mathcal{H}^{N-1},
\end{aligned} \tag{6.28}$$

for all $w \in BV(\Omega) \cap L^2(\Omega)$.

Theorem 6.7. *Let $\varphi \in L^1(\partial\Omega)$. Assume we are under assumptions (H), then the operator $\partial\Phi_\varphi$ has dense domain in $L^2(\Omega)$ and*

$$\partial\Phi_\varphi = \mathcal{B}_\varphi.$$

We note that, in the particular case of the nonparametric area integrand $f(x,\xi) = \sqrt{1+\|\xi\|^2}$, the characterization of the subdifferential of Φ_φ given in Theorem 6.7 coincides with the one given by F. Demengel and R. Temam in [90], Theorem 3.1, where they use a different approach. More precisely, they characterize the subdifferential by means of the duality method of convex optimization introduced by R. T. Rockafellar in [170]. To prove Theorem 6.7 we need the following proposition.

Proposition 6.8. *Let $\varphi \in L^1(\partial\Omega)$. Assume we are under assumptions (H), then $L^\infty(\Omega) \subset R(I + \mathcal{B}_\varphi)$ and $D(\mathcal{B}_\varphi)$ is dense in $L^2(\Omega)$.*

To prove Proposition 6.8 we need to introduce the following sequence of auxiliary operators. For $\varphi \in W^{\frac{1}{2},2}(\Omega)$, let

$$W^{1,2}_\varphi(\Omega) := \left\{u \in W^{1,2}(\Omega) \ : \ u|_{\partial\Omega} = \varphi \ \ \mathcal{H}^{N-1} - \text{a.e.}\right\}.$$

For every $n \in \mathbb{N}$, consider $\mathbf{a}_n(x,\xi) := \mathbf{a}(x,\xi) + \frac{1}{n}\xi$. We define the operator $A_{n,\varphi}$ in $L^2(\Omega)$:

$$(u,v) \in A_{n,\varphi} \iff u \in W^{1,2}_\varphi(\Omega) \cap L^\infty(\Omega), v \in L^2(\Omega), \text{ and}$$

$$\int_\Omega (w-u)v \, dx \leq \int_\Omega \mathbf{a}_n(x, \nabla u) \cdot \nabla(w-u) \, dx \qquad \forall\, w \in W^{1,2}_\varphi(\Omega).$$

A similar proof to the one given in Proposition 5.9 gives us the following result.

Lemma 6.9. *Let $\varphi \in W^{\frac{1}{2},2}(\partial\Omega) \cap L^\infty(\partial\Omega)$. Then for every $n \in \mathbb{N}$ the operator $A_{n,\varphi}$ satisfies*

$$L^\infty(\Omega) \subset R(I + A_{n,\varphi}).$$

We also need an approximation lemma similar to the one given by Anzellotti in [27]. The proof of this lemma will be given in Section 6.6

Lemma 6.10. *Let Ω be an open bounded set in $\mathbb{R}^N$, $N \geq 2$, and assume that $\partial\Omega$ is of class C^1. If $v, u \in BV(\Omega)$ and $g \in L^1(\partial\Omega)$, then there exists a sequence of functions $v_j \in C^1(\overline{\Omega})$ such that*

$$v_j \to g \quad \text{in} \quad L^1(\partial\Omega), \tag{6.29}$$

$$v_j \to v \quad \text{in} \quad L^{N/(N-1)}(\Omega), \tag{6.30}$$

$$\int_\Omega \sqrt{1 + |\nabla v_j(x)|^2} dx \to \int_\Omega \sqrt{1 + |Dv|^2} + \int_{\partial\Omega} |g - v| d\mathcal{H}^{N-1}, \tag{6.31}$$

$$\nabla v_j(x) \to \nabla v(x) \quad \mathcal{L}^N\text{-a.e. in } \Omega, \tag{6.32}$$

$$|\nabla v_j(x)| \to \infty \text{ and } \frac{\nabla v_j(x)}{|\nabla v_j(x)|} \to \frac{Dv(x)}{|Dv(x)|} \quad |Dv|^s \text{ a.e. in } \Omega, \tag{6.33}$$

$$|\nabla v_j(x)| \to \infty \text{ and } \frac{\nabla v_j(x)}{|\nabla v_j(x)|} \to \frac{Du(x)}{|Du(x)|} \quad |Du|^{ss} \text{ a.e. in } \Omega, \tag{6.34}$$

where $|Du|^{ss}$ denotes the part of the singular measure $|Du|^s$ which is singular with respect to $|Dv|^s$,

$$|\nabla v_j(x)| \to \infty \text{ and } \frac{\nabla v_j(x)}{|\nabla v_j(x)|} \to \frac{g(x)-v(x)}{|g(x)-v(x)|}\nu(x) \tag{6.35}$$

$$\mathcal{H}^{N-1} \text{ a.e. in } \{x \in \partial\Omega : g(x) \neq v(x)\},$$

$$|\nabla v_j(x)| \to \infty \text{ and } \frac{\nabla v_j(x)}{|\nabla v_j(x)|} \to \frac{v(x)-u(x)}{|v(x)-u(x)|}\nu(x) \tag{6.36}$$

$$\mathcal{H}^{N-1} \text{ a.e. in } \{x \in \partial\Omega : g(x) = v(x), u(x) \neq v(x)\}.$$

Next three Lemmas will be used to prove Proposition 6.7 and Theorem 6.8.

Lemma 6.11. *Let $\varphi, \varphi_n \in L^1(\partial\Omega)$, $\varphi_n \to \varphi$ in $L^1(\partial\Omega)$. Let $u_n, u \in BV(\Omega)$ and $z \in X(\Omega)_1$ with $\operatorname{div}(z) \in L^2(\Omega)$. We assume that*

$$\Phi_\varphi(u_n) \to \Phi_\varphi(u), \tag{6.37}$$

$$\mathbf{a}(x, \nabla u_n) \rightharpoonup z \quad \text{weakly}^* \text{ in } \quad L^\infty(\Omega), \tag{6.38}$$

$$|[z, \nu(x)]| \leq f^0(x, \nu(x)) \quad \text{a.e. in } \partial\Omega, \tag{6.39}$$

$$|z \cdot D^s u| \leq f^0(x, D^s u) \quad \text{as measures in} \quad \Omega, \tag{6.40}$$

$$\begin{aligned} &\lim_{n\to\infty} \int_\Omega h(x, Du_n) + \int_{\partial\Omega} |u_n - \varphi_n| f^0(x, \nu(x))\, d\mathcal{H}^{N-1} \\ &= \int_\Omega h(x, Du) + \int_{\partial\Omega} |u - \varphi| f^0(x, \nu(x))\, d\mathcal{H}^{N-1} \end{aligned} \tag{6.41}$$

and

$$\begin{aligned} &\int_\Omega h(x, Du) + \int_{\partial\Omega} |u - \varphi| f^0(x, \nu(x))\, d\mathcal{H}^{N-1} \\ &\leq \int_\Omega (z, Du) + \int_{\partial\Omega} [z, \nu](\varphi - u)\, d\mathcal{H}^{N-1}. \end{aligned} \tag{6.42}$$

Then

$$\int_\Omega z \cdot \nabla u\, dx = \int_\Omega h(x, \nabla u)\, dx = \int_\Omega \mathbf{a}(x, \nabla u) \cdot \nabla u\, dx, \tag{6.43}$$

$$z \cdot D^s u = f^0(x, D^s u), \tag{6.44}$$

$$[z, \nu] \in \operatorname{sign}\,(\varphi - u) f^0(x, \nu(x)) \qquad \mathcal{H}^{N-1} - \text{a.e.} \tag{6.45}$$

Proof. By the convexity of f, we have

$$\begin{aligned}
&\int_\Omega \mathbf{a}(x,\nabla u_n)\cdot\nabla u\,dx \\
&\leq \int_\Omega \mathbf{a}(x,\nabla u_n)\cdot\nabla u_n\,dx + \int_\Omega f(x,\nabla u)\,dx - \int_\Omega f(x,\nabla u_n)\,dx \\
&\leq \int_\Omega \mathbf{a}(x,\nabla u_n)\cdot\nabla u_n\,dx + \int_\Omega f^0(x,D^s u_n) + \int_{\partial\Omega} |u_n-\varphi_n| f^0(x,\nu(x))\,d\mathcal{H}^{N-1} \\
&\quad + \int_{\partial\Omega} |\varphi_n-\varphi| f^0(x,\nu(x))\,d\mathcal{H}^{N-1} + \int_\Omega f(x,\nabla u)\,dx \\
&\quad - \left(\int_\Omega f(x,\nabla u_n)\,dx + \int_\Omega f^0(x,D^s u_n) + \int_{\partial\Omega} |u_n-\varphi| f^0(x,\nu(x))d\mathcal{H}^{N-1} \right) \\
&= \int_\Omega h(x,Du_n)\,dx + \int_{\partial\Omega} |u_n-\varphi_n| f^0(x,\nu(x))\,d\mathcal{H}^{N-1} \\
&\quad + \int_{\partial\Omega} |\varphi_n-\varphi| f^0(x,\nu(x))\,d\mathcal{H}^{N-1} + \int_\Omega f(x,\nabla u)\,dx - \Phi_\varphi(u_n).
\end{aligned}$$

Letting $n\to\infty$, and using (6.37), (6.38) and (6.41), we obtain

$$\begin{aligned}
\int_\Omega z\cdot\nabla u\,dx \leq &\int_\Omega h(x,Du) + \int_{\partial\Omega} |u-\varphi| f^0(x,\nu(x))\,d\mathcal{H}^{N-1} \\
&+ \int_\Omega f(x,\nabla u)\,dx - \Phi_\varphi(u) = \int_\Omega \mathbf{a}(x,\nabla u)\cdot\nabla u\,dx.
\end{aligned}$$

Now, using (6.39) and (6.40), we have

$$|[z,\nu](\varphi-u)| \leq |u-\varphi| f^0(x,\nu(x))$$

and

$$|z\cdot D^s u| \leq f^0(x,D^s u).$$

Hence from (6.42), we obtain (6.43), (6.44) and (6.45). □

Lemma 6.12. (i) *Let $u_n \in BV(\Omega)\cap L^2(\Omega)$ and $z\in X(\Omega)_1$. Suppose that*

$$\mathbf{a}(x,\nabla u_n) \rightharpoonup z \qquad \text{weakly}^* \text{ in } L^\infty(\Omega,\mathbb{R}^N) \tag{6.46}$$

and

$$\operatorname{div}(\mathbf{a}(x,\nabla u_n)) \rightharpoonup \operatorname{div}(z) \qquad \text{weakly in } L^2(\Omega). \tag{6.47}$$

Then

$$[\mathbf{a}(x,\nabla u_n),\nu(x)] \rightharpoonup [z,\nu(x)] \quad \text{weakly in } L^2(\partial\Omega) \text{ and} \tag{6.48}$$

$$|z(x)\cdot\nu(x)| \leq f^0(x,\nu(x)) \quad \text{a.e. in } \partial\Omega. \tag{6.49}$$

(ii) *Let* $u_n \in W^{1,2}(\Omega)$. *Let* $\mathbf{a}_n(x,\xi) = \mathbf{a}(x,\xi) + \frac{1}{n}\xi$. *Suppose that*

$$\|u_n\|_2 \quad \text{is bounded in} \quad L^2(\Omega), \tag{6.50}$$

$$\frac{1}{n}|\nabla u_n| \to 0 \quad \text{in} \quad L^2(\Omega), \tag{6.51}$$

$$\mathbf{a}_n(x,\nabla u_n) \rightharpoonup z \qquad \text{weakly in} \quad L^2(\Omega,\mathbb{R}^N) \tag{6.52}$$

and

$$\operatorname{div}(\mathbf{a}_n(x,\nabla u_n)) \rightharpoonup \operatorname{div}(z) \qquad \text{weakly in} \quad L^2(\Omega). \tag{6.53}$$

Then

$$[\mathbf{a}_n(x,\nabla u_n),\nu(x)] \rightharpoonup [z,\nu(x)] \quad \text{weakly in } W^{1/2,2}(\partial\Omega)^* \quad \text{and} \tag{6.54}$$

$$|[z(x),\nu(x)]| \le f^0(x,\nu(x)) \quad \text{a.e. in } \partial\Omega. \tag{6.55}$$

Proof. Since both proofs are based on similar arguments, we shall only prove (ii). Observe that, if $\sigma \in L^2(\Omega,\mathbb{R}^N)$ and $\operatorname{div}(\sigma) \in L^2(\Omega)$, we can define $[\sigma,\nu]$ using the integration by parts formula

$$\int_{\partial\Omega} [\sigma,\nu]\psi \, d\mathcal{H}^{N-1} = \int_\Omega \operatorname{div}(\sigma)\psi \, dx + \int_\Omega \sigma\cdot\nabla\psi \, dx \tag{6.56}$$

for all $\psi \in W^{1,2}(\Omega)$. This is consistent with the classical notion of trace at the boundary and it defines $[\sigma,\nu]$ as an element of $W^{1/2,2}(\partial\Omega)^*$. According to the assumptions (6.52), (6.53) we have that $[\mathbf{a}_n(x,\nabla u_n),\nu(x)] \to [z,\nu(x)]$ weakly in $W^{1/2,2}(\partial\Omega)^*$. In i), the analogous conclusion (6.48) follows from the results in [25] and the fact that $\mathbf{a}(x,\nabla u_n)$ is uniformly bounded in $L^\infty(\Omega)$. In this case, the traces $[\mathbf{a}(x,\nabla u_n),\nu(x)]$ are in $L^\infty(\partial\Omega)$.

To prove (6.55), again, we observe that (see [141]) if $\sigma \in L^2(\Omega,\mathbb{R}^N)$ and $\operatorname{div}(\sigma) \in L^2(\Omega)$, then there is a sequence $\sigma_k \in C^\infty(\overline{\Omega},\mathbb{R}^N)$ satisfying

$$\sigma_k \to \sigma \quad \text{in } L^2(\Omega,\mathbb{R}^N), \tag{6.57}$$

$$\operatorname{div}(\sigma_k) \to \operatorname{div}(\sigma) \quad \text{in } L^2(\Omega). \tag{6.58}$$

We recall the construction in [141]. We use a partition of unity θ_j, $j = 1,2,\ldots,p$, in $\overline{\Omega}$ with $0 \le \theta_j \le 1$, $\theta_j \in C_0^\infty(\mathbb{R}^N)$, such that if the support of θ_j intersects $\partial\Omega$, then for some bounded open cone K_j with vertex 0, every $x \in \partial\Omega \cap \operatorname{supp}(\theta_j)$ satisfies $(x+K_j)\cap\overline{\Omega} = \emptyset$, and for some $r > 0$, every $x \in \partial\Omega\cap(\operatorname{supp}(\theta_j)+B(0,r))$ satisfies $(x-K_j) \subset \Omega$. For each j, we choose $\rho_j \in C_0^\infty(\mathbb{R}^N)$, $0 \le \rho_j \le 1$, with $\int_{\mathbb{R}^N} \rho_j dx = 1$, and let $\rho_{j,k}(x) = k^N\rho_j(kx)$. If j is such that the support of θ_j intersects $\partial\Omega$, we choose ρ_j such that $\operatorname{supp}(\rho_j) \subseteq K_j$. Then we define

$$\sigma_k = \sum_{j=1}^{p} \rho_{j,k} * (\theta_j\sigma\chi_\Omega).$$

As it was proved in [141], σ_k satisfies (6.57) and (6.58). As in the first part of the proof, we have that

$$\int_{\partial\Omega} [\sigma_k, \nu]\psi \, d\mathcal{H}^{N-1} \to \int_{\partial\Omega} [\sigma, \nu]\psi \, d\mathcal{H}^{N-1}$$

for all $\psi \in W^{1,2}(\Omega)$. We shall use this observation for $\sigma = \mathbf{a}_n(x, \nabla u_n)$. Previously, we extended u_n as a function in $W^{1,2}(\mathbb{R}^N)$ such that $\|u_n\|_{W^{1,2}(\mathbb{R}^N)} \leq C\|u_n\|_{W^{1,2}(\Omega)}$, for some constant $C > 0$ depending only on Ω ([2]). Then we defined

$$\mathbf{a}_{n,k}(x, \nabla u_n) = \sum_{j=1}^{p} \rho_{j,k} * \left(\theta_j \mathbf{a}(x, \nabla u_n)\chi_\Omega + \theta_j \frac{\nabla u_n}{n}\right).$$

Now, since $\mathbf{a}_{n,k}(x, \nabla u_n) \in C^\infty(\overline{\Omega})$, $[\mathbf{a}_{n,k}(x, \nabla u_n), \nu(x)]$ can be understood in a classical sense. For a given function $\psi \in W^{1/2,2}(\partial\Omega)$, we may write

$$\begin{aligned}
&\int_{\partial\Omega} [\mathbf{a}_{n,k}(x, \nabla u_n), \nu(x)]\psi \, d\mathcal{H}^{N-1} \\
&= \sum_{j=1}^{p} \int_{\partial\Omega} [\rho_{j,k} * (\theta_j \mathbf{a}(x, \nabla u_n)\chi_\Omega), \nu(x)]\psi(x) \, d\mathcal{H}^{N-1} \\
&\quad + \sum_{j=1}^{p} \frac{1}{n} \int_{\partial\Omega} [\rho_{j,k} * (\theta_j \nabla u_n), \nu(x)]\psi(x) \, d\mathcal{H}^{N-1}.
\end{aligned}$$

By taking k sufficiently large, we may assume that all θ_j used in the above expression are such that $\operatorname{supp}(\theta_j)$ intersects $\partial\Omega$. We observe that

$$\begin{aligned}
&\int_{\partial\Omega} |[\rho_{j,k} * (\theta_j \mathbf{a}(x, \nabla u_n)\chi_\Omega), \nu(x)]||\psi(x)| \, d\mathcal{H}^{N-1} \\
&\leq \int_{\partial\Omega} \int_\Omega \rho_{j,k}(x - y)\theta_j(y)|\mathbf{a}(y, \nabla u_n(y)) \cdot \nu(x)||\psi(x)| \, dy d\mathcal{H}^{N-1}(x) \\
&\leq \int_{\partial\Omega} \int_\Omega \rho_{j,k}(x - y)\theta_j(y) f^0(y, \nu(x))|\psi(x)| \, dy d\mathcal{H}^{N-1}(x).
\end{aligned}$$

Since

$$\nabla u_n \theta_j = \nabla(u_n \theta_j) - u_n \nabla \theta_j,$$

we may write

$$\begin{aligned}
&\int_{\partial\Omega} [\rho_{j,k} * (\theta_j \nabla u_n), \nu(x)]\psi(x) \, d\mathcal{H}^{N-1} \\
&= \int_{\partial\Omega} [\nabla \rho_{j,k} * (\theta_j u_n), \nu(x)]\psi(x) \, d\mathcal{H}^{N-1} - \int_{\partial\Omega} [\rho_{j,k} * (\nabla \theta_j u_n), \nu(x)]\psi(x) \, d\mathcal{H}^{N-1}.
\end{aligned}$$

We estimate both integrals in the right-hand side of the above expression. First,

$$\left|\int_{\partial\Omega}[\nabla\rho_{j,k}*(\theta_j u_n),\nu(x)]\psi(x)\right| \leq \left\|\frac{\partial}{\partial\nu}(\rho_{j,k}*(u_n\theta_j))\right\|_{W^{1/2,2}(\partial\Omega)^*}\|\psi\|_{W^{1/2,2}(\partial\Omega)}$$
$$\leq C\|\rho_{j,k}*(u_n\theta_j)\|_{W^{1,2}(\Omega)}\|\psi\|_{W^{1/2,2}(\partial\Omega)} \leq C\|\rho_{j,k}*(u_n\theta_j)\|_{W^{1,2}(\mathbb{R}^N)}\|\psi\|_{W^{1/2,2}(\partial\Omega)}$$
$$\leq C\|u_n\theta_j\|_{W^{1,2}(\mathbb{R}^N)}\|\psi\|_{W^{1/2,2}(\partial\Omega)} \leq C\|u_n\|_{W^{1,2}(\mathbb{R}^N)}\|\psi\|_{W^{1/2,2}(\partial\Omega)}$$
$$\leq C\|u_n\|_{W^{1,2}(\Omega)}\|\psi\|_{W^{1,2}(\Omega)}$$

for some constant $C > 0$ (which may change from line to line). A similar analysis proves that

$$\left|\int_{\partial\Omega}[\rho_{j,k}*(\nabla\theta_j u_n),\nu(x)]\psi(x)\right| \leq C\|u_n\|_{W^{1,2}(\Omega)}\|\psi\|_{W^{1,2}(\Omega)}$$

for some constant $C > 0$. Taking all the above into account , we obtain

$$\left|\int_{\partial\Omega}[\mathbf{a}_{n,k}(x,\nabla u_n),\nu(x)]\psi\, d\mathcal{H}^{N-1}\right|$$
$$\leq \sum_{j=1}^{p}\int_{\partial\Omega}\int_{\Omega}\rho_{j,k}(x-y)\theta_j(y)f^0(y,\nu(x))|\psi(x)|\,dy\,d\mathcal{H}^{N-1}(x)$$
$$+\frac{C}{n}\|u_n\|_{W^{1,2}(\Omega)}\|\psi\|_{W^{1,2}(\Omega)}.$$

Letting $k \to \infty$, and taking into account the fact that θ_j is a partition of unity in $\overline{\Omega}$ and our assumptions on θ_j and K_j, we obtain

$$\left|\int_{\partial\Omega}[\mathbf{a}_n(x,\nabla u_n),\nu(x)]\psi\, d\mathcal{H}^{N-1}\right|$$
$$\leq \int_{\partial\Omega}f^0(x,\nu(x))|\psi(x)|\,d\mathcal{H}^{N-1} + \frac{C}{n}\|u_n\|_{W^{1,2}(\Omega)}\|\psi\|_{W^{1,2}(\Omega)}.$$

Now, letting $n \to \infty$, and using (6.50), (6.51), we obtain

$$\left|\int_{\partial\Omega}[z,\nu(x)]\psi\, d\mathcal{H}^{N-1}\right| \leq \int_{\partial\Omega}f^0(x,\nu(x))|\psi(x)|\,d\mathcal{H}^{N-1} \tag{6.59}$$

for all $\psi \in W^{1,2}(\Omega)$. Now, since $z \in L^\infty(\Omega)$ and $\operatorname{div}(z) \in L^2(\Omega)$, $[z,\nu]$ coincides with the trace given in Section C.1, and, therefore, $[z,\nu] \in L^\infty(\partial\Omega)$. Hence, from (6.59), we conclude that $|[z(x),\nu(x)]| \leq f^0(x,\nu(x))$. □

Lemma 6.13. *Suppose that any of the assumptions of Lemma 6.12 hold. Moreover we assume that*

$$\mathbf{a}(x,\nabla u_n)\cdot D^s u_n = f^0(x,D^s u_n), \tag{6.60}$$

$$u_n \to u \text{ in } L^2(\Omega) \text{ and } \|u_n\|_{BV} \text{ is bounded.} \tag{6.61}$$

Then

$$z(x) = \mathbf{a}(x, \nabla u(x)) \qquad \text{a.e. } x \in \Omega. \tag{6.62}$$

Proof. Again, since both proofs are based on similar arguments, we shall only prove (6.62) under the assumptions given in (i) of Lemma 6.12. Let $0 \leq \phi \in C_0^1(\Omega)$ and $g \in C^1(\overline{\Omega})$. We observe that

$$\begin{aligned}
&\int_\Omega \phi[(\mathbf{a}(x, \nabla u_n), D(u_n - g)) - \mathbf{a}(x, \nabla g) D(u_n - g)] \\
&= \int_\Omega \phi[\mathbf{a}(x, \nabla u_n) - \mathbf{a}(x, \nabla g)) \cdot \nabla(u_n - g)]\, dx \\
&\quad + \int_\Omega \phi[\mathbf{a}(x, \nabla u_n) - \mathbf{a}(x, \nabla g)] \cdot D^s(u_n - g)).
\end{aligned}$$

Since by (H_5), (6.17) and (6.60) both terms at the right-hand side of the above expression are positive, we have

$$\int_\Omega \phi[(\mathbf{a}(x, \nabla u_n), D(u_n - g)) - \mathbf{a}(x, \nabla g) D(u_n - g)] \geq 0.$$

Since

$$\begin{aligned}
&\int_\Omega \phi(\mathbf{a}(x, \nabla u_n), D(u_n - g)) \\
&= - \int_\Omega \operatorname{div}(\mathbf{a}(x, \nabla u_n)) \phi(u_n - g)\, dx - \int_\Omega (u_n - g) \mathbf{a}(x, \nabla u_n) \cdot \nabla \phi\, dx,
\end{aligned}$$

we get

$$\begin{aligned}
&\lim_{n \to \infty} \int_\Omega \phi(\mathbf{a}(x, \nabla u_n), D(u_n - g)) \\
&= - \int_\Omega \operatorname{div}(z) \phi(u - g)\, dx - \int_\Omega (u - g) z \cdot \nabla \phi\, dx = \int_\Omega \phi(z, D(u - g)).
\end{aligned}$$

On the other hand,

$$\lim_{n \to \infty} \int_\Omega \phi\, \mathbf{a}(x, \nabla g) D(u_n - g) = \int_\Omega \phi\, \mathbf{a}(x, \nabla g) D(u - g).$$

Hence, we obtain

$$\int_\Omega \phi[(z, D(u - g)) - \mathbf{a}(x, \nabla g) D(u - g)] \geq 0, \qquad \forall\, 0 \leq \phi \in C_0^1(\Omega).$$

Thus, the measure $(z, D(u-g)) - \mathbf{a}(x, \nabla g)D(u-g) \geq 0$, and, therefore, its absolutely continuous part

$$(z - \mathbf{a}(x, \nabla g)) \cdot \nabla(u-g) \geq 0 \quad \text{a.e. in } \Omega.$$

Since we may take a countable set dense in $C^1(\overline{\Omega})$ we have that the above inequality holds for all $x \in \tilde{\Omega}$, where $\tilde{\Omega} \subset \Omega$ is such that $\mathcal{L}^N(\Omega \setminus \tilde{\Omega}) = 0$, and all $g \in C^1(\overline{\Omega})$. Now, fixed $x \in \tilde{\Omega}$ and given $\xi \in \mathbb{R}^N$, there is $g \in C^1(\overline{\Omega})$ such that $\nabla g(x) = \xi$. Then

$$(z(x) - \mathbf{a}(x, \xi)) \cdot (\nabla u(x) - \xi) \geq 0, \quad \forall\, \xi \in R^N.$$

These inequalities imply (6.62) by an application of Minty–Browder's method in $\mathbb{R}^N$. □

Proof of Proposition 6.8. We divide the proof in three steps.
Step 1. Suppose first that $\varphi \in C^1(\overline{\Omega})$. Let $v \in L^\infty(\Omega)$. We shall find $u \in BV(\Omega) \cap L^2(\Omega)$ such that $(u, v-u) \in \mathcal{B}_\varphi$. That is, there is $\mathbf{a}(x, \nabla u) \in X(\Omega)_1$ satisfying

$$(v-u) = -\text{div } \mathbf{a}(x, \nabla u) \qquad \text{in } \mathcal{D}'(\Omega), \tag{6.63}$$

$$\mathbf{a}(x, \nabla u) \cdot D^s u = f^0(x, D^s u) \quad \text{and} \tag{6.64}$$

$$[\mathbf{a}(x, \nabla u), \nu] \in \text{sign } (\varphi - u) f^0(x, \nu(x)) \qquad \mathcal{H}^{N-1} - \text{a.e.} \tag{6.65}$$

By Lemma 6.9, we know that for any $n \in \mathbb{N}$ there exists $u_n \in W^{1,2}_\varphi(\Omega) \cap L^\infty(\Omega)$ such that $(u_n, v-u_n) \in A_{n,\varphi}$. Hence

$$\int_\Omega (w-u_n)(v-u_n)\, dx \leq \int_\Omega \mathbf{a}_n(x, \nabla u_n) \cdot \nabla(w-u_n)\, dx \tag{6.66}$$

for all $w \in W^{1,2}_\varphi(\Omega)$. Let $M_1 := \sup\{\|\varphi\|_\infty, \|v\|_\infty\}$. Then, taking $w = u_n - (u_n - M_1)^+$ as test function in (6.66), we obtain

$$\int_\Omega (u_n - M_1)^+ (u_n - v)\, dx \leq 0.$$

Hence,

$$\begin{aligned} \int_{\{u_n > M_1\}} (u_n - M_1)^2\, dx &\leq \int_{\{u_n > M_1\}} (u_n - M_1)(u_n - v)\, dx \\ &= \int_\Omega (u_n - M_1)^+ (u_n - v)\, dx \leq 0, \end{aligned}$$

and, thus, $u_n \leq M_1$ a.e. in Ω. In a similar way, taking $w = u_n + (u_n + M_1)^-$ as test function, we get $-M_1 \leq u_n$ a.e. in Ω. Therefore,

$$\|u_n\|_\infty \leq M_1 \qquad \text{for all } n \in \mathbb{N}. \tag{6.67}$$

Taking $w = w_0 \in W^{1,2}_{\varphi}(\Omega) \cap L^\infty(\Omega)$ in (6.66), applying Young's inequality, and using (6.67) we get

$$\begin{aligned}
&\int_\Omega \mathbf{a}(x, \nabla u_n) \cdot \nabla u_n \, dx + \frac{1}{n} \int_\Omega |\nabla u_n|^2 \, dx \\
&\leq \int_\Omega \mathbf{a}(x, \nabla u_n) \cdot \nabla w_0 \, dx + \frac{1}{n} \int_\Omega \nabla u_n \cdot \nabla w_0 \, dx + \int_\Omega (w_0 - u_n)(u_n - v) \, dx \\
&\leq M_2 \left(\int_\Omega |\nabla w_0|^2 \, dx \right)^{\frac{1}{2}} + \frac{1}{2n} \int_\Omega |\nabla u_n|^2 \, dx + \frac{1}{2n} \int_\Omega |\nabla w_0|^2 \, dx + M_3 \\
&\leq M_4 + \frac{1}{2n} \int_\Omega |\nabla u_n|^2 \, dx.
\end{aligned}$$

Hence, by (6.19), we obtain

$$\int_\Omega |\nabla u_n| \, dx \leq M_5 \quad \forall\, n \in \mathbb{N} \tag{6.68}$$

and

$$\frac{1}{n} \int_\Omega |\nabla u_n|^2 \, dx \leq M_6 \quad \forall n \in \mathbb{N}. \tag{6.69}$$

Thus, $\{u_n : n \in \mathbb{N}\}$ is bounded in $W^{1,1}(\Omega)$ and, by extracting a subsequence if necessary, we may assume that u_n converges in $L^1(\Omega)$ and almost everywhere to some function $u \in L^1(\Omega)$ as $n \to +\infty$. Now, by (6.67) and (6.68), we have that $u_n \to u$ in $L^2(\Omega)$ and $u \in BV(\Omega) \cap L^\infty(\Omega)$.

Observe that by (6.18) and (6.69), $\{\mathbf{a}_n(x, \nabla u_n) : n \in \mathbb{N}\}$ is bounded in $L^2(\Omega, \mathbb{R}^N)$. Hence, we may assume that

$$\mathbf{a}_n(x, \nabla u_n) \rightharpoonup z \quad \text{as } n \to \infty, \text{ weakly in } L^2(\Omega, \mathbb{R}^N). \tag{6.70}$$

Given $\psi \in C_0^\infty(\Omega)$, taking $w = u_n \pm \psi$ in (6.66) we obtain

$$\int_\Omega \psi(v - u_n) \, dx = \int_\Omega \mathbf{a}_n(x, \nabla u_n) \cdot \nabla \psi \, dx.$$

Letting $n \to +\infty$, we obtain

$$\int_\Omega (v - u)\psi \, dx = \int_\Omega z \cdot \nabla \psi \, dx,$$

that is,

$$v - u = -\operatorname{div}(z), \qquad \text{in } \mathcal{D}'(\Omega) \tag{6.71}$$

and

$$\operatorname{div} \mathbf{a}_n(x, \nabla u_n) \rightharpoonup \operatorname{div}(z) \quad \text{weakly in } L^2(\Omega). \tag{6.72}$$

Since, by (6.69),

$$\frac{1}{n}|\nabla u_n| \to 0 \quad \text{in } L^2(\Omega), \tag{6.73}$$

as a consequence of (6.70), it follows that

$$\mathbf{a}(x, \nabla u_n) \rightharpoonup z \quad \text{as } n \to \infty, \text{ weakly in } L^2(\Omega, \mathbb{R}^N). \tag{6.74}$$

Moreover, by (6.18) we may assume that

$$\mathbf{a}(x, \nabla u_n) \rightharpoonup z \quad \text{as } n \to \infty, \text{ weakly* in } L^\infty(\Omega, \mathbb{R}^N). \tag{6.75}$$

Let us prove that

$$\lim_{n\to\infty} \int_\Omega \mathbf{a}(x, \nabla u_n) \cdot \nabla u_n \, dx = \int_\Omega (z, Du) - \int_{\partial\Omega} [z, \nu](u - \varphi) \, d\mathcal{H}^{N-1}. \tag{6.76}$$

By (6.66), we have

$$\begin{aligned} &\int_\Omega (w - u_n)(v - u_n) \, dx + \int_\Omega \mathbf{a}(x, \nabla u_n) \cdot \nabla u_n \, dx \\ &\leq \int_\Omega \mathbf{a}(x, \nabla u_n) \cdot \nabla w \, dx + \frac{1}{n} \int_\Omega \nabla u_n \cdot \nabla w \, dx \end{aligned} \tag{6.77}$$

for all $w \in W^{1,2}_\varphi(\Omega)$. By Lemma 6.10, there exists $v_j \in C^1(\overline{\Omega})$ such that $v_j|_{\partial\Omega} = \varphi$, $v_j \to u$ in $L^1(\Omega)$. If we set $w = v_j$ in (6.77), taking the upper limit when $n \to \infty$, we get

$$\int_\Omega (v_j - u)(v - u) \, dx + \limsup_{n\to\infty} \int_\Omega \mathbf{a}(x, \nabla u_n) \cdot \nabla u_n \, dx \leq \int_\Omega z \cdot \nabla v_j \, dx. \tag{6.78}$$

Now, by Green's formula (C.10) we have

$$\begin{aligned} \int_\Omega z \cdot \nabla v_j \, dx &= - \int_\Omega \operatorname{div}(z) v_j \, dx + \int_{\partial\Omega} [z, \nu]\varphi \, d\mathcal{H}^{N-1} \\ &= \int_\Omega (v - u) v_j \, dx + \int_{\partial\Omega} [z, \nu]\varphi \, d\mathcal{H}^{N-1}. \end{aligned}$$

Hence, taking limit as $j \to \infty$ and applying again Green's formula we obtain that

$$\lim_{j\to\infty} \int_\Omega z \cdot \nabla v_j \, dx = \int_\Omega (z, Du) - \int_{\partial\Omega} [z, \nu](u - \varphi) \, d\mathcal{H}^{N-1}. \tag{6.79}$$

Letting $j \to \infty$ in (6.78) , we have

$$\limsup_{n\to\infty} \int_\Omega \mathbf{a}(x, \nabla u_n) \cdot \nabla u_n \, dx \leq \int_\Omega (z, Du) - \int_{\partial\Omega} [z, \nu](u - \varphi) \, d\mathcal{H}^{N-1}. \tag{6.80}$$

On the other hand,

$$\int_\Omega \mathbf{a}(x,\nabla u_n)\cdot\nabla u_n\,dx = \int_\Omega (\mathbf{a}(x,\nabla u_n)-\mathbf{a}(x,\nabla v_j))\cdot\nabla(u_n-v_j)\,dx$$
$$+\int_\Omega (\mathbf{a}(x,\nabla u_n)-\mathbf{a}(x,\nabla v_j))\cdot\nabla v_j\,dx + \int_\Omega \mathbf{a}(x,\nabla v_j)\cdot\nabla u_n\,dx$$
$$\geq \int_\Omega (\mathbf{a}(x,\nabla u_n)-\mathbf{a}(x,\nabla v_j))\cdot\nabla v_j\,dx + \int_\Omega \mathbf{a}(x,\nabla v_j)\cdot\nabla u_n\,dx.$$

Hence

$$\liminf_{n\to\infty}\int_\Omega \mathbf{a}(x,\nabla u_n)\cdot\nabla u_n\,dx$$
$$\geq \lim_{n\to\infty}\left(\int_\Omega \mathbf{a}(x,\nabla u_n)\cdot\nabla v_j\,dx - \int_\Omega \mathbf{a}(x,\nabla v_j)\cdot\nabla v_j\,dx + \int_\Omega \mathbf{a}(x,\nabla v_j)\cdot\nabla u_n\,dx\right).$$

If we consider the $\mathbb{R}^N$-valued measures μ_n, μ on $\overline{\Omega}$ which are defined as

$$\mu_n(B) := \int_{B\cap\Omega}\nabla u_n\,dx,$$

$$\mu(B) := \int_{B\cap\Omega} Du + \int_{B\cap\partial\Omega}(\varphi-u)\nu\,d\mathcal{H}^{N-1}$$

for all Borel sets $B\subset\overline{\Omega}$, we have

$$\mu_n \rightharpoonup \mu \qquad \text{weakly as measures in } \overline{\Omega}.$$

Then, since $\mathbf{a}(x,\nabla v_j(x))\in C(\overline{\Omega},\mathbb{R}^N)$, we have

$$\lim_{n\to\infty}\int_\Omega \mathbf{a}(x,\nabla v_j)\cdot\nabla u_n\,dx = \int_\Omega \mathbf{a}(x,\nabla v_j)\,dDu + \int_{\partial\Omega}\mathbf{a}(x,\nabla v_j)\cdot\nu(\varphi-u)\,d\mathcal{H}^{N-1}.$$

Therefore, we have

$$\liminf_{n\to\infty}\int_\Omega \mathbf{a}(x,\nabla u_n)\cdot\nabla u_n\,dx \geq \int_\Omega z\cdot\nabla v_j\,dx - \int_\Omega \mathbf{a}(x,\nabla v_j)\cdot\nabla v_j\,dx$$
$$+\int_\Omega \mathbf{a}(x,\nabla v_j)\,dDu + \int_{\partial\Omega}\mathbf{a}(x,\nabla v_j)\cdot\nu(\varphi-u)\,d\mathcal{H}^{N-1}.$$

Now, by Theorem 7.4 of [27], we have

$$\lim_{j\to\infty}\int_\Omega \mathbf{a}(x,\nabla v_j)\cdot\nabla v_j\,dx = \int_\Omega \mathbf{a}(x,\nabla u)\cdot\nabla u\,dx$$
$$+\int_\Omega \mathbf{a}^\infty(x,\overrightarrow{D^s u})\cdot D^s u + \int_{\partial\Omega}\mathbf{a}^\infty(x,(\varphi-u)\nu)\cdot\nu(\varphi-u)\,d\mathcal{H}^{N-1}.$$

On the other hand, as a consequence of Lemma 6.10, we have

$$\begin{aligned}\lim_{j\to\infty}\int_\Omega \mathbf{a}(x,\nabla v_j)\,dDu &= \lim_{j\to\infty}\left(\int_\Omega \mathbf{a}(x,\nabla v_j)\cdot\nabla u\,dx + \int_\Omega \mathbf{a}(x,\nabla v_j)dD^s u\right)\\ &= \int_\Omega \mathbf{a}(x,\nabla u)\cdot\nabla u\,dx + \int_\Omega \mathbf{a}^\infty(x,\overrightarrow{D^s u})\cdot D^s u\end{aligned}$$

and

$$\begin{aligned}\lim_{j\to\infty}\int_{\partial\Omega}\mathbf{a}(x,\nabla v_j)\cdot\nu(\varphi-u)\,d\mathcal{H}^{N-1} &= \int_{\partial\Omega}\mathbf{a}^\infty\left(x,\frac{\varphi-u}{|\varphi-u|}\nu\right)\cdot\nu(\varphi-u)\,d\mathcal{H}^{N-1}\\ &= \int_{\partial\Omega}\mathbf{a}^\infty(x,(\varphi-u)\nu)\cdot\nu(\varphi-u)\,d\mathcal{H}^{N-1}.\end{aligned}$$

Collecting all these facts, we obtain

$$\begin{aligned}\liminf_{n\to\infty}\int_\Omega \mathbf{a}(x,\nabla u_n)\cdot\nabla u_n\,dx &\geq \lim_{j\to\infty}\int_\Omega z\cdot\nabla v_j\,dx\\ &= \int_\Omega (z,Du) - \int_{\partial\Omega}[z,\nu](u-\varphi)\,d\mathcal{H}^{N-1}.\end{aligned}$$

Combining this inequality with (6.80), we obtain (6.76).

Our next purpose will be to show that

$$\begin{aligned}&\int_\Omega h(x,Du) + \int_{\partial\Omega}|\varphi-u|f^0(x,\nu(x))\,d\mathcal{H}^{N-1}\\ &= \int_\Omega (z,Du) - \int_{\partial\Omega}[z,\nu](u-\varphi)\,d\mathcal{H}^{N-1}.\end{aligned}\tag{6.81}$$

According to Lemma 6.2, there exists a sequence $\{w_j\}\subset C^1(\Omega)\cap BV(\Omega)$ such that $w_j|_{\partial\Omega}=\varphi$,

$$w_j\to u \text{ in } L^1(\Omega), \quad \text{and} \quad \Phi_\varphi(w_j)\to\Phi_\varphi(u).$$

Now, by the convexity of f, we have

$$\int_\Omega f(x,\nabla u_n)\,dx \leq \int_\Omega \mathbf{a}(x,\nabla u_n)\cdot\nabla u_n\,dx - \int_\Omega \mathbf{a}(x,\nabla u_n)\cdot\nabla w_j\,dx + \int_\Omega f(x,\nabla w_j)\,dx.$$

Thus,

$$\Phi_\varphi(u_n) \leq \int_\Omega \mathbf{a}(x,\nabla u_n)\cdot\nabla u_n\,dx - \int_\Omega \mathbf{a}(x,\nabla u_n)\cdot\nabla w_j\,dx + \Phi_\varphi(w_j).$$

Using (6.76), it follows that

$$\begin{aligned}
&\limsup_{n\to\infty} \Phi_\varphi(u_n) \\
&\leq \int_\Omega (z, Du) - \int_{\partial\Omega} [z,\nu](u-\varphi)\, d\mathcal{H}^{N-1} - \lim_{n\to\infty} \int_\Omega \mathbf{a}(x, \nabla u_n)\cdot \nabla w_j\, dx + \Phi_\varphi(w_j) \\
&= \int_\Omega (z, Du) - \int_{\partial\Omega} [z,\nu](u-\varphi)\, d\mathcal{H}^{N-1} - \int_\Omega z\cdot \nabla w_j\, dx + \Phi_\varphi(w_j).
\end{aligned}$$

Since

$$\begin{aligned}
\lim_{j\to\infty} \int_\Omega z\cdot\nabla w_j\, dx &= \lim_{j\to\infty}\left(-\int_\Omega \operatorname{div}(z) w_j\, dx + \int_{\partial\Omega} [z,\nu]\varphi\, d\mathcal{H}^{N-1}\right) \\
&= -\int_\Omega \operatorname{div}(z) u\, dx + \int_{\partial\Omega} [z,\nu]\varphi\, d\mathcal{H}^{N-1} \\
&= \int_\Omega (z, Du) - \int_{\partial\Omega} [z,\nu](u-\varphi)\, d\mathcal{H}^{N-1},
\end{aligned}$$

letting $j\to\infty$ in the above inequality, we obtain

$$\limsup_{n\to\infty} \Phi_\varphi(u_n) \leq \lim_{j\to\infty} \Phi_\varphi(w_j) = \Phi_\varphi(u).$$

Thus, by the lower-semi-continuity of Φ_φ, we get

$$\Phi_\varphi(u) = \lim_{n\to\infty} \Phi_\varphi(u_n). \tag{6.82}$$

Now,

$$\Phi_\varphi(u) = \int_{\overline{\Omega}} \tilde{f}(x,\tilde{\mu}) \qquad \text{and} \qquad \Phi_\varphi(u_n) = \int_{\overline{\Omega}} \tilde{f}(x,\tilde{\mu}_n).$$

Hence, (6.82) yields

$$\lim_{n\to\infty} \int_{\overline{\Omega}} \tilde{f}(x,\tilde{\mu}_n) = \int_{\overline{\Omega}} \tilde{f}(x,\tilde{\mu}).$$

Then, applying Theorem 3 of [169], it follows that

$$\int_{\overline{\Omega}} \tilde{h}(x,\tilde{\mu}) = \lim_{n\to\infty} \int_{\overline{\Omega}} \tilde{h}(x,\tilde{\mu}_n) = \lim_{n\to\infty} \int_\Omega \mathbf{a}(x,\nabla u_n)\cdot\nabla u_n. \tag{6.83}$$

Since

$$\int_{\overline{\Omega}} \tilde{h}(x,\tilde{\mu}) = \int_{\overline{\Omega}} \tilde{h}(x,\tilde{\mu}^a(x))\,dx + \int_{\overline{\Omega}} \tilde{h}\left(x, \frac{d\tilde{\mu}^s}{d|\tilde{\mu}^s|}(x)\right)\,d|\tilde{\mu}^s|$$

$$= \int_{\overline{\Omega}} \tilde{h}(x,\mu^a(x),1)\,dx + \int_{\overline{\Omega}} \tilde{h}\left(x, \frac{d\mu^s}{d|\mu^s|}(x),0\right)\,d|\mu^s|$$

$$= \int_{\overline{\Omega}} h(x,\mu^a(x))\,dx + \int_{\overline{\Omega}} h^0\left(x, \frac{d\mu^s}{d|\mu^s|}(x)\right)\,d|\mu^s|$$

$$= \int_{\Omega} h(x,\nabla u(x))\,dx + \int_{\Omega} h^0(x,\overrightarrow{D^s u}(x))\,d|D^s u| + \int_{\partial\Omega} h^0\left(x, \frac{(\varphi-u)\cdot\nu}{|(\varphi-u)\cdot\nu|}\right)\,d\mathcal{H}^{N-1}$$

$$= \int_{\Omega} h(x,Du) + \int_{\partial\Omega} |\varphi-u| f^0(x,\nu(x))\,d\mathcal{H}^{N-1},$$

(6.81) follows from (6.76) and (6.83).

By (6.73), (6.74) and (6.72), applying Lemma 6.12 (ii), we get

$$|[z(x),\nu(x)]| \le f^0(x,\nu(x)) \quad \text{a.e. in } \partial\Omega. \tag{6.84}$$

Let $v_j \in C^1(\overline{\Omega})$ be a sequence such that $v_j \to u$ in $L^2(\Omega)$ and $\int_{\Omega} |\nabla v_j| \to \|Du\|$. According to ($H_5$), we have

$$|a(x,\nabla u_n)\cdot \nabla v_j| \le f^0(x,\nabla v_j).$$

Then, if $\psi,\phi \in C^1(\Omega)$, with $0 \le \psi \le \phi$, we have

$$\left|\int_{\Omega} \mathbf{a}(x,\nabla u_n)\cdot\nabla v_j\,\psi\,dx\right| \le \int_{\Omega} f^0(x,\nabla v_j)\psi\,dx,$$

and, letting $n \to \infty$, we get

$$\left|\int_{\Omega} z\cdot\nabla v_j\,\psi\,dx\right| \le \int_{\Omega} f^0(x,\nabla v_j)\psi\,dx.$$

Now, since

$$\left|\int_{\Omega} z\cdot\nabla v_j\,\psi\,dx\right| = \left|-\int_{\Omega} \operatorname{div}(z) v_j\psi\,dx - \int_{\Omega} v_j z\cdot\nabla\psi\,dx\right|,$$

letting $j \to \infty$ we obtain that

$$|\langle (z,Du),\psi\rangle| = \left|-\int_{\Omega} \operatorname{div}(z) u\psi\,dx - \int_{\Omega} uz\cdot\nabla\psi\,dx\right|$$
$$\le \int_{\Omega} \psi f^0(x,Du) \le \int_{\Omega} \phi f^0(x,Du).$$

Hence
$$\langle |(z, Du)|, \phi\rangle \leq \int_\Omega \phi f^0(x, Du).$$

Thus, we have
$$|(z, Du)| \leq f^0(x, Du) \qquad \text{as measures in } \Omega.$$

Then, the singular parts also satisfy a similar inequality,
$$|z \cdot D^s u| \leq f^0(x, D^s u) \qquad \text{as measures in } \Omega. \tag{6.85}$$

Now, by (6.82), (6.75), (6.84) and (6.85), the assumptions of Lemma 6.11 are satisfied, and we have
$$\int_\Omega z \cdot \nabla u \, dx = \int_\Omega h(x, \nabla u) \, dx = \int_\Omega \mathbf{a}(x, \nabla u) \cdot \nabla u \, dx, \tag{6.86}$$
$$z \cdot D^s u = f^0(x, D^s u), \tag{6.87}$$
$$[z, \nu] \in \text{sign } (\varphi - u) f^0(x, \nu(x)) \qquad \mathcal{H}^{N-1} - \text{a.e.} \tag{6.88}$$

Moreover, since the assumptions of Lemma 6.13 hold, we have that
$$z(x) = \mathbf{a}(x, \nabla u(x)) \qquad \text{a.e. } \ x \in \Omega. \tag{6.89}$$

Observe that (6.63) follows from (6.71) and (6.89); (6.64) is a consequence of (6.86), (6.87) and (6.89); and (6.65) follows from (6.88) and (6.89). This concludes the proof in the case $\varphi \in C^1(\overline{\Omega})$.

Step 2. Suppose now we are in the general case, that is, $\varphi \in L^1(\partial\Omega)$. Take $\varphi_j \in C^1(\overline{\Omega})$ such that $\varphi_j \to \varphi$ in $L^1(\partial\Omega)$. Given $v \in L^\infty(\Omega)$, from Step 1, there exists $u_j \in D(\mathcal{B}_{\varphi_j})$ such that $(u_j, v - u_j) \in \mathcal{B}_{\varphi_j}$. Hence, we have
$$-\text{div}(\mathbf{a}(x, \nabla u_j)) = v - u_j, \qquad \text{in} \quad \mathcal{D}'(\Omega), \tag{6.90}$$
$$\mathbf{a}(x, \nabla u_j) \cdot D^s u_j = f^0(x, D^s u_j), \tag{6.91}$$
$$[\mathbf{a}(x, \nabla u_j), \nu] \in \text{sign}(\varphi_j - u_j) f^0(x, \nu(x)) \qquad \mathcal{H}^{N-1} - \text{a.e.} \tag{6.92}$$

By (6.90), (6.91) and (6.92), we get
$$\begin{aligned}\int_\Omega \mathbf{a}(x, \nabla u_j) \cdot \nabla u_j \, dx + \int_\Omega f^0(x, D^s u_j) + \int_{\partial\Omega} |\varphi_j - u_j| f^0(x, \nu(x)) \, d\mathcal{H}^{N-1} \\ + \int_\Omega u_j^2 \, dx = \int_\Omega u_j v \, dx + \int_{\partial\Omega} (\mathbf{a}(x, \nabla u_j) \cdot \nu) \varphi_j \, d\mathcal{H}^{N-1}.\end{aligned} \tag{6.93}$$

From (6.93), using Young's inequality and (6.19), we obtain that
$$C_0 \|Du_j\| + C_0 \int_{\partial\Omega} |\varphi_j - u_j| f^0(x, \nu(x)) \, d\mathcal{H}^{N-1} + \frac{1}{2} \int_\Omega u_j^2 \, dx \leq C \qquad \forall\, j \in \mathbb{N},$$

for some constant $C > 0$. It follows that there exists $u \in BV(\Omega) \cap L^2(\Omega)$, such that

$$u_j \rightharpoonup u \quad \text{weakly in } L^2(\Omega), \qquad u_j \to u \quad \text{in } L^q(\Omega)\ \forall\, 1 \le q < \frac{N}{N-1}. \tag{6.94}$$

Hence,

$$\int_\Omega u^2\, dx \le \limsup_{j\to\infty} \int_\Omega u_j^2\, dx. \tag{6.95}$$

After passing to a subsequence, if necessary, we may assume that

$$\mathbf{a}(x, \nabla u_j) \rightharpoonup z \quad \text{as } j \to \infty, \text{ weakly* in } L^\infty(\Omega, \mathbb{R}^N) \tag{6.96}$$

and

$$-\mathrm{div}(z) = v - u \qquad \text{in} \quad \mathcal{D}'(\Omega). \tag{6.97}$$

By Lemma 6.2, there exists a sequence $\{w_k\} \subset C^1(\Omega) \cap BV(\Omega)$ such that $w_k|_{\partial\Omega} = \varphi$,

$$w_k \to u \ \text{ in } L^2(\Omega) \quad \text{and} \quad \Phi_\varphi(w_k) \to \Phi_\varphi(u). \tag{6.98}$$

Now, by the convexity of f we have

$$\int_\Omega f(x, \nabla u_j)\, dx$$
$$\le \int_\Omega \mathbf{a}(x, \nabla u_j) \cdot \nabla u_j\, dx - \int_\Omega \mathbf{a}(x, \nabla u_j) \cdot \nabla w_k\, dx + \int_\Omega f(x, \nabla w_k)\, dx.$$

Thus, having in mind (6.90), (6.91) and (6.92), we get

$$\begin{aligned}
\Phi_{\varphi_j}(u_j) &= \int_\Omega f(x, \nabla u_j)\, dx + \int_\Omega f^0(x, D^s u_j) + \int_{\partial\Omega} |u_j - \varphi_j| f^0(x, \nu(x))\, d\mathcal{H}^{N-1} \\
&\le \int_\Omega f(x, \nabla w_k)\, dx + \int_\Omega \mathbf{a}(x, \nabla u_j) \cdot \nabla u_j\, dx + \int_\Omega f^0(x, D^s u_j) \\
&\quad + \int_{\partial\Omega} |u_j - \varphi_j| f^0(x, \nu(x))\, d\mathcal{H}^{N-1} - \int_\Omega \mathbf{a}(x, \nabla u_j) \cdot \nabla w_k\, dx \\
&\le \int_\Omega f(x, \nabla w_k)\, dx + \int_\Omega (v - u_j) u_j\, dx \\
&\quad + \int_{\partial\Omega} [\mathbf{a}(x, \nabla u_j), \nu] \varphi_j\, d\mathcal{H}^{N-1} - \int_\Omega \mathbf{a}(x, \nabla u_j) \cdot \nabla w_k\, dx.
\end{aligned}$$

Using (6.95) and (6.96), it follows that

$$\limsup_{j\to\infty} \Phi_\varphi(u_j) = \limsup_{j\to\infty} \Phi_{\varphi_j}(u_j)$$
$$\le \int_\Omega f(x, \nabla w_k)\, dx + \int_\Omega uv\, dx - \int_\Omega u^2\, dx + \int_{\partial\Omega} [z, \nu]\varphi\, d\mathcal{H}^{N-1} - \int_\Omega z \cdot \nabla w_k\, dx$$
$$\le \int_\Omega f(x, \nabla w_k)\, dx + \int_\Omega (v - u) u\, dx + \int_\Omega \mathrm{div}(z) w_k\, dx.$$

Hence, by (6.97), letting $k \to \infty$, we arrive to

$$\limsup_{j\to\infty} \Phi_\varphi(u_j) \leq \lim_{k\to\infty} \Phi_\varphi(w_k) = \Phi_\varphi(u).$$

Thus, by the lower-semi-continuity of Φ_φ, we get

$$\Phi_\varphi(u) = \lim_{j\to\infty} \Phi_\varphi(u_j). \tag{6.99}$$

Applying Theorem 3 of [169] as in Step 1, it follows that

$$\begin{aligned}
&\lim_{j\to\infty} \int_\Omega h(x, Du_j) + \int_{\partial\Omega} |u_j - \varphi_j| f^0(x, \nu(x)) \, d\mathcal{H}^{N-1} \\
&= \int_\Omega h(x, Du) + \int_{\partial\Omega} |u - \varphi| f^0(x, \nu(x)) \, d\mathcal{H}^{N-1}.
\end{aligned} \tag{6.100}$$

On the other hand, by Green's formula, (6.90), (6.91) and (6.92), we have

$$\begin{aligned}
&\int_\Omega h(x, Du_j) + \int_{\partial\Omega} |u_j - \varphi_j| f^0(x, \nu(x)) \, d\mathcal{H}^{N-1} \\
&= \int_\Omega (\mathbf{a}(x, \nabla u_j), Du_j) + \int_{\partial\Omega} [\mathbf{a}(x, \nabla u_j), \nu](\varphi_j - u_j) \, d\mathcal{H}^{N-1} \\
&= \int_\Omega u_j(v - u_j) \, dx + \int_{\partial\Omega} [\mathbf{a}(x, \nabla u_j), \nu]\varphi_j \, d\mathcal{H}^{N-1}.
\end{aligned}$$

Since $[\mathbf{a}(x, \nabla u_j), \nu] \rightharpoonup [z, \nu]$ weakly* in $L^\infty(\partial\Omega)$, letting $j \to +\infty$, and using (6.100), it follows that

$$\begin{aligned}
&\int_\Omega h(x, Du) + \int_{\partial\Omega} |u - \varphi| f^0(x, \nu(x)) \, d\mathcal{H}^{N-1} \\
&\leq \int_\Omega u(v - u) \, dx + \int_{\partial\Omega} [z, \nu]\varphi \, d\mathcal{H}^{N-1} = \int_\Omega (z, Du) + \int_{\partial\Omega} [z, \nu](\varphi - u) \, d\mathcal{H}^{N-1}.
\end{aligned}$$

Now, by Lemma 6.12 (i), we have

$$|[z(x), \nu(x)]| \leq f^0(x, \nu(x)) \qquad \mathcal{H}^{N-1} - \text{a.e. in } \partial\Omega.$$

Moreover, as in the Step 1, we get

$$|z \cdot D^s u| \leq f^0(x, D^s u) \quad \text{as measures in } \Omega.$$

With this, and using Lemma 6.11, we obtain

$$\int_\Omega z \cdot \nabla u \, dx = \int_\Omega h(x, \nabla u) \, dx = \int_\Omega \mathbf{a}(x, \nabla u) \cdot \nabla u \, dx, \tag{6.101}$$

$$z \cdot D^s u = f^0(x, D^s u), \tag{6.102}$$

$$[z, \nu] \in \text{sign}\,(\varphi - u) f^0(x, \nu(x)) \qquad \mathcal{H}^{N-1} - \text{a.e.} \tag{6.103}$$

As in Step 1, to get that $(u, v-u) \in \mathcal{B}_\varphi$, we only need to prove that

$$\text{div}(z) = \text{div}\,\mathbf{a}(x, \nabla u) \qquad \text{in} \quad \mathcal{D}'(\Omega), \tag{6.104}$$

and

$$[z, \nu] = [\mathbf{a}(x, \nabla u), \nu] \qquad \mathcal{H}^{N-1} - \text{a.e. on } \partial\Omega. \tag{6.105}$$

Now, by (6.90), (6.94) and using Fatou's lemma, we are able to adapt the proof of Lemma 6.13 obtaining that $z(x) = a(x, \nabla u(x))$ a.e. in Ω and this implies both (6.104) and (6.105).

Step 3. To prove the density of $D(\mathcal{B}_\varphi)$ in $L^2(\Omega)$, we prove that $C_0^\infty(\Omega) \subseteq \overline{D(\mathcal{B}_\varphi)}^{L^2(\Omega)}$. Let $v \in C_0^\infty(\Omega)$. By the above, $v \in R(I + \frac{1}{n}\mathcal{B}_\varphi)$ for all $n \in \mathbb{N}$. Thus, for each $n \in \mathbb{N}$, there exists $u_n \in D(\mathcal{B}_\varphi)$ such that $(u_n, n(v - u_n)) \in \mathcal{B}_\varphi$. Consequently, we have $\mathbf{a}(x, \nabla u_n) \in X(\Omega)_1$, $n(v - u_n) = -\text{div}(\mathbf{a}(x, \nabla u_n))$ in $\mathcal{D}'(\Omega)$ and

$$\begin{aligned}
&\int_\Omega (w - u_n) n(v - u_n)\, dx \\
&= \int_\Omega (\mathbf{a}(x, \nabla u_n), Dw) - \int_{\partial\Omega} [\mathbf{a}(x, \nabla u_n), \nu](w - \varphi)\, d\mathcal{H}^{N-1} \\
&\quad - \int_\Omega h(x, Du_n) - \int_{\partial\Omega} |\varphi - u_n| f^0(x, \nu(x))\, d\mathcal{H}^{N-1}
\end{aligned}$$

for every $w \in BV(\Omega) \cap L^2(\Omega)$. Taking $w = v$, we get

$$\begin{aligned}
\int_\Omega (v - u_n)^2\, dx &= \frac{1}{n}\Bigg(\int_\Omega \mathbf{a}(x, \nabla u_n) \cdot \nabla v\, dx - \int_{\partial\Omega} [\mathbf{a}(x, \nabla u_n), \nu](v - \varphi)\, d\mathcal{H}^{N-1} \\
&\qquad - \int_\Omega h(x, Du_n) - \int_{\partial\Omega} |\varphi - u_n| f^0(x, \nu(x))\, d\mathcal{H}^{N-1} \Bigg) \\
&\le \frac{1}{n}\Bigg(\int_\Omega \mathbf{a}(x, \nabla u_n) \cdot \nabla v\, dx - \int_{\partial\Omega} [\mathbf{a}(x, \nabla u_n), \nu](v - \varphi)\, d\mathcal{H}^{N-1} \Bigg) \\
&\le \frac{M}{n}\Bigg(\int_\Omega |\nabla v|\, dx + \int_{\partial\Omega} |v - \varphi|\, d\mathcal{H}^{N-1} \Bigg).
\end{aligned}$$

Letting $n \to \infty$, it follows that $u_n \to v$ in $L^2(\Omega)$. Therefore $v \in \overline{D(\mathcal{B}_\varphi)}^{L^2(\Omega)}$ and the proof is complete. □

Proof of Theorem 6.7. First, we prove that $\mathcal{B}_\varphi \subset \partial\Phi_\varphi$. Let $(u,v) \in \mathcal{B}_\varphi$ and $w \in W^{1,2}_\varphi(\Omega)$. Then, by (6.16), and applying Green's formula (C.10) we get

$$\int_\Omega (w-u)v\,dx = -\int_\Omega (w-u)\text{ div }\mathbf{a}(x,\nabla u)\,dx$$
$$= \int_\Omega (\mathbf{a}(x,\nabla u), Dw - Du) - \int_{\partial\Omega} [\mathbf{a}(x,\nabla u),\nu](\varphi - u)\,d\mathcal{H}^{N-1}$$
$$= \int_\Omega \mathbf{a}(x,\nabla u)\cdot\nabla w\,dx - \int_\Omega \mathbf{a}(x,\nabla u)\cdot\nabla u\,dx - \int_\Omega \mathbf{a}(x,\nabla u)\cdot D^s u$$
$$- \int_{\partial\Omega} [\mathbf{a}(x,\nabla u),\nu](\varphi-u)\,d\mathcal{H}^{N-1}$$
$$\leq \int_\Omega f(x,\nabla w)\,dx - \int_\Omega f(x,Du)\,dx - \int_{\partial\Omega} |\varphi - u| f^0(x,\nu(x))\,d\mathcal{H}^{N-1}$$
$$= \Phi_\varphi(w) - \Phi_\varphi(u).$$

Suppose that $w \in BV(\Omega)\cap L^2(\Omega)$. According to Lemma 6.2, there exists a sequence $w_n \in W^{1,2}_\varphi(\Omega)$, with $w_n \to w$ in $L^2(\Omega)$ and $\Phi_\varphi(w_n) \to \Phi_\varphi(w)$. Then, by the above inequality, we have

$$\int_\Omega (w_n - u)v\,dx \leq \Phi_\varphi(w_n) - \Phi_\varphi(u).$$

Now, letting $n\to\infty$, we get

$$\int_\Omega (w-u)v\,dx \leq \Phi_\varphi(w) - \Phi_\varphi(u),$$

and therefore, $(u,v)\in\partial\Phi_\varphi$.

Since $\mathcal{B}_\varphi \subset \partial\Phi_\varphi$, and, by Proposition 6.8, $L^\infty(\Omega)\subset R(I+\mathcal{B}_\varphi)$, we have $\partial\Phi_\varphi = \overline{\mathcal{B}_\varphi}^{L^2(\Omega)}$. To finish the proof we only need to prove that the operator $\mathcal{B}_\varphi$ is closed. Let $(u_n,v_n)\in\mathcal{B}_\varphi$ and assume that $(u_n,v_n)\to(u,v)$ in $L^2(\Omega)\times L^2(\Omega)$. Let us prove that $(u,v)\in\mathcal{B}_\varphi$. Since $(u_n,v_n)\in\mathcal{B}_\varphi$, we know that $\mathbf{a}(x,\nabla u_n)\in X(\Omega)_1$ is such that

$$-v_n = \text{div }\mathbf{a}(x,\nabla u_n) \qquad \text{in } \mathcal{D}'(\Omega), \tag{6.106}$$

$$\mathbf{a}(x,\nabla u_n)\cdot D^s u_n = f^0(x, D^s u_n), \tag{6.107}$$

$$[\mathbf{a}(x,\nabla u_n),\nu] \in \text{sign }(\varphi - u_n) f^0(x,\nu(x)) \qquad \mathcal{H}^{N-1} - \text{a.e.} \tag{6.108}$$

Multiplying (6.106) by u_n and applying Green's formula we obtain

$$-\int_\Omega u_n v_n \, dx$$

$$= \int_{\partial\Omega} [\mathbf{a}(x,\nabla u_n),\nu]\varphi \, dH^{N-1} - \int_\Omega h(x,Du_n) - \int_{\partial\Omega} |\varphi - u_n| f^0(x,\nu(x)) \, d\mathcal{H}^{N-1}.$$

Hence,

$$\int_\Omega h(x,Du_n) \le \int_\Omega u_n v_n \, dx + \int_{\partial\Omega} [\mathbf{a}(x,\nabla u_n),\nu]\varphi \, d\mathcal{H}^{N-1}. \tag{6.109}$$

From (6.19) and (6.109), we have

$$\begin{aligned} C_0 \int_\Omega \|Du_n\| \, dx - D_1 \mathcal{L}^N(\Omega) &\le \int_\Omega h(x,Du_n) \, dx \\ &\le \int_\Omega u_n v_n \, dx + \int_{\partial\Omega} [\mathbf{a}(x,\nabla u_n),\nu]\varphi \, d\mathcal{H}^{N-1}. \end{aligned}$$

Hence,

$$\int_\Omega \|Du_n\| \, dx \le M_1 \qquad \forall\, n \in \mathbb{N}. \tag{6.110}$$

Therefore, $u \in BV(\Omega) \cap L^2(\Omega)$. On the other hand, since $\|\mathbf{a}(x,\nabla u_n)\|_\infty \le M$, we may assume that

$$\mathbf{a}(x,\nabla u_n) \rightharpoonup z \quad \text{in the weak}^* \text{ topology of } L^\infty(\Omega,\mathbb{R}^N), \tag{6.111}$$

with $\|z\|_\infty \le M$. Moreover, since $v_n \to v$ in $L^2(\Omega)$, we have that $v = -\mathrm{div}(z)$ in $\mathcal{D}'(\Omega)$. By the definition of the weak trace on $\partial\Omega$ of the normal component of z, it is easy to see that

$$[\mathbf{a}(x,\nabla u_n),\nu] \rightharpoonup [z,\nu] \qquad \text{weakly}^* \text{ in } L^\infty(\partial\Omega). \tag{6.112}$$

Now, we prove the convergence of the energies. According to Lemma 6.2, there exists a sequence $w_j \in C^1(\Omega) \cap BV(\Omega)$, with $w_j|_{\partial\Omega} = \varphi$, $w_j \to u$ in $L^1(\Omega)$ and $\Phi_\varphi(w_j) \to \Phi_\varphi(u)$. Moreover, looking at the proof of Lemma 6.2, we have that, $w_j = w_j^1 + w_j^2$ with $w_j^1|_{\partial\Omega} = u|_{\partial\Omega}$ and $w_j^1 \to u$ in $L^1(\Omega)$, $w_j^2|_{\partial\Omega} = \varphi - u|_{\partial\Omega}$, $w_j^2 \to 0$ in $L^1(\Omega)$, and, using Lemma C.8, we have that

$$\int_\Omega (z, Dw_j^1) \to \int_\Omega (z, Du).$$

By the convexity of f and taking (6.107) and (6.108) into account we have

$$\Phi_\varphi(u_n) = \int_\Omega f(x, \nabla u_n)\, dx + \int_\Omega f^0(x, D^s u_n) + \int_{\partial\Omega} |u_n - \varphi| f^0(x, \nu(x))\, d\mathcal{H}^{N-1}$$

$$\leq \int_\Omega \mathbf{a}(x, \nabla u_n) \cdot \nabla u_n\, dx - \int_\Omega \mathbf{a}(x, \nabla u_n) \cdot \nabla w_j\, dx + \int_\Omega f(x, \nabla w_j)\, dx$$

$$+ \int_\Omega \mathbf{a}(x, \nabla u_n) \cdot D^s u_n + \int_{\partial\Omega} [\mathbf{a}(x, \nabla u_n), \nu](\varphi - u_n)\, d\mathcal{H}^{N-1}$$

$$= \int_\Omega (\mathbf{a}(x, \nabla u_n), Du_n) - \int_\Omega \mathbf{a}(x, \nabla u_n) \cdot \nabla w_j\, dx + \Phi_\varphi(w_j)$$

$$+ \int_{\partial\Omega} [\mathbf{a}(x, \nabla u_n), \nu](\varphi - u_n)\, d\mathcal{H}^{N-1}$$

$$= \Phi_\varphi(w_j) - \int_\Omega \mathbf{a}(x, \nabla u_n) \cdot \nabla w_j\, dx - \int_\Omega \operatorname{div}(\mathbf{a}(x, \nabla u_n)) u_n\, dx$$

$$+ \int_{\partial\Omega} [\mathbf{a}(x, \nabla u_n), \nu]\varphi\, d\mathcal{H}^{N-1}$$

$$= \Phi_\varphi(w_j) - \int_\Omega \mathbf{a}(x, \nabla u_n) \cdot \nabla w_j\, dx + \int_\Omega v_n u_n\, dx + \int_{\partial\Omega} [\mathbf{a}(x, \nabla u_n), \nu]\varphi\, d\mathcal{H}^{N-1}.$$

Hence, by (6.111) and (6.112), it follows that

$$\limsup_{n\to\infty} \Phi_\varphi(u_n) \leq \Phi_\varphi(w_j) - \int_\Omega (z, Dw_j) + \int_\Omega uv\, dx + \int_{\partial\Omega} [z, \nu]\varphi\, d\mathcal{H}^{N-1}$$

$$= \Phi_\varphi(w_j) - \int_\Omega (z, Dw_j^1) - \int_\Omega (z, Dw_j^2) - \int_\Omega \operatorname{div}(z) u\, dx + \int_{\partial\Omega} [z, \nu]\varphi\, d\mathcal{H}^{N-1}$$

$$= \Phi_\varphi(w_j) - \int_\Omega (z, Dw_j^1) + \int_\Omega \operatorname{div}(z) w_j^2\, dx + \int_{\partial\Omega} [z, \nu] u\, d\mathcal{H}^{N-1} - \int_\Omega \operatorname{div}(z) u\, dx.$$

Letting $j \to \infty$, we have that

$$\limsup_{n\to\infty} \Phi_\varphi(u_n) \leq \Phi_\varphi(u) - \int_\Omega (z, Du) + \int_{\partial\Omega} [z, \nu] u\, d\mathcal{H}^{N-1} - \int_\Omega \operatorname{div}(z) u\, dx$$

$$= \Phi_\varphi(u).$$

Finally, by the lower-semi-continuity of Φ_φ, we obtain

$$\Phi_\varphi(u) = \lim_{n\to\infty} \Phi_\varphi(u_n). \tag{6.113}$$

If we consider the $\mathbb{R}^N$-valued measures μ_n, μ on $\overline{\Omega}$ which are defined as

$$\mu_n(B) := \int_{B\cap\Omega} Du_n + \int_{B\cap\partial\Omega} (\varphi - u_n)\nu\, d\mathcal{H}^{N-1},$$

$$\mu(B) := \int_{B\cap\Omega} Du + \int_{B\cap\partial\Omega} (\varphi - u)\nu \, d\mathcal{H}^{N-1}$$

for all Borel sets $B \subset \overline{\Omega}$, we have

$$\mu_j \rightharpoonup \mu \qquad \text{weakly as measures in} \quad \overline{\Omega}.$$

Moreover,

$$\Phi_\varphi(u) = \int_{\overline{\Omega}} \tilde{f}(x, \tilde{\mu}) \qquad \text{and} \qquad \Phi_\varphi(u_n) = \int_{\overline{\Omega}} \tilde{f}(x, \tilde{\mu}_n).$$

Hence, (6.113) yields

$$\lim_{n\to\infty} \int_{\overline{\Omega}} \tilde{f}(x, \tilde{\mu}_n) = \int_{\overline{\Omega}} \tilde{f}(x, \tilde{\mu}).$$

Then, applying [169], Theorem 3, it follows that

$$\int_{\overline{\Omega}} \tilde{h}(x, \tilde{\mu}) = \lim_{n\to\infty} \int_{\overline{\Omega}} \tilde{h}(x, \tilde{\mu}_n)$$

$$= \lim_{n\to\infty} \int_\Omega h(x, Du_n) + \int_{\partial\Omega} |u_n - \varphi| f^0(x, \nu(x)) \, d\mathcal{H}^{N-1}.$$

Since

$$\int_{\overline{\Omega}} \tilde{h}(x, \tilde{\mu}) = \int_\Omega h(x, Du) + \int_{\partial\Omega} |\varphi - u| f^0(x, \nu(x)) \, d\mathcal{H}^{N-1},$$

we have

$$\begin{aligned} &\int_\Omega h(x, Du) + \int_{\partial\Omega} |\varphi - u| f^0(x, \nu(x)) \, d\mathcal{H}^{N-1} \\ &= \lim_{n\to\infty} \int_\Omega h(x, Du_n) + \int_{\partial\Omega} |u_n - \varphi| f^0(x, \nu(x)) \, d\mathcal{H}^{N-1}. \end{aligned} \tag{6.114}$$

Now, since

$$\begin{aligned} &\lim_{n\to\infty} \int_\Omega h(x, Du_n) + \int_{\partial\Omega} |u_n - \varphi| f^0(x, \nu(x)) \, d\mathcal{H}^{N-1} \\ &= \lim_{n\to\infty} \int_\Omega \mathbf{a}(x, \nabla u_n)\cdot\nabla u_n \, dx + \int_\Omega f^0(x, D^s u_n) + \int_{\partial\Omega} [\mathbf{a}(x, \nabla u_n), \nu](\varphi - u_n) \, d\mathcal{H}^{N-1} \\ &= \lim_{n\to\infty} \int_\Omega (\mathbf{a}(x, \nabla u_n), Du_n) + \int_{\partial\Omega} [\mathbf{a}(x, \nabla u_n), \nu](\varphi - u_n) \, d\mathcal{H}^{N-1} \\ &= \lim_{n\to\infty} \int_{\partial\Omega} [\mathbf{a}(x, \nabla u_n), \nu]\varphi \, d\mathcal{H}^{N-1} - \int_\Omega \operatorname{div}(\mathbf{a}(x, \nabla u_n)) u_n \, dx \\ &= \int_{\partial\Omega} [z, \nu]\varphi \, d\mathcal{H}^{N-1} - \int_\Omega \operatorname{div}(z) u \, dx = \int_\Omega (z, Du) + \int_{\partial\Omega} [z, \nu](\varphi - u) \, d\mathcal{H}^{N-1}, \end{aligned}$$

we finally obtain

$$\begin{aligned}&\int_\Omega h(x, Du) + \int_{\partial\Omega} |\varphi - u| f^0(x, \nu(x)) \, d\mathcal{H}^{N-1} \\ &= \int_\Omega (z, Du) + \int_{\partial\Omega} [z, \nu](\varphi - u) \, d\mathcal{H}^{N-1}.\end{aligned} \tag{6.115}$$

Again, by (6.111) and (6.112) we can apply Lemma 6.12 obtaining that

$$|[z(x), \nu(x)]| \leq f^0(x, \nu(x)) \quad \text{a.e. in } \partial\Omega. \tag{6.116}$$

Moreover, acting as in the proof of Proposition 6.8, we get that

$$|z \cdot D^s u| \leq f^0(x, D^s u) \quad \text{as measures in } \Omega. \tag{6.117}$$

Hence by (6.113), (6.114), (6.115), (6.111) (6.117) and (6.116), we can apply Lemma 6.11, to obtain

$$\int_\Omega z \cdot \nabla u \, dx = \int_\Omega h(x, \nabla u) \, dx = \int_\Omega \mathbf{a}(x, \nabla u) \cdot \nabla u \, dx, \tag{6.118}$$

$$z \cdot D^s u = f^0(x, D^s u), \tag{6.119}$$

$$[z, \nu] \in \text{sign } (\varphi - u) f^0(x, \nu(x)) \qquad \mathcal{H}^{N-1} - \text{a.e.} \tag{6.120}$$

Now, using Lemma 6.13, we have

$$\text{div}(z) = \text{div } \mathbf{a}(x, \nabla u) \qquad \text{in } \mathcal{D}'(\Omega), \tag{6.121}$$

and

$$[z, \nu] = [\mathbf{a}(x, \nabla u), \nu] \qquad \mathcal{H}^{N-1} - \text{a.e. on } \partial\Omega. \tag{6.122}$$

Since $v = -\text{div}(z)$ in $\mathcal{D}'(\Omega)$, taking (6.121) into account, we get

$$v = -\text{div}(\mathbf{a}(x, \nabla u)) \qquad \text{in } \mathcal{D}'(\Omega),$$

and, using (6.118), (6.119) and (6.121), we obtain

$$\mathbf{a}(x, \nabla u) \cdot D^s u = f^0(x, D^s u).$$

Finally, by (6.120) and (6.122)

$$[\mathbf{a}(x, \nabla u), \nu] \in \text{sign } (\varphi - u) f^0(x, \nu(x)) \qquad \mathcal{H}^{N-1} - \text{a.e.}$$

Therefore, $(u, v) \in \mathcal{B}_\varphi$. □

Proof of Theorem 6.6. Let $(S(t))_{t \geq 0}$ be the semigroup in $L^2(\Omega)$ generated by the subdifferential of Φ_φ. Then by the theory of nonlinear semigroups (see Appendix A), given $u_0 \in L^2(\Omega) = \overline{D(\partial\Phi_\varphi)}$, $u(t) = S(t)u_0$ is the only strong solution of problem (6.24). Thus, by Theorem 6.7, we have that for almost all $t \in [0, +\infty[$, $u(t) \in D(\mathcal{B}_\varphi)$ and $-u'(t) \in \mathcal{B}_\varphi(u(t))$. This concludes the proof. □

We have the following weak form of the maximum principle.

Theorem 6.14. *Suppose u_1 and u_2 are two solutions of (6.1) corresponding to initial data $u_{1,0}$ and $u_{2,0}$ in $L^2(\Omega)$ and boundary data φ_1 and φ_2 in $L^1(\partial\Omega)$, respectively. If*

$$u_{1,0} \geq u_{2,0} \quad \text{and} \quad \varphi_1 \geq \varphi_2,$$

then $u_1 \geq u_2$.

Proof. For almost all $t \in [0, +\infty[$, we have $u_i'(t) \in L^2(\Omega)$, $u_i(t) \in BV(\Omega) \cap L^2(\Omega)$, $\mathbf{a}(x, \nabla u_i(t)) \in X(\Omega)_1$, and

$$u_2'(t) - u_1'(t) = \operatorname{div}[\mathbf{a}(x, \nabla u_2(t)) - \mathbf{a}(x, \nabla u_1(t))] \quad \text{in } \mathcal{D}'(\Omega), \tag{6.123}$$

$$\mathbf{a}(x, \nabla u_i(t)) \cdot D^s u_i(t) = f^0(x, D^s u_i(t)), \tag{6.124}$$

$$[\mathbf{a}(x, \nabla u_i(t)), \nu] \in \operatorname{sign}(\varphi_i - u_i(t)) f^0(x, \nu(x)) \quad \mathcal{H}^{N-1} - \text{a.e. on } \partial\Omega. \tag{6.125}$$

Multiplying in (6.123) by $\left(u_2(t) - u_1(t)\right)^+$, integrating in Ω, and using Green's formula, we get

$$\begin{aligned}
&\frac{1}{2} \int_\Omega \frac{d}{dt} \left[\left(u_2(t) - u_1(t)\right)^+\right]^2 dx \\
&= \int_\Omega \operatorname{div}\left[\mathbf{a}(x, \nabla u_2(t)) - \mathbf{a}(x, \nabla u_1(t))\right] \left(u_2(t) - u_1(t)\right)^+ dx \\
&= - \int_\Omega \left(\mathbf{a}(x, \nabla u_2(t)) - \mathbf{a}(x, \nabla u_1(t)), D\left(\left(u_2(t) - u_1(t)\right)^+\right)\right) \\
&\quad + \int_{\partial\Omega} [\mathbf{a}(x, \nabla u_2(t)) - \mathbf{a}(x, \nabla u_1(t)), \nu] \left(u_2(t) - u_1(t)\right)^+ d\mathcal{H}^{N-1}.
\end{aligned} \tag{6.126}$$

Now, by the chain rule for BV-functions ([10], [129], Lemma 1.2), there exists a scalar function $\eta(t)$, with $0 \leq \eta(t) \leq 1$, such that

$$\begin{aligned}
&\int_\Omega \left(\mathbf{a}(x, \nabla u_2(t)) - \mathbf{a}(x, \nabla u_1(t)), D\left(\left(u_2(t) - u_1(t)\right)^+\right)\right) \\
&= \int_{\{u_2 \geq u_1\}} \left(\mathbf{a}(x, \nabla u_2(t)) - \mathbf{a}(x, \nabla u_1(t)\right) \cdot \left(\nabla u_2(t) - \nabla u_1(t)\right) dx \\
&\quad + \int_\Omega \eta(t) \left(\mathbf{a}(x, \nabla u_2(t)) - \mathbf{a}(x, \nabla u_1(t)\right) \cdot D^s\left(u_2(t) - u_1(t)\right).
\end{aligned}$$

Observe that, by the monotonicity of $\mathbf{a}$, (H_5) and (6.124), we have that

$$\int_\Omega \left(\mathbf{a}(x, \nabla u_2(t)) - \mathbf{a}(x, \nabla u_1(t)), D\left(\left(u_2(t) - u_1(t)\right)^+\right)\right) \geq 0. \tag{6.127}$$

On the other hand, since $\varphi_1 \geq \varphi_2$, from (6.125), it is easy to prove that

$$\int_{\partial\Omega} [\mathbf{a}(x, \nabla u_2(t)) - \mathbf{a}(x, \nabla u_1(t)), \nu] \big(u_2(t) - u_1(t)\big)^+ \, d\mathcal{H}^{N-1} \leq 0. \tag{6.128}$$

From (6.126), (6.127) and (6.128), it follows that

$$\frac{1}{2} \int_\Omega \frac{d}{dt} \left[\big(u_2(t) - u_1(t)\big)^+ \right]^2 dx \leq 0.$$

Since $u_{1,0} \geq u_{2,0}$, we have $u_1 \geq u_2$, and the proof is concluded. □

6.5 Asymptotic Behaviour

We shall now prove that the solution $u(t)$ of problem (6.1) stabilizes as $t \to +\infty$ by converging to a solution of the steady-state problem. To do that, we follow the proof of Theorem 4.2 in [145].

Theorem 6.15. *Suppose $u_0 \in L^2(\Omega) \cap BV(\Omega)$ and $\varphi \in L^\infty(\partial\Omega)$. Assume that $0 \leq f(x, \xi)$ for all $x \in \Omega$ and $\xi \in \mathbb{R}^N$. Then the solution $u(t)$ of (6.1) converges as $t \to +\infty$ to some limit $w \in \mathcal{B}_\varphi^{-1}(0)$ in the following sense:*

$$u(t) \to w \quad \text{strongly in} \quad L^1(\Omega) \quad \text{and weakly in} \quad L^2(\Omega).$$

Proof. Since $\mathcal{B}_\varphi$ is the subdifferential of Φ_φ, by a classical result of Bruck ([60], Theorem 4), to prove the weak convergence in $L^2(\Omega)$, it is sufficient to prove that Φ_φ attains its minimun in $L^2(\Omega)$. In fact, let $\{u_n\}$ be a minimizing sequence for Φ_φ. Without loss of generality, we may assume that $u_n \in BV(\Omega) \cap L^2(\Omega)$. Now, by approximation, we may assume that $u_n \in W^{1,1}(\Omega) \cap L^2(\Omega)$. Denote by $J : \mathbb{R} \to \mathbb{R}$ the truncation function

$$J(r) := \begin{cases} -\|\varphi\|_\infty & \text{if} \quad r < -\|\varphi\|_\infty, \\ r & \text{if} \quad |r| \leq \|\varphi\|_\infty, \\ \|\varphi\|_\infty & \text{if} \quad x > \|\varphi\|_\infty. \end{cases}$$

If we take $w_n := J \circ u_n$, $w_n \in W^{1,1}(\Omega) \cap L^\infty(\Omega)$, and using that $|J'| \leq 1$, we have

$$\begin{aligned} \Phi_\varphi(w_n) &= \int_\Omega f(x, \nabla w_n) \, dx + \int_{\partial\Omega} |w_n - \varphi| f^0(x, \nu(x)) \, d\mathcal{H}^{N-1} \\ &= \int_{\{|u_n| \leq \|\varphi\|_\infty\}} f(x, \nabla u_n) \, dx + \int_{\partial\Omega} |J \circ u_n - J \circ \varphi| f^0(x, \nu(x)) \, d\mathcal{H}^{N-1} \\ &\leq \int_\Omega f(x, \nabla u_n) \, dx + \int_{\partial\Omega} |u_n - \varphi| f^0(x, \nu(x)) \, d\mathcal{H}^{N-1}. \end{aligned}$$

Thus, $\{w_n\}$ is still a minimizing sequence for Φ_φ. Moreover, this sequence is bounded in $W^{1,1}(\Omega)\cap L^\infty(\Omega)$, hence, relatively compact in $L^1(\Omega)$. We may extract a subsequence converging in $L^1(\Omega)$ to some $\overline{u} \in L^1(\Omega) \cap BV(\Omega)$. Therefore,

$$\Phi_\varphi(\overline{u}) = \inf_{u\in L^2(\Omega)} \Phi_\varphi(u).$$

Then, by Bruck's result ([60], Theorem 4), there exists $w \in \mathcal{B}_\varphi^{-1}(0)$, such that $u(t) \to w$ weakly in $L^2(\Omega)$. Finally, we prove the strong convergence in $L^1(\Omega)$. Since $(u(t), -u'(t)) \in \partial\Phi_\varphi$, by (A.35), we have

$$\frac{d}{ds}\Phi_\varphi(u(s)) = -\int_\Omega u'(s)^2\,dx \leq 0,$$

hence,

$$\Phi_\varphi(u(t)) \leq \Phi_\varphi(u_0) \qquad \forall\, t > 0.$$

Thus, $\{u(t) \ : \ t \geq 0\}$ is bounded in $BV(\Omega)$, and therefore relatively compact in $L^1(\Omega)$. The result follows. □

6.6 Proof of the Approximation Lemma

In this section we give the proof of Lemma 6.10. Before giving the proof, let us construct a substitute for the distance function to the boundary $d(.,\partial\Omega)$. That construction would be unnecessary if $\partial\Omega$ would be of class $W^{2,\infty}$ ([27]). We follow the proof of Lemma 5.1 in [27] for C^2 domains.

If $\partial\Omega$ is a manifold of class C^1, then there is some $\epsilon > 0$ such that for all points $y \in \Omega$ such that $d(y,\Omega) < \epsilon$ there is $z \in \partial\Omega$ and $t \in (0,\epsilon)$ such that $y = z - t\nu(z)$, $\nu(z)$ being the outer unit normal to $\partial\Omega$ at z ([88]). In other words, $\Omega^\epsilon := \{x \in \Omega : x = y - t\nu(y),\, y \in \partial\Omega,\, t \in (0,\epsilon)\}$ is open. Then there is a function $D \in C^1(\overline{\Omega})$ such that $D = 0$ on $\partial\Omega$, $D > 0$ on Ω and $\nabla D(x) = -\nu(x)$ for all $x \in \partial\Omega$. This is a consequence of Withney's extension theorem ([133], p.48, [110], p.245). Indeed, since

$$\nu(y)\frac{x-y}{|x-y|} \to 0 \text{ as } x, y \to p,\ x \neq y,\ x, y \in \partial\Omega,$$

by Withney's theorem , we know that there exists a function $\tilde{D} \in C^1(\overline{\Omega})$ such that $\tilde{D} = 0$ on $\partial\Omega$ and $\nabla\tilde{D}(x) = -\nu(x)$ for all $x \in \partial\Omega$. Now, let $y \in \partial\Omega$ and $t \in (0,\epsilon)$. Using the mean value theorem, we know that

$$\tilde{D}(y - t\nu(y)) = \tilde{D}(y) - t\nabla\tilde{D}(y - s\nu(y))\cdot\nu(y) = -t\nabla\tilde{D}(y - s\nu(y))\cdot\nu(y).$$

Since $\tilde{D} \in C^1(\overline{\Omega})$, we have

$$\tilde{D}(y - t\nu(y)) = t(1 + \omega(t)),$$

where $\omega(t) = o(1)$ as $t \to 0+$ and is the modulus of continuity of $\nabla\tilde{D}$. Without loss of generality we may assume that $\epsilon > 0$ is such that $\omega(t) < \frac{1}{2}$ for all $t \in (0, \epsilon)$. In particular, we have that

$$\tilde{D}(x) > 0 \qquad \text{for all } x \in \Omega^\epsilon. \tag{6.129}$$

We shall modify $\tilde{D}$ so that the modified function is strictly positive in Ω. Let $\eta \in C([0,\infty))$, $\eta(t) > 0$, for all $t \in (0,\infty)$, $\eta(t) = o(t)$ as $t \to 0+$. Let Ω_1 be an open set, $\overline{\Omega_1} \subset \Omega$, with smooth boundary $\partial\Omega_1 \subset \Omega^\epsilon$ such that $0 < \delta - \eta(\delta) < \tilde{D}(x) < \delta + \eta(\delta)$ for all $x \in \partial\Omega_1$ for some $\delta > 0$. Let Ω_2' be an open set with smooth boundary such that $\overline{\Omega_2'} \subset \Omega_1$ and $\eta(\delta) < d(\partial\Omega_1, \partial\Omega_2') < 2\eta(\delta)$, where $d(\partial\Omega_1, \partial\Omega_2') = \inf\{|x-y| \,:\, x \in \partial\Omega_1,\ y \in \partial\Omega_2'\}$. Let $d_{\partial\Omega_2'}$ be the distance function to $\partial\Omega_2'$, $d_{\partial\Omega_2'} > 0$ in Ω_2', negative outside. Let $d_{\partial\Omega_2',n} = \rho_n * d_{\partial\Omega_2'}$, ρ_n being a positive regularizing kernel. Observe that $\|\nabla d_{\partial\Omega_2',n}\|_\infty \le 1$. We may choose n large enough, and Ω_2 such that $\overline{\Omega_2} \subset \Omega_2'$, $\eta(\delta) < d(\partial\Omega_1, \partial\Omega_2) < 2\eta(\delta)$, $0 < d_{\partial\Omega_2',n} < \eta(\delta)$ in $\partial\Omega_2$, and $d_{\partial\Omega_2',n} > 0$ in Ω_2. Let $B_{1,2} = \Omega_1 \setminus \overline{\Omega_2}$. Then, using again Withney's extension theorem, there is a function $R \in C^1(\overline{B_{1,2}})$ such that $R = \tilde{D} - \delta$ and $\nabla R = \nabla\tilde{D}$ on $\partial\Omega_1$, and $R = d_{\partial\Omega_2',n}$, $\nabla R = \nabla d_{\partial\Omega_2',n}$ on $\partial\Omega_2$. Moreover, $\|\nabla R\|_\infty$ is bounded by a constant depending on $\|\tilde{D}\|_{\infty,\partial\Omega_1}$, $\|d_{\partial\Omega_2',n}\|_{\infty,\partial\Omega_2}$, $\|\nabla\tilde{D}\|_{\infty,\partial\Omega_1}$, $\|\nabla d_{\partial\Omega_2',n}\|_{\infty,\partial\Omega_2}$ and

$$\sup_{x\in\partial\Omega_1, y\in\partial\Omega_2} \frac{|\tilde{D}(x) - \delta - d_{\partial\Omega_2',n}(y)|}{|x-y|} \le \frac{2\eta(\delta)}{\eta(\delta)} = 2.$$

We define $D : \overline{\Omega} \to \mathbb{R}$ by

$$D = \tilde{D} \quad \text{in } \overline{\Omega_1},$$
$$D = R + \delta \quad \text{in } B_{1,2},$$
$$D = d_{\partial\Omega_2',n} + \delta \quad \text{in } \overline{\Omega_2}.$$

Then $D \in C^1(\overline{\Omega})$, $D = 0$ on $\partial\Omega$, $D > 0$ on Ω and $\nabla D(x) = -\nu(x)$ for all $x \in \partial\Omega$.

Proof of Lemma 6.10. We may conclude that u and v are extended as BV functions in $\mathbb{R}^N$ in such a way that

$$\int_{\partial\Omega} \|Du\| = \int_{\partial\Omega} \|Dv\| = 0. \tag{6.130}$$

We consider a family of radially symmetric positive mollifiers $\eta_j = \frac{1}{\tau_j^N}\eta(\frac{x}{\tau_j})$, $\eta \ge 0$, $\int_{\mathbb{R}^N} \eta(x)dx = 1$, $\tau_j \downarrow 0+$, and we set

$$z_j = \eta_j * \left(v + \frac{u}{j}\right). \tag{6.131}$$

Clearly, we have $z_j \in C^1(\overline{\Omega})$ and obviously we have

$$z_j \to v \quad \text{in } L^{N/(N-1)}(\Omega). \tag{6.132}$$

Also, from (6.130) it follows that

$$\int_\Omega \sqrt{1+|\nabla z_j(x)|^2 dx} \to \int_\Omega \sqrt{1+|Dv|^2}. \tag{6.133}$$

This implies, by the theorem of convergence of traces for BV functions (Theorem B.11) that

$$z_j|_{\partial\Omega} \to v|_{\partial\Omega} \quad \text{in } L^1(\partial\Omega). \tag{6.134}$$

By the theorem of differentiation of measures ([27]), we obtain

$$\nabla z_j(x) \to \nabla v(x) \quad \mathcal{L}^N \text{ a.e. in } \Omega. \tag{6.135}$$

Indeed, since $Dz_j = \eta_j * Dv + \frac{1}{j}\eta_j * Du$, this is a consequence of the four following limits:

$$\lim_j [\eta_j * \nabla v](x) = \nabla v(x) \quad \mathcal{L}^N \text{ a.e. in } \Omega, \tag{6.136}$$

$$\lim_j [\eta_j * (Dv)^s](x) = (Dv)^s(x) = 0 \quad \mathcal{L}^N \text{ a.e. in } \Omega, \tag{6.137}$$

$$\lim_j \frac{1}{j}[\eta_j * \nabla u](x) = \nabla u(x) \lim_j \frac{1}{j} = 0 \quad \mathcal{L}^N \text{ a.e. in } \Omega, \tag{6.138}$$

$$\lim_j \frac{1}{j}[\eta_j * (Du)^s](x) = 0 \quad \mathcal{L}^N \text{ a.e. in } \Omega, \tag{6.139}$$

since $(Du)^s$, $(Dv)^s$ are singular with respect to $\mathcal{L}^N$ and $|\nabla u(x)| < \infty$ $\mathcal{L}^N$ a.e. in Ω. In the same way, using the theorem of differentiation of measures, we have

$$\lim_j [\eta_j * \nabla v](x) = 0 \quad |Dv|^s \text{ a.e. in } \Omega, \tag{6.140}$$

$$\lim_j [\eta_j * \nabla u](x) = 0 \quad |Dv|^s \text{ a.e. in } \Omega, \tag{6.141}$$

$$\lim_j [\eta_j * (Du)^{ss}](x) = 0 \quad |Dv|^s \text{ a.e. in } \Omega, \tag{6.142}$$

$$\lim_j \frac{1}{j}[\eta_j * (Du)^{sa}](x) = (Du)^{sa}(x) \lim \frac{1}{j} = 0 \quad |Dv|^s \text{ a.e. in } \Omega, \tag{6.143}$$

where $(Du)^{sa}$, $(Du)^{ss}$ denote the absolutely continuous and singular part of $(Du)^s$ with respect to $(Dv)^s$, and we obtain

$$\lim_j \frac{Dz_j(x)}{|Dz_j(x)|} = \lim_j \frac{Dz_j(x)}{|[\eta_j * (Dv)^s]|(x)} = \frac{Dv}{|Dv|}(x) \quad |Dv|^s \text{ a.e. in } \Omega. \tag{6.144}$$

Similarly

$$\lim_j |Dz_j(x)| = \lim_j |[\eta_j * |Dv|^s](x)| = \infty \quad |Dv|^s \text{ a.e. in } \Omega. \tag{6.145}$$

Next, we prove that for a suitable choice of the numbers τ_j one has

$$\lim_j \frac{Dz_j(x)}{(1/j)[\eta_j * |Du|^{ss}](x)} = \frac{Du}{|Du|}(x) \quad |Du|^{ss} \text{ a.e.} \tag{6.146}$$

Assuming this, it is easy to prove that

$$\lim_j \frac{Dz_j(x)}{|Dz_j(x)|} = \frac{Du}{|Du|}(x) \quad |Du|^{ss} \text{ a.e.} \tag{6.147}$$

Indeed,

$$\begin{aligned} Dz_j &= \eta_j * Dv(x) + \frac{1}{j}\eta_j * Du(x) = \eta_j * \nabla v(x) + \eta_j * (Dv)^s(x) \\ &+\frac{1}{j}\eta_j * \nabla u(x) + \frac{1}{j}\eta_j * (Du)^{sa}(x) + \frac{1}{j}\eta_j * (Du)^{ss}(x). \end{aligned}$$

Since $\eta_j * \nabla v(x) \to 0$, $\eta_j * (Dv)^s(x) \to 0$, $\frac{1}{j}\eta_j * \nabla u(x) \to 0$, $\frac{1}{j}\eta_j * (Du)^{sa}(x) \to 0$ $|Du|^{ss}$-a.e., we see that (6.147) follows from (6.146). To prove (6.146) we observe that

$$\frac{Dz_j(x)}{(1/j)[\eta_j * |Du|^{ss}](x)} = \frac{[\eta_j * Dv](x)}{(1/j)[\eta_j * |Du|^{ss}](x)} + \frac{[\eta_j * Du](x)}{[\eta_j * |Du|^{ss}](x)}. \tag{6.148}$$

Since

$$\frac{[\eta_j * Du](x)}{[\eta_j * |Du|^{ss}](x)} \to \frac{Du}{|Du|}(x) \qquad |Du|^{ss} \text{ a.e.,} \tag{6.149}$$

it is sufficient to prove that

$$\frac{[\eta_j * Dv](x)}{\frac{1}{j}[\eta_j * |Du|^{ss}](x)} \to 0 \qquad |Du|^{ss} \text{ a.e.} \tag{6.150}$$

To prove (6.150), we define

$$a_\tau(x) = \frac{[\eta_\tau * Dv](x)}{[\eta_\tau * |Du|^{ss}](x)}. \tag{6.151}$$

Since Dv and $|Du|^{ss}$ are mutually singular, then

$$a_\tau(x) \to 0 \qquad |Du|^{ss} \text{ a.e.}$$

Thus, if we consider the sets

$$E(\tau, j) = \left\{x \in \Omega : |a_\tau(x)| > \frac{1}{j^2}\right\},$$

for any fixed $j \in \mathbb{N}$ we have

$$\lim_{\tau \to 0} |Du|^{ss}(E(\tau, j)) = 0.$$

For each $j \in \mathbb{N}$, there is some τ_j such that

$$|Du|^{ss}(E(\tau_j, j)) < \frac{1}{2^j},$$

that is

$$|Du|^{ss}\left(\left\{x \in \Omega : j|a_\tau(x)| > \frac{1}{j}\right\}\right) < \frac{1}{2^j}.$$

This easily implies that

$$\lim_{j\to\infty} j a_{\tau_j}(x) = 0 \qquad |Du|^{ss} \text{ a.e.},$$

which is exactly (6.150). Moreover, we may choose τ_j such that

$$\frac{1}{j}[\eta_j * |Du|^{ss}](x) \to \infty \qquad |Du|^{ss}\text{a.e.}$$

From this, and (6.146), it follows that

$$|Dz_j|(x) \to \infty \qquad |Du|^{ss} \text{ a.e.} \tag{6.152}$$

We observe that up to now we have used neither the hypothesis on the regularity of $\partial\Omega$ nor the regularity of g.

On the other hand, the functions z_j that we have constructed satisfy some of the requirements of the lemma but not all of them, in particular, (6.29), (6.31), (6.35), (6.36) have yet to be satisfied. For that, we construct suitable correction functions σ_j and ρ_j around the boundary and we shall define

$$v_j = z_j + \sigma_j + \rho_j.$$

Let $g_j \in C^1(\partial\Omega)$ be such that $g_j \to g$ in $L^1(\partial\Omega)$. We shall construct the sequence of functions $\sigma_j \in C^1(\overline{\Omega})$ such that

$$\sigma_j = g_j - z_j \quad \text{on } \partial\Omega, \tag{6.153}$$

$$\int_\Omega |\sigma_j|^{\frac{N}{N-1}} \to 0, \tag{6.154}$$

$$\sigma_j(x) = 0 \qquad \text{if } D(x) > \epsilon_j + \epsilon_j^2, \tag{6.155}$$

$$\int_\Omega \psi \cdot D\sigma_j \to \int_{\partial\Omega} \psi \cdot \nu(g - v)\,d\mathcal{H}^{N-1} \tag{6.156}$$

for all $\psi \in C(\overline{\Omega}, \mathbb{R}^N)$,

$$\int_\Omega |D\sigma_j| \to \int_{\partial\Omega} |v - g|\,d\mathcal{H}^{N-1}, \tag{6.157}$$

$$|D(\sigma_j + z_j)(x)| \to \infty \quad \mathcal{H}^{N-1} \text{ a.e. in } T = \{x \in \partial\Omega : g(x) \neq v(x)\}, \tag{6.158}$$

$$\frac{D(\sigma_j + z_j)(x)}{|D(\sigma_j + z_j)(x)|} \to \frac{g(x) - v(x)}{|g(x) - v(x)|}\nu(x) \tag{6.159}$$

H^{N-1} a.e. in $T = \{x \in \partial\Omega : g(x) \neq v(x)\}$.

Construction of σ_j.

For each number $\epsilon \in (0, \epsilon_0)$ we consider a function $h_\epsilon(t) : [0, \infty) \to [0, \infty)$ such that

$$h_\epsilon \in C^1([0, \infty)),$$

$$h'_\epsilon(t) \leq 0, \quad h'_\epsilon(0) = -\frac{1}{\epsilon},$$

$$h'_\epsilon(t) \text{ is not decreasing,}$$

$$h_\epsilon(0) = 1, \; h_\epsilon(t) = 0 \text{ for } t \geq \epsilon + \epsilon^2.$$

Let $\{\epsilon_n\}_{n=1}^{\infty}$ be a decreasing sequence of numbers such that

$$2\epsilon_1 < \epsilon_0 < 1, \quad \lim_j \epsilon_j = 0.$$

Now, let $G \in W^{1,1}(\Omega)$ such that $G_{|\partial\Omega} = g$. Since $g_j \in C^1(\partial\Omega)$, we may consider a function $G_j \in C^1(\overline{\Omega})$ which is an extension of g_j. We may assume that $G_j \to G$ in $L^1(\Omega)$ and $\int_\Omega |\nabla G_j| \to \int_\Omega |\nabla G|$. We define

$$\sigma_j = [G_j(x) - z_j(x)]h_{\epsilon_j}(D(x)). \tag{6.160}$$

Clearly, $\sigma_j \in C^1(\overline{\Omega})$,

$$\sigma_j = g_j - z_j \quad \text{on } \partial\Omega,$$

and, if $D(x) > \epsilon_j + \epsilon_j^2$, then $h_{\epsilon_j}(D(x)) = 0$, and, therefore

$$\sigma_j(x) = 0.$$

Now,

$$\int_\Omega |\sigma_j|^{N/(N-1)} = \int_{\Omega_{2\epsilon_j}} |\sigma_j|^{N/(N-1)} \leq \int_{\Omega_{2\epsilon_j}} |G_j(x) - z_j(x)|^{N/(N-1)},$$

where, for any $\epsilon > 0$, we denote

$$\Omega_\epsilon = \{x \in \Omega : D(x) < \epsilon\}.$$

The functions G_j, z_j being independent of ϵ_j, we may choose $\epsilon_j > 0$ small enough such that

$$\int_{\Omega_{2\epsilon_j}} |G_j(x) - z_j(x)|^{N/(N-1)} < \frac{1}{j}.$$

Hence

$$\int_\Omega |\sigma_j|^{N/(N-1)} \to 0 \quad \text{as } j \to \infty.$$

Let $\psi \in C(\overline{\Omega}, \mathbb{R}^N)$. Since

$$\nabla \sigma_j(x) = \nabla(G_j - z_j)(x) h_{\epsilon_j}(D(x)) + (G_j - z_j)(x) h'_{\epsilon_j}(D(x)) \nabla D(x),$$

we have

$$\begin{aligned}
\int_\Omega \psi(x) \cdot \nabla \sigma_j(x) dx &= \int_\Omega \psi(x) \cdot (\nabla G_j(x) - \nabla z_j(x)) h_{\epsilon_j}(D(x)) dx \\
&\quad + \int_\Omega (G_j(x) - z_j(x)) \psi(x) \cdot \nabla (h_{\epsilon_j}(D(x))) dx \\
&= \int_{\Omega_{\epsilon_j + \epsilon_j^2}} \psi(x) \cdot (\nabla G_j(x) - \nabla z_j(x)) h_{\epsilon_j}(D(x)) dx \\
&\quad + \int_{\Omega_{\epsilon_j + \epsilon_j^2}} (G_j(x) - z_j(x)) \psi(x) \cdot h'_{\epsilon_j}(D(x)) \nabla D(x) dx.
\end{aligned}$$

Again, since $|h_\epsilon| \leq 1$ for all $\epsilon > 0$, a proper choice of ϵ_j guarantees that

$$\int_{\Omega_{\epsilon_j + \epsilon_j^2}} \psi(x) \cdot (\nabla G_j(x) - \nabla z_j(x)) h_{\epsilon_j}(D(x)) dx \to 0$$

as $j \to \infty$. Now, by our choice of G_j, (6.133), and a proper choice of ϵ_j, we have that

$$\lim_j \int_{\Omega_{\epsilon_j + \epsilon_j^2}} (G_j(x) - z_j(x)) \psi(x) \cdot h'_{\epsilon_j}(D(x)) \nabla D(x) dx = \int_{\partial\Omega} \psi \cdot \nu (g - v) d\mathcal{H}^{N-1}. \tag{6.161}$$

Indeed, using the change of variable formula ([110], p. 118, [179], p. 96),

$$\begin{aligned}
&\int_{\Omega_{\epsilon_j + \epsilon_j^2}} (G_j(x) - z_j(x)) \psi(x) \cdot h'_{\epsilon_j}(D(x)) \nabla D(x) dx \\
&= \int_0^{\epsilon_j + \epsilon_j^2} \int_{[D=\lambda]} (G_j(y) - z_j(y)) \psi(y) \cdot h'_{\epsilon_j}(D(y)) \frac{\nabla D(y)}{|\nabla D(y)|} d\mathcal{H}^{N-1}(y) d\lambda \\
&= (\epsilon_j + \epsilon_j^2) \int_{[D=\lambda_j]} (G_j(y) - z_j(y)) \psi(y) \cdot h'_{\epsilon_j}(D(y)) \frac{\nabla D(y)}{|\nabla D(y)|} d\mathcal{H}^{N-1}(y)
\end{aligned}$$

for some $\lambda_j \in (0, \epsilon_j + \epsilon_j^2)$ by the intermediate value theorem. Now, since G_j, z_j do not depend on our choice of ϵ_j, by choosing $\epsilon_j \to 0+$ sufficiently fast, we obtain(6.161). Hence

$$\int_\Omega \psi(x) \cdot \nabla \sigma_j(x) dx \to \int_{\partial\Omega} \psi \cdot \nu (g - v) d\mathcal{H}^{N-1}$$

as $j \to \infty$. In particular, we have

$$\liminf_j \int_\Omega |\nabla \sigma_j(x)| dx \geq \int_{\partial\Omega} |g - v| d\mathcal{H}^{N-1}. \tag{6.162}$$

On the other hand, we have

$$\int_\Omega |\nabla \sigma_j(x)| dx \leq \int_{\Omega_{\epsilon_j + \epsilon_j^2}} |\nabla G_j(x) - \nabla z_j(x)| dx + \int_{\Omega_{\epsilon_j + \epsilon_j^2}} |G_j(x) - z_j(x)| |h'_{\epsilon_j}(D(x))| |\nabla D(x)| dx.$$

Again, a suitable choice of ϵ_j guarantees that

$$\int_{\Omega_{\epsilon_j + \epsilon_j^2}} |\nabla G_j(x) - \nabla z_j(x)| dx \to 0$$

as $j \to \infty$. Similarly, the properties of G_j, z_j and a choice of ϵ_j imply that

$$\int_{\Omega_{\epsilon_j + \epsilon_j^2}} |G_j(x) - z_j(x)| |h'_{\epsilon_j}(D(x))| |\nabla D(x)| dx \to \int_{\partial\Omega} |g(x) - v(x)| d\mathcal{H}^{N-1}.$$

Hence

$$\limsup_j \int_\Omega |\nabla \sigma_j(x)| dx \leq \int_{\partial\Omega} |g(x) - v(x)| d\mathcal{H}^{N-1}.$$

This, together with (6.162) proves that

$$\lim_j \int_\Omega |\nabla \sigma_j(x)| dx = \int_{\partial\Omega} |g - v| d\mathcal{H}^{N-1}. \tag{6.163}$$

Finally, since

$$D\sigma_j + Dz_j = \nabla G_j h_{\epsilon_j}(D) + \nabla z_j (1 - h_{\epsilon_j}(D)) + (G_j - z_j) h'_{\epsilon_j}(D) \nabla D,$$

we may write on $\partial\Omega$,

$$D\sigma_j + Dz_j = \nabla G_j - (G_j - z_j) h'_{\epsilon_j}(0) \nu(x). \tag{6.164}$$

Hence, on $\partial\Omega$, we have

$$\frac{D\sigma_j + Dz_j}{|D\sigma_j + Dz_j|} = \frac{\nabla G_j - (G_j - z_j) h'_{\epsilon_j}(0)\nu}{|\nabla G_j - (G_j - z_j) h'_{\epsilon_j}(0)\nu|} = \frac{\epsilon_j \nabla G_j + (G_j - z_j)\nu}{|\epsilon_j \nabla G_j + (G_j - z_j)\nu|}.$$

Now, choosing ϵ_j such that $\epsilon_j \nabla G_j \to 0$ as $j \to \infty$, we obtain that

$$\frac{D\sigma_j + Dz_j}{|D\sigma_j + Dz_j|} \to \frac{g(x) - v(x)}{|g(x) - v(x)|} \nu(x) \tag{6.165}$$

H^{N-1} a.e. in $T = \{x \in \partial\Omega : g(x) \neq v(x)\}$. Next, a proper choice of ϵ_j in (6.164) guarantees that

$$|D\sigma_j(x) + Dz_j(x)| \to \infty \tag{6.166}$$

H^{N-1} a.e. in T.

Next, we construct a sequence of functions $\rho_j \in C^1(\overline{\Omega})$ such that

$$\rho_j = 0 \qquad \text{on } \partial\Omega, \tag{6.167}$$

$$\int_\Omega |\rho_j|^{\frac{N}{N-1}} \to 0, \tag{6.168}$$

$$\rho_j = 0 \qquad \text{if } D(x) > \delta_j^2 \tag{6.169}$$

for some $\delta_j > 0$,

$$\int_\Omega |D\rho_j| \to 0, \tag{6.170}$$

$$\text{for } H^{N-1}\text{-a.e. } x \in T,\ \exists\ j_0(x) \text{ such that } D\rho_j(x) = 0\ \ \forall\ j \geq j_0(x). \tag{6.171}$$

$$\text{If we set } v_j = z_j + \sigma_j + \rho_j \text{ then (6.36) holds.} \tag{6.172}$$

Construction of ρ_j.

For all $\delta > 0$ consider a function $\psi_\delta : [0,\infty) \to [0,\infty)$ such that

$$\psi_\delta \in C^1([0,\infty)),$$

$$\psi_\delta(0) = 0, \quad \psi_\delta(t) = 0 \text{ for } t \geq \delta^2,$$

$$|\psi'_\delta(t)| \leq \frac{4}{\delta}, \text{ for } t \in (0, \delta^2),$$

$$\psi'_\delta(0) \geq \frac{1}{\delta},$$

$$\int_0^\infty |\psi'_\delta(t)|dt \leq 2\delta.$$

We define $\zeta : \partial\Omega \to \mathbb{R}$.

$$\zeta(x) = \begin{cases} \dfrac{u(x) - g(x)}{|u(x) - g(x)|} & \text{if } g(x) = v(x) \text{ and } g(x) \neq u(x) \\ 0 & \text{elsewhere.} \end{cases} \tag{6.173}$$

Let ζ_j be a sequence of functions in $C^1(\partial\Omega)$ converging to ζ in $L^1(\partial\Omega)$. Now, we may assume that ζ is the trace of a function $\Theta \in W^{1,1}(\Omega)$ and ζ_j are traces of functions $\Theta_j \in C^1(\overline{\Omega})$ such that $\Theta_j \to \Theta$ in $L^1(\Omega)$ and $\int_\Omega |D\Theta_j| \to \int_\Omega |D\Theta|$. Let δ_j be a decreasing sequence of positive numbers that converges to 0 and consider the functions

$$\rho_j(x) = \Theta_j(x)\psi_{\delta_j}(D(x)). \tag{6.174}$$

Clearly, $\rho_j \in C^1(\overline{\Omega})$, $\rho_j(x) = 0$ if $x \in \partial\Omega$. Also (6.169) holds. Since, by our choice of the functions ψ_δ, we have

$$|\psi_\delta(t)| \le 2\delta. \tag{6.175}$$

Now,

$$\int_\Omega |\rho_j|^{N/(N-1)} \le 2\delta_j \int_\Omega |\Theta_j|^{N/(N-1)}$$

which tends to 0 as $j \to \infty$, which proves (6.168).

Our purpose now is to choose the functions ζ_j such that (6.171) holds. For that, we consider the sets

$$N^+ = \{x \in \partial\Omega : \zeta(x) = 1\},$$

$$N^- = \{x \in \partial\Omega : \zeta(x) = -1\},$$

$$N = N^+ \cup N^-.$$

We consider increasing sequences of compact sets $K_j^+ \subseteq N^+$, $K_j^- \subseteq N^-$ such that

$$\lim_j \mathcal{H}^{N-1}(N^+ \setminus K_j^+) = \lim_j \mathcal{H}^{N-1}(N^- \setminus K_j^-) = 0.$$

We consider also decreasing sequences of open sets $G_j^+ \supseteq N^+$, $G_j^- \supseteq N^-$ such that

$$\lim_j \mathcal{H}^{N-1}(G_j^+ \setminus N^+) = \lim_j \mathcal{H}^{N-1}(G_j^- \setminus N^-) = 0.$$

Now, we take functions $\zeta_j^+, \zeta_j^- \in C^1(\partial\Omega)$ with values in $[0,1]$ such that

$$\zeta_j^+(x) = \begin{cases} 1 & \text{in } K_j^+, \\ 0 & \text{in } \partial\Omega \setminus G_j^+, \end{cases} \tag{6.176}$$

$$\zeta_j^-(x) = \begin{cases} 1 & \text{in } K_j^-, \\ 0 & \text{in } \partial\Omega \setminus G_j^-, \end{cases} \tag{6.177}$$

and we set

$$\zeta_j = \zeta_j^+ - \zeta_j^-.$$

The functions ζ_j satisfy

$$\zeta_j(x) = \begin{cases} 1 & \text{in } K_j^+ \setminus G_j^-, \\ 0 & \text{in } \partial\Omega \setminus (G_j^+ \cup G_j^-), \\ -1 & \text{in } K_j^- \setminus G_j^+. \end{cases} \tag{6.178}$$

Moreover

$$\lim_j \mathcal{H}^{N-1}(T \cap (G_j^+ \cup G_j^-)) = 0. \tag{6.179}$$

Recall that the functions Θ_j are extensions of ζ_j to Ω. Now,

$$\nabla \rho_j(x) = \nabla \Theta_j(x)\psi_{\delta_j}(D(x)) + \Theta_j(x)\psi'_{\delta_j}(D(x))\nabla D(x).$$

If $x \in \partial\Omega$, then

$$\nabla \rho_j(x) = -\Theta_j(x)\psi'_{\delta_j}(0)\nu(x).$$

Now, using (6.179), for almost all $x \in T = \{x \in \partial\Omega : g(x) \neq v(x)\}$, there exists $j_0(x) \in \mathbb{N}$ such that $\zeta_j(x) = 0$ for all $j \geq j_0(x)$. Hence, also

$$\nabla \rho_j(x) = 0 \quad \text{for all } j \geq j_0(x).$$

Next

$$\begin{aligned}
\int_\Omega |\nabla \rho_j| &\leq 2\delta_j \int_\Omega |\nabla \Theta_j| + \|\Theta_j\|_\infty \|\nabla D\|_\infty \int_{\Omega_{\delta_j^2}} |\psi'_{\delta_j}(D(x))| dx \\
&\leq 2\delta_j \int_\Omega |\nabla \Theta_j| + \|\Theta_j\|_\infty \|\nabla D\|_\infty \frac{4}{\delta_j} \int_{\Omega_{\delta_j^2}} dx.
\end{aligned}$$

Now, for j large enough

$$\int_{\Omega_{\delta_j^2}} dx = \int_0^{\delta_j^2} \int_{[D=\lambda]} \frac{d\mathcal{H}^{N-1}(z)}{|\nabla D(z)|} d\lambda \leq 2 \int_0^{\delta_j^2} \int_{[D=\lambda]} d\mathcal{H}^{N-1}(z) d\lambda \leq C\delta_j^2,$$

where C depends on $\mathrm{Per}(\partial\Omega)$. Hence

$$\int_\Omega |\nabla \rho_j| \leq 2\delta_j \int_\Omega |\nabla \Theta_j| + \|\Theta_j\|_\infty \|\nabla D\|_\infty 4C\delta_j.$$

Now, choosing δ_j we may guarantee that

$$\int_\Omega |\nabla \rho_j| \to 0 \quad \text{as } j \to \infty.$$

Since

$$\nabla v_j = \nabla G_j h_{\epsilon_j} + \nabla z_j(1 - h_{\epsilon_j}) + (G_j - z_j)h'_{\epsilon_j}(D)\nabla D + \nabla \Theta_j \psi_{\delta_j} + \Theta_j \psi'_{\delta_j} \nabla D,$$

on $\partial\Omega$, we have

$$\nabla v_j(x) = \nabla G_j(x) - (G_j(x) - z_j(x))h'_{\epsilon_j}(0)\nu(x) - \Theta_j(x)\psi'_{\delta_j}(0)\nu(x) \tag{6.180}$$

and we may write on $\partial\Omega$,

$$\begin{aligned}\frac{\nabla v_j}{|\nabla v_j|} &= \frac{\nabla G_j - (G_j - z_j)h'_{\epsilon_j}(0)\nu - \Theta_j\psi'_{\delta_j}(0)\nu}{|\nabla G_j - (G_j - z_j)h'_{\epsilon_j}(0)\nu - \Theta_j\psi'_{\delta_j}(0)\nu|} \\ &= \frac{\delta_j\nabla G_j - \delta_j(G_j - z_j)h'_{\epsilon_j}(0)\nu - \delta_j\Theta_j\psi'_{\delta_j}(0)\nu}{|\delta_j\nabla G_j - \delta_j(G_j - z_j)h'_{\epsilon_j}(0)\nu - \delta_j\Theta_j\psi'_{\delta_j}(0)\nu|}.\end{aligned}$$

Now, we choose δ_j such that $\delta_j\nabla G_j \to 0$, $\delta_j(G_j - z_j)h'_{\epsilon_j}(0) \to 0$, as $j \to \infty$, and we obtain that

$$\frac{\nabla v_j}{|\nabla v_j|} \to \frac{g(x) - u(x)}{|g(x) - u(x)|}\nu(x)$$

H^{N-1} a.e. on $\{x \in \partial\Omega : g(x) = v(x), u(x) \neq v(x)\}$. By choosing δ_j to converge sufficiently fast to 0, from (6.180), we obtain that

$$|\nabla v_j(x)| \to \infty$$

H^{N-1} a.e. on $\{x \in \partial\Omega : g(x) = v(x), u(x) \neq v(x)\}$.

Let us now check that $v_j = z_j + \sigma_j + \rho_j$ satisfies the required properties. Since $v_j = g_j$ on $\partial\Omega$, (6.29) follows immediately. The property (6.30) follows from (6.132), (6.154) and (6.168). To check (6.31), let $\psi \in C^1(\overline{\Omega}, \mathbb{R}^N)$, $\psi = (\psi_1, \ldots, \psi_n)$ and $\psi_{N+1} \in C^1(\overline{\Omega}, \mathbb{R})$. Using (6.29) and (6.30), we obtain

$$\begin{aligned}&\lim_j \int_\Omega [\sum_{i=1}^N \psi_i(x)D_iv_j(x) + \psi_{N+1}(x)]dx \\ &= -\lim_j \int_\Omega \operatorname{div}\psi(x)v_j(x)dx + \int_{\partial\Omega} g_j\psi\cdot\nu\, d\mathcal{H}^{N-1} + \int_\Omega \psi_{N+1}(x)dx \\ &= -\int_\Omega \operatorname{div}\psi(x)v(x)dx + \int_{\partial\Omega} g\psi\cdot\nu\, d\mathcal{H}^{N-1} + \int_\Omega \psi_{N+1}(x)dx \\ &= \int_\Omega [\psi(x)\cdot Dv(x) + \psi_{N+1}(x)]dx + \int_{\partial\Omega}(g - v)\psi\cdot\nu\, d\mathcal{H}^{N-1}.\end{aligned}$$

Now, because of the lower semi-continuity of the total variation with respect to weak convergence, we have

$$\int_\Omega \sqrt{1 + |Dv|^2} + \int_{\partial\Omega} |g - v| d\mathcal{H}^{N-1} \leq \liminf_j \int_\Omega \sqrt{1 + |Dv_j|^2}dx.$$

On the other hand, since

$$\int_\Omega \sqrt{1 + |\nabla v_j|^2}dx \leq \int_\Omega \sqrt{1 + |\nabla z_j|^2}dx + \int_\Omega |\nabla\sigma_j|dx + \int_\Omega |\nabla\rho_j|dx,$$

using (6.133), (6.157) and (6.170) we obtain that

$$\limsup_j \int_\Omega \sqrt{1 + |Dv_j|^2}dx \leq \int_\Omega \sqrt{1 + |Dv|^2} + \int_{\partial\Omega} |g - v| d\mathcal{H}^{N-1}.$$

This proves (6.31). Now, using (6.135), (6.155) and (6.169) we obtain (6.32). Next, we observe that (6.33) is a consequence of (6.144), (6.155) and (6.169). In the same way, (6.34) is a consequence of (6.147), (6.155) and (6.169). We observe that (6.35) follows from (6.159) and (6.171). Finally, (6.36) has already been proved. □

Chapter 7

Parabolic Equations Minimizing Linear Growth Functionals: L^1-Theory

7.1 Introduction

Let Ω be an open bounded set in $\mathbb{R}^N$ with boundary $\partial\Omega$ of class C^1. We are interested in the Dirichlet problem

$$\begin{cases} \dfrac{\partial u}{\partial t} = \operatorname{div} \mathbf{a}(x, Du) & \text{in} \quad Q = (0,\infty) \times \Omega, \\ u(t,x) = \varphi(x) & \text{on} \quad S = (0,\infty) \times \partial\Omega, \\ u(0,x) = u_0(x) & \text{in} \quad x \in \Omega, \end{cases} \tag{7.1}$$

where $\varphi \in L^1(\partial\Omega)$, $u_0 \in L^1(\Omega)$ and $\mathbf{a}(x,\xi) = \nabla_\xi f(x,\xi)$, f being a function with linear growth in $\|\xi\|$ as $\|\xi\| \to \infty$. In the previous chapter we proved existence and uniqueness of solutions of problem (7.1) for initial data in $L^2(\Omega)$. Our aim here is to solve this problem for initial and boundary data in $L^1(\Omega)$ using the technique introduced in Chapter 5 to solve the Dirichlet problem for the total variation flow. To do that we use some techniques introduced by Bénilan et al. in [41] to get an existence and uniqueness L^1-theory of solutions of nonlinear elliptic equations in divergence form when the associated variational energy has growth in $|\nabla u|$ of order p with $p > 1$, and also the doubling variables technique introduced by Kruzhkov to prove uniqueness of scalar conservation laws. Let us give a brief description of these ideas.

Let $\Omega \subset \mathbb{R}^N$ an open set, $1 < p < \infty$ and assume that $\mathbf{a} : \Omega \times \mathbb{R}^N \to \mathbb{R}^N$ is a Caratheodory function which is monotone satisfying:

$$\langle \mathbf{a}(x,\xi), \xi \rangle \geq \lambda \|\xi\|^p \qquad \text{for every } \xi \in \mathbb{R}^N \text{ and a.e. } x \in \Omega \tag{7.2}$$

and

$$|\mathbf{a}(x,\xi)| \leq \Lambda(j(x) + \|\xi\|^{p-1}) \qquad \text{for every } \xi \in \mathbb{R}^N \text{ with } j \in L^{p'}(\Omega). \tag{7.3}$$

Associated with $\mathbf{a}$ we may consider the Dirichlet problem

$$\begin{cases} -\text{div}\, \mathbf{a}(x, \nabla u) = f & \text{in } \Omega, \\ u = 0 & \text{on } \partial\Omega. \end{cases} \tag{7.4}$$

When $f \in W^{-1,p'}(\Omega)$, J. Leray and J.L. Lions ([144]), using variational methods, solved problem (7.4). They proved that there exists a function $u \in W_0^{1,p}(\Omega)$ satisfying

$$\int_\Omega \langle \mathbf{a}(x, \nabla u), \nabla w \rangle = \langle f, w \rangle \qquad \forall\, w \in W_0^{1,p}(\Omega).$$

Hence, if $f \in L^1(\Omega)$ and $p > N$, problem (7.4) has a solution. The case $1 < p \leq N$ presents several difficulties. The first one is to give a sense to the notion of solution for p close to 1, more precisely for $1 < p \leq 2 - (\frac{1}{N})$. In fact, we cannot expect the solution to be in $W_{loc}^{1,1}(\Omega)$ (see [41]) and consequently, the gradient appearing in the divergence operator can not be taken in the usual distributional sense. This difficulty is solved in [41] by introducing a new space $\mathcal{T}_{loc}^{1,1}(\Omega)$ in which the gradient of u can be defined by pasting together the gradients of the truncations $T_k(u)$, which turn out to be locally integrable. The second difficulty appears with the question of uniqueness. This question was open even in the linear elliptic case. Indeed, as it is known from the work of J. Serrin [183], in the general linear elliptic case, there is no uniqueness of distributional solutions u in $W_0^{1,q}(\Omega)$ (with $q < 2$).

To overcome these difficulties Bénilan et al. introduced in [41] the concept of *entropy solution*. To avoid technical difficulties we assume that Ω is bounded. In this case, the space $\mathcal{T}^{1,p}(\Omega)$ is defined as the set of all measurable functions $u : \Omega \to \mathbb{R}$ such that $T_k(u) \in W^{1,p}(\Omega)$ for all $k > 0$; and the space $\mathcal{T}_0^{1,p}(\Omega)$ is the set of all $u \in \mathcal{T}^{1,p}(\Omega)$ such that for every $k > 0$ and every smooth cutoff function $\eta \in C_0^\infty(\mathbb{R}^N)$, $\eta T_k(u) \in W_0^{1,p}(\Omega)$. Then, a function $u \in \mathcal{T}_0^{1,p}(\Omega)$ is said to be an *entropy solution* of problem (7.4) if

$$\int_\Omega \langle \mathbf{a}(x, Du), DT_k(u - \phi) \rangle \, dx \leq \int_\Omega T_k(u - \phi) f \, dx$$

for all test functions $\phi \in C_0^\infty(\Omega)$ (or, equivalently, $\phi \in \mathcal{T}_0^{1,p}(\Omega) \cap L^\infty(\Omega)$) and all $k > 0$. It is proved in [41] that for every $f \in L^1(\Omega)$ there exists a unique entropy solution u of problem (7.4). For linear elliptic equations, the uniqueness of distributional solutions u such that $T_k(u) \in H_0^1(\Omega)$ for all $k > 0$ is an open question (see [42]).

Let us mention some parallel developments. The notion of *renormalized solution* was introduced by R. J. DiPerna and P.L. Lions [99] in the context of Boltzmann equations. This notion was adapted to the study of some nonlinear elliptic problems by L. Boccardo, D. Giachetti, J.I. Diaz and F. Murat [55] when the right-hand side is in $W^{-1,p'}(\Omega)$, and by P. L. Lions and F. Murat [147] (see also [160]) when the right-hand side is in $L^1(\Omega)+W^{-1,p'}(\Omega)$. The concept of renormalized solution can be proved to be equivalent to the concept of entropy solution. In [20], the notion of entropy solution was extended to elliptic problems of the form (7.4) with nonlinear boundary condition of the form $-\langle \mathbf{a}(x, Du), \eta\rangle \in \beta(u)$, β being a maximal monotone graph in $\mathbb{R}\times\mathbb{R}$. Existence and uniqueness of entropy solutions for the corresponding parabolic problem was obtained in [21] (see also [168], for similar results in the case of Dirichlet boundary conditions). In parallel with the elliptic case, D. Blanchard and F. Murat [49] introduced the concept of renormalized solution for nonlinear parabolic problems with L^1 data and proved existence and uniqueness of these solutions. Nowadays there is an extensive literature dealing with entropy/renormalized solutions.

As it is well known, for initial value problems for scalar conservation laws of the form

$$\begin{cases} u_t + f(u)_x = 0, \\ u(0,\cdot) = u_0, \end{cases} \tag{7.5}$$

the concept of weak solution is usually not stringent enough to single out a unique solution (see for instance [57]). In some cases, as it happens for Burger's equation, an entire family of weak solutions can be found. Consequently, in order to achieve uniqueness, the notion of weak solution must be supplemented with further admissibility conditions. Motivated by physical considerations, the concept of solution is the following: A locally integrable function $u : [0,\infty) \to \mathbb{R}$ is an *entropy solution* of (7.5) if $u(0,\cdot) = u_0$ and

$$\int\!\!\int \{|u-k|\varphi_t + \mathrm{sign}(u-k)(f(u)-f(k))\varphi_x\}\, dx\, dt \geq 0 \tag{7.6}$$

for every constant $k \in \mathbb{R}$ and every C^1 function $\varphi \geq 0$ with compact support in $[0,\infty)\times\mathbb{R}$.

In [143], S. Kruzhkov proved an estimate of the L^1 distance between any two bounded entropy solutions of (7.5), and, in particular, that the entropy solution of (7.6) is unique, within the class of L^∞ functions. To do that he introduced the doubling variables technique. This method has been very useful to prove uniqueness in different contexts. Carrillo [61] was the first to apply Kruzhkov's method to parabolic equations. In this monograph we use Kruzhkov's method several times to prove uniqueness.

7.2 The Main Result

In this section we give the concept of solution for the Dirichlet problem (7.1) and we state an existence and uniqueness result for this type of solutions.

Here we assume that Ω is an open bounded set in $\mathbb{R}^N$, $N \geq 2$, with boundary $\partial\Omega$ of class C^1, and the Lagrangian $f : \overline{\Omega} \times \mathbb{R}^N \to \mathbb{R}$ satisfies the assumptions (H) given in Section 6.3. Moreover, additionally we need to consider here the following assumption:

(H_6) We assume that

$$|a(x, \xi) - a(y, \xi)| \leq \omega(\|x - y\|) \tag{7.7}$$

for all $x, y \in \Omega$, and all $\xi \in \mathbb{R}^N$, where $\omega(r)$ is a modulus of continuity.

From the definition of $\mathbf{a}^\infty$,

$$|\mathbf{a}^\infty(x, \xi) - \mathbf{a}^\infty(y, \xi)| \leq \omega(\|x - y\|).$$

Hence

$$|h^0(x, \xi) - h^0(y, \xi)| = |\mathbf{a}^\infty(x, \xi) \cdot \xi - \mathbf{a}^\infty(y, \xi) \cdot \xi| \leq \omega(\|x - y\|)\|\xi\|. \tag{7.8}$$

Remark 7.1. We note that assumption (H_6) is only needed to prove uniqueness. The Lipschitz continuity in x of the flux is a common assumption to prove uniqueness of Kruzhkov's solutions of scalar conservation laws ([143]).

Recall that in Chapter 5 we have introduced the set

$$\mathcal{T} = \left\{T_k, T_k^+, T_k^- \ : \ k > 0\right\}.$$

where $T_k(r) = [k - (k - |r|)^+]\text{sign}_0(r)$, $k \geq 0$, $r \in \mathbb{R}$.

Our concept of solution for the Dirichlet problem (7.1) is the following.

Definition 7.2. A measurable function $u : (0, T) \times \Omega \to \mathbb{R}$ is an *entropy solution* of (7.1) in $Q_T = (0, T) \times \Omega$ if $u \in C([0, T]; L^1(\Omega))$, $p(u(\cdot)) \in L^1_w(0, T, BV(\Omega))$ for all $p \in \mathcal{T}$, and there exists $\xi \in \left(L^1(0, T, BV(\Omega)_2)\right)^*$ such that:

(i) $(\mathbf{a}(x, \nabla u(t)), \xi(t)) \in Z(\Omega)$ a.e. $t \in [0, T]$.

(ii) ξ is the time derivative of u in $\left(L^1(0, T, BV(\Omega)_2)\right)^*$ in the sense of Definition 5.3.

(iii) $\xi = \text{div}(\mathbf{a}(x, \nabla u(t)))$ in $\left(L^1(0, T, BV(\Omega))\right)^*$ a.e. $t \in [0, T]$ in the sense of Definition 5.4.

(iv) $[\mathbf{a}(x, \nabla u(t)), \nu] \in \text{sign}\big(p(\varphi) - p(u(t))\big) f^0(x, \nu(x))$ a.e. in $t \in [0, T]$ for all $p \in \mathcal{T}$.

(v) The following inequality is satisfied

$$-\int_0^T \int_\Omega j(u(t)-l)\eta_t \, dxdt + \int_0^T \int_\Omega \eta(t) h(x, Dp(u(t)-l)) \, dt$$

$$+ \int_0^T \int_\Omega \mathbf{a}(x, \nabla u(t)) \cdot \nabla \eta(t) p(u(t)-l) \, dxdt$$

$$\leq \int_0^T \int_{\partial\Omega} [\mathbf{a}(x, \nabla u(t)), \nu] \eta(t) p(u(t)-l) \, d\mathcal{H}^{N-1} dt,$$

for all $l \in \mathbb{R}$, for all $\eta \in C^\infty(\overline{Q_T})$, with $\eta \geq 0$, $\eta(t,x) = \phi(t)\psi(x)$, being $\phi \in \mathcal{D}(]0,T[)$, $\psi \in C^\infty(\overline{\Omega})$, and $p \in \mathcal{T}$, where $j(r) = \displaystyle\int_0^r p(s) \, ds$.

The main result of this chapter is the following existence and uniqueness theorem.

Theorem 7.3. *Assume we are under assumptions (H) and (H_6) of Chapter 6, and let $u_0 \in L^1(\Omega)$ and $\varphi \in L^1(\partial\Omega)$. Then there exists a unique entropy solution of (7.1) in $(0,T) \times \Omega$ for every $T > 0$ such that $u(0) = u_0$. Moreover, if $u(t), \hat{u}(t)$ are the entropy solutions corresponding to initial data $u_0, \hat{u}_0$, respectively, then*

$$\| (u(t) - \hat{u}(t))^+ \|_1 \leq \| (u_0 - \hat{u}_0)^+ \|_1 \quad \text{and} \quad \|u(t) - \hat{u}(t)\|_1 \leq \|u_0 - \hat{u}_0\|_1 \tag{7.9}$$

for all $t \geq 0$.

7.3 The Semigroup Solution

To prove the existence part of Theorem 7.3 we shall use the techniques of completely accretive operators and Crandall–Liggett's semigroup generation theorem. In Chapter 6, using nonlinear semigroups, we studied the Dirichlet problem (7.1) for initial data in L^2. For that we have considered the energy functional $\Phi_\varphi : L^2(\Omega) \to [0, +\infty]$ defined by

$$\Phi_\varphi(u) := \int_\Omega f(x, Du) + \int_{\partial\Omega} f^0(x, \nu(x)[\varphi - u]) \, d\mathcal{H}^{N-1},$$

if $u \in BV(\Omega) \cap L^2(\Omega)$ and

$$\Phi_\varphi(u) := +\infty \quad \text{if} \quad u \in L^2(\Omega) \setminus BV(\Omega).$$

Since Φ_φ is lower-semi-continuous, the subdifferential $\partial\Phi_\varphi$ of Φ_φ is a maximal monotone operator in $L^2(\Omega)$. Thus, existence and uniqueness of a solution of the abstract Cauchy problem

$$\begin{cases} u'(t) + \partial\Phi_\varphi(u(t)) \ni 0 & t \in]0, \infty[, \\ u(0) = u_0 & u_0 \in L^2(\Omega), \end{cases} \tag{7.10}$$

follows immediately from the nonlinear semigroup theory (see Appendix A.1). Now, to get the full strength of the abstract result derived from semigroup theory, $\partial\Phi_\varphi(u)$ was characterized. To get this characterization, the operator $\mathcal{B}_\varphi$ in $L^2(\Omega)$ was introduced in Chapter 6, Section 6.4. Let us recall its definition.

$$(u,v) \in \mathcal{B}_\varphi \iff u \in BV(\Omega)\cap L^2(\Omega), v \in L^2(\Omega) \text{ and } \mathbf{a}(x,\nabla u) \in X(\Omega)_1 \text{ satisfies :}$$

$$-v = \operatorname{div} \mathbf{a}(x,\nabla u) \quad \text{in } \mathcal{D}'(\Omega), \tag{7.11}$$

$$\mathbf{a}(x,\nabla u)\cdot D^s u = f^0(x, D^s u) = f^0(x,\overrightarrow{D^s u})|D^s u|, \tag{7.12}$$

$$[\mathbf{a}(x,\nabla u),\nu] \in \operatorname{sign}\,(p(\varphi)-p(u))f^0(x,\nu(x)), \ \forall\, p\in\mathcal{P} \quad \mathcal{H}^{N-1}-\text{a.e.} \tag{7.13}$$

Proposition 7.4. *Let $\varphi \in L^1(\partial\Omega)$. Assume we are under assumptions (H). Then the operator $\mathcal{B}_\varphi = \partial\Phi_\varphi$ is completely accretive.*

Proof. Let us consider the functional $\Gamma_\varphi : L^2(\Omega) \to]-\infty,+\infty]$ given by

$$\Gamma_\varphi(u) := \int_\Omega f(x,\nabla u)\,dx + \int_{\partial\Omega} f^0(x,\nu(x)[\varphi-u])\,d\mathcal{H}^{N-1},$$

if $u \in W^{1,1}(\Omega)\cap L^2(\Omega)$ and

$$\Gamma_\varphi(u) := +\infty \quad \text{if} \quad u \in L^2(\Omega)\setminus W^{1,1}(\Omega).$$

Let us prove that Γ_φ satisfies

$$\Gamma_\varphi(u+p(\hat u-u)) + \Gamma_\varphi(\hat u - p(\hat u-u)) \le \Gamma_\varphi(u)+\Gamma_\varphi(\hat u) \tag{7.14}$$

for all $u,\hat u \in L^1(\Omega)$, $p\in P_0$. In fact, we may assume $u,\hat u \in W^{1,1}(\Omega)$. If $v := u+p(\hat u-u)$ and $\hat v := \hat u - p(\hat u - u)$, then

$$\begin{aligned}\Gamma_\varphi(v) &= \int_\Omega f(x,\lambda\nabla\hat u+(1-\lambda)\nabla u)\,dx \\ &\quad + \int_{\partial\Omega} f^0(x,\nu(x))|\alpha(\varphi-\hat u)+(1-\alpha)(\varphi-u)|\,d\mathcal{H}^{N-1}\end{aligned}$$

and

$$\begin{aligned}\Gamma_\varphi(\hat v) &= \int_\Omega f(x,\lambda\nabla u+(1-\lambda)\nabla\hat u)\,dx \\ &\quad + \int_{\partial\Omega} f^0(x,\nu(x))|\alpha(\varphi-u)+(1-\alpha)(\varphi-\hat u)|\,d\mathcal{H}^{N-1},\end{aligned}$$

where

$$\lambda = p'(\hat u-u) \quad \text{and} \quad \alpha = \chi_{\{u\neq\hat u\}}\frac{p(\hat u-u)}{\hat u-u}.$$

By the convexity of $f(x, \xi)$ with respect to ξ, we have

$$\begin{aligned}
& f(x, \lambda \nabla \hat{u} + (1-\lambda)\nabla u) + f(x, \lambda \nabla u + (1-\lambda)\nabla \hat{u}) \\
& \leq \lambda f(x, \nabla \hat{u}) + (1-\lambda) f(x, \nabla u) + \lambda f(x, \nabla u) + (1-\lambda) f(x, \nabla \hat{u}) \\
& = f(x, \nabla u) + f(x, \nabla \hat{u}).
\end{aligned}$$

In the same way

$$|\alpha(\varphi - \hat{u}) + (1-\alpha)(\varphi - u)| + |\alpha(\varphi - u) + (1-\alpha)(\varphi - \hat{u})| \leq |\varphi - u| + |\varphi - \hat{u}|.$$

Then, integrating, we obtain $\Gamma_\varphi(v) + \Gamma_\varphi(\hat{v}) \leq \Gamma_\varphi(u) + \Gamma_\varphi(\hat{u})$.

Since Γ_φ satisfies (7.14), by Lemma A.48, we have that the operator $\partial\Gamma_\varphi$ is completely accretive. Now, by Theorem 6.4, the lower semi-continuous envelope of the functional Γ_φ is the functional Φ_φ. Hence, by Lemma A.49 and Theorem A.50, we get that $\partial\Phi_\varphi$ is completely accretive. □

To associate an m-completely accretive operator in $L^1(\Omega)$ to problem (7.1) we need to consider the function space

$$TBV(\Omega) := \left\{ u \in L^1(\Omega) \ : \ T_k(u) \in BV(\Omega), \ \ \forall\, k > 0 \right\},$$

and to give a sense to the Radon–Nikodym derivative ∇u of a function $u \in TBV(\Omega)$. A similar problem was treated in [41] where the authors had to give a sense to the derivative of functions whose truncations are in a Sobolev space (in their notation, for functions in $\mathcal{T}^{1,p}_{loc}(\Omega)$, $p \geq 1$). Notice that the function space $TBV(\Omega)$ is closely related to the space $GBV(\Omega)$ of generalized functions of bounded variation introduced by E. Di Giorgi and L. Ambrosio ([86], see also [10]). Using the chain rule for BV-functions (see for instance [10]), with a similar proof to the one given in Lemma 2.1 of [41], we obtain the following result.

Lemma 7.5. *For every $u \in TBV(\Omega)$ there exists a unique measurable function $v : \Omega \to \mathbb{R}^N$ such that*

$$\nabla T_k(u) = v \chi_{\{|u|<k\}} \qquad \mathcal{L}^N - \text{a.e.} \tag{7.15}$$

Thanks to this result we define ∇u for a function $u \in TBV(\Omega)$ as the unique function v which satisfies (7.15). This notation will be used throughout in the sequel.

Lemma 7.6. *If $u \in TBV(\Omega)$, then $p(u) \in BV(\Omega)$ for every Lipschitz continuous function $p : \mathbb{R} \to \mathbb{R}$ satisfying $p'(s) = 0$ for $|s|$ large enough. Moreover, $\nabla p(u) = p'(u)\nabla u$ $\mathcal{L}^N$-a.e.*

Proof. The proof of this lemma is straightforward since $p(u) = p(T_k(u))$ for k large enough. Hence, $p(u) \in BV(\Omega)$ and by the chain rule,

$$\nabla p(u) = \nabla p(T_k(u)) = p'(T_k(u)) \nabla T_k(u) = p'(u) \nabla u \, \chi_{\{|u|<k\}} = p'(u)\nabla u. \qquad \square$$

Let $\mathcal{P}$ be the set of truncation functions defined in (3.16).

We define the operator $\mathcal{A}_\varphi$ in $L^1(\Omega)$ by

$$(u,v) \in \mathcal{A}_\varphi \iff u, v \in L^1(\Omega), p(u) \in BV(\Omega) \text{ for all } p \in \mathcal{P}$$
$$\text{and } \mathbf{a}(x, \nabla u) \in X(\Omega)_1 \text{ satisfies :}$$

$$-v = \operatorname{div} \mathbf{a}(x, \nabla u) \qquad \text{in } \mathcal{D}'(\Omega), \tag{7.16}$$

$$\mathbf{a}(x, \nabla u) \cdot D^s p(u) = f^0(x, D^s p(u)) \qquad \forall\, p \in \mathcal{P}, \tag{7.17}$$

$$[\mathbf{a}(x, \nabla u), \nu] \in \text{sign }(p(\varphi) - p(u)) f^0(x, \nu(x)), \;\; \forall\, p \in \mathcal{P} \;\; \mathcal{H}^{N-1} - \text{a.e.} \tag{7.18}$$

Taking into account Lemma 7.6, we have that if $(u,v) \in \mathcal{A}_\varphi$, then

$$\begin{aligned} \int_\Omega (w - p(u))v \, dx = & \int_\Omega (\mathbf{a}(x, \nabla u), Dw) - \int_\Omega h(x, Dp(u)) \\ & - \int_{\partial\Omega} [\mathbf{a}(x, \nabla u), \nu](w - p(\varphi)) \, d\mathcal{H}^{N-1} \\ & - \int_{\partial\Omega} |p(u) - p(\varphi)| f^0(x, \nu(x)) \, d\mathcal{H}^{N-1}, \end{aligned} \tag{7.19}$$

for all $w \in BV(\Omega) \cap L^\infty(\Omega)$ and for all $p \in \mathcal{P}$.

Lemma 7.7. *If $(u,v) \in \mathcal{A}_\varphi$, for $a, b > 0$, we have*

$$\begin{aligned} & \frac{1}{b} \int_{\{a<|u|<a+b\}} \mathbf{a}(x, \nabla u) \cdot \nabla u \, dx \\ & \leq \int_{\{|u|>a\}} |v| \, dx + \int_{\{|\varphi|>a\}} |[\mathbf{a}(x, \nabla u), \nu]| \, d\mathcal{H}^{N-1}. \end{aligned} \tag{7.20}$$

In particular,

$$\lim_{a\to\infty} \int_{\{a<|u|<a+b\}} \mathbf{a}(x, \nabla u) \cdot \nabla u \, dx = \lim_{a\to\infty} \int_{\{a<|u|<a+b\}} |\nabla u| \, dx = 0. \tag{7.21}$$

Proof. We consider the truncation function $T_{a,b}(s) := T_b(s - T_a(s))$. Since $T_{a,b}$ is an element of $\mathcal{P}$, applying Green's formula (C.10), we have

$$\begin{aligned}
-\int_\Omega vT_{a,b}(u)\,dx &= \int_\Omega \operatorname{div}(\mathbf{a}(x,\nabla u))T_{a,b}(u)\,dx \\
&= -\int_\Omega (\mathbf{a}(x,\nabla u), DT_{a,b}(u)) + \int_{\partial\Omega} [\mathbf{a}(x,\nabla u),\nu]T_{a,b}(u)\,d\mathcal{H}^{N-1} \\
&= -\int_\Omega \mathbf{a}(x,\nabla u)\cdot\nabla T_{a,b}(u)\,dx - \int_\Omega \mathbf{a}(x,\nabla u)\cdot D^sT_{a,b}(u) \\
&\quad + \int_{\partial\Omega} [\mathbf{a}(x,\nabla u),\nu]T_{a,b}(u)\,d\mathcal{H}^{N-1} \\
&= -\int_{\{a<|u|<a+b\}} \mathbf{a}(x,\nabla u)\cdot\nabla u\,dx - \int_\Omega f^0(x, D^sT_{a,b}(u)) \\
&\quad + \int_{\partial\Omega} [\mathbf{a}(x,\nabla u),\nu]T_{a,b}(u)\,d\mathcal{H}^{N-1}.
\end{aligned}$$

Since

$$\begin{aligned}
&\int_{\partial\Omega} [\mathbf{a}(x,\nabla u),\nu]T_{a,b}(u)\,d\mathcal{H}^{N-1} \\
&= \int_{\partial\Omega} [\mathbf{a}(x,\nabla u),\nu](T_{a,b}(u)-T_{a,b}(\varphi))\,d\mathcal{H}^{N-1} + \int_{\partial\Omega} [\mathbf{a}(x,\nabla u),\nu]T_{a,b}(\varphi)\,d\mathcal{H}^{N-1} \\
&= -\int_{\partial\Omega} |T_{a,b}(u)-T_{a,b}(\varphi)| f^0(x,\nu(x))\,d\mathcal{H}^{N-1} + \int_{\partial\Omega} [\mathbf{a}(x,\nabla u),\nu]T_{a,b}(\varphi)\,d\mathcal{H}^{N-1} \\
&\le \int_{\partial\Omega} [\mathbf{a}(x,\nabla u),\nu]T_{a,b}(\varphi)\,d\mathcal{H}^{N-1},
\end{aligned}$$

it follows that

$$\begin{aligned}
&\int_{\{a<|u|<a+b\}} \mathbf{a}(x,\nabla u)\cdot\nabla u\,dx \\
&\le \int_\Omega vT_{a,b}(u)\,dx + \int_{\partial\Omega} [\mathbf{a}(x,\nabla u),\nu]T_{a,b}(\varphi)\,d\mathcal{H}^{N-1} \\
&\le b\left(\int_{\{|u|>a\}} |v|\,dx + \int_{\{|\varphi|>a\}} |[\mathbf{a}(x,\nabla u),\nu]|\,d\mathcal{H}^{N-1}\right),
\end{aligned}$$

and the result follows. □

The main result of this section is the following.

Theorem 7.8. *Let $\varphi \in L^1(\partial\Omega)$. Assume we are under assumptions (H). Then, $\mathcal{B}_\varphi \subset \mathcal{A}_\varphi$ and the operator $\mathcal{A}_\varphi$ is m-completely accretive in $L^1(\Omega)$ with dense domain. Moreover, if $(T(t))_{t\ge 0}$ is the semigroup of order preserving contractions in $L^1(\Omega)$ generated by the operator $\mathcal{A}_\varphi$, then its restriction to $L^2(\Omega)$ coincides with the semigroup generated by the operator $\mathcal{B}_\varphi$.*

Proof. First, let us prove that $\mathcal{B}_\varphi \subset \mathcal{A}_\varphi$. If $(u,v) \in \mathcal{B}_\varphi$, then $u \in BV(\Omega) \cap L^2(\Omega)$, $v \in L^2(\Omega)$ and $\mathbf{a}(x, \nabla u) \in X(\Omega)_1$ satisfies

$$-v = \operatorname{div} \mathbf{a}(x, \nabla u) \qquad \text{in } \mathcal{D}'(\Omega), \tag{7.22}$$

$$\mathbf{a}(x, \nabla u) \cdot D^s u = f^0(x, D^s u) = f^0(x, \overrightarrow{D^s u})|D^s u|, \tag{7.23}$$

$$[\mathbf{a}(x, \nabla u), \nu] \in \text{sign } (\varphi - u) f^0(x, \nu(x)), \quad \forall\, p \in \mathcal{P} \quad \mathcal{H}^{N-1} - \text{a.e.} \tag{7.24}$$

Since the functions of $\mathcal{P}$ are nondecreasing, (7.18) follows from (7.24). On the other hand, by (7.23), we have

$$\theta(\mathbf{a}(x, \nabla u), Du, x) = f^0(x, \overrightarrow{D^s u}) \qquad |D^s u| - \text{a.e.} \tag{7.25}$$

Then, given $p \in \mathcal{P}$, by Corollary C.16, for every Borel set $B \subset \Omega$, we have

$$\begin{aligned}
&\int_B \mathbf{a}(x, \nabla u) \cdot \nabla p(u)\, dx + \int_B \mathbf{a}(x, \nabla u) \cdot D^s p(u) = \int_B (\mathbf{a}(x, \nabla u), Dp(u)) \\
&= \int_B \theta(\mathbf{a}(x, \nabla u), Dp(u), x) \|Dp(u)\| = \int_B \theta(\mathbf{a}(x, \nabla u), Du, x) \|Dp(u)\| \\
&= \int_B \theta(\mathbf{a}(x, \nabla u), Du, x) |\nabla p(u)| + \int_B \theta(\mathbf{a}(x, \nabla u), Du, x) |D^s p(u)| \\
&= \int_B \theta(\mathbf{a}(x, \nabla u), Du, x) \cdot p'(u) |\nabla u|\, dx + \int_B f^0(x, \overrightarrow{D^s u}) |D^s p(u)| \\
&= \int_B \mathbf{a}(x, \nabla u) \cdot \nabla p(u)\, dx + \int_B f^0(x, \overrightarrow{D^s u}) |D^s p(u)|.
\end{aligned}$$

Hence, $\mathbf{a}(x, \nabla u) \cdot D^s p(u) = f^0(x, \overrightarrow{D^s u}) |D^s p(u)| = f^0(x, D^s p(u))$. Thus, (7.17) holds and, therefore, we have proved that $\mathcal{B}_\varphi \subset \mathcal{A}_\varphi$. Moreover, since the domain of $\mathcal{B}_\varphi$ is dense in $L^2(\Omega)$ (Theorem 6.7), we also obtain that the domain of $\mathcal{A}_\varphi$ is dense in $L^1(\Omega)$.

Next we are going to prove that the operator $\mathcal{A}_\varphi$ is accretive in $L^1(\Omega)$. We must show that

$$\int_\Omega |u - \hat{u}| \le \int_\Omega |f - \hat{f}| \tag{7.26}$$

whenever $f \in u + \mathcal{A}_\varphi u$, $\hat{f} \in \hat{u} + \mathcal{A}_\varphi \hat{u}$. Indeed, for $r, k > 0$, we have

$$\begin{aligned}
&\int_\Omega \left[(f - u) - (\hat{f} - \hat{u})\right] T_r(T_k(u) - T_k(\hat{u}))\, dx \\
&= \int_\Omega (\mathbf{a}(x, \nabla u) - \mathbf{a}(x, \nabla \hat{u}), DT_r(T_k(u) - T_k(\hat{u}))) \\
&\quad - \int_{\partial\Omega} [\mathbf{a}(x, \nabla u) - \mathbf{a}(x, \nabla \hat{u}), \nu] T_r(T_k(u) - T_k(\hat{u}))\, d\mathcal{H}^{N-1}
\end{aligned}$$

$$= \int_\Omega (\mathbf{a}(x,\nabla u) - \mathbf{a}(x,\nabla \hat{u})) \cdot \nabla T_r(T_k(u) - T_k(\hat{u}))\, dx$$
$$+ \int_\Omega (\mathbf{a}(x,\nabla u) - \mathbf{a}(x,\nabla \hat{u})) \cdot D^s T_r(T_k(u) - T_k(\hat{u}))$$
$$- \int_{\partial\Omega} [\mathbf{a}(x,\nabla u) - \mathbf{a}(x,\nabla \hat{u}), \nu] T_r(T_k(u) - T_k(\hat{u}))\, d\mathcal{H}^{N-1}.$$

If we write

$$I_{r,k} := \int_\Omega (\mathbf{a}(x,\nabla u) - \mathbf{a}(x,\nabla \hat{u})) \cdot \nabla T_r(T_k(u) - T_k(\hat{u}))\, dx,$$

$$J_{r,k} := \int_\Omega (\mathbf{a}(x,\nabla u) - \mathbf{a}(x,\nabla \hat{u})) \cdot D^s T_r(T_k(u) - T_k(\hat{u})),$$

since

$$\int_{\partial\Omega} [\mathbf{a}(x,\nabla u) - \mathbf{a}(x,\nabla \hat{u}), \nu] T_r(T_k(u) - T_k(\hat{u}))\, d\mathcal{H}^{N-1} \leq 0,$$

we have

$$\begin{aligned} &\int_\Omega (u - \hat{u}) T_r(T_k(u) - T_k(\hat{u}))\, dx \\ &\leq -I_{r,k} - J_{r,k} + \int_\Omega (f - \hat{f}) T_r(T_k(u) - T_k(\hat{u}))\, dx. \end{aligned} \tag{7.27}$$

Now, if $\Omega_{k,r} := \{|T_k(u) - T_k(\hat{u})| < r\}$, then

$$\begin{aligned} I_{r,k} &= \int_{\Omega_{k,r}} (\mathbf{a}(x,\nabla u) - \mathbf{a}(x,\nabla \hat{u})) \cdot \nabla (T_k(u) - T_k(\hat{u}))\, dx \\ &= \int_{\Omega_{k,r}\cap\{|u|<k,|\hat{u}|<k\}} (\mathbf{a}(x,\nabla u) - \mathbf{a}(x,\nabla \hat{u})) \cdot (\nabla u - \nabla \hat{u})\, dx \\ &\quad + \int_{\Omega_{k,r}\cap\{|u|<k,|\hat{u}|\geq k\}} (\mathbf{a}(x,\nabla u) - \mathbf{a}(x,\nabla \hat{u})) \cdot \nabla u\, dx \\ &\quad - \int_{\Omega_{k,r}\cap\{|u|\geq k,|\hat{u}|<k\}} (\mathbf{a}(x,\nabla u) - \mathbf{a}(x,\nabla \hat{u})) \cdot \nabla \hat{u}\, dx \\ &\geq \int_{\Omega_{k,r}\cap\{|u|<k,|\hat{u}|\geq k\}} (\mathbf{a}(x,\nabla u) - \mathbf{a}(x,\nabla \hat{u})) \cdot \nabla u\, dx \\ &\quad - \int_{\Omega_{k,r}\cap\{|u|\geq k,|\hat{u}|<k\}} (\mathbf{a}(x,\nabla u) - \mathbf{a}(x,\nabla \hat{u})) \cdot \nabla \hat{u}\, dx. \end{aligned}$$

On the other hand, by Lemma 1.2 of [129], there exists a nonnegative function ξ_r

such that

$$\begin{aligned} J_{r,k} &= \int_\Omega (\mathbf{a}(x,\nabla u) - \mathbf{a}(x,\nabla \hat{u})) \cdot \xi_r (D^s T_k(u) - D^s T_k(\hat{u})) \\ &= \int_\Omega \xi_r \left(f^0(x, D^s T_k(u)) - \mathbf{a}(x,\nabla \hat{u}) \cdot D^s T_k(u) \right) \\ &\quad + \int_\Omega \xi_r \left(f^0(x, D^s T_k(\hat{u})) - \mathbf{a}(x,\nabla u) \cdot D^s T_k(\hat{u}) \right). \end{aligned}$$

Then, by (H_5), we have $J_{r,k} \geq 0$. Hence, letting $k \to \infty$ in (7.27), it follows that

$$\begin{aligned} \int_\Omega (u-\hat{u}) T_r(u-\hat{u})\, dx \leq & \int_\Omega (f - \hat{f}) T_r(u - \hat{u})\, dx \\ & - \lim_{k\to\infty} \int_{\Omega_{k,r} \cap \{|u|<k, |\hat{u}|\geq k\}} (\mathbf{a}(x,\nabla u) - \mathbf{a}(x,\nabla \hat{u})) \cdot \nabla u \, dx \\ & + \lim_{k\to\infty} \int_{\Omega_{k,r} \cap \{|u|\geq k, |\hat{u}|< k\}} (\mathbf{a}(x,\nabla u) - \mathbf{a}(x,\nabla \hat{u})) \cdot \nabla \hat{u} \, dx. \end{aligned}$$

Now, by Lemma 7.7,

$$\begin{aligned} & \lim_{k\to\infty} \left| \int_{\Omega_{k,r} \cap \{|u|<k, |\hat{u}|\geq k\}} (\mathbf{a}(x,\nabla u) - \mathbf{a}(x,\nabla \hat{u})) \cdot \nabla u \, dx \right| \\ & \leq \lim_{k\to\infty} \int_{\{k-r<|u|<k\}} |(\mathbf{a}(x,\nabla u) - \mathbf{a}(x,\nabla \hat{u})) \cdot \nabla u| \, dx = 0 \end{aligned}$$

and

$$\begin{aligned} & \lim_{k\to\infty} \left| \int_{\Omega_{k,r} \cap \{|u|\geq k, |\hat{u}|< k\}} (\mathbf{a}(x,\nabla u) - \mathbf{a}(x,\nabla \hat{u})) \cdot \nabla \hat{u} \, dx \right| \\ & \leq \lim_{k\to\infty} \int_{\{k-r<|\hat{u}|<k\}} |(\mathbf{a}(x,\nabla u) - \mathbf{a}(x,\nabla \hat{u})) \cdot \nabla \hat{u}| \, dx = 0. \end{aligned}$$

Consequently, we get

$$\frac{1}{r} \int_\Omega (u - \hat{u}) T_r(u - \hat{u})\, dx \leq \frac{1}{r} \int_\Omega (f - \hat{f}) T_r(u - \hat{u})\, dx \leq \int_\Omega |f - \hat{f}|\, dx.$$

Passing to the limit as $r \to 0^+$, (7.26) follows.

Having in mind Theorem 6.7 and Proposition 7.4, to finish the proof, we only need to prove that $\overline{\mathcal{B}_\varphi}^{L^1(\Omega)} \subset \mathcal{A}_\varphi$. Let $(u_n, v_n) \in \mathcal{B}_\varphi$ be such that $(u_n, v_n) \to (u, v)$ in $L^1(\Omega) \times L^1(\Omega)$. Let us prove that $(u, v) \in \mathcal{A}_\varphi$. Since $(u_n, v_n) \in \mathcal{B}_\varphi \subset \mathcal{A}_\varphi$, we have $\mathbf{a}(x, \nabla u_n) \in X(\Omega)_1$ satisfying

$$-v_n = \operatorname{div} \mathbf{a}(x, \nabla u_n) \qquad \text{in } \mathcal{D}'(\Omega), \tag{7.28}$$

$$\mathbf{a}(x,\nabla u_n)\cdot D^s p(u_n) = f^0(x, D^s p(u_n)) \qquad \forall\, p\in\mathcal{P}, \tag{7.29}$$

$$[\mathbf{a}(x,\nabla u_n),\nu] \in \text{sign}\,(p(\varphi)-p(u_n))f^0(x,\nu(x)), \;\; \forall\, p\in\mathcal{P} \quad \mathcal{H}^{N-1}-\text{a.e.} \tag{7.30}$$

Then, given $p\in\mathcal{P}$, we have

$$\begin{aligned}
\int_\Omega v_n p(u_n)\,dx &= \int_\Omega (\mathbf{a}(x,\nabla u_n), Dp(u_n)) - \int_{\partial\Omega} [\mathbf{a}(x,\nabla u_n),\nu]p(u_n)\,d\mathcal{H}^{N-1} \\
&= \int_\Omega \mathbf{a}(x,\nabla u_n)\cdot\nabla p(u_n)\,dx + \int_\Omega f^0(x,D^s p(u_n)) \\
&\quad - \int_{\partial\Omega} [\mathbf{a}(x,\nabla u_n),\nu]p(u_n)\,d\mathcal{H}^{N-1} \\
&= \int_\Omega \mathbf{a}(x,\nabla u_n)\cdot\nabla p(u_n)\,dx + \int_\Omega f^0(x,D^s p(u_n)) \\
&\quad + \int_{\partial\Omega} |p(\varphi)-p(u_n)| f^0(x,\nu(x))\,d\mathcal{H}^{N-1} - \int_{\partial\Omega} [\mathbf{a}(x,\nabla u_n),\nu]p(\varphi)\,d\mathcal{H}^{N-1}.
\end{aligned}$$

Hence, by (6.19) and (6.20), it follows that

$$\begin{aligned}
\int_\Omega \|Dp(u_n)\| &= \int_\Omega |\nabla p(u_n)|\,dx + \int_\Omega |D^s p(u_n)| \\
&\le \frac{1}{C_0}\left(\int_\Omega \mathbf{a}(x,\nabla p(u_n))\cdot\nabla p(u_n)\,dx + \int_\Omega f^0(x,D^s p(u_n)) + D_1\mathcal{L}^N(\Omega)\right) \\
&\le \frac{1}{C_0}\left(\int_\Omega v_n p(u_n)\,dx + \int_{\partial\Omega} [\mathbf{a}(x,\nabla u_n),\nu]p(\varphi)\,d\mathcal{H}^{N-1} + D_1\mathcal{L}^N(\Omega)\right).
\end{aligned}$$

Thus,

$$\int_\Omega \|Dp(u_n)\| \le M_1\|p\|_\infty \qquad \forall\, n\in\mathbb{N}. \tag{7.31}$$

Therefore, $p(u)\in BV(\Omega)$ for any $p\in\mathcal{P}$. On the other hand, since

$$\|\mathbf{a}(x,\nabla u_n)\|_\infty \le M,$$

we may assume that $\mathbf{a}(x,\nabla u_n) \rightharpoonup z$ in the weak* topology of $L^\infty(\Omega,\mathbb{R}^N)$, with $\|z\|_\infty \le M$. Moreover, since $v_n \to v$ in $L^1(\Omega)$, we have that $v = -\text{div}(z)$ in $\mathcal{D}'(\Omega)$. By the definition of the weak trace on $\partial\Omega$ of the normal component of z, it is easy to see that

$$[\mathbf{a}(x,\nabla u_n),\nu] \rightharpoonup [z,\nu] \qquad \text{weakly}^* \text{ in } L^\infty(\partial\Omega). \tag{7.32}$$

On the other hand,

$$\begin{aligned}
&\lim_{n\to\infty}\left(\int_\Omega h(x,Dp(u_n)) + \int_{\partial\Omega} |p(\varphi)-p(u_n)| f^0(x,\nu(x))\, d\mathcal{H}^{N-1}\right)\\
&= \lim_{n\to\infty}\left(\int_\Omega v_n p(u_n)\, dx + \int_{\partial\Omega} [\mathbf{a}(x,\nabla u_n),\nu] p(\varphi)\, d\mathcal{H}^{N-1}\right)\\
&= \int_\Omega v p(u)\, dx + \int_{\partial\Omega} [z,\nu] p(\varphi)\, d\mathcal{H}^{N-1}\\
&= -\int_\Omega \operatorname{div}(z) p(u) + \int_{\partial\Omega} [z,\nu] p(\varphi)\, d\mathcal{H}^{N-1}.
\end{aligned}$$

Now, applying Green's formula (C.10) we get

$$\begin{aligned}
&\lim_{n\to\infty}\left(\int_\Omega h(x,Dp(u_n)) + \int_{\partial\Omega} |p(\varphi)-p(u_n)| f^0(x,\nu(x))\, d\mathcal{H}^{N-1}\right)\\
&= \int_\Omega (z,Dp(u)) + \int_{\partial\Omega} [z,\nu](p(\varphi)-p(u))\, d\mathcal{H}^{N-1}.
\end{aligned} \tag{7.33}$$

It is not difficult to see that

$$\lim_{n\to\infty}\int_\Omega (\mathbf{a}(x,\nabla u_n) - \mathbf{a}(x,\nabla p(u_n)))\cdot \nabla p(u)\, dx = 0$$

for all $p \in \mathcal{P}$. Consequently,

$$\lim_{n\to\infty}\int_\Omega \mathbf{a}(x,\nabla p(u_n))\cdot\nabla p(u)\, dx = \int_\Omega z\cdot\nabla p(u)\, dx \quad \forall\, p\in\mathcal{P}. \tag{7.34}$$

Let us now prove the convergence of the energies. We consider the energy functional $\Psi_\varphi : L^1(\Omega) \to [0,+\infty]$ defined by

$$\Psi_\varphi(v) := \begin{cases} \displaystyle\int_\Omega f(x,Dv) + \int_{\partial\Omega} |\varphi - v| f^0(x,\nu(x))\, d\mathcal{H}^{N-1} & \text{if } v\in BV(\Omega),\\ +\infty & \text{if } v\in L^1(\Omega)\setminus BV(\Omega).\end{cases}$$

As a consequence of Theorem 6.4, the functional Ψ_φ is convex and lower-semi-continuous. By the convexity of f, we have

$$\begin{aligned}
&\Psi_{p(\varphi)}(p(u_n))\\
&=\int_\Omega f(x,\nabla p(u_n))\, dx + \int_\Omega f^0(x,D^s p(u_n)) + \int_{\partial\Omega} |p(u_n)-p(\varphi)| f^0(x,\nu(x))\, d\mathcal{H}^{N-1}
\end{aligned}$$

$$
\begin{aligned}
&\leq \int_\Omega f(x,\nabla p(u))\,dx + \int_\Omega \mathbf{a}(x,\nabla p(u_n))\cdot\nabla p(u_n)\,dx \\
&\quad - \int_\Omega \mathbf{a}(x,\nabla p(u_n))\cdot\nabla p(u)\,dx + \int_\Omega \mathbf{a}(x,\nabla u_n)\cdot D^s p(u_n) \\
&\quad + \int_{\partial\Omega} [\mathbf{a}(x,\nabla u_n),\nu](p(\varphi)-p(u_n))\,d\mathcal{H}^{N-1} \\
&= \int_\Omega f(x,\nabla p(u))\,dx + \int_\Omega (\mathbf{a}(x,\nabla u_n), Dp(u_n)) \\
&\quad - \int_\Omega \mathbf{a}(x,\nabla p(u_n))\cdot\nabla p(u)\,dx + \int_{\partial\Omega} [\mathbf{a}(x,\nabla u_n),\nu](p(\varphi)-p(u_n))\,d\mathcal{H}^{N-1} \\
&= \int_\Omega f(x,\nabla p(u))\,dx - \int_\Omega \operatorname{div}(\mathbf{a}(x,\nabla u_n))p(u_n)\,dx \\
&\quad - \int_\Omega \mathbf{a}(x,\nabla p(u_n))\cdot\nabla p(u)\,dx + \int_{\partial\Omega} [\mathbf{a}(x,\nabla u_n),\nu]p(\varphi)\,d\mathcal{H}^{N-1}.
\end{aligned}
$$

Then, by (7.34), letting $n\to\infty$, we obtain

$$
\begin{aligned}
&\limsup_{n\to\infty} \Psi_{p(\varphi)}(p(u_n)) \leq \int_\Omega f(x,\nabla p(u))\,dx - \int_\Omega \operatorname{div}(z)p(u)\,dx \\
&\quad - \int_\Omega z\cdot\nabla p(u)\,dx + \int_{\partial\Omega} [z,\nu]p(\varphi)\,d\mathcal{H}^{N-1} = \int_\Omega f(x,\nabla p(u))\,dx \\
&\quad + \int_\Omega (z, Dp(u)) - \int_\Omega z\cdot\nabla p(u)\,dx + \int_{\partial\Omega} [z,\nu](p(\varphi)-p(u))\,d\mathcal{H}^{N-1} \\
&= \int_\Omega f(x,\nabla p(u))\,dx + \int_\Omega z\cdot D^s p(u) + \int_{\partial\Omega} [z,\nu](p(\varphi)-p(u))\,d\mathcal{H}^{N-1}.
\end{aligned}
$$

Now, by (H_5)

$$
\int_\Omega z\cdot D^s p(u) = \lim_{n\to\infty}\int_\Omega \mathbf{a}(x,\nabla u_n)\cdot D^s p(u) \leq \int_\Omega f^0(x, D^s p(u)).
$$

Moreover, by (7.30)

$$
\begin{aligned}
\int_{\partial\Omega} [z,\nu](p(\varphi)-p(u))\,d\mathcal{H}^{N-1} &= \lim_{n\to\infty}\int_{\partial\Omega} [\mathbf{a}(x,\nabla u_n),\nu](p(\varphi)-p(u))\,d\mathcal{H}^{N-1} \\
&\leq \int_{\partial\Omega} f^0(x,\nu(x))|p(\varphi)-p(u)|\,d\mathcal{H}^{N-1}.
\end{aligned}
$$

Hence, we have

$$
\begin{aligned}
\limsup_{n} \Psi_{p(\varphi)}(p(u_n)) &\leq \int_\Omega f(x,\nabla p(u))\,dx + \int_\Omega f^0(x, D^s p(u))\,dx \\
&\quad + \int_{\partial\Omega} f^0(x,\nu(x))|p(\varphi)-p(u)|\,d\mathcal{H}^{N-1} = \Psi_{p(\varphi)}(p(u)),
\end{aligned}
$$

and, having in mind the lower-semi-continuity of $\Psi_{p(\varphi)}$, this yields

$$\lim_{n\to\infty} \Psi_{p(\varphi)}(p(u_n)) = \Psi_{p(\varphi)}(p(u)). \tag{7.35}$$

If we consider the $\mathbb{R}^N$-valued measures μ_n, μ on $\overline{\Omega}$ which are defined by

$$\mu_n(B) := \int_{B\cap\Omega} Dp(u_n) + \int_{B\cap\partial\Omega} (p(\varphi) - p(u_n))\nu \, d\mathcal{H}^{N-1}, \tag{7.36}$$

$$\mu(B) := \int_{B\cap\Omega} Dp(u) + \int_{B\cap\partial\Omega} (p(\varphi) - p(u))\nu \, d\mathcal{H}^{N-1} \tag{7.37}$$

for all Borel sets $B \subset \overline{\Omega}$, we have

$$\mu_n \to \mu \qquad \text{weakly as measures in} \quad \overline{\Omega}.$$

Moreover,

$$\Psi_{p(\varphi)}(p(u)) = \int_{\overline{\Omega}} \tilde{f}(x,\tilde{\mu}) \qquad \text{and} \qquad \Psi_{p(\varphi)}(p(u_n)) = \int_{\overline{\Omega}} \tilde{f}(x,\tilde{\mu}_n).$$

Hence, (7.35) yields

$$\lim_{n\to\infty} \int_{\overline{\Omega}} \tilde{f}(x,\tilde{\mu}_n) = \int_{\overline{\Omega}} \tilde{f}(x,\tilde{\mu}).$$

Then, applying Theorem 3 of [169], it follows that

$$\int_{\overline{\Omega}} \tilde{h}(x,\tilde{\mu}) = \lim_{n\to\infty} \int_{\overline{\Omega}} \tilde{h}(x,\tilde{\mu}_n)$$

$$= \lim_{n\to\infty} \int_{\Omega} h(x, Dp(u_n)) + \int_{\partial\Omega} |p(u_n) - p(\varphi)| f^0(x,\nu(x)) \, d\mathcal{H}^{N-1}.$$

Now, it is easy to prove that

$$\int_{\overline{\Omega}} \tilde{h}(x,\tilde{\mu}) = \int_{\Omega} h(x, Dp(u)) + \int_{\partial\Omega} |p(\varphi) - p(u)| f^0(x,\nu(x)) \, d\mathcal{H}^{N-1},$$

hence, we obtain

$$\begin{aligned} &\lim_{n\to\infty} \int_{\Omega} h(x, Dp(u_n)) + \int_{\partial\Omega} |p(u_n) - p(\varphi)| f^0(x,\nu(x)) \, d\mathcal{H}^{N-1} \\ &= \int_{\Omega} h(x, Dp(u)) + \int_{\partial\Omega} |p(\varphi) - p(u)| f^0(x,\nu(x)) \, d\mathcal{H}^{N-1}. \end{aligned} \tag{7.38}$$

By (7.33) and (7.38), we get

$$\begin{aligned} &\int_{\Omega} h(x, Dp(u)) + \int_{\partial\Omega} |p(\varphi) - p(u)| f^0(x,\nu(x)) \, d\mathcal{H}^{N-1} \\ &= \int_{\Omega} (z, Dp(u)) + \int_{\partial\Omega} [z,\nu](p(\varphi) - p(u)) \, d\mathcal{H}^{N-1}. \end{aligned} \tag{7.39}$$

Now, (7.39) can be written as

$$\int_\Omega \mathbf{a}(x,\nabla u)\cdot\nabla p(u)\,dx + \int_\Omega f^0(x,D^sp(u)) + \int_{\partial\Omega}|p(\varphi)-p(u)|f^0(x,\nu(x))\,d\mathcal{H}^{N-1}$$
$$= \int_\Omega z\cdot\nabla p(u)\,dx + \int_\Omega z\cdot D^sp(u) + \int_{\partial\Omega}[z,\nu](p(\varphi)-p(u))\,d\mathcal{H}^{N-1}. \tag{7.40}$$

On the other hand, by Lemma 6.12 (i), we get

$$|[z,\nu]| \le f^0(x,\nu(x)) \qquad \mathcal{H}^{N-1}-\text{a.e. on } \partial\Omega. \tag{7.41}$$

Let $v_j \in C^1(\overline{\Omega})$ be a sequence such that $v_j \to p(u)$ in $L^2(\Omega)$ and $f^0(x,\nabla v_j) \to f^0(x,Dp(u))$ weakly as measures. According to (H_5), we have

$$|a(x,\nabla u_n)\cdot\nabla v_j| \le f^0(x,\nabla v_j).$$

Then, if $\psi,\phi \in C_c^1(\Omega)$, with $0\le\psi\le\phi$, we have

$$\left|\int_\Omega \mathbf{a}(x,\nabla u_n)\cdot\nabla v_j\,\psi\,dx\right| \le \int_\Omega f^0(x,\nabla v_j)\psi\,dx,$$

and, letting $n\to\infty$, we get

$$\left|\int_\Omega z\cdot\nabla v_j\,\psi\,dx\right| \le \int_\Omega f^0(x,\nabla v_j)\psi\,dx.$$

Now, since

$$\left|\int_\Omega z\cdot\nabla v_j\,\psi\,dx\right| = \left|-\int_\Omega \mathrm{div}(z)v_j\psi\,dx - \int_\Omega v_j z\cdot\nabla\psi\,dx\right|,$$

letting $j\to\infty$ we obtain that

$$\begin{aligned}|\langle (z,Dp(u)),\psi\rangle| &= \left|-\int_\Omega \mathrm{div}(z)p(u)\psi\,dx - \int_\Omega p(u)z\cdot\nabla\psi\,dx\right|\\ &\le \int_\Omega \psi f^0(x,Dp(u)) \le \int_\Omega \phi f^0(x,Dp(u)).\end{aligned}$$

Hence

$$\langle |(z,Dp(u))|,\phi\rangle \le \int_\Omega \phi f^0(x,Dp(u)).$$

Thus, we have

$$|(z,Dp(u))| \le f^0(x,Dp(u)) \qquad \text{as measures in } \Omega.$$

Then, the singular parts also satisfy a similar inequality,

$$|z\cdot D^sp(u)| \le f^0(x,D^sp(u)) \qquad \text{as measures in } \Omega. \tag{7.42}$$

By the convexity of f, we have

$$\begin{aligned}
&\int_\Omega \mathbf{a}(x,\nabla p(u_n))\cdot\nabla p(u)\,dx \\
&\leq \int_\Omega \mathbf{a}(x,\nabla p(u_n))\cdot\nabla p(u_n)\,dx + \int_\Omega f(x,\nabla p(u))\,dx - \int_\Omega f(x,\nabla p(u_n))\,dx \\
&\leq \int_\Omega \mathbf{a}(x,\nabla p(u_n))\cdot\nabla p(u_n)\,dx + \int_\Omega h^0(x,D^s p(u_n)) \\
&\quad + \int_{\partial\Omega} |p(u_n)-p(\varphi)| h^0(x,\nu(x))\,d\mathcal{H}^{N-1} + \int_\Omega f(x,\nabla p(u))\,dx \\
&\quad - \left(\int_\Omega f(x,\nabla p(u_n))\,dx + \int_\Omega f^0(x,D^s p(u_n)) + \int_{\partial\Omega} |p(u_n)-p(\varphi)| f^0(x,\nu(x))\,d\mathcal{H}^{N-1}\right) \\
&= \int_\Omega h(x,Dp(u_n)) + \int_{\partial\Omega} |p(u_n)-p(\varphi)| f^0(x,\nu(x))\,d\mathcal{H}^{N-1} \\
&\quad + \int_\Omega f(x,\nabla p(u))\,dx - \Psi_{p(\varphi)}(p(u_n)).
\end{aligned}$$

Letting $n\to\infty$, and using (7.34), (7.35) and (7.38), we obtain

$$\int_\Omega z\cdot\nabla p(u)\,dx \leq \int_\Omega \mathbf{a}(x,\nabla u)\cdot\nabla p(u)\,dx. \tag{7.43}$$

Now, from (7.40), (7.41), (7.42) and (7.43), we finally obtain

$$\int_\Omega z\cdot\nabla p(u)\,dx = \int_\Omega h(x,\nabla p(u))\,dx = \int_\Omega \mathbf{a}(x,\nabla u)\cdot\nabla p(u)\,dx, \tag{7.44}$$

$$z\cdot D^s p(u) = f^0(x,D^s p(u)), \tag{7.45}$$

$$[z,\nu]\in \operatorname{sign}\,(p(\varphi)-p(u)) f^0(x,\nu(x)) \qquad \mathcal{H}^{N-1}-\text{a.e. on }\partial\Omega. \tag{7.46}$$

Let us see now that

$$z(x) = \mathbf{a}(x,\nabla u(x)) \qquad \text{a.e. } x\in\Omega. \tag{7.47}$$

Let $0\leq\phi\in C_0^1(\Omega)$ and $g\in C^1(\overline{\Omega})$. We observe that

$$\begin{aligned}
&\int_\Omega \phi[(\mathbf{a}(x,\nabla u_n),Dp(u_n-g)) - \mathbf{a}(x,\nabla g)Dp(u_n-g)] \\
&= \int_\Omega \phi[\mathbf{a}(x,\nabla u_n)-\mathbf{a}(x,\nabla g))\cdot\nabla p(u_n-g)]\,dx \\
&\quad + \int_\Omega \phi[\mathbf{a}(x,\nabla u_n)-\mathbf{a}(x,\nabla g)]\cdot D^s p(u_n-g)).
\end{aligned}$$

Since both terms at the right-hand side of the above expression are positive, we have

$$\int_\Omega \phi[(\mathbf{a}(x,\nabla u_n), Dp(u_n-g)) - \mathbf{a}(x,\nabla g)Dp(u_n-g)] \geq 0.$$

Since

$$\int_\Omega \phi(\mathbf{a}(x,\nabla u_n), Dp(u_n-g))$$

$$= -\int_\Omega \operatorname{div}(\mathbf{a}(x,\nabla u_n))\phi p(u_n-g)\, dx - \int_\Omega p(u_n-g)\mathbf{a}(x,\nabla u_n)\cdot\nabla\phi\, dx,$$

we get

$$\lim_{n\to\infty}\int_\Omega \phi(\mathbf{a}(x,\nabla u_n), Dp(u_n-g))$$

$$= -\int_\Omega \operatorname{div}(z)\phi p(u-g)\, dx - \int_\Omega p(u-g)z\cdot\nabla\phi\, dx = \int_\Omega \phi(z, Dp(u-g)).$$

On the other hand,

$$\lim_{n\to\infty}\int_\Omega \phi\,\mathbf{a}(x,\nabla g)Dp(u_n-g) = \int_\Omega \phi\,\mathbf{a}(x,\nabla g)Dp(u-g).$$

Hence, we obtain

$$\int_\Omega \phi[(z, Dp(u-g)) - \mathbf{a}(x,\nabla g)Dp(u-g)] \geq 0, \qquad \forall\, 0\leq\phi\in C_0^1(\Omega).$$

Thus the measure $(z, Dp(u-g)) - \mathbf{a}(x,\nabla g)Dp(u-g) \geq 0$. Then its absolutely continuous part

$$(z-\mathbf{a}(x,\nabla g))\cdot\nabla p(u-g)) \geq 0 \quad \text{a.e. in } \Omega.$$

Hence,

$$(z-\mathbf{a}(x,\nabla g))\cdot\nabla(u-g)) \geq 0 \quad \text{a.e. in } \Omega.$$

Since we may take a countable set dense in $C^1(\overline{\Omega})$, we have that the above inequality holds for all $x\in\tilde{\Omega}$, where $\tilde{\Omega}\subset\Omega$ is such that $\mathcal{L}^N(\Omega\setminus\tilde{\Omega})=0$, and all $g\in C^1(\overline{\Omega})$. Now, fixed $x\in\tilde{\Omega}$ and given $\xi\in\mathbb{R}^N$, there is $g\in C^1(\overline{\Omega})$ such that $\nabla g(x)=\xi$. Then

$$(z(x)-\mathbf{a}(x,\xi))\cdot(\nabla u(x)-\xi)\geq 0, \quad \forall\,\xi\in\mathbb{R}^N.$$

These inequalities imply (7.47) by an application of the Minty–Browder method in $\mathbb{R}^N$. Now, since $v=-\operatorname{div}(z)$ in $\mathcal{D}'(\Omega)$, by (7.47) we get

$$v = -\operatorname{div}\mathbf{a}(x,\nabla u) \quad \text{in } \mathcal{D}'(\Omega).$$

Finally, by (7.42), (7.47), (7.45) and (7.46), we get

$$\mathbf{a}(x, \nabla u) \cdot D^s p(u) = f^0(x, D^s p(u)) \quad \forall\, p \in \mathcal{P}$$

and

$$[\mathbf{a}(x, \nabla u), \nu] \in \text{sign}\,(p(\varphi) - p(u)) f^0(x, \nu(x)) \qquad \mathcal{H}^{N-1} - \text{a.e. on } \partial\Omega, \ \ \forall\, p \in \mathcal{P}.$$

Therefore, $(u, v) \in \mathcal{A}_\varphi$ and the proof concludes. □

7.4 Existence and Uniqueness for Data in $L^1(\Omega)$

In this section we are going to prove Theorem 7.3.

7.4.1 Proof of Theorem 7.3. Existence

Let $u_0 \in L^1(\Omega)$ and $(T(t))_{t \geq 0}$ be the contraction semigroup in $L^1(\Omega)$ generated by $\mathcal{A}_\varphi$. We shall prove that $u(t) := T(t)u_0$ is an entropy solution of problem (7.1). We divide the proof in several steps.

Step 1. Since $\mathcal{D}(\mathcal{A}_\varphi) \cap L^\infty(\Omega)$ is dense in $L^1(\Omega)$, given $u_0 \in L^1(\Omega)$ there exists a sequence $u_{0,n} \in \mathcal{D}(\mathcal{A}_\varphi) \cap L^\infty(\Omega)$ such that $u_{0,n} \to u_0$ in $L^1(\Omega)$. Then, if $u_n(t) := T(t)u_{0,n}$, we have that $u_n \to u$ in $C([0,T]; L^1(\Omega))$ for every $T > 0$. As a consequence of Theorem 7.8, $u_n(t), u_n'(t) \in L^2(\Omega)$, $u_n(t) \in BV(\Omega)$, $z_n(t) := \mathbf{a}(x, \nabla u_n(t)) \in X(\Omega)_1$ a.e. $t \in [0,T]$, and for almost all $t \in [0,T]$ $u_n(t)$ satisfies

$$u_n'(t) = \operatorname{div}(z_n(t)) \quad \text{in } \mathcal{D}'(\Omega), \tag{7.48}$$

$$\begin{cases} z_n(t) \cdot D^s u_n(t) = f^0(x, D^s u_n(t)), \\ z_n(t) \cdot D^s p(u_n(t)) = f^0(x, D^s p(u_n(t))) \qquad \forall\, p \in \mathcal{P}, \end{cases} \tag{7.49}$$

$$[z_n(t), \nu] \in \text{sign}\,(p(\varphi) - p(u_n(t))) f^0(x, \nu(x)) \ \ \forall\, p \in \mathcal{P} \ \ \mathcal{H}^{N-1} - \text{a.e.} \tag{7.50}$$

From (7.48), (7.49) and (7.50), it follows that

$$\begin{aligned} -\int_\Omega (w - u_n(t)) u_n'(t)\, dx &= \int_\Omega (z_n(t), Dw) - \int_\Omega h(x, Du_n(t)) \\ &- \int_{\partial\Omega} [z_n(t), \nu](w - \varphi)\, d\mathcal{H}^{N-1} - \int_{\partial\Omega} |u_n(t) - \varphi| f^0(x, \nu(x))\, d\mathcal{H}^{N-1} \end{aligned} \tag{7.51}$$

for every $w \in BV(\Omega) \cap L^2(\Omega)$.

On the other hand, since $\| [z_n(t), \nu] \|_\infty \leq \| z_n(t) \|_\infty \leq M$, we may assume (up to extraction of a subsequence, if necessary) that

$$[z_n(\cdot), \nu] \to \rho \quad \sigma(L^\infty(S_T), L^1(S_T)).$$

Step 2. Convergence of the derivatives and identification of the limit. Since the map $t \mapsto u_n'(t)$ is strongly measurable from $[0,T]$ into $L^2(\Omega)$, and, by (5.6),

$$\|u_n'(t)\|_{BV(\Omega)_2^*} \leq \|u_n'(t)\|_{L^2(\Omega)},$$

it follows that this map is strongly measurable from $[0,T]$ into $BV(\Omega)_2^*$. Moreover, for every $w \in BV(\Omega)_2$, if we take $u_n(t) - w$ as test function in (7.51), since

$$\int_\Omega h(x, Du_n(t)) = \int_\Omega (z_n(t), Du_n(t)),$$

we get

$$\int_\Omega u_n'(t) w \, dx = - \int_\Omega (z_n(t), Dw) + \int_{\partial\Omega} [z_n(t), \nu] w \, d\mathcal{H}^{N-1}.$$

Hence

$$\left| \int_\Omega u_n'(t) w \, dx \right| \leq M \int_\Omega \|Dw\| + M \int_{\partial\Omega} |w| \, dH^{N-1} \leq M_1 \|w\|_{BV(\Omega)_2} \quad \forall \, n \in \mathbb{N}.$$

Thus,

$$\|u_n'(t)\|_{BV(\Omega)_2^*} \leq M_1 \qquad \forall \, n \in \mathbb{N} \text{ and } t \in [0,T].$$

Therefore, $\{u_n'\}_{n\in\mathbb{N}}$ is a bounded sequence in $L^\infty(0,T; BV(\Omega)_2^*)$. Now, since the space $L^\infty(0,T; BV(\Omega)_2^*)$ is a vector subspace of the dual space $\left(L^1(0,T; BV(\Omega)_2)\right)^*$, we can find a subnet $\{u_\alpha'\}$ such that

$$u_\alpha' \to \xi \in \left(L^1(0,T; BV(\Omega)_2)\right)^* \quad \text{weakly}^*. \tag{7.52}$$

Since $\|z_n(t)\|_\infty \leq M$ for all $n \in \mathbb{N}$ and a.e. $t \in [0,T]$, we may also assume that

$$z_n \to z \in L^\infty(Q_T, \mathbb{R}^N) \qquad \text{weakly}^*. \tag{7.53}$$

Given $\eta \in \mathcal{D}(Q_T)$, since $\eta \in L^1(0,T; BV(\Omega)_2)$, we have

$$\begin{aligned}
\langle \xi, \eta \rangle &= \lim_\alpha \langle u_\alpha', \eta \rangle = \lim_\alpha \int_0^T \langle u_\alpha'(t), \eta(t) \rangle \, dt \\
&= \lim_\alpha \int_0^T \int_\Omega u_\alpha'(t) \eta(t) \, dx dt = \lim_\alpha \int_0^T \int_\Omega \operatorname{div}(z_\alpha(t)) \eta(t) \, dx dt \\
&= - \lim_\alpha \int_0^T \int_\Omega z_\alpha(t) \cdot \nabla \eta(t) \, dx dt = - \int_{Q_T} z \cdot \nabla \eta \, dx dt = \langle \operatorname{div}_x(z), \eta \rangle.
\end{aligned}$$

Hence,

$$\xi = \operatorname{div}_x(z) \qquad \text{in } \mathcal{D}'(Q_T). \tag{7.54}$$

On the other hand, if we take $\eta(t,x) = \phi(t)\psi(x)$ with $\phi \in \mathcal{D}(]0,T[)$ and $\psi \in \mathcal{D}(\Omega)$, the same calculation as above shows that

$$\xi(t) = \operatorname{div}_x(z(t)) \qquad \text{in } \mathcal{D}'(\Omega) \text{ a.e. } t \in [0,T]. \tag{7.55}$$

Hence, $(z(t), \xi(t)) \in Z(\Omega)$ for almost all $t \in [0,T]$, and the trace $[z(t),\nu]$ defined as in Section 2 has a sense.

With a similar proof to the one given for Lemma 5.18, we get the following result.

Lemma 7.9. *ξ is the time derivative of u in the sense of Definition* 5.3.

Step 3. *The boundary condition.* Let us now prove that

$$\rho(t) = [z(t),\nu] \qquad \mathcal{H}^{N-1} - \text{a.e. on } \partial\Omega, \quad \text{a.e. } t \in [0,T]. \tag{7.56}$$

Indeed, if $w \in BV(\Omega) \cap L^\infty(\Omega)$ and $v \in R(\Omega)$ are such that $v|_{\partial\Omega} = w|_{\partial\Omega}$, we have that

$$\int_0^t \langle z_\alpha(s), w\rangle_{\partial\Omega}\, ds = \int_0^t \langle \operatorname{div}(z_\alpha(s)), v\rangle\, ds + \int_0^t \int_\Omega z_\alpha(s) \cdot \nabla v\, dxds.$$

Hence

$$\begin{aligned} \lim_\alpha \int_0^t \langle z_\alpha(s), w\rangle_{\partial\Omega}\, ds &= \int_0^t \langle \xi(s), v\rangle\, ds + \int_0^t \int_\Omega z(s) \cdot \nabla v\, dxds \\ &= \int_0^t \langle z(s), w\rangle_{\partial\Omega}\, ds = \int_0^t \int_{\partial\Omega} [z(s),\nu] w\, d\mathcal{H}^{N-1} ds. \end{aligned} \tag{7.57}$$

On the other hand, since $z_\alpha(s) \in X(\Omega)_1$, if we apply Green's formula we have that

$$\int_0^t \langle \operatorname{div}(z_\alpha(s)), v\rangle ds = -\int_0^t \int_\Omega z_\alpha(s) \cdot \nabla v\, dxds + \int_0^t \int_{\partial\Omega} [z_\alpha(s),\nu] w\, d\mathcal{H}^{N-1} ds.$$

Hence,

$$\int_0^t \langle z_\alpha(s), w\rangle_{\partial\Omega}\, ds = \int_0^t \int_{\partial\Omega} [z_\alpha(s),\nu] w\, d\mathcal{H}^{N-1} ds.$$

Taking limits in α, we get

$$\int_0^t \int_{\partial\Omega} \rho(s) w\, d\mathcal{H}^{N-1} ds = \int_0^t \int_{\partial\Omega} [z(s),\nu] w\, d\mathcal{H}^{N-1} ds \tag{7.58}$$

for all $w \in BV(\Omega) \cap L^\infty(\Omega)$ and $t \in [0,T]$. Now, if $w \in L^1(\partial\Omega)$, we take $w_k \in BV(\Omega) \cap L^\infty(\Omega)$ such that $w_k|_{\partial\Omega} = T_k(w)$. By (7.58), we have

$$\int_0^t \int_{\partial\Omega} \rho(s) w_k\, d\mathcal{H}^{N-1} ds = \int_{\partial\Omega} [z(s),\nu] w_k\, d\mathcal{H}^{N-1} ds.$$

Letting $k \to \infty$, it follows that

$$\int_0^t \int_{\partial\Omega} \rho(s) w\, d\mathcal{H}^{N-1} ds = \int_0^t \int_{\partial\Omega} [z(s),\nu] w\, d\mathcal{H}^{N-1} ds$$

for all $w \in L^1(\partial\Omega)$ and $t \in [0,T]$. Therefore, (7.56) holds.

Step 4. Next, we prove that $\xi = \operatorname{div}(z)$ in $\left(L^1(0,T,BV(\Omega)_2)\right)^*$ in the sense of the Definition 5.4. To do that, let us first observe that (z, Dw), defined by (5.7), is a Radon measure in Q_T for all $w \in L^1_w(0,T,BV(\Omega)) \cap L^\infty(Q_T)$. Let $\phi \in \mathcal{D}(Q_T)$, then

$$\begin{aligned}\langle (z,Dw),\phi\rangle &= -\int_0^T \langle \xi(t) - u'_\alpha(t), w(t)\phi(t)\rangle \, dt \\ &\quad - \int_{Q_T} w(z - z_\alpha)\cdot \nabla_x \phi \, dxdt + \int_0^T \langle (z_\alpha(t), Dw(t)), \phi(t)\rangle \, dt.\end{aligned}$$

Then, taking limits in α, and using (7.52), we get

$$\langle (z,Dw),\phi\rangle = \lim_\alpha \int_0^T \langle (z_\alpha(t), Dw(t)), \phi(t)\rangle \, dt. \tag{7.59}$$

Therefore

$$|\langle (z,Dw),\phi\rangle| \le M\|\phi\|_\infty \int_0^T \int_\Omega \|Dw(t)\| \, dt.$$

Hence, (z, Dw) is a Radon measure in Q_T. Moreover, from (7.59), and Green's formula we obtain that

$$\begin{aligned}&\int_{Q_T} (z,Dw) = \lim_\alpha \int_0^T (z_\alpha(t), Dw(t)) \, dt \\ &= \lim_\alpha \left(-\int_0^T \int_\Omega \operatorname{div}(z_\alpha(t))w(t) \, dxdt + \int_0^T \int_{\partial\Omega} [z_\alpha(t),\nu] w(t) \, d\mathcal{H}^{N-1} dt \right) \\ &= -\int_0^T \langle \xi(t), w(t)\rangle \, dt + \int_0^T \int_{\partial\Omega} [z(t),\nu] w(t) \, d\mathcal{H}^{N-1} dt,\end{aligned}$$

that is,

$$\int_{Q_T} (z,Dw) + \int_0^T \langle \xi(t), w(t)\rangle \, dt = \int_0^T \int_{\partial\Omega} [z(t),\nu] w(t) \, d\mathcal{H}^{N-1} dt. \tag{7.60}$$

As a consequence of the boundedness of $\{u'_n\}$, (7.52) and the above statement, we have

$$u'_n \to \xi \in \left(L^1(0,T;BV(\Omega)_2)\right)^* \quad \text{weakly}^*. \tag{7.61}$$

Step 5. *Convergence of the energy.*

Multiplying (7.48) by $w - p(u_n(t))$ and integrating in Ω, using (7.49) and (7.50), we have that

$$
\begin{aligned}
&-\int_\Omega (w - p(u_n(t)))u_n'(t)\, dx = \int_\Omega (z_n(t), Dw) \\
&-\int_\Omega h(x, Dp(u_n(t))) - \int_{\partial\Omega} [z_n(t), \nu](w - p(\varphi))\, d\mathcal{H}^{N-1} \\
&-\int_{\partial\Omega} |p(u_n(t)) - p(\varphi)| f^0(x, \nu(x))\, d\mathcal{H}^{N-1}
\end{aligned}
\tag{7.62}
$$

for every $w \in BV(\Omega) \cap L^\infty(\Omega)$ and all $p \in \mathcal{P}$.

First, we observe that setting $w = 0$ in (7.62), and integrating in $(0, T)$, we obtain

$$
\begin{aligned}
&\int_\Omega J_p(u_n(T))dx + \int_0^T \int_\Omega h(x, Dp(u_n(t)))\, dxdt \\
&+ \int_0^T \int_{\partial\Omega} |p(\varphi) - p(u_n(t))| f^0(x, \nu(x))\, d\mathcal{H}^{N-1}dt \\
&= \int_0^T \int_{\partial\Omega} [z_n(t), \nu] p(\varphi)\, d\mathcal{H}^{N-1}dt + \int_\Omega J_p(u_{0,n})\, dx,
\end{aligned}
$$

where $J_p'(r) = p(r)$. In particular, we have

$$
\int_0^T \int_\Omega J_p(u_n(t))\, dxdt \leq C, \tag{7.63}
$$

and

$$
\begin{aligned}
&\int_0^T \int_\Omega h(x, Dp(u_n(t)))\, dxdt \\
&+ \int_0^T \int_{\partial\Omega} |p(\varphi) - p(u_n(t))| f^0(x, \nu(x))\, d\mathcal{H}^{N-1}dt \leq C
\end{aligned}
$$

where C is a constant depending on $\|u_0\|_1$, $\|\varphi\|_1$ and $\|p\|_\infty$. Hence

$$
\int_0^T \|Dp(u_n(t))\|\, dt + \int_0^T \int_{\partial\Omega} |p(\varphi) - p(u_n(t))| f^0(x, \nu(x))\, d\mathcal{H}^{N-1}dt \leq C \tag{7.64}
$$

where C is a constant depending on $\|u_0\|_1$, $\|\varphi\|_1$, $\|p\|_\infty$ and the constants in (6.19). Since the functional $\Phi_p : L^1(\Omega) \to]-\infty, +\infty]$, defined by

$$
\Phi_p(w) = \begin{cases} \displaystyle\int_\Omega \|Dw\| + \int_{\partial\Omega} |w - p(\varphi)| f^0(x, \nu(x)) & \text{if } \; w \in BV(\Omega), \\ +\infty & \text{if } \; w \in L^1(\Omega) \setminus BV(\Omega), \end{cases} \tag{7.65}
$$

is lower semi-continuous in $L^1(\Omega)$, we have that

$$
\begin{aligned}
&\Phi_p(p(u(t))) \leq \liminf_{n\to\infty} \Phi_p(p(u_n(t)) \\
&= \liminf_{n\to\infty} \left(\int_\Omega \|Dp(u_n(t))\| + \int_{\partial\Omega} |p(u_n(t)) - p(\varphi)| f^0(x, \nu(x))\, d\mathcal{H}^{N-1} \right).
\end{aligned}
$$

On the other hand, by Lemma 5.19, the map $t \mapsto \|p(u_n(t))\|_{BV(\Omega)}$ is measurable, then by Fatou's lemma and (7.64), it follows that

$$\int_0^T \liminf_{n\to\infty} \left(\int_\Omega \|Dp(u_n(t))\| + \int_{\partial\Omega}\int_\Omega |p(u_n(t)) - p(\varphi)| f^0(x,\nu(x))\, d\mathcal{H}^{N-1} \right) dt$$
$$\leq \liminf_{n\to\infty} \int_0^T \left(\int_\Omega \|Dp(u_n(t))\| + \int_{\partial\Omega} |p(u_n(t)) - p(\varphi)| f^0(x,\nu(x)) \right) dt \leq C. \tag{7.66}$$

As a consequence of (7.66), we obtain that $p(u(t)) \in BV(\Omega)$ for almost all $t \in [0,T]$.

From Lemma 5.19, if $0 \leq \eta \in \mathcal{D}(]0,T[)$, the map $t \mapsto p(u(t))\eta(t)$, from $[0,T]$ into $BV(\Omega)$ is weakly measurable.

Using the same technique as in the proofs of Lemmas 5.20 and 5.21, we obtain the following two results.

Lemma 7.10. *For any $\tau > 0$, we define the function ψ^τ, as the Dunford integral (see [98])*

$$\psi^\tau(t) := \frac{1}{\tau}\int_{t-\tau}^t \eta(s)p(u(s))\, ds \in BV(\Omega)^{**},$$

that is,

$$\langle \psi^\tau(t), w\rangle = \frac{1}{\tau}\int_{t-\tau}^t \langle \eta(s)p(u(s)), w\rangle\, ds$$

for any $w \in BV(\Omega)^*$. Then $\psi^\tau \in C([0,T]; BV(\Omega))$. Moreover, $\psi^\tau(t) \in L^2(\Omega)$, and, thus, $\psi^\tau(t) \in BV(\Omega)_2$, and ψ^τ admits a weak derivative in $L^1_w(0,T,BV(\Omega)) \cap L^\infty(Q_T)$.

Lemma 7.11. *For $\tau > 0$ small enough, we have*

$$\int_0^T \langle \psi^\tau(t), \xi(t)\rangle\, dt \leq -\int_0^T \int_\Omega \frac{\eta(t-\tau) - \eta(t)}{-\tau} J_p(u(t))\, dxdt. \tag{7.67}$$

We need the following result.

Lemma 7.12. *Let*

$$A_n := \int_0^T \int_\Omega h(x, Dp(u_n))\, dt + \int_0^T \int_{\partial\Omega} |p(\varphi) - p(u_n)| f^0(x,\nu(x))\, d\mathcal{H}^{N-1} dt,$$

$\eta \in \mathcal{D}(0,T)$, and

$$(\eta p(u))^\tau(t) = \frac{1}{\tau}\int_{t-\tau}^t \eta(s)p(u(s))ds.$$

Then

$$\limsup_{n\to\infty} A_n \leq \int_0^T\int_\Omega z(t)\cdot\nabla p(u(t))\;dxdt$$
$$+\liminf_{\eta\uparrow 1}\liminf_{\tau\to 0}\textstyle\int_0^T\langle z(t), D^s(\eta p(u))^\tau(t)\rangle\;dt$$
$$+\int_0^T\int_{\partial\Omega}[z(t),\nu](p(\varphi)-p(u(t)))\;d\mathcal{H}^{N-1}dt.$$

Proof. Let $w\in W^{1,1}((0,T)\times\Omega)$ be such that $w(t)|_{\partial\Omega}=\varphi$. We use as test function in (7.62) $(\eta p(w(t)))^\tau$ and integrate in $(0,T)$ to obtain

$$-\int_0^T\int_\Omega(\eta p(w(t)))^\tau u_n'(t)\;dxdt+\int_0^T\int_\Omega p(u_n(t))u_n'(t)\;dxdt$$
$$+\int_0^T\int_\Omega h(x,Dp(u_n(t)))\;dt+\int_0^T\int_{\partial\Omega}|p(\varphi)-p(u_n(t))|f^0(x,\nu(x))\;d\mathcal{H}^{N-1}dt$$
$$=\int_0^T\int_\Omega(z_n(t),D(\eta p(w(t)))^\tau)\;dt-\int_0^T\int_{\partial\Omega}[z_n(t),\nu](\eta^\tau(t)-1)p(\varphi)\;d\mathcal{H}^{N-1}dt,$$

where $\eta^\tau(t)=\frac{1}{\tau}\int_{t-\tau}^t\eta(s)ds$. Our purpose is to take limits in the above expression as $n\to\infty$, $w\to u$ in $L^1((0,T)\times\Omega)$, $\tau\to 0$ and $\eta\uparrow\chi_\Omega$. We take $\tau>0$ small enough. Let us analyze the first term

$$-\int_0^T\int_\Omega(\eta p(w(t)))^\tau u_n'(t)\;dxdt=\int_0^T\int_\Omega(\eta p(w(t)))_t^\tau u_n(t)\;dxdt$$
$$\to\int_0^T\int_\Omega(\eta p(w(t)))_t^\tau u(t)\;dxdt\quad\text{as }n\to\infty.$$

Now, using (5.63) and Lemma 7.11,

$$\int_0^T\int_\Omega(\eta p(w(t)))_t^\tau u=\int_0^T\int_\Omega\frac{\eta(t)p(w(t))-\eta(t-\tau)p(w(t-\tau))}{\tau}u(t)\;dxdt$$
$$\to\int_0^T\int_\Omega\frac{\eta(t)p(u(t))-\eta(t-\tau)p(u(t-\tau))}{\tau}u(t)\;dxdt,\quad\text{as }w\to u\text{ in }L^1,$$
$$=-\int_0^T\langle\xi(t),(\eta p(u(t)))^\tau\rangle\;dt\geq\int_0^T\int_\Omega\frac{\eta(t-\tau)-\eta(t)}{-\tau}J_p(u(t))\;dxdt$$
$$\to\int_0^T\int_\Omega\eta_t J_p(u(t))\;dxdt,\quad\text{as }\tau\to 0$$
$$\to\int_\Omega(J_p(u(0))-J_p(u(T)))\;dx\quad\text{as }\eta\uparrow\chi_\Omega.$$

The analysis of the second term is easy. Letting $n\to\infty$ we have

$$\int_0^T \int_\Omega p(u_n(t))u_n'(t)\, dxdt = \int_0^T \frac{d}{dt}\int_\Omega J_p(u_n(t))\, dx$$
$$= \int_\Omega (J_p(u_n)(T) - J_p(u_n(0)))\, dx \to \int_\Omega (J_p(u(T)) - J_p(u(0)))\, dx.$$

Let us analyze the fifth term. Having in mind Steps 3 and 4, and (7.61), taking limits as $n \to \infty$, $w \to u$ in L^1 and $\tau \to 0$, we get:

$$\int_0^T \int_\Omega (z_n(t), D(\eta p(w))^\tau)\, dt$$
$$\to -\int_0^T \langle \xi(t), (\eta p(w))^\tau \rangle\, dt + \int_0^T \int_{\partial\Omega} [z(t),\nu]\eta^\tau p(\varphi)\, d\mathcal{H}^{N-1} dt$$
$$= \int_0^T \int_\Omega u(t)(\eta p(w))_t^\tau\, dxdt + \int_0^T \int_{\partial\Omega} [z(t),\nu]\eta^\tau p(\varphi)\, d\mathcal{H}^{N-1} dt$$
$$= \int_0^T \int_\Omega u(t)\frac{\eta(t)p(w)(t) - \eta(t-\tau)p(w)(t-\tau)}{\tau}\, dxdt$$
$$+ \int_0^T \int_{\partial\Omega} [z(t),\nu]\eta^\tau p(\varphi)\, d\mathcal{H}^{N-1} dt$$
$$\to \int_0^T \int_\Omega u(t)\frac{\eta(t)p(u)(t) - \eta(t-\tau)p(u)(t-\tau)}{\tau}\, dxdt$$
$$+ \int_0^T \int_{\partial\Omega} [z(t),\nu]\eta^\tau p(\varphi)\, d\mathcal{H}^{N-1} dt$$
$$= \int_0^T \int_\Omega u(t)(\eta p(u))_t^\tau\, dxdt + \int_0^T \int_{\partial\Omega} [z(t),\nu]\eta^\tau p(\varphi)\, d\mathcal{H}^{N-1} dt$$
$$= -\int_0^T \langle \xi(t), (\eta p(u))^\tau \rangle\, dt + \int_0^T \int_{\partial\Omega} [z(t),\nu]\eta^\tau p(\varphi)\, d\mathcal{H}^{N-1} dt$$
$$= \int_0^T \int_\Omega z(t)\cdot \nabla(\eta p(u))^\tau\, dxdt + \int_0^T \int_\Omega z(t)\cdot D^s(\eta p(u))^\tau\, dt$$
$$+ \int_0^T \int_{\partial\Omega} [z(t),\nu](\eta^\tau p(\varphi) - (\eta p(u))^\tau)\, d\mathcal{H}^{N-1} dt$$
$$\to \int_0^T \int_\Omega z(t)\cdot \nabla(\eta p(u))\, dxdt + \liminf_{\tau\to 0} \int_0^T \int_\Omega z(t)\cdot D^s(\eta p(u))^\tau\, dt$$
$$+ \int_0^T \int_{\partial\Omega} [z(t),\nu]\eta(p(\varphi) - p(u(t)))\, d\mathcal{H}^{N-1} dt$$
$$\to \int_0^T \int_\Omega z(t)\cdot \nabla p(u(t))\, dxdt + \liminf_{\eta\uparrow 1}\liminf_{\tau\to 0} \int_0^T \int_\Omega z(t)\cdot D^s(\eta p(u))^\tau\, dt$$
$$+ \int_0^T \int_{\partial\Omega} [z(t),\nu](p(\varphi) - p(u(t)))\, d\mathcal{H}^{N-1} dt, \quad \text{as } \eta\uparrow \chi_\Omega.$$

Finally let us analyze the last term. Using Step 3, we have

$$\begin{aligned}
&\int_0^T \int_{\partial\Omega} [z_n(t), \nu](\eta^\tau(t) - 1)p(\varphi)\, d\mathcal{H}^{N-1} dt \\
&\to \int_0^T \int_{\partial\Omega} [z(t), \nu](\eta^\tau(t) - 1)p(\varphi)\, d\mathcal{H}^{N-1} dt, \quad \text{as } n \to \infty \\
&\to \int_0^T \int_{\partial\Omega} [z(t), \nu](\eta(t) - 1)p(\varphi)\, d\mathcal{H}^{N-1} dt, \quad \text{as } \tau \to 0 \\
&\to 0, \quad \text{as } \eta \uparrow \chi_\Omega.
\end{aligned}$$

The lemma follows by collecting all these facts. □

Lemma 7.13. *Let*

$$\Psi_{p(\varphi)}(p(u(t))) = \int_\Omega f(x, Dp(u(t))) + \int_{\partial\Omega} |p(\varphi) - p(u(t))| f^0(x, \nu(x))\, d\mathcal{H}^{N-1}.$$

Then

$$\int_0^T \Psi_{p(\varphi)}(p(u(t)))\, dt = \lim_{n\to\infty} \int_0^T \Psi_{p(\varphi)}(p(u_n(t)))\, dt. \tag{7.68}$$

As a consequence, we also have that

$$\Psi_{p(\varphi)}(p(u(t))) = \lim_{n\to\infty} \Psi_{p(\varphi)}(p(u_n(t))) \quad \text{a.e. in } t. \tag{7.69}$$

Proof. From the convexity of f, we have

$$\begin{aligned}
&\int_0^T \Psi_{p(\varphi)}(p(u_n(t))\, dt = \int_0^T \int_\Omega f(x, \nabla p(u_n))\, dxdt \\
&+ \int_0^T \int_\Omega f^0(x, D^s p(u_n))\, dt + \int_0^T \int_{\partial\Omega} |p(\varphi) - p(u_n)| f^0(x, \nu(x))\, d\mathcal{H}^{N-1} dt \\
&\leq \int_0^T \int_\Omega f(x, \nabla p(u))\, dxdt + \int_0^T \int_\Omega a(x, \nabla u_n) \cdot \nabla p(u_n)\, dxdt \\
&- \int_0^T \int_\Omega a(x, \nabla p(u_n)) \nabla p(u)\, dxdt + \int_0^T \int_\Omega a(x, \nabla u_n) D^s p(u_n)\, dt \\
&+ \int_0^T \int_{\partial\Omega} |p(\varphi) - p(u_n)| f^0(x, \nu(x))\, dH^{N-1} dt \\
&= \int_0^T \int_\Omega f(x, \nabla p(u))\, dxdt + \int_0^T \int_\Omega (a(x, \nabla u_n), Dp(u_n))\, dt \\
&- \int_0^T \int_\Omega a(x, \nabla p(u_n)) \cdot \nabla p(u)\, dxdt + \int_0^T \int_{\partial\Omega} |p(\varphi) - p(u_n)| f^0(x, \nu(x))\, d\mathcal{H}^{N-1} dt.
\end{aligned}$$

Taking limits as $n \to \infty$ and using Lemma 7.12, we obtain

$$\begin{aligned}
&\limsup_{n\to\infty} \int_0^T \Psi_{p(\varphi)}(p(u_n(t))\, dt \\
&\leq \int_0^T \int_\Omega f(x, \nabla p(u))\, dxdt + \limsup_{n\to\infty} A_n - \int_0^T \int_\Omega z(t) \cdot \nabla p(u(t))\, dxdt \\
&\leq \int_0^T \int_\Omega f(x, \nabla p(u))\, dxdt + \liminf_{\eta\uparrow 1} \liminf_{\tau\to 0} \int_0^T z(t) \cdot D^s(\eta p(u))^\tau\, dt \\
&+ \int_0^T \int_{\partial\Omega} [z(t), \nu](p(\varphi) - p(u))\, d\mathcal{H}^{N-1} dt.
\end{aligned}$$

Now, having in mind that $f^0(x, \xi)$ is positively homogeneous in ξ, and applying Jensen's inequality,

$$\int_\Omega f^0(x, D^s(\eta p(u))^\tau) \leq \frac{1}{\tau} \int_{t-\tau}^t \eta(r) \int_\Omega f^0(x, D^s p(u(r)))\, dr. \tag{7.70}$$

Since $f^0(x, D^s(\eta p(u))^\tau)$ is a measure, by (7.70), using (H_5), we have that

$$\begin{aligned}
&\liminf_{\eta\uparrow 1} \liminf_{\tau\to 0} \int_0^T \int_\Omega z(t) \cdot D^s(\eta p(u))^\tau(t)\, dt \\
&\leq \liminf_{\eta\uparrow 1} \liminf_{\tau\to 0} \int_0^T \int_\Omega f^0(x, D^s(\eta p(u))^\tau)\, dt \\
&\leq \liminf_{\eta\uparrow 1} \liminf_{\tau\to 0} \int_0^T \frac{1}{\tau} \int_{t-\tau}^t \eta(r) \int_\Omega f^0(x, D^s p(u(r)))\, drdt \\
&= \int_0^T \int_\Omega f^0(x, D^s p(u(t)))\, dt.
\end{aligned} \tag{7.71}$$

Hence

$$\begin{aligned}
&\limsup_{n\to\infty} \int_0^T \Psi_{p(\varphi)}(p(u_n(t))\, dt \\
&\leq \int_0^T \int_\Omega f(x, \nabla p(u))\, dxdt + \int_0^T \int_\Omega f^0(x, D^s p(u(t)))\, dt \\
&+ \int_0^T \int_{\partial\Omega} [z(t), \nu](p(\varphi) - p(u))\, d\mathcal{H}^{N-1} dt.
\end{aligned}$$

Since $[z, \nu] = \lim_n [z_n, \nu]$, we have that $|[z, \nu]| \leq f^0(x, \nu(x))$; from the above inequalities, we conclude that

$$\limsup_{n\to\infty}\int_0^T \Psi_{p(\varphi)}(p(u_n(t))\,dt$$
$$\leq \int_0^T\int_\Omega f(x,Dp(u(t)))\,dt + \int_0^T\int_{\partial\Omega}|p(\varphi)-p(u(t))|f^0(x,\nu(x))\,d\mathcal{H}^{N-1}dt$$
$$= \int_0^T \Psi_{p(\varphi)}(u)\,dt.$$

Now, from the lower semi-continuity of $\Psi_{p(\varphi)}$, we obtain

$$\int_0^T \Psi_{p(\varphi)}(p(u(t)))\,dt \leq \liminf_n \int_0^T \Psi_{p(\varphi)}(p(u_n(t)))\,dt$$
$$\leq \limsup_n \int_0^T \Psi_{p(\varphi)}(p(u_n(t)))\,dt \leq \int_0^T \Psi_{p(\varphi)}(p(u(t)))\,dt.$$

The proof of (7.68) is concluded. To prove (7.69) we observe that, since $u_n(t) \to u(t)$ in $L^1(\Omega)$ for all $t\in[0,T]$, we have that

$$\Psi_{p(\varphi)}(p(u(t))) \leq \liminf_n \Psi_{p(\varphi)}(p(u_n(t)) \quad \text{for all } t\in[0,T].$$

Using Fatou's lemma and (7.68), we have

$$\int_0^T \Psi_{p(\varphi)}(p(u(t)))\,dt \leq \int_0^T \liminf_n \Psi_{p(\varphi)}(p(u_n(t))\,dt$$
$$\leq \liminf_n \int_0^T \Psi_{p(\varphi)}(p(u_n(t))\,dt = \int_0^T \Psi_{p(\varphi)}(p(u(t)))\,dt,$$

and, therefore, (7.69) holds. □

Remark 7.14. Let $\eta(t,x)=\phi(t)\psi(x)$ with $\eta\geq 0$, $\phi\in\mathcal{D}(]0,T[)$, $\psi\in C^\infty(\overline{\Omega})$. Multiplying (7.48) by $\hat{w}-p(u_n(t))\eta$ and integrating in Ω, using (7.49) and (7.50), we have that

$$\begin{aligned}-\int_\Omega(\hat{w}-p(u_n(t))\eta)u_n'(t)\,dx &= \int_\Omega(z_n(t),D\hat{w}) - \int_\Omega h(x,Dp(u_n(t)))\eta\\ &-\int_\Omega p(u_n(t))z_n(t)\cdot\nabla_x\eta\,dx - \int_{\partial\Omega}[z_n(t),\nu](\hat{w}-p(\varphi)\eta)\,d\mathcal{H}^{N-1}\\ &-\int_{\partial\Omega}|p(u_n(t))-p(\varphi)|f^0(x,\nu(x))\eta\,d\mathcal{H}^{N-1}\end{aligned} \tag{7.72}$$

for every $\hat{w}\in BV(\Omega)\cap L^\infty(\Omega)$ and all $p\in\mathcal{P}$. Now, let $w\in W^{1,1}((0,T)\times\Omega)$ be such that $w(t)|_{\partial\Omega}=\varphi$. Using $(\eta p(w(t)))^\tau$ as test function in (7.72), integrating in $(0,T)$, and proceeding as in the proof of Lemma 7.12 (this time we do not let $\eta \uparrow \chi_\Omega$), we obtain that

$$\begin{aligned}\limsup_{n\to\infty} A_n(\eta) \leq & \int_0^T \int_\Omega z(t)\cdot \nabla p(u(t))\eta \, dxdt \\ & + \liminf_{\tau\to 0} \int_0^T \int_\Omega z(t) \cdot D^s(\eta p(u))^\tau(t)\, dt \\ & + \int_0^T \int_{\partial\Omega} [z(t),\nu](p(\varphi) - p(u(t)))\eta \, d\mathcal{H}^{N-1} dt,\end{aligned}$$

where

$$\begin{aligned}A_n(\eta) := & \int_0^T \int_\Omega h(x, Dp(u_n))\eta \, dt \\ & + \int_0^T \int_{\partial\Omega} |p(\varphi) - p(u_n)| f^0(x,\nu(x))\eta \, d\mathcal{H}^{N-1} dt.\end{aligned}$$

Now, proceeding as in Lemma 7.13, we prove that

$$\begin{aligned}& \int_0^T \int_\Omega f(x, Dp(u(t)))\eta \, dxdt + \int_0^T \int_{\partial\Omega} |p(\varphi) - p(u(t))| f^0(x,\nu(x))\eta \, d\mathcal{H}^{N-1} dt \\ & = \lim_n \int_0^T \int_\Omega f(x, Dp(u_n(t)))\eta \, dxdt \\ & + \int_0^T \int_{\partial\Omega} |p(\varphi) - p(u_n(t))| f^0(x,\nu(x))\eta \, d\mathcal{H}^{N-1} dt,\end{aligned}$$

and

$$\begin{aligned}& \int_\Omega f(x, Dp(u(t)))\eta \, dx + \int_{\partial\Omega} |p(\varphi) - p(u(t))| f^0(x,\nu(x))\eta \, d\mathcal{H}^{N-1} \\ & = \lim_n \int_\Omega f(x, Dp(u_n(t)))\eta \, dx + \int_{\partial\Omega} |p(\varphi) - p(u_n(t))| f^0(x,\nu(x))\eta \, d\mathcal{H}^{N-1},\end{aligned}$$

for almost all $t \in (0,T)$.

From Lemma 7.13, and arguing as in the proof of Theorem 7.8, it follows that

$$\begin{aligned}& \int_\Omega h(x, Dp(u(t))) + \int_{\partial\Omega} |p(\varphi) - p(u(t))| f^0(x,\nu(x)) \, d\mathcal{H}^{N-1} \\ & = \lim_{n\to\infty} \int_\Omega h(x, Dp(u_n(t))) + \lim_{n\to\infty} \int_{\partial\Omega} |p(\varphi) - p(u_n(t))| f^0(x,\nu(x)) \, d\mathcal{H}^{N-1},\end{aligned} \tag{7.73}$$

a.e. in $t \in (0,T)$.

Let us now prove that

$$\int_0^T \int_\Omega z(t) \cdot \nabla p(u(t))\, dxdt \leq \int_0^T \int_\Omega a(x, \nabla u(t)) \cdot \nabla p(u(t))\, dxdt. \tag{7.74}$$

In fact, from the convexity of f in ξ, we have

$$\begin{aligned}
&\int_0^T \int_\Omega a(x, \nabla p(u_n(t))) \cdot \nabla p(u(t))\, dxdt \leq \int_0^T \int_\Omega a(x, \nabla p(u_n(t))) \cdot \nabla p(u_n(t))\, dxdt \\
&\quad + \int_0^T \int_\Omega f(x, \nabla p(u(t)))\, dxdt - \int_0^T \int_\Omega f(x, \nabla p(u_n(t)))\, dxdt \\
&= \int_0^T \int_\Omega h(x, \nabla p(u_n(t)))\, dxdt + \int_0^T \int_\Omega f^0(x, D^s p(u_n(t)))\, dt \\
&\quad + \int_0^T \int_{\partial\Omega} |p(\varphi) - p(u_n(t))| f^0(x, \nu(x))\, d\mathcal{H}^{N-1} dt - \int_0^T \int_\Omega f^0(x, D^s p(u_n(t)))\, dt \\
&\quad - \int_0^T \int_{\partial\Omega} |p(\varphi) - p(u_n(t))| f^0(x, \nu(x))\, d\mathcal{H}^{N-1} dt + \int_0^T \int_\Omega f(x, \nabla p(u(t)))\, dxdt \\
&\quad - \int_0^T \int_\Omega f(x, \nabla p(u_n(t)))\, dxdt \\
&= \int_0^T \int_\Omega h(x, Dp(u_n(t)))\, dt \\
&\quad + \int_0^T \int_{\partial\Omega} |p(\varphi) - p(u_n(t))| f^0(x, \nu(x))\, d\mathcal{H}^{N-1} dt - \int_0^T \int_\Omega f(x, Dp(u_n(t)))\, dt \\
&\quad - \int_0^T \int_{\partial\Omega} |p(\varphi) - p(u_n(t))| f^0(x, \nu(x))\, d\mathcal{H}^{N-1} dt + \int_0^T \int_\Omega f(x, \nabla p(u(t)))\, dxdt.
\end{aligned}$$

Now, since

$$\lim_{n\to\infty} \int_0^T \int_\Omega \mathbf{a}(x, \nabla u_n(t)) - \mathbf{a}(x, \nabla p(u_n(t))) \cdot \nabla p(u(t))\, dxdt = 0,$$

we have

$$\lim_{n\to\infty} \int_0^T \int_\Omega \mathbf{a}(x, \nabla p(u_n(t))) \cdot \nabla p(u(t))\, dxdt = \int_0^T \int_\Omega z(t) \cdot \nabla p(u(t))\, dxdt.$$

Then, letting $n \to \infty$, and using Lemma 7.13 and (7.73), we deduce that

$$\begin{aligned}
&\int_0^T \int_\Omega z(t) \cdot \nabla p(u(t))\, dxdt \leq \int_0^T \int_\Omega h(x, Dp(u(t)))\, dt \\
&\quad + \int_0^T \int_{\partial\Omega} |p(\varphi) - p(u(t))| f^0(x, \nu(x))\, d\mathcal{H}^{N-1} dt - \int_0^T \int_\Omega f(x, Dp(u(t)))\, dt
\end{aligned}$$

$$-\int_0^T\int_{\partial\Omega}|p(\varphi)-p(u(t))|f^0(x,\nu(x))\,d\mathcal{H}^{N-1}dt+\int_0^T\int_{\Omega}f(x,\nabla p(u(t)))\,dt$$
$$=\int_0^T\int_{\Omega}a(x,\nabla u(t))\cdot\nabla p(u(t))\,dxdt,$$

and (7.74) holds.

Lemma 7.15. *We have*

$$\begin{aligned}&\int_0^T\int_{\partial\Omega}[z(t),\nu(x)](p(\varphi)-p(u(t)))\,dt\\&=\int_0^T\int_{\partial\Omega}|p(\varphi)-p(u(t))|f^0(x,\nu(x))\,dt.\end{aligned}\tag{7.75}$$

In particular, since $|[z,\nu]|\le f^0(x,\nu(x))$, (7.75) *implies that*

$$[z,\nu]\in\operatorname{sign}(p(\varphi)-p(u))f^0(x,\nu(x)).\tag{7.76}$$

Proof. Using (7.73), Fatou's lemma, Lemma 7.12, (7.71) and (7.74), we have

$$\begin{aligned}&\int_0^T\int_{\Omega}h(x,Dp(u(t)))\,dt+\int_0^T\int_{\partial\Omega}|p(\varphi)-p(u(t))|f^0(x,\nu(x))\,d\mathcal{H}^{N-1}dt\\
&\le\liminf_{n\to\infty}\int_0^T\int_{\Omega}h(x,Dp(u_n(t)))\,dt+\int_0^T\int_{\partial\Omega}|p(\varphi)-p(u_n(t))|f^0(x,\nu(x))\,d\mathcal{H}^{N-1}dt\\
&\le\int_0^T\int_{\Omega}z(t)\cdot\nabla p(u(t))\,dxdt+\liminf_{\eta\uparrow 1}\liminf_{\tau\to 0}\int_0^T\int_{\Omega}z(t)\cdot D^s(\eta p(u(t)))^\tau\,dt\\
&\quad+\int_0^T\int_{\partial\Omega}[z(t),\nu(x)](p(\varphi)-p(u(t)))\,d\mathcal{H}^{N-1}dt\\
&\le\int_0^T\int_{\Omega}z(t)\cdot\nabla p(u(t))\,dxdt+\int_0^T\int_{\Omega}f^0(x,D^sp(u(t)))\,dt\\
&\quad+\int_0^T\int_{\partial\Omega}[z(t),\nu(x)](p(\varphi)-p(u(t)))\,d\mathcal{H}^{N-1}\,dt\\
&\le\int_0^T\int_{\Omega}a(x,\nabla u(t))\cdot\nabla p(u(t))\,dxdt\\
&\quad+\int_0^T\int_{\Omega}f^0(x,D^sp(u(t)))\,dt+\int_0^T\int_{\partial\Omega}[z(t),\nu(x)](p(\varphi)-p(u(t)))\,d\mathcal{H}^{N-1}dt\\
&=\int_0^T\int_{\Omega}h(x,Dp(u(t)))\,dt+\int_0^T\int_{\partial\Omega}[z(t),\nu(x)](p(\varphi)-p(u(t)))\,d\mathcal{H}^{N-1}dt\\
&\le\int_0^T\int_{\Omega}h(x,Dp(u(t)))\,dt+\int_0^T\int_{\partial\Omega}|p(\varphi)-p(u(t))|f^0(x,\nu(x))\,d\mathcal{H}^{N-1}dt.\end{aligned}$$

From this series of inequalities, the identity (7.75) follows. □

Remark 7.16. From the last series of inequalities, we also have the following identities:

$$\int_0^T \int_\Omega h(x, Dp(u(t)))\, dt + \int_0^T \int_{\partial\Omega} |p(\varphi) - p(u(t))| f^0(x, \nu(x))\, d\mathcal{H}^{N-1} dt$$

$$= \lim_{n\to\infty} \left(\int_0^T \int_\Omega h(x, Dp(u_n(t)))\, dt + \int_0^T \int_{\partial\Omega} |p(\varphi) - p(u_n(t))| f^0(x, \nu(x))\, d\mathcal{H}^{N-1} dt \right), \tag{7.77}$$

$$\int_0^T \int_\Omega z(t) \cdot \nabla p(u(t))\, dxdt = \int_0^T \int_\Omega a(x, \nabla u(t)) \cdot \nabla p(u(t))\, dxdt, \tag{7.78}$$

$$\liminf_{\eta\uparrow 1} \liminf_{\tau\to 0} \int_0^T \int_\Omega z(t) \cdot D^s(\eta p(u))^\tau\, dt = \int_0^T \int_\Omega f^0(x, D^s p(u(t)))\, dt. \tag{7.79}$$

Step 6. Identification of the limit. Let us now prove that

$$z(t,x) = \mathbf{a}(x, \nabla u(t,x)) \qquad \text{a.e. } (t,x) \in (0,T) \times \Omega. \tag{7.80}$$

Let $0 \le \phi \in C_0^1((0,T) \times \Omega)$ and $g \in C^1([0,T] \times \overline{\Omega})$. We observe that

$$\int_0^T \int_\Omega \phi[(\mathbf{a}(x, \nabla u_n(t)), Dp(u_n(t) - g)) - \mathbf{a}(x, \nabla g) Dp(u_n(t) - g)]\, dt$$

$$= \int_0^T \int_\Omega \phi[\mathbf{a}(x, \nabla u_n(t)) - \mathbf{a}(x, \nabla g)] \cdot \nabla p(u_n(t) - g)\, dxdt$$

$$+ \int_0^T \int_\Omega \phi[\mathbf{a}(x, \nabla u_n(t)) - \mathbf{a}(x, \nabla g)] \cdot D^s p(u_n(t) - g)\, dt.$$

Since both terms at the right hand side of the above expression are positive, we have

$$\int_0^T \int_\Omega \phi[(\mathbf{a}(x, \nabla u_n(t)), Dp(u_n(t) - g)) - \mathbf{a}(x, \nabla g) Dp(u_n(t) - g)]\, dt \ge 0. \tag{7.81}$$

Our purpose is to take limits as $n \to \infty$ in the above inequality. We assume that $\phi(t,x) = \eta(t)\psi(x)$, where $\eta \in \mathcal{D}(0,T)$, $\psi \in \mathcal{D}(\Omega)$, $\eta \ge 0$, $\psi \ge 0$. First, integrating by parts in the first term, we have

$$\int_0^T \int_\Omega \phi(\mathbf{a}(x, \nabla u_n(t)), Dp(u_n(t) - g))\, dt$$

$$= -\int_0^T \int_\Omega p(u_n(t) - g) \nabla_x \phi \cdot \mathbf{a}(x, \nabla u_n(t))\, dxdt$$

$$- \int_0^T \int_\Omega \phi \operatorname{div}(\mathbf{a}(x, \nabla u_n(t))) p(u_n(t) - g)\, dxdt$$

$$= -\int_0^T \int_\Omega p(u_n(t) - g)\nabla_x \phi \cdot \mathbf{a}(x, \nabla u_n(t)) \, dxdt$$

$$- \int_0^T \int_\Omega \phi u_n'(t) p(u_n(t) - g) \, dxdt$$

$$= -\int_0^T \int_\Omega p(u_n(t) - g)\nabla_x \phi \cdot \mathbf{a}(x, \nabla u_n) \, dxdt - \int_0^T \int_\Omega \phi \frac{d}{dt} J_p(u_n(t) - g) \, dxdt$$

$$- \int_0^T \int_\Omega \phi g_t p(u_n(t) - g) \, dxdt$$

$$= -\int_0^T \int_\Omega p(u_n(t) - g)\nabla_x \phi \cdot \mathbf{a}(x, \nabla u_n(t)) \, dxdt + \int_0^T \int_\Omega \phi_t J_p(u_n(t) - g) \, dxdt$$

$$- \int_0^T \int_\Omega \phi g_t p(u_n(t) - g) \, dxdt.$$

Letting $n \to \infty$ in (7.81), taking into account the above equalities, we obtain

$$\begin{aligned} &- \int_0^T \int_\Omega p(u(t) - g)\nabla_x \phi \cdot z \, dxdt + \int_0^T \int_\Omega \phi_t J_p(u(t) - g) \, dxdt \\ &- \int_0^T \int_\Omega \phi g_t p(u(t) - g) \, dxdt - \int_0^T \int_\Omega \phi(a(x, \nabla g), Dp(u(t) - g)) \, dt \geq 0. \end{aligned} \tag{7.82}$$

Now,

$$\begin{aligned} &\int_0^T \int_\Omega \phi_t J_p(u(t) - g) \, dxdt \\ &= \lim_{\tau \to 0} \int_0^T \int_\Omega \frac{\phi(t - \tau) - \phi(t)}{-\tau} J_p(u(t) - g) \, dxdt \\ &= \lim_{\tau \to 0} \int_0^T \int_\Omega \psi(x) \frac{\eta(t - \tau) - \eta(t)}{-\tau} J_p(u(t) - g) \, dxdt. \end{aligned} \tag{7.83}$$

For simplicity, let us write $v = u - g$. Since $J_p(v(t)) - J_p(v(t + \tau)) \leq (v(t) - v(t + \tau))p(v(t))$, for τ small enough, we have

$$\begin{aligned} &\int_0^T \int_\Omega \frac{v(t + \tau) - v(t)}{\tau} \eta(t)\psi(x) p(v(t)) \, dxdt \\ &\leq \int_0^T \int_\Omega \frac{J_p(v(t + \tau)) - J_p(v(t))}{\tau} \eta(t)\psi(x) \, dxdt \\ &= \int_0^T \int_\Omega \frac{\eta(t - \tau) - \eta(t)}{\tau} \psi(x) J_p(v(t)) \, dxdt. \end{aligned} \tag{7.84}$$

By Lemma 7.9, we have

$$\begin{aligned}
&\int_0^T \int_\Omega \frac{v(t+\tau)-v(t)}{\tau}\eta(t)\psi(x)p(v(t))\,dxdt \\
&= -\int_0^T \int_\Omega v(t)\frac{d}{dt}(\eta p(v))^\tau(t)\psi(x)\,dxdt \\
&= \int_0^T \int_\Omega (\xi - g_t)(t)(\eta p(v))^\tau(t)\psi(x)\,dxdt.
\end{aligned} \tag{7.85}$$

Collecting these inequalities, we obtain

$$\begin{aligned}
&\int_0^T \int_\Omega \frac{\eta(t-\tau)-\eta(t)}{-\tau}\psi J_p(v(t))\,dxdt \\
&\leq -\int_0^T \int_\Omega (\xi - g_t)(t)(\eta p(v))^\tau(t)\psi(x)\,dxdt \\
&= -\lim_n \int_0^T \langle u_n'(t) - g_t, (\eta p(v))^\tau(t)\psi\rangle dt \\
&= -\lim_n \int_0^T \frac{1}{\tau}\int_{t-\tau}^t \eta(s)\langle \operatorname{div}(z_n(t)) - g_t(t), p(v(s))\psi\rangle\; dsdt \\
&= \lim_n \int_0^T \frac{1}{\tau}\int_{t-\tau}^t \eta(s)\left[\int_\Omega (z_n(t), D(p(v(s))\psi)) + \langle g_t, p(v(s))\psi\rangle\right]\; dsdt \\
&= \lim_n \int_0^T \frac{1}{\tau}\int_{t-\tau}^t \eta(s)\int_\Omega (z_n(t), Dp(v(s)))\psi\; dsdt \\
&\quad + \lim_n \int_0^T \frac{1}{\tau}\int_{t-\tau}^t \eta(s)p(v(s))\int_\Omega z_n(t)\cdot\nabla\psi\; dsdt \\
&\quad + \int_0^T \frac{1}{\tau}\int_{t-\tau}^t \eta(s)\langle g_t, p(v(s))\psi\rangle\; dsdt.
\end{aligned}$$

Since

$$Dp(v(s)) = \nabla p(u(s) - g(s)) + D^s p(u(s) - g(s))$$

and

$$\begin{aligned}
z_n(t)\cdot D^s p(u(s)-g(s)) &= \mathbf{a}(x, \nabla u_n(t,x))\cdot D^s p(u(s)-g(s)) \\
&\leq h^0(x, D^s p(u(s)-g(s))),
\end{aligned}$$

from the above inequality, it follows that

$$\begin{aligned}
&\int_0^T \int_\Omega \frac{\eta(t-\tau)-\eta(t)}{-\tau} \psi J_p(u(t)-g)\, dxdt \\
&\leq \int_0^T \frac{1}{\tau} \int_{t-\tau}^t \eta(s) \int_\Omega z(t) \cdot \nabla p(u(s)-g(s)) \psi \, dxdsdt \\
&+ \int_0^T \frac{1}{\tau} \int_{t-\tau}^t \eta(s) \int_\Omega \psi h^0(x, D^s p(u(s)-g(s)))\, dsdt \\
&+ \int_0^T \frac{1}{\tau} \int_{t-\tau}^t \eta(s) \int_\Omega p(u(s)-g(s)) z(t) \cdot \nabla \psi \, dxdsdt \\
&+ \int_0^T \frac{1}{\tau} \int_{t-\tau}^t \eta(s) \int_\Omega g_t(t) p(u(s)-g(s)) \psi(x)\, dxdsdt.
\end{aligned} \tag{7.86}$$

Hence, letting $\tau \to 0$ in (7.86), we obtain

$$\begin{aligned}
\int_0^T \int_\Omega \phi_t J_p(u(t)-g) &\leq \int_0^T \int_\Omega \eta(t) z(t) \cdot \nabla p(u(t)-g(t)) \psi \, dxdt \\
&+ \int_0^T \eta(t) \int_\Omega \psi h^0(x, D^s p(u(t)-g(t)))\, dt \\
&+ \int_0^T \int_\Omega \eta(t) p(u(t)-g(t)) z(t) \cdot \nabla \psi \, dxdt \\
&+ \int_0^T \int_\Omega \eta(t) g_t(t) p(u(t)-g(t)) \psi(x)\, dxds\, dt.
\end{aligned} \tag{7.87}$$

Taking into account (7.82) and (7.87), we obtain

$$\begin{aligned}
&\int_0^T \int_\Omega \phi \left([z(t) - \mathbf{a}(x, \nabla g)] \cdot \nabla p(u(t)-g) \right) dtdx \\
&+ \int_0^T \int_\Omega \phi \left(h^0(x, D^s p(u(t)-g)) - \mathbf{a}(x, \nabla g) \cdot D^s p(u(t)-g) \right) dt \geq 0
\end{aligned}$$

for all $\phi(t,x) = \eta(t)\psi(x)$, $\eta \in \mathcal{D}(0,T)$, $\psi \in \mathcal{D}(\Omega)$, $\eta, \psi \geq 0$. Thus, the measure

$$\left([z - \mathbf{a}(x, \nabla g)] \cdot \nabla p(u-g) + h^0(x, D^s p(u-g)) - \mathbf{a}(x, \nabla g) \cdot D^s p(u-g) \right) \geq 0.$$

Then its absolutely continuous part

$$[z(t) - \mathbf{a}(x, \nabla g)] \cdot \nabla p(u(t)-g) \geq 0 \quad \text{a.e. in } \Omega.$$

Hence

$$[z(t) - \mathbf{a}(x, \nabla g)] \cdot \nabla (u(t)-g) \geq 0 \quad \text{a.e. in } \Omega.$$

Since we may take a countable set dense in $C^1([0,T]\times\overline{\Omega})$, we have that the above inequality holds for all $(t,x)\in S$ where $S\subseteq(0,T)\times\Omega$ is such that $\mathcal{L}^N((0,T)\times\Omega\setminus S)=0$, and all $g\in C^1([0,T]\times\overline{\Omega})$. Now, fix $(t,x)\in S$, and, given $y\in\mathbb{R}^N$, there is $g\in C^1([0,T]\times\overline{\Omega})$ such that $\nabla g(t,x)=y$. Then

$$(z(t,x)-\mathbf{a}(x,y))\cdot(\nabla u(t,x)-y)\geq 0 \quad \forall y\in\mathbb{R}^N,$$

and we get that

$$z(t,x)=\mathbf{a}(x,\nabla u(t,x)) \qquad \text{a.e.} \quad (t,x)\in Q_T. \tag{7.88}$$

Then, we have

$$\operatorname{div}(z(t))=\operatorname{div}(\mathbf{a}(x,\nabla u(t)) \quad \text{in } \mathcal{D}'(\Omega), \quad \text{a.e. } t\in[0,T],$$

and, since

$$|[z,\nu]|\leq f^0(x,\nu(x)) \;\; \mathcal{H}^{N-1}-\text{a.e. on } \partial\Omega,$$

we also get

$$|[\mathbf{a}(x,\nabla u(t)),\nu]|\leq f^0(x,\nu(x)) \quad \mathcal{H}^{N-1}\text{-a.e. on } \partial\Omega, \text{ a.e. in } t\in(0,T).$$

Finally, from (7.88) and (7.76), we obtain that

$$[\mathbf{a}(x,\nabla u(t)),\nu]\in\operatorname{sign}(p(\varphi)-p(u(t)))f^0(x,\nu(x))$$

$\mathcal{H}^{N-1}$-a.e. on $\partial\Omega$, a.e. in $t\in(0,T)$, and all $p\in\mathcal{P}$.

Step 7. Conclusion. Finally, we are going to prove that u verifies:

$$\begin{aligned}
&-\int_0^T\int_\Omega j(u(t)-l)\eta_t\,dxdt+\int_0^T\int_\Omega \eta(t)h(x,Dp(u(t)-l))\,dt\\
&+\int_0^T\int_\Omega z(t)\cdot\nabla\eta(t)p(u(t)-l)\,dxdt\\
&\leq\int_0^T\int_{\partial\Omega}[z(t),\nu]\eta(t)p(u(t)-l)\,d\mathcal{H}^{N-1}dt,
\end{aligned} \tag{7.89}$$

for all $\eta\in C^\infty(\overline{Q_T})$, with $\eta\geq 0$, $\eta(t,x)=\phi(t)\psi(x)$, being that $\phi\in\mathcal{D}(]0,T[)$, $\psi\in C^\infty(\overline{\Omega})$, and $p\in\mathcal{T}$, where $j(r)=\displaystyle\int_0^r p(s)\,ds$.

Let $\eta\in C^\infty(\overline{Q_T})$, with $\eta\geq 0$, $\eta(t,x)=\phi(t)\psi(x)$, $\phi\in\mathcal{D}(]0,T[)$, $\psi\in C^\infty(\overline{\Omega})$, $p\in\mathcal{P}$ and $a\in\mathbb{R}$. Let $G_p(r)=\displaystyle\int_a^r p(s)\,ds$. Since $u_n'(t)=\operatorname{div}(z_n(t))$, multiplying by $p(u_n(t))\eta(t)$ and integrating, we obtain that

$$
\begin{aligned}
&\int_0^T \int_\Omega \frac{d}{dt} G_p(u_n(t))\eta(t)\,dxdt = \int_0^T \int_\Omega p(u_n(t))u_n'(t)\eta(t)\,dxdt \\
&= \int_0^T \int_\Omega \operatorname{div}(z_n(t))p(u_n(t))\eta(t)\,dxdt \\
&= -\int_0^T \int_\Omega (z_n(t), D(p(u_n(t))\eta(t)))\,dt + \int_0^T \int_{\partial\Omega} [z_n(t),\nu]p(u_n(t))\eta(t)\,d\mathcal{H}^{N-1}dt \\
&= -\int_0^T \int_\Omega \eta(t)h(x, Dp(u_n(t)))\,dt - \int_0^T \int_\Omega z_n(t)\cdot\nabla\eta(t)\;p(u_n(t))\,dxdt \\
&\quad + \int_0^T \int_{\partial\Omega} [z_n(t),\nu]p(u_n(t))\eta(t)\,d\mathcal{H}^{N-1}dt \\
&= -\int_0^T \int_\Omega \eta(t)h(x, Dp(u_n(t)))\,dt - \int_0^T \int_\Omega z_n(t)\cdot\nabla\eta(t)\;p(u_n(t))\,dxdt \\
&\quad - \int_0^T \int_{\partial\Omega} |p(u_n(t)) - p(\varphi)| f^0(x,\nu(x))\eta(t)\,d\mathcal{H}^{N-1}dt \\
&\quad + \int_0^T \int_{\partial\Omega} [z_n(t),\nu]p(\varphi)\eta(t)\,d\mathcal{H}^{N-1}dt.
\end{aligned}
$$

Hence, having in mind that $\eta(0) = \eta(T) = 0$, we get

$$
\begin{aligned}
&\int_0^T \int_\Omega \eta(t)h(x, Dp(u_n(t)))\,dt + \int_0^T \int_{\partial\Omega} |p(u_n(t)) - p(\varphi)| f^0(x,\nu(x))\eta(t)\,d\mathcal{H}^{N-1}dt \\
&= -\int_0^T \int_\Omega z_n(t)\cdot\nabla\eta(t)\;p(u_n(t))\,dxdt + \int_0^T \int_{\partial\Omega} [z_n(t),\nu]p(\varphi)\eta(t)\,d\mathcal{H}^{N-1}dt \\
&\quad - \int_0^T \int_\Omega \frac{d}{dt} G_p(u_n(t))\eta(t)\,dxdt \\
&= -\int_0^T \int_\Omega z_n(t)\cdot\nabla\eta(t)\;p(u_n(t))\,dxdt + \int_0^T \int_{\partial\Omega} [z_n(t),\nu]p(\varphi)\eta(t)\,d\mathcal{H}^{N-1}dt \\
&\quad + \int_0^T \int_\Omega G_p(u_n(t))\;\eta_t\,dxdt.
\end{aligned}
$$

Now, observe that, by Remark 7.14, we have that

$$
\begin{aligned}
&\int_\Omega \eta(t,x)f(x, Dp(u_n)) + \int_{\partial\Omega} |p(\varphi) - p(u_n)| f^0(x,\nu(x))\eta(t,x)\;d\mathcal{H}^{N-1} \\
&\to \int_\Omega \eta(t,x)f(x, Dp(u)) + \int_{\partial\Omega} |p(\varphi) - p(u)| f^0(x,\nu(x))\eta(t,x)\;d\mathcal{H}^{N-1}
\end{aligned}
$$

a.e. in $t \in (0,T)$, and, therefore,

$$\int_\Omega \eta(t,x)h(x,Dp(u_n)) + \int_{\partial\Omega} |p(\varphi)-p(u_n)|f^0(x,\nu(x))\eta(t,x)\,d\mathcal{H}^{N-1}$$
$$\to \int_\Omega \eta(t,x)h(x,Dp(u)) + \int_{\partial\Omega} |p(\varphi)-p(u)|f^0(x,\nu(x))\eta(t,x)\,d\mathcal{H}^{N-1},$$

a.e. in $t \in (0,T)$. Hence, integrating in $(0,T)$ and using Fatou's lemma, it follows that

$$\int_0^T \int_\Omega \eta(t)h(x,Dp(u(t)))\,dt + \int_0^T \int_{\partial\Omega} |p(u(t))-p(\varphi)|f^0(x,\nu(x))\eta(t)\,d\mathcal{H}^{N-1}dt$$
$$\le \lim_{n\to\infty}\Bigg[\int_0^T \int_\Omega \eta(t)h(x,Dp(u_n(t)))\,dt$$
$$+\int_0^T \int_{\partial\Omega} |p(u_n(t))-p(\varphi)|f^0(x,\nu(x))\eta(t)\,d\mathcal{H}^{N-1}dt\Bigg]$$
$$= \lim_{n\to\infty}\Bigg[-\int_0^T\int_\Omega z_n(t)\cdot\nabla\eta(t)\;p(u_n(t))\,dxdt + \int_0^T\int_{\partial\Omega}[z_n(t),\nu]p(\varphi)\eta(t)\,d\mathcal{H}^{N-1}dt$$
$$+\int_0^T\int_\Omega G_p(u_n(t))\;\eta_t\,dxdt\Bigg]$$
$$= -\int_0^T\int_\Omega z(t)\cdot\nabla\eta(t)\;p(u(t))\,dxdt + \int_0^T\int_{\partial\Omega}[z(t),\nu]p(\varphi)\eta(t)\,d\mathcal{H}^{N-1}dt$$
$$+\int_0^T\int_\Omega G_p(u(t))\;\eta_t\,dxdt.$$

Now, using that $|p(u(t))-p(\varphi)|f^0(x,\nu(x)) = [z(t),\nu](p(\varphi)-p(u(t)))$, we have

$$-\int_0^T\int_\Omega G_p(u(t))\eta_t\,dxdt + \int_0^T\int_\Omega \eta(t)h(x,Dp(u(t)))\,dt$$
$$+\int_0^T\int_\Omega z(t)\cdot\nabla\eta(t)\;p(u(t))\,dxdt \le \int_0^T\int_{\partial\Omega}[z(t),\nu]p(u(t))\eta(t)\,d\mathcal{H}^{N-1}dt. \tag{7.90}$$

Finally, given $l \in \mathbb{R}$ and $p \in \mathcal{T}$, since $q(r) := p(r-l)$ is an element of $\mathcal{P}$, and taking $a = l$, we obtain (7.89) as a consequence of (7.90). The proof of the existence is finished.

7.4.2 Proof of Theorem 7.3: Uniqueness

To prove uniqueness of entropy solutions, we shall use the doubling variables technique introduced by Kruzhkov [143] (see also [61] and [115]) to prove the L^1-contraction property for entropy solutions of scalar conservation laws.

Since the operator $\mathcal{A}_\varphi$ is m-completely accretive in $L^1(\Omega)$, if we prove that the entropy solution coincides with the semigroup solution, then, by Proposition A.44, (7.9) holds. So we only need to prove that any entropy solution is a semigroup solution.

Let $u(t)$ be an entropy solution with initial datum $u_0 \in L^1(\Omega)$ and $\overline{u}(t) = T(t)\overline{u}_0$ the semigroup solution with initial datum $\overline{u}_0 \in L^\infty(\Omega)$. Then, there exist $\xi, \overline{\xi} \in \left(L^1(0,T,BV(\Omega)_2)\right)^*$ such that, letting $z(t) := \mathbf{a}(x, \nabla u(t))$ and $\overline{z}(t) := \mathbf{a}(x, \nabla \overline{u}(t))$, we have $(z(t), \xi(t)), (\overline{z}(t), \overline{\xi}(t)) \in Z(\Omega)$ for almost all $t \in [0,T]$,

$$[z(t), \nu] \in \operatorname{sign}\left(T_k^+(\varphi) - T_k^+(u(t))\right) f^0(x, \nu(x)) \quad \text{a.e. in } t \in [0,T], \tag{7.91}$$

$$[\overline{z}(t), \nu] \in \operatorname{sign}\left(T_k^+(\varphi) - T_k^+(\overline{u}(t))\right) f^0(x, \nu(x)) \quad \text{a.e. in } t \in [0,T], \tag{7.92}$$

and such that, if $r, \overline{r} \in \mathbb{R}^N$ and $l_1, l_2 \in \mathbb{R}$, then

$$\begin{aligned} &-\int_0^T \int_\Omega j_k^+(u(t) - l_1)\eta_t + \int_0^T \int_\Omega \eta(t) h(x, DT_k^+(u(t) - l_1)) \\ &+ \int_0^T \int_\Omega (z(t) - r) \cdot \nabla \eta(t)\, T_k^+(u(t) - l_1) + \int_0^T \int_\Omega r \cdot \nabla \eta(t)\, T_k^+(u(t) - l_1) \\ &\leq \int_0^T \int_{\partial\Omega} [z(t), \nu] \eta(t) T_k^+(u(t) - l_1) \end{aligned} \tag{7.93}$$

and

$$\begin{aligned} &-\int_0^T \int_\Omega j_k^-(\overline{u}(t) - l_2)\eta_t + \int_0^T \int_\Omega \eta(t) h(x, DT_k^-(\overline{u}(t) - l_2)) \\ &+ \int_0^T \int_\Omega (\overline{z}(t) - \overline{r}) \cdot \nabla \eta(t)\, T_k^-(\overline{u}(t) - l_2) + \int_0^T \int_\Omega \overline{r} \cdot \nabla \eta(t)\, T_k^-(\overline{u}(t) - l_2) \\ &\leq \int_0^T \int_{\partial\Omega} [\overline{z}(t), \nu] \eta(t) T_k^-(\overline{u}(t) - l_2), \end{aligned} \tag{7.94}$$

for all $\eta \in C^\infty(\overline{Q_T})$, with $\eta \geq 0$, $\eta(t,x) = \phi(t)\psi(x)$, being that $\phi \in \mathcal{D}(]0,T[)$, $\psi \in C^\infty(\overline{\Omega})$, and $j_k^+(r) = \int_0^r T_k^+(s)\, ds$, $j_k^-(r) = \int_0^r T_k^-(s)\, ds$.

We choose two different pairs of variables (t,x), (s,y) and consider u, z as functions in (t,x); $\overline{u}$, $\overline{z}$ in (s,y). Let $0 \leq \phi \in \mathcal{D}(]0,T[)$, $0 \leq \psi \in \mathcal{D}(\Omega)$, ρ_n a classical sequence of mollifiers in $\mathbb{R}^N$ and $\tilde{\rho}_n$ a sequence of mollifiers in $\mathbb{R}$. Define

$$\eta_n(t,x,s,y) := \rho_n(x-y)\tilde{\rho}_n(t-s)\phi\left(\frac{t+s}{2}\right)\psi\left(\frac{x+y}{2}\right).$$

Note that, for n sufficiently large,

$$(t,x) \mapsto \eta_n(t,x,s,y) \in \mathcal{D}(]0,T[\times\Omega) \quad \forall\, (s,y) \in Q_T,$$

$$(s,y) \mapsto \eta_n(t,x,s,y) \in \mathcal{D}(]0,T[\times\Omega) \quad \forall\, (t,x) \in Q_T.$$

Hence, for (s,y) fixed, if we take $l_1 = \overline{u}(s,y)$ and $r = \overline{z}(s,y)$ in (7.93), we get

$$\begin{aligned}
&-\int_0^T\int_\Omega j_k^+(u(t,x)-\overline{u}(s,y))(\eta_n)_t\,dxdt \\
&+\int_0^T\int_\Omega \eta_n h(x, D_x T_k^+(u(t,x)-\overline{u}(s,y)))\,dt \\
&+\int_0^T\int_\Omega (z(t,x)-\overline{z}(s,y))\cdot\nabla_x\eta_n\; T_k^+(u(t,x)-\overline{u}(s,y))\,dxdt \\
&+\int_0^T\int_\Omega \overline{z}(s,y)\cdot\nabla_x\eta_n\; T_k^+(u(t,x)-\overline{u}(s,y))\,dxdt \le 0.
\end{aligned} \tag{7.95}$$

Similarly, for (t,x) fixed, if we take $l_2 = u(t,x)$ and $\overline{r} = z(t,x)$ in (7.94), we get

$$\begin{aligned}
&-\int_0^T\int_\Omega j_k^-(\overline{u}(s,y)-u(t,x))(\eta_n)_s\,dyds \\
&+\int_0^T\int_\Omega \eta_n h(y, D_y T_k^-(\overline{u}(s,y)-u(t,x)))\,dyds \\
&+\int_0^T\int_\Omega (\overline{z}(s,y)-z(t,x))\cdot\nabla_y\eta_n\; T_k^-(\overline{u}(s,y)-u(t,x))\,dyds \\
&+\int_0^T\int_\Omega z(t,x)\cdot\nabla_y\eta_n\; T_k^-(\overline{u}(s,y)-u(t,x))\,dyds \le 0.
\end{aligned}$$

Now, since $T_k^-(r) = -T_k^+(-r)$, $j_k^-(r) = j_k^+(-r)$ and $h(x,-\xi) = h(x,\xi)$, we can rewrite the last inequality as

$$\begin{aligned}
&-\int_0^T\int_\Omega j_k^+(u(t,x)-\overline{u}(s,y))(\eta_n)_s\,dyds \\
&+\int_0^T\int_\Omega \eta_n h(y, D_y T_k^+(u(t,x)-\overline{u}(s,y)))\,ds \\
&+\int_0^T\int_\Omega (z(t,x)-\overline{z}(s,y))\cdot\nabla_y\eta_n\; T_k^+(u(t,x)-\overline{u}(s,y))\,dyds \\
&-\int_0^T\int_\Omega z(t,x)\cdot\nabla_y\eta_n T_k^+(u(t,x)-\overline{u}(s,y))\,dyds \le 0.
\end{aligned} \tag{7.96}$$

Integrating (7.95) in (s,y), (7.96) in (t,x) and taking their sum yields

$$
\begin{aligned}
&-\int_{Q_T\times Q_T} j_k^+(u(t,x)-\overline{u}(s,y))\,((\eta_n)_t+(\eta_n)_s)\\
&+\int_{Q_T\times Q_T} \eta_n h(x, D_x T_k^+(u(t,x)-\overline{u}(s,y)))\\
&+\int_{Q_T\times Q_T} \eta_n h(y, D_y T_k^+(u(t,x)-\overline{u}(s,y)))\\
&+\int_{Q_T\times Q_T} (z(t,x)-\overline{z}(s,y))\cdot(\nabla_x\eta_n+\nabla_y\eta_n)\,T_k^+(u(t,x)-\overline{u}(s,y))\\
&+\int_{Q_T\times Q_T} \overline{z}(s,y)\cdot\nabla_x\eta_n T_k^+(u(t,x)-\overline{u}(s,y))\\
&-\int_{Q_T\times Q_T} z(t,x)\cdot\nabla_y\eta_n T_k^+(u(t,x)-\overline{u}(s,y))\le 0.
\end{aligned}
\tag{7.97}
$$

Now, by Green's formula (C.10) and the identities $z(t,x) = \mathbf{a}(x,\nabla u(t,x))$, $\overline{z}(s,y) = a(y,\nabla\overline{u}(s,y))$, we have

$$
\begin{aligned}
J_n := &\int_{Q_T\times Q_T} \overline{z}(s,y)\cdot\nabla_x\eta_n\; T_k^+(u(t,x)-\overline{u}(s,y))\\
&+\int_{Q_T\times Q_T} \eta_n h(x, D_x T_k^+(u(t,x)-\overline{u}(s,y)))\\
&-\int_{Q_T\times Q_T} z(t,x)\cdot\nabla_y\eta_n\; T_k^+(u(t,x)-\overline{u}(s,y))\\
&+\int_{Q_T\times Q_T} \eta_n h(y, D_y T_k^+(u(t,x)-\overline{u}(s,y)))\\
= &-\int_{Q_T\times Q_T} \eta_n\left(\overline{z}(s,y), D_x T_k^+(u(t,x)-\overline{u}(s,y))\right)\\
&+\int_{Q_T\times Q_T} \eta_n h(x, D_x T_k^+(u(t,x)-\overline{u}(s,y)))\\
&+\int_{Q_T\times Q_T} \eta_n\left(z(t,x), D_y T_k^+(u(t,x)-\overline{u}(s,y))\right)\\
&+\int_{Q_T\times Q_T} \eta_n h(y, D_y T_k^+(u(t,x)-\overline{u}(s,y)))\\
= &\int_{Q_T\times Q_T} \eta_n (T_k^+)'(u(t,x)-\overline{u}(s,y))[z(t,x)-\overline{z}(s,y)]\cdot(\nabla_x u(t,x)-\nabla_y\overline{u}(s,y))\\
&-\int_{Q_T\times Q_T} \eta_n\overline{z}(s,y)\cdot D_x^s T_k^+(u(t,x)-\overline{u}(s,y))\\
&+\int_{Q_T\times Q_T} \eta_n h^0(x, D_x^s T_k^+(u(t,x)-\overline{u}(s,y)))
\end{aligned}
$$

$$+\int_{Q_T\times Q_T}\eta_n z(t,x)\cdot D_y^s T_k^+(u(t,x)-\overline{u}(s,y))$$
$$+\int_{Q_T\times Q_T}\eta_n h^0(y,D_y^s T_k^+(u(t,x)-\overline{u}(s,y))).$$

We claim that

$$J_n\geq o(1), \tag{7.98}$$

where $o(1)$ is an expression that tends to 0 as $n\to\infty$. Indeed, let us analyze the term

$$-\int_{Q_T\times Q_T}\eta_n \overline{z}(s,y)\cdot D_x^s T_k^+(u(t,x)-\overline{u}(s,y))$$
$$+\int_{Q_T\times Q_T}\eta_n h^0(x,D_x^s T_k^+(u(t,x)-\overline{u}(s,y))).$$

By assumption (H_5) we have

$$-\int_{Q_T\times Q_T}\eta_n \overline{z}(s,y)\cdot D_x^s T_k^+(u(t,x)-\overline{u}(s,y))$$
$$+\int_{Q_T\times Q_T}\eta_n h^0(x,D_x^s T_k^+(u(t,x)-\overline{u}(s,y)))$$
$$\geq\int_{Q_T\times Q_T}\eta_n h^0(x,D_x^s T_k^+(u(t,x)-\overline{u}(s,y)))$$
$$-\int_{Q_T\times Q_T}\eta_n h^0(y,D_x^s T_k^+(u(t,x)-\overline{u}(s,y))).$$

Let us prove that the term in the right-hand side tends to 0 as $n\to\infty$. Indeed, let $l=k+\|\overline{u}\|_\infty$, using (7.8), and having in mind that

$$\int_\Omega\|D_x^s T_k^+(u(t,x)-\overline{u}(s,y))\|=\int_\Omega\|D_x^s T_k^+(T_l(u(t,x))-\overline{u}(s,y))\|$$
$$\leq\int_\Omega\|D_x^s(T_l(u(t,x))-\overline{u}(s,y))\|=\int_\Omega\|D_x^s T_l(u(t,x))\|,$$

we have

$$\left|\int_{Q_T\times Q_T}\eta_n h^0(x,D_x^s T_k^+(u(t,x)-\overline{u}(s,y)))\right.$$
$$\left.-\int_{Q_T\times Q_T}\eta_n h^0(y,D_x^s T_k^+(u(t,x)-\overline{u}(s,y)))\right|$$
$$\leq\int_{Q_T\times Q_T}\eta_n\omega(\|x-y\|)\|D_x^s T_k^+(u(t,x)-\overline{u}(s,y))\|$$
$$\leq\int_{Q_T}dy\int_{Q_T}\eta_n\omega(\|x-y\|)\|D_x^s T_l(u(t,x))\|$$
$$=\int_{Q_T}\|D_x^s T_l(u(t,x))\|\int_{Q_T}\eta_n\omega(\|x-y\|)\,dy.$$

Now, we observe that

$$\int_{Q_T} dy \eta_n \omega(\|x-y\|)$$
$$= \int_0^T \tilde{\rho}_n(t-s)\phi\left(\frac{t+s}{2}\right) ds \int_\Omega \rho_n(x-y)\psi\left(\frac{x+y}{2}\right)\omega(\|x-y\|)\, dy$$

and

$$\left|\int_\Omega \rho_n(x-y)\psi\left(\frac{x+y}{2}\right)\omega(\|x-y\|)dy\right|$$
$$\leq \|\psi\|_\infty \int_{\mathbb{R}^N} \rho_n(x-y)\omega(\|x-y\|)dy$$
$$= \|\psi\|_\infty \int_{\mathbb{R}^N} \rho_n(z)\omega(\|z\|)dz \to 0, \quad \text{as } n \to \infty.$$

Hence

$$\int_{Q_T\times Q_T} \eta_n h^0(x, D_x^s T_k^+(u(t,x) - \overline{u}(s,y)))$$
$$- \int_{Q_T\times Q_T} \eta_n h^0(y, D_x^s T_k^+(u(t,x) - \overline{u}(s,y))) \to 0, \quad \text{as } n \to \infty.$$

In a similar way, we prove that

$$\int_{Q_T\times Q_T} \eta_n z(t,x) \cdot D_y^s T_k^+(u(t,x) - \overline{u}(s,y))$$
$$+ \int_{Q_T\times Q_T} \eta_n h^0(y, D_y^s T_k^+(u(t,x) - \overline{u}(s,y))) \geq o(1).$$

Let us prove that

$$\int_{Q_T\times Q_T} \eta_n (T_k^+)'(u(t,x) - \overline{u}(s,y))[z(t,x) - \overline{z}(s,y)]\cdot(\nabla_x u(t,x) - \nabla_y \overline{u}(s,y)) \geq o(1).$$

Indeed, let $\overline{z}(s,x,y) = a(x, \nabla_y \overline{u}(s,y))$, then

$$\int_{Q_T\times Q_T} \eta_n (T_k^+)'(u(t,x) - \overline{u}(s,y))[z(t,x) - \overline{z}(s,y)] \cdot (\nabla_x u(t,x) - \nabla_y \overline{u}(s,y))$$
$$= \int_{Q_T\times Q_T} \eta_n (T_k^+)'(u(t,x) - \overline{u}(s,y))[z(t,x) - \overline{z}(s,x,y)]\cdot(\nabla_x u(t,x) - \nabla_y \overline{u}(s,y))$$
$$+ \int_{Q_T\times Q_T} \eta_n (T_k^+)'(u(t,x) - \overline{u}(s,y))[\overline{z}(s,x,y) - \overline{z}(s,y)]\cdot(\nabla_x u(t,x) - \nabla_y \overline{u}(s,y))$$

$$\geq \int_{Q_T\times Q_T} \eta_n (T_k^+)'(u(t,x)-\overline{u}(s,y))[\overline{z}(s,x,y)-\overline{z}(s,y)]\cdot(\nabla_x u(t,x)-\nabla_y \overline{u}(s,y))$$
$$= \int_{Q_T\times Q_T} \eta_n[\overline{z}(s,x,y)-\overline{z}(s,y)]\cdot \nabla_x T_k^+(u(t,x)-\overline{u}(s,y))$$
$$+ \int_{Q_T\times Q_T} \eta_n[\overline{z}(s,x,y)-\overline{z}(s,y)]\cdot \nabla_y T_k^+(u(t,x)-\overline{u}(s,y)).$$

Now, we observe that using the same argument as above, both terms in the right-hand side in the last inequality tend to zero as $n\to\infty$. With this we finish the proof of (7.98).

From (7.98) and (7.97), it follows that

$$\begin{aligned} &-\int_{Q_T\times Q_T} j_k^+(u(t,x)-\overline{u}(s,y))\,((\eta_n)_t+(\eta_n)_s) \\ &+\int_{Q_T\times Q_T} (z(t,x)-\overline{z}(s,y))\cdot(\nabla_x\eta_n+\nabla_y\eta_n)\,T_k^+(u(t,x)-\overline{u}(s,y)) \\ &\leq o(1). \end{aligned} \tag{7.99}$$

Since

$$(\eta_n)_t+(\eta_n)_s = \rho_n(x-y)\tilde{\rho}_n(t-s)\phi'\left(\frac{t+s}{2}\right)\psi\left(\frac{x+y}{2}\right),$$

and

$$\nabla_x\eta_n+\nabla_y\eta_n = \rho_n(x-y)\tilde{\rho}_n(t-s)\phi\left(\frac{t+s}{2}\right)\nabla\psi\left(\frac{x+y}{2}\right),$$

passing to the limit in (7.99), it yields

$$\begin{aligned} &-\int_{Q_T} j_k^+(u(t,x)-\overline{u}(t,x))\phi'(t)\psi(x) \\ &+\int_{Q_T} (z(t,x)-\overline{z}(t,x))\cdot\nabla\psi(x)\;\phi(t)T_k^+(u(t,x)-\overline{u}(t,x))\leq 0. \end{aligned} \tag{7.100}$$

We have to prove that

$$\lim_n \int_{Q_T} (z(t,x)-\overline{z}(t,x))\cdot\nabla\psi_n(x)\;\phi(t)T_k^+(u(t,x)-\overline{u}(t,x))\geq 0$$

for any sequence $\psi_n\uparrow\chi_\Omega$. Since $\xi=\operatorname{div}(z)$, $\overline{\xi}=\operatorname{div}(\overline{z})$ in $\left(L^1(0,T,BV(\Omega)_2)\right)^*$, the following integration by parts formula holds:

$$\int_{Q_T}(z-\overline{z},Dw)+\int_0^T\langle\xi(t)-\overline{\xi}(t),w(t)\rangle dt$$
$$=\int_0^T\int_{\partial\Omega}[z(t,x)-\overline{z}(t,x),\nu]w(t,x)d\mathcal{H}^{N-1}dt,$$

for all $w \in L^1(0,T,BV(\Omega)) \cap L^\infty(Q_T)$. Set

$$w(t) = ((\psi - 1)\phi T_k^+(u - \overline{u}))^\tau(t,x) = (\psi(x) - 1)(\phi T_k^+(u - \overline{u}))^\tau(t,x),$$

where $\psi \in \mathcal{D}(\Omega)$, $\quad 0 \leq \psi \leq 1$ and

$$(\phi T_k^+(u - \overline{u}))^\tau(t,x) = \frac{1}{\tau}\int_t^{t+\tau} \phi(s)T_k^+(u(s,x) - \overline{u}(s,x))ds,$$

in the above formula to obtain

$$\begin{aligned}
&\int_{Q_T} (z(t,x) - \overline{z}(t,x)) \cdot \nabla(\psi(x) - 1)\ (\phi T_k^+(u - \overline{u}))^\tau(t,x)dxdt \\
&= -\int_0^T \int_\Omega (\psi - 1)\left(z(t) - \overline{z}(t), D(\phi T_k^+(u - \overline{u}))^\tau(t)\right) dt \\
&\quad - \int_{Q_T} (\xi(t) - \overline{\xi}(t))(\psi(x) - 1)(\phi T_k^+(u - \overline{u}))^\tau(t,x) \\
&\quad + \int_0^T \int_{\partial\Omega} [z(t,x) - \overline{z}(t,x), \nu](\psi(x) - 1)(\phi T_k^+(u - \overline{u}))^\tau(t,x)d\mathcal{H}^{N-1}dt.
\end{aligned}$$

Since

$$\begin{aligned}
&\int_{Q_T} (z - \overline{z}) \cdot \nabla(\psi - 1)\phi T_k^+(u - \overline{u})dxdt \\
&= \lim_{\tau \to 0+} \int_{Q_T} (z - \overline{z}) \cdot \nabla(\psi - 1)\ (\phi T_k^+(u - \overline{u}))^\tau dxdt,
\end{aligned}$$

and, using that $\psi|_{\partial\Omega} = 0$, also

$$\begin{aligned}
&-\int_0^T \int_{\partial\Omega} [z(t) - \overline{z}(t), \nu]\phi T_k^+(u(t) - \overline{u}(t))d\mathcal{H}^{N-1}dt \\
&= \lim_{\tau \to 0+} \int_0^T \int_{\partial\Omega} [z(t) - \overline{z}(t), \nu](\psi - 1)(\phi T_k^+(u - \overline{u}))^\tau(t)d\mathcal{H}^{N-1}dt,
\end{aligned}$$

we may write

$$\begin{aligned}
&\int_{Q_T} (z - \overline{z})\nabla\psi\phi T_k^+(u - \overline{u}) = \int_{Q_T} (z - \overline{z})\nabla(\psi - 1)\phi T_k^+(u - \overline{u}) \\
&= \lim_{\tau \to 0+} \left(\int_0^T \int_\Omega (1 - \psi)\left(z(t) - \overline{z}(t), D(\phi T_k^+(u - \overline{u}))^\tau(t)\right) dt \right. \\
&\qquad \left. + \int_{Q_T} (\xi - \overline{\xi})(1 - \psi)(\phi T_k^+(u - \overline{u}))^\tau \right) \\
&\quad - \int_0^T \int_{\partial\Omega} [z(t) - \overline{z}(t), \nu]\phi T_k^+(u(t) - \overline{u}(t))d\mathcal{H}^{N-1}dt.
\end{aligned}$$

Now, since $\xi, \overline{\xi}$ are the time derivatives in $\left(L^1(0,T,BV(\Omega)_2)\right)^*$ of u and $\overline{u}$, respectively, we have that

$$\int_0^T \int_\Omega (\xi - \overline{\xi})(1-\psi)(\phi T_k^+(u-\overline{u}))^\tau = \int_0^T \int_\Omega (\xi - \overline{\xi})((1-\psi)\phi T_k^+(u-\overline{u}))^\tau$$
$$= \int_0^T \int_\Omega (1-\psi)\phi T_k^+(u-\overline{u})\frac{1}{\tau}\Delta_\tau^-(u-\overline{u}),$$

where $\Delta_\tau^-(u-\overline{u}) = (u-\overline{u})(t) - (u-\overline{u})(t-\tau)$. Let $v = u - \overline{u}$. Since

$$T_k^+(v(t))(v(t) - v(t-\tau)) \geq j_k^+(v(t)) - j_k^+(v(t-\tau))$$

(j_k^+ being the primitive of T_k^+), and $\phi, (1-\psi) \geq 0$, we have for τ small enough that

$$\int_0^T \int_\Omega (\xi - \overline{\xi})(1-\psi)(\phi T_k^+(u-\overline{u}))^\tau$$
$$\geq \int_0^T \int_\Omega (1-\psi)\phi \frac{j_k^+(v(t)) - j_k^+(v(t-\tau))}{\tau}$$
$$= -\int_0^T \int_\Omega \frac{\phi(t+\tau) - \phi(t)}{\tau}(1-\psi)j_k^+(u-\overline{u}).$$

Thus, we have

$$\int_{Q_T} (z - \overline{z})\nabla\psi\phi T_k^+(u-\overline{u})\, dxdt$$
$$\geq \lim_{\tau\to 0+}\left(\int_0^T \int_\Omega (1-\psi)\left(z(t) - \overline{z}(t), D(\phi T_k^+(u-\overline{u}))^\tau(t)\right) dt \right.$$
$$\left. - \int_0^T \int_\Omega \frac{\phi(t+\tau) - \phi(t)}{\tau}(1-\psi)j_k^+(u(t) - \overline{u}(t))\, dxdt\right)$$
$$- \int_0^T \int_{\partial\Omega} [z(t) - \overline{z}(t), \nu]\phi T_k^+(u(t) - \overline{u}(t)) d\mathcal{H}^{N-1} dt.$$

Finally, we observe that

$$\lim_{\tau\to 0+}\left|\int_0^T \int_\Omega (1-\psi)\left(z(t) - \overline{z}(t), D(\phi T_k^+(u-\overline{u}))^\tau(t)\right) dt\right|$$
$$\leq 2M \int_{Q_T} (1-\psi)\phi \| DT_k^+(u(t) - \overline{u}(t))\|\, dt,$$

which enables us to write that

$$\int_{Q_T} (z - \overline{z})\nabla\psi\phi T_k^+(u - \overline{u})\, dxdt \geq -2M \int_{Q_T} (1 - \psi)\phi \|DT_k^+(u(t) - \overline{u}(t))\|\, dt$$
$$- \int_0^T \int_\Omega \phi'(t)(1 - \psi)j_k^+(u(t) - \overline{u}(t))\, dxdt$$
$$- \int_0^T \int_{\partial\Omega} [z(t) - \overline{z}(t), \nu]\phi T_k^+(u(t) - \overline{u}(t)) d\mathcal{H}^{N-1} dt.$$

Let $\psi = \psi_n$ where $\psi_n \uparrow \chi_\Omega$ in the last expression. Using that $\|DT_k^+(u(t) - \overline{u}(t))\|$ is a Radon measure a.e. in t with $\|DT_k^+(u(t) - \overline{u}(t))\| \in L^1(0,T)$, which follows from Lemma 5.19, and letting $n \to \infty$, we obtain

$$\lim_n \int_{Q_T} (z - \overline{z})\nabla\psi_n\phi T_k^+(u - \overline{u})\, dxdt$$
$$\geq - \int_0^T \int_{\partial\Omega} [z(t) - \overline{z}(t), \nu]\phi T_k^+(u(t) - \overline{u}(t)) d\mathcal{H}^{N-1} dt.$$

Thus, using (7.100), we get

$$\int_{Q_T} j_k^+(u(t,x) - \overline{u}(t,x))\phi'(t)\, dxdt$$
$$\geq - \int_0^T \int_{\partial\Omega} [z(t) - \overline{z}(t), \nu]\phi T_k^+(u(t) - \overline{u}(t)) d\mathcal{H}^{N-1} dt \geq 0.$$

Since this is true for all $0 \leq \phi \in \mathcal{D}(]0,T[)$, we have

$$\frac{d}{dt} \int_\Omega j_k^+(u(t,x) - \overline{u}(t,x))\, dx \leq 0.$$

Hence

$$\int_\Omega j_k^+(u(t,x) - \overline{u}(t,x))\, dx \leq \int_\Omega j_k^+(u_0 - \overline{u}_0)\, dx.$$

Then, dividing by k the last inequality and letting $k \to 0$, we obtain

$$\int_\Omega (u(t,x) - \overline{u}(t,x))^+\, dx \leq \int_\Omega (u_0 - \overline{u}_0)^+\, dx.$$

From this, we deduce that

$$\|u(t) - \overline{u}(t)\|_1 \leq \|u_0 - \overline{u}_0\|_1, \qquad \forall\, t \geq 0.$$

Hence, taking $u_n(t) = T(t)u_{0,n}$, $u_{0,n} \in L^\infty(\Omega)$ and $u_{0,n} \to u_0$ in $L^1(\Omega)$, we have

$$\|u(t) - u_n(t)\|_1 \leq \|u_0 - u_{0,n}\|_1, \qquad \forall\, t \geq 0.$$

Letting $n \to \infty$, we obtain that $u(t) = T(t)u_0$. We have that entropy solutions coincide with semigroup solutions. Thus, entropy solutions are unique, and the proof is concluded. □

Remark 7.17. Let us sketch the proof of uniqueness of the Dirichlet problem for the total variation flow. Let $u(t)$ be an entropy solution of (5.1) with initial datum $u_0 \in L^1(\Omega)$ and $\overline{u}(t) = S(t)\overline{u}_0$ the semigroup solution with initial datum $\overline{u}_0 \in L^\infty(\Omega)$. Then, there exist $z(t), \overline{z}(t) \in Z(\Omega)$ with $\|z(t)\|_\infty \leq 1$, $\|\overline{z}(t)\|_\infty \leq 1$,

$$[z(t), \nu] \in \operatorname{sign}\big(T_k^+(\varphi) - T_k^+(u(t))\big) \quad \text{a.e. in } t \in [0,T], \tag{7.101}$$

$$[\overline{z}(t), \nu] \in \operatorname{sign}\left(T_k^+(\varphi) - T_k^+(\overline{u}(t))\right) \quad \text{a.e. in } t \in [0,T], \tag{7.102}$$

and such that, if $r, \overline{r} \in \mathbb{R}^N$, with $\|r\| \leq 1$, $\|\overline{r}\| \leq 1$ and $l_1, l_2 \in \mathbb{R}$, then

$$\begin{aligned} &-\int_0^T \int_\Omega j_k^+(u(t) - l_1)\eta_t + \int_0^T \int_\Omega \eta(t)\, d\|DT_k^+(u(t) - l_1)\| \\ &+ \int_0^T \int_\Omega (z(t) - r) \cdot D\eta(t)\ T_k^+(u(t) - l_1) + \int_0^T \int_\Omega r \cdot D\eta(t)\ T_k^+(u(t) - l_1) \\ &\leq \int_0^T \int_{\partial\Omega} [z(t), \nu]\eta(t) T_k^+(u(t) - l_1), \end{aligned} \tag{7.103}$$

and

$$\begin{aligned} &-\int_0^T \int_\Omega j_k^-(\overline{u}(t) - l_2)\eta_t + \int_0^T \int_\Omega \eta(t)\, d\|DT_k^-(\overline{u}(t) - l_2)\| \\ &+ \int_0^T \int_\Omega (\overline{z}(t) - \overline{r}) \cdot D\eta(t)\ T_k^-(\overline{u}(t) - l_2) + \int_0^T \int_\Omega \overline{r} \cdot D\eta(t)\ T_k^-(\overline{u}(t) - l_2) \\ &\leq \int_0^T \int_{\partial\Omega} [\overline{z}(t), \nu]\eta(t) T_k^-(\overline{u}(t) - l_2), \end{aligned} \tag{7.104}$$

for all $\eta \in C^\infty(\overline{Q_T})$, with $\eta \geq 0$, $\eta(t,x) = \phi(t)\psi(x)$, being that $\phi \in \mathcal{D}(]0,T[)$, $\psi \in C^\infty(\overline{\Omega})$, and $j_k^+(r) = \displaystyle\int_0^r T_k^+(s)\, ds$, $j_k^-(r) = \displaystyle\int_0^r T_k^-(s)\, ds$.

As before, we choose two different pairs of variables (t,x), (s,y) and consider u, z as functions in (t,x), $\overline{u}$, $\overline{z}$ in (s,y). Let $0 \leq \phi \in \mathcal{D}(]0,T[)$, $0 \leq \psi \in \mathcal{D}(\Omega)$, ρ_n, $\tilde{\rho}_n$, and $\eta_n(t,x,s,y)$ be as in the previous proof. Hence, for (s,y) fixed, if we take in (7.103) $l_1 = \overline{u}(s,y)$ and $r = \overline{z}(s,y)$, we get

$$\begin{aligned} &-\int_0^T \int_\Omega j_k^+(u(t,x) - \overline{u}(s,y))(\eta_n)_t + \int_0^T \int_\Omega \eta_n \|D_x T_k^+(u(t,x) - \overline{u}(s,y))\| \\ &+ \int_0^T \int_\Omega (z(t,x) - \overline{z}(s,y)) \cdot \nabla_x \eta_n\ T_k^+(u(t,x) - \overline{u}(s,y)) \\ &+ \int_0^T \int_\Omega \overline{z}(s,y) \cdot \nabla_x \eta_n\ T_k^+(u(t,x) - \overline{u}(s,y)) \leq 0. \end{aligned} \tag{7.105}$$

Similarly, for (t,x) fixed, if we take in (7.104) $l_2 = u(t,x)$ and $\overline{r} = z(t,x)$, we get

$$\begin{aligned}
&-\int_0^T\int_\Omega j_k^-(\overline{u}(s,y)-u(t,x))(\eta_n)_s + \int_0^T\int_\Omega \eta_n \, d\|D_y T_k^-(\overline{u}(s,y)-u(t,x))\| \\
&+\int_0^T\int_\Omega (\overline{z}(s,y)-z(t,x))\cdot\nabla_y\eta_n \; T_k^-(\overline{u}(s,y)-u(t,x)) \\
&+\int_0^T\int_\Omega z(t,x)\cdot\nabla_y\eta_n \; T_k^-(\overline{u}(s,y)-u(t,x)) \le 0.
\end{aligned} \tag{7.106}$$

Now, since $T_k^-(r) = -T_k^+(-r)$ and $j_k^-(r) = j_k^+(-r)$, we can rewrite (7.106) as

$$\begin{aligned}
&-\int_0^T\int_\Omega j_k^+(u(t,x)-\overline{u}(s,y))(\eta_n)_s + \int_0^T\int_\Omega \eta_n \, d\|D_y T_k^+(u(t,x)-\overline{u}(s,y))\| \\
&+\int_0^T\int_\Omega (z(t,x)-\overline{z}(s,y))\cdot\nabla_y\eta_n \; T_k^+(u(t,x)-\overline{u}(s,y)) \\
&-\int_0^T\int_\Omega z(s,y)\cdot\nabla_y\eta_n T_k^+(u(t,x)-\overline{u}(s,y)) \le 0.
\end{aligned} \tag{7.107}$$

Integrating (7.105) in (s,y), (7.107) in (t,x) and taking their sum yields

$$\begin{aligned}
&-\int_{Q_T\times Q_T} j_k^+(u(t,x)-\overline{u}(s,y))\big((\eta_n)_t+(\eta_n)_s\big) \\
&+\int_{Q_T\times Q_T} \eta_n \, d\|D_x T_k^+(u(t,x)-\overline{u}(s,y))\| \\
&+\int_{Q_T\times Q_T} \eta_n \, d\|D_y T_k^+(u(t,x)-\overline{u}(s,y))\| \\
&+\int_{Q_T\times Q_T} \big(z(t,x)-\overline{z}(s,y)\big)\cdot\big(\nabla_x\eta_n+\nabla_y\eta_n\big)T_k^+(u(t,x)-\overline{u}(s,y)) \\
&+\int_{Q_T\times Q_T} \overline{z}(s,y)\cdot\nabla_x\eta T_k^+(u(t,x)-\overline{u}(s,y)) \\
&-\int_{Q_T\times Q_T} z(t,x)\cdot\nabla_y\eta_n T_k^+(u(t,x)-\overline{u}(s,y)) \le 0.
\end{aligned} \tag{7.108}$$

Now, by Green's formula (C.10) we have

$$\begin{aligned}
&\int_{Q_T\times Q_T} \overline{z}(s,y)\cdot\nabla_x\eta_n \; T_k^+(u(t,x)-\overline{u}(s,y)) \\
&\quad+\int_{Q_T\times Q_T} \eta_n \, d\|D_x T_k^+(u(t,x)-\overline{u}(s,y))\|
\end{aligned}$$

$$= - \int_{Q_T \times Q_T} \eta_n \overline{z}(s,y) \cdot D_x T_k^+(u(t,x) - \overline{u}(s,y)) + \int_{Q_T \times Q_T} \eta_n \, d\|D_x T_k^+(u(t,x) - \overline{u}(s,y))\| \geq 0,$$

and

$$- \int_{Q_T \times Q_T} z(t,x) \cdot \nabla_y \eta_n \; T_k^+(u(t,x) - \overline{u}(s,y)) + \int_{Q_T \times Q_T} \eta_n \, d\|D_y T_k^+(u(t,x) - \overline{u}(s,y))\|$$
$$= \int_{Q_T \times Q_T} \eta_n z(t,x) \cdot D_y T_k^+(u(t,x) - \overline{u}(s,y)) + \int_{Q_T \times Q_T} \eta_n \, d\|D_y T_k^+(u(t,x) - \overline{u}(s,y))\| \geq 0.$$

Hence, from (7.108), it follows that

$$- \int_{Q_T \times Q_T} j_k^+(u(t,x) - \overline{u}(s,y))\big((\eta_n)_t + (\eta_n)_s\big) + \int_{Q_T \times Q_T} \big(z(t,x) - \overline{z}(s,y)\big) \cdot \big(\nabla_x \eta_n + \nabla_y \eta_n\big) T_k^+(u(t,x) - \overline{u}(s,y)) \leq 0. \tag{7.109}$$

Now, passing to the limit in (7.109), we obtain

$$- \int_{Q_T} j_k^+(u(t,x) - \overline{u}(t,x))\phi'(t)\psi(x) + \int_{Q_T} \big(z(t,x) - \overline{z}(t,x)\big) \cdot \nabla\psi(x) \; \phi(t) T_k^+(u(t,x) - \overline{u}(t,x)) \leq 0. \tag{7.110}$$

The last steps of the proof are similar to the ones in the general case and we shall omit the details.

7.5 A Remark for Strictly Convex Lagrangians

As it is well known, if $u_n, u \in BV(\Omega)$ and $u_n \to u$ in $L^1(\Omega)$, $\|Du_n\| \to \|Du\|$, and $\nabla u_n \to F$ a.e in Ω, then F does not coincide with ∇u, in general, as the following example shows. Consider $\Omega =]0,1[$ and $u_n \in BV(\Omega)$, defined by

$$u_n := \sum_{i=1}^{n} \frac{i}{n} \chi_{[\frac{i-1}{n}, \frac{i}{n}[}.$$

Then, $u_n(x) \to u(x) = x$ for almost all $x \in \Omega$, but $\nabla u_n = 0$ for all $n \in \mathbb{N}$, and $\nabla u = 1$.

Now, in the proof of Theorem 7.8, we have seen that if $(u_n, v_n) \in \mathcal{B}_\varphi$ is such that $(u_n, v_n) \to (u, v)$ in $L^1(\Omega) \times L^1(\Omega)$, then $(u, v) \in \mathcal{A}_\varphi$. Thus, a natural question is when $\nabla u_n \to \nabla u$ a.e. in Ω. We are going to prove that, if we assume that the Lagrangian f is strictly convex, then the answer to this question is positive.

First, observe that from the strict convexity of f, we deduce the following strict monotonicity condition on $\mathbf{a}$:

$$(\mathbf{a}(x,\eta) - \mathbf{a}(x,\xi)) \cdot (\eta - \xi) > 0 \quad \text{if } \xi \neq \eta. \tag{7.111}$$

Let us prove that $\{\nabla u_n\}$ is a Cauchy sequence in measure. To do that, we follow the same technique as in [54]. Let $t, \epsilon > 0$. For $a > 1$, we set

$$C(x,a,t) := \inf\{(\mathbf{a}(x,\xi) - \mathbf{a}(x,\eta)) \cdot (\xi - \eta) \ : \ \|\xi\| \le a, \ \|\eta\| \le a, \ \|\xi - \eta\| \ge t\}.$$

Having in mind that the function $\xi \mapsto \mathbf{a}(x,\xi)$ is continuous for almost all $x \in \Omega$, and the set $\{(\xi,\eta) \ : \ \|\xi\| \le a, \ \|\eta\| \le a, \ \|\xi - \eta\| \ge t\}$ is compact, the infimum in the definition of $C(x,a,t)$ is a minimum. Hence, from (7.111) it follows that

$$C(x,a,t) > 0 \quad \text{for almost all } x \in \Omega. \tag{7.112}$$

For $n, m \in \mathbb{N}$, and any $k > 0$, we have

$$\begin{aligned} &\{\|\nabla u_n - \nabla u_m\| > t\} \subset \{\|\nabla T_a u_n\| \ge a^2\} \cup \{\|\nabla T_a u_m\| \ge a^2\} \\ &\cup \{|u_n| \ge a\} \cup \{|u_m| \ge a\} \cup \{|u_n - u_m| \ge k^2\} \cup \{C(x,a^2,t) \le k\} \\ &\cup \{|u_n - u_m| < k^2, |u_n| < a, |u_m| < a, C(x,a^2,t) \ge k, \\ &\qquad \|\nabla T_a u_n\| \le a^2, \|\nabla T_a u_m\| \le a^2, \|\nabla u_n - \nabla u_m\| > t\}. \end{aligned} \tag{7.113}$$

Since $\{u_n\}$ is bounded in $L^1(\Omega)$ we can choose a large enough in order to have

$$\mathcal{L}^N(\{|u_n| \ge a\} \cup \{|u_m| \ge a\}) \le \frac{\epsilon}{5} \qquad \text{for all } n, m \in \mathbb{N}. \tag{7.114}$$

Similarly, by (7.31), we can choose a large enough in order to have

$$\mathcal{L}^N(\{\|\nabla T_a u_n\| \ge a^2\} \cup \{\|\nabla T_a u_m\| \ge a^2\}) \le \frac{\epsilon}{5} \qquad \text{for all } n, m \in \mathbb{N}. \tag{7.115}$$

Fixing a satisfying (7.114) and (7.115), by (7.112), taking k small enough, we have

$$\mathcal{L}^N(\{C(x,a^2,t) \le k\}) \le \frac{\epsilon}{5}. \tag{7.116}$$

On the other hand, since $v_n = -\operatorname{div} \mathbf{a}(x, \nabla u_n)$, using Green's formula (C.10), we have

$$\begin{aligned} &\int_\Omega (\mathbf{a}(x,\nabla u_n) - \mathbf{a}(x,\nabla u_m), DT_r(u_n - u_m)) \\ &= \int_\Omega (v_n - v_m) T_r(u_n - u_m)\, dx + \int_{\partial\Omega} [\mathbf{a}(x,\nabla u_n) - \mathbf{a}(x,\nabla u_m), \nu] T_r(u_n - u_m)\, d\mathcal{H}^{N-1} \\ &\le 2Qr \quad \forall\, n, m \in \mathbb{N}. \end{aligned}$$

Now,

$$\begin{aligned}\int_\Omega (\mathbf{a}(x,\nabla u_n) - \mathbf{a}(x,\nabla u_m), DT_r(u_n-u_m))\\ = \int_\Omega (\mathbf{a}(x,\nabla u_n) - \mathbf{a}(x,\nabla u_m))\cdot \nabla T_r(u_n-u_m)\,dx\\ + \int_\Omega (\mathbf{a}(x,\nabla u_n) - \mathbf{a}(x,\nabla u_m))\cdot D^sT_r(u_n-u_m).\end{aligned}$$

Moreover, by the chain rule, there exists a positive function η such that

$$\begin{aligned}&\int_\Omega (\mathbf{a}(x,\nabla u_n) - \mathbf{a}(x,\nabla u_m))\cdot D^sT_r(u_n-u_m)\\ &= \int_\Omega \eta[(\mathbf{a}(x,\nabla u_n) - \mathbf{a}(x,\nabla u_m))\cdot D^s(u_n-u_m)]\\ &= \int_\Omega \eta[f^0(x,D^su_n) - \mathbf{a}(x,\nabla u_m)\cdot D^su_n + f^0(x,D^su_m) - \mathbf{a}(x,\nabla u_m)\cdot D^su_m]\\ &\geq 0,\end{aligned}$$

by (H_5). Therefore, we obtain

$$\int_\Omega (\mathbf{a}(x,\nabla u_n) - \mathbf{a}(x,\nabla u_m))\cdot \nabla T_r(u_n-u_m)\,dx \leq 2Qr. \tag{7.117}$$

If

$$\begin{aligned}S := \{\ &|u_n-u_m| < k^2, |u_n| < a, |u_m| < a, C(x,a^2,t) \geq k,\\ &\|\nabla T_a u_n\| \leq a^2, \|\nabla T_a u_m\| \leq a^2, \|\nabla u_n - \nabla u_m\| > t\},\end{aligned}$$

since $\nabla T_a u_n = \nabla u_n$ a.e. in S, by (7.117), we get

$$\begin{aligned}\mathcal{L}^N(S) &\leq \mathcal{L}^N\left(\{|u_n-u_m| < k^2, (\mathbf{a}(x,\nabla u_n) - \mathbf{a}(x,\nabla u_m))\cdot(\nabla u_n - \nabla u_m) \geq k\}\right)\\ &\leq \frac{1}{k}\int_{|u_n-u_m|<k^2} (\mathbf{a}(x,\nabla u_n) - \mathbf{a}(x,\nabla u_m))\cdot(\nabla u_n-\nabla u_m)\,dx \leq 2Qk.\end{aligned}$$

Hence, for k small enough, we have

$$\mathcal{L}^N(S) \leq \frac{\epsilon}{5}. \tag{7.118}$$

Since a and k have already been chosen, if n_0 is large enough, we have for $n,m \geq n_0$ the estimate $\mathcal{L}^N(\{|u_n-u_m| \geq k^2\}) \leq \frac{\epsilon}{5}$. Now, using (7.113), (7.114), (7.115), (7.116) and (7.118), it follows that

$$\mathcal{L}^N\left(\{\|\nabla u_n - \nabla u_m\| > t\}\right) \leq \epsilon \qquad \text{for } n,m \geq n_0.$$

Hence, $\{\nabla u_n\}$ is a Cauchy sequence in measure. Then, up to extraction of a subsequence, we have convergence a.e., and we can say that there exists a measurable function F, such that

$$\nabla u_n \to F \quad \text{a.e. in } \Omega. \tag{7.119}$$

Now, $\mathbf{a}(x, \nabla u_n) \rightharpoonup \mathbf{a}(x, \nabla u)$ in the weak* topology of $L^\infty(\Omega, \mathbb{R}^N)$ and, by (7.119), $\mathbf{a}(x, \nabla u_n) \to \mathbf{a}(x, F)$ a.e. in Ω. Hence, $\mathbf{a}(x, F) = \mathbf{a}(x, \nabla u)$ a.e. in Ω. Therefore, by (7.111), we deduce that

$$\nabla u_n \to \nabla u \quad \text{a.e. in } \Omega.$$

Remark 7.18. Coming back to the observation made at the beginning of this section, the only thing we can expect is that there exists $\lambda : \Omega \to \mathbb{R}$, $0 \le \lambda(x) \le 1$ a.e., such that $F = \lambda(x)\nabla u(x)$ a.e.. Indeed, we have: if μ_k, μ are vector measures in Ω (with values in $\mathbb{R}^N$) such that $\mu_k \to \mu$, $|\mu_k| \to |\mu|$ weakly* as measures in Ω, and $\mu_k^{ac} \to F$ in measure in Ω, then there is $\lambda : \Omega \to \mathbb{R}$, $0 \le \lambda(x) \le 1$ a.e., such that $F = \lambda(x)\mu^{ac}(x)$ a.e.. This can be proved using Reshetnyak's theorem ([119], Thm. 19). Indeed, Reshetnyak's theorem implies that

$$\int_\Omega N_u(x, \mu_k)\phi \to \int_\Omega N_u(x, \mu)\phi$$

for any $\phi \in C_0(\Omega)$, where $C_0(\Omega)$ denotes the space of continuous functions with compact support in Ω, and $N_u^+(x, v) = (\langle u, v\rangle)^+$, $u \in \mathbb{R}^N$. Now, for any $k > 0$ we have

$$\begin{aligned}
\int_\Omega (N_u(x,\mu) \wedge k)\phi &= \int_\Omega N_u(x,\mu)\phi - \int_\Omega (N_u(x,\mu) - k)^+\phi \\
&\ge \lim_n \int_\Omega N_u(x,\mu_n)\phi - \liminf_n \int_\Omega (N_u(x,\mu_n) - k)^+\phi \\
&\ge \liminf_n \int_\Omega (N_u(x,\mu_n) \wedge k)\phi = \liminf_n \int_\Omega (N_u(x,\mu_n^{ac}) \wedge k)\phi \\
&= \int_\Omega N_u(x, F)\phi.
\end{aligned}$$

Since

$$\int_\Omega (N_u(x,\mu) \wedge k)\phi = \int_\Omega (N_u(x,\mu^{ac}) \wedge k)\phi$$

and the previous inequality holds for any $k \in \mathbb{R}$, any $u \in \mathbb{R}^N$, and any $\phi \in C_0(\Omega)$, and all these spaces are separable we obtain that

$$N_u(x, F(x)) \le N_u(x, \mu^{ac}(x))$$

for all $x \in Q$, where $\mathcal{L}^N(\Omega \setminus Q) = 0$, and all $u \in \mathbb{R}^N$. Now, we observe that if $v, w \in \mathbb{R}^N$ are such that

$$(\langle u, v\rangle)^+ \le (\langle u, w\rangle)^+ \tag{7.120}$$

for all $u \in \mathbb{R}^N$, then there is $\lambda \in [0,1]$ such that $v = \lambda w$. If we fix $x \in Q$, applying the last observation, we conclude that there is $\lambda(x) \in [0,1]$ such that $F(x) = \lambda(x)\mu^{ac}(x)$. These observations can be used to prove that there exists $\lambda : \Omega \to \mathbb{R}$, $0 \le \lambda(x) \le 1$ a.e., such that $F = \lambda(x)\nabla u(x)$ a.e., once we know that $\nabla u_n(x) \to F(x)$ in measure. Then, we need structural assumptions on $\mathbf{a}(x,\xi)$ to obtain either that $\mathbf{a}(x,F) = \mathbf{a}(x,\nabla u)$, or $F(x) = \nabla u(x)$. Since, to prove that $\nabla u_n(x) \to F(x)$ in measure, we need to use the strict convexity of f, and this also gives that $F = \nabla u$, we do not need the more involved approach of this remark.

7.6 The Cauchy Problem

The Cauchy problem

$$\begin{cases} \dfrac{\partial u}{\partial t} = \operatorname{div} \mathbf{a}(x, Du) & \text{in} \quad Q = (0,\infty) \times \mathbb{R}^N, \\ u(0,x) = u_0(x) & \text{in} \quad x \in \mathbb{R}^N, \end{cases} \tag{7.121}$$

where $u_0 \in L^1_{loc}(\mathbb{R}^N)$ and $\mathbf{a}(x,\xi) = \nabla_\xi f(x,\xi)$, $f : \mathbb{R}^N \times \mathbb{R}^N \to \mathbb{R}$ being a function with linear growth as $\|\xi\| \to \infty$ satisfying the assumptions (H_1–H_6) was studied in [18].

The concept of solution for the Cauchy problem (7.121) is the following.

Definition 7.19. A measurable function $u : (0,T) \times \mathbb{R}^N \to \mathbb{R}$ is an *entropy solution* of (7.121) in Q_T if $u \in C([0,T]; L^1_{loc}(\mathbb{R}^N))$, $u(t)$ converges to u_0 in $L^1_{\rm loc}(\mathbb{R}^N)$ as $t \to 0^+$, $p(u(\cdot)) \in L^1_w(0,T; BV_{\rm loc}(\mathbb{R}^N))$ for all $p \in \mathcal{P}$, and there exists $\xi \in (L^1(0,T; BV(\mathbb{R}^N)_2))^*$ such that:

(i) $(\mathbf{a}(x,\nabla u(t)), \xi(t)) \in Z(\mathbb{R}^N)$ a.e. $t \in [0,T]$,

(ii) ξ is the time derivative of u in $(L^1(0,T; BV(\mathbb{R}^N)_2))^*$ in the sense of Definition 5.3,

(iii) $\xi = \operatorname{div}(\mathbf{a}(x,\nabla u(t)))$ in $(L^1(0,T; BV(\mathbb{R}^N)_2))^*$ in the sense of Definition 5.4,

(iv) the following inequality is satisfied,

$$-\int_0^T \int_{\mathbb{R}^N} j(u(t)-l)\eta_t \, dxdt + \int_0^T \int_{\mathbb{R}^N} \eta(t) h(x, Dp(u(t)-l)) \, dt$$
$$+ \int_0^T \int_{\mathbb{R}^N} \mathbf{a}(x,\nabla u(t)) \cdot \nabla \eta(t) p(u(t)-l) \, dxdt \le 0$$

for all $l \in \mathbb{R}$, all $\eta \in C^\infty(]0,T[\times \mathbb{R}^N)$, with $\eta \ge 0$, $\eta(t,x) = \phi(t)\psi(x)$, being $\phi \in C_0^\infty(]0,T[)$, $\psi \in C_0^\infty(\mathbb{R}^N)$, and all $p \in \mathcal{P}$, where $j(r) := \displaystyle\int_0^r p(s)\, ds$.

We have the following existence and uniqueness result.

Theorem 7.20. *([18]) Assume we are under assumptions* $(\mathrm{H}_1\text{–}\mathrm{H}_6)$. *Let* $u_0 \in L^1_{\mathrm{loc}}(\mathbb{R}^N)$. *Then there exists a unique entropy solution of* (7.121) *in* $[0,T] \times \mathbb{R}^N$ *for all* $T > 0$.

The proof of uniqueness is similar to the one given in subsection 7.4.2 for the Dirichlet problem. To prove the existence part of Theorem 7.20 we can not apply the method used for the Dirichlet problem since in this case the energy functional can be infinity for functions in $BV(\mathbb{R}^N)$. To overcome this difficulty, one can approximate (7.121) by problems of the form

$$\begin{cases} \dfrac{\partial u}{\partial t} = \operatorname{div}\,(\varphi \mathbf{a}(x, Du)) & \text{in} \quad Q = (0,\infty) \times \mathbb{R}^N, \\ u(0,x) = u_0(x) & \text{in} \quad x \in \mathbb{R}^N, \end{cases} \tag{7.122}$$

where $u_0 \in L^2(\mathbb{R}^N)$, $\varphi \in \mathcal{S}(\mathbb{R}^N)$ and $\mathcal{S}(\mathbb{R}^N)$ denotes the space of rapidly decreasing C^∞ functions in $\mathbb{R}^N$. The techniques used in Chapter 6 can be used to prove the well-posedness of problem (7.122). We have:

Theorem 7.21. *([18]) Assume we are under assumptions* $(\mathrm{H}_1\text{–}\mathrm{H}_6)$, $\varphi \in \mathcal{S}(\mathbb{R}^N)$, $\varphi(x) > 0$ *for all* $x \in \mathbb{R}^N$, *and satisfies the property*

$$|\varphi(x) - \varphi(y)| \le C\varphi(y)\|x - y\| \tag{7.123}$$

for all $x, y \in \mathbb{R}^N$ *such that* $\|x - y\| \le 1$, *for some constant* $C > 0$. *Given* $u_0 \in L^2(\mathbb{R}^N)$, *there exists a unique solution* u *of* (7.122) *in* Q_T *for every* $T > 0$ *such that* $u(0) = u_0$.

Using the above result, the existence part of Theorem 7.20 is proved in the following way. Given $u_0 \in L^1_{\mathrm{loc}}(\mathbb{R}^N)$, we take $u_{0n} \in L^2(\mathbb{R}^N) \cap L^\infty(\mathbb{R}^N)$ such that $u_{0n} \to u_0$ in $L^1_{\mathrm{loc}}(\mathbb{R}^N)$. Let $\varphi_n \in \mathcal{S}(\mathbb{R}^N)$ satisfying (7.123), $0 < \varphi_n \le 1$ and $\varphi_n(x) = 1$ for all $x \in B(0,n)$. By Theorem 7.21, for every $n \in \mathbb{N}$ there exists a solution u_n of (7.122) for $\varphi = \varphi_n$, corresponding to the initial condition u_{0n}. Then one proves that $\{u_n\}$ is a Cauchy sequence in $C([0,T]; L^1_{\mathrm{loc}}(\mathbb{R}^N))$, and, thus, $u_n \to u$ in $C([0,T]; L^1_{\mathrm{loc}}(\mathbb{R}^N))$. Finally, one proves that u is a solution of problem (7.121).

Appendix A

Nonlinear Semigroups

A.1 Introduction

In this appendix we outline some of the main points of the theory of nonlinear semigroups and evolution equations governed by accretive operators. We refer the reader to: [32], [40], [45], [58], [83], [84], [85].

The linear part of this theory started in the 1930s with the works of E. Hille, Y. Yosida and R. Phillips on semigroups of linear operators in Banach spaces. Now, one of the first ideas came from a paper of G. Peano from 1887 where he wrote the system of differential equations

$$\begin{cases} \dfrac{du_1}{dt} = a_{11}u_1 + \cdots + a_{1n}u_n + f_1(t), \\ \vdots \\ \dfrac{du_n}{dt} = a_{n1}u_1 + \cdots + a_{nn}u_n + f_n(t) \end{cases} \tag{A.1}$$

in a matrix form as

$$\mathbf{u}'(t) = A\mathbf{u}(t) + \mathbf{f}(t) \tag{A.2}$$

where $\mathbf{u}(t) = (u_1(t), u_2(t), \ldots, u_n(t))$, $\mathbf{f}(t) = (f_1(t), f_2(t), \ldots, f_n(t))$ and $A = (a_{ij})$, and solved it by means of the explicit formula

$$\mathbf{u}(t) = e^{tA}\mathbf{u}(0) + \int_0^t e^{(t-s)A}\mathbf{f}(s)\, ds,$$

where

$$e^{tA} = \sum_{k=0}^{\infty} \frac{1}{k!} t^k A^k.$$

So, he transformed a complicated problem in one dimension to a formally simpler one in higher dimension. That is the essence of the nonlinear semigroups theory.

Now, since we want to apply this abstract theory to solve partial differential equations, we must work in infinite dimensions. So our main object will be the study of evolution problems of the form:

$$\begin{cases} u'(t) + Au(t) = f(t) & \text{on } (0,T), \\ u(0) = u_0, \end{cases} \tag{A.3}$$

where X is a Banach space, $f : (0,T) \to X$ and $A : D(A) \to X$ is an operator.

Let us give one example about how to write a PDE problem as a problem in the form (A.3).

Example A.1. Let Ω be a bounded domain in $\mathbb{R}^N$ with smooth boundary $\partial\Omega$. Consider the classical initial-boundary problem for the heat equation, that is the problem

$$\begin{cases} \dfrac{\partial w}{\partial t} = \Delta w & \text{for } (t,x) \in [0,\infty[\times\Omega, \\ w(t,x) = 0 & \text{for } x \in \partial\Omega,\ t \geq 0, \\ w(0,x) = f(x) & \text{for } x \in \Omega. \end{cases} \tag{A.4}$$

Write $u(t) = w(t,\cdot)$, regarded as a function of x, and take X to be a space of functions on Ω, for example, $X = L^p(\Omega)$ for some $p \geq 1$ or $X = C(\overline{\Omega})$. Suppose we are in this last case. Let A be the operator with domain

$$D(A) := \left\{ v \in C(\overline{\Omega}) \ : \ \Delta v \in C(\overline{\Omega}) \ \text{ and } \ v(x) = 0 \ \ \forall\, x \in \partial\Omega \right\}$$

and defined by $Av := -\Delta v$, for $v \in D(A)$. Then we can write problem (A.4) in the form (A.3). Note that the boundary condition of (A.4) is absorbed into the domain of the definition of the operator A and into the requirement that $u(t) \in D(A)$ for all $t \geq 0$.

A.2 Abstract Cauchy Problems

From now on, X will be a real Banach space with norm $\| \ \|$ and dual X^*.

We will use multivalued nonlinear operators not only since they permit a coherent theory but also because it is often necessary in applications. So let us recall some notation and basic facts concerning multivalued operators .

A mapping $A : X \to 2^X$ from X into 2^X (the collection of subsets of X) will be called an *operator* in X. For $x \in X$, Ax denotes the value of A at x, $D(A) := \{x \in X \ : Ax \neq \emptyset\}$ will be called the *effective domain* of A, and $R(A) := \cup\{Ax \ : \ x \in D(A)\}$ its *range.*

If A is an operator in X, it determines the subset $G(A) = \{(x,y) \in X \times X \ : \ y \in Ax\}$, called the graph of A; conversely, a subset G of $X \times X$ determines a unique operator A whose graph is G; the operator A is given by $Ax := \{y : (x,y) \in G\}$. Whenever it is convenient we will identify an operator with its graph.

Given two operators A and B in X and $\alpha \in \mathbb{R}$, we define new operators $A+B$, αA and A^{-1}, according to:

$$\begin{aligned}(A+B)x &:= Ax + Bx\\ (\alpha A)x &:= \alpha(Ax)\\ A^{-1}x &:= \{y \in X : x \in Ay\}.\end{aligned}$$

The *closure* of the operator A, denoted by $\overline{A}$, is defined to be the closure of the graph of A in $X \times X$, that is:

$$y \in \overline{A}\text{ iff } \exists\, y_n \in Ax_n : x_n \to x,\ y_n \to y.$$

Before proceeding we fix some notation: By $L^1(a,b;X)$ we denote the vector space of all Bochner integrable functions $f : [a,b] \to X$ with respect to the Lebesgue measure (i.e., the strong measurable functions f such that $\int_a^b \|f(t)\|\, dt < +\infty$). If I is an interval in $\mathbb{R}$, $L^1_{loc}(I;X)$ is the space of those functions $f : I \to X$ which are Bochner integrable on compact subintervals of I. As in the case of real functions, if $f \in L^1(a,b;X)$, for almost all $t \in]a,b[$ one has

$$\lim_{h\downarrow 0} \frac{1}{h}\int_{t-h}^{t+h} \|f(s) - f(t)\|\, ds = 0. \tag{A.5}$$

If (A.5) holds, t is called a *Lebesgue point* of f.

The space $W^{1,1}(a,b;X)$ consists of those functions f which have the form

$$f(t) = f(0) + \int_0^t h(s)\, ds \tag{A.6}$$

for some $h \in L^1(a,b;X)$. It is well known that $W^{1,1}(a,b;X)$ consists of exactly those absolutely continuous functions $f : [a,b] \to X$ which are differentiable a.e. on $[a,b]$ and if (A.6) holds, then $f'(t) = h(t)$ a.e.

In a general Banach space X, the absolute continuity of a function $f : [a,b] \to X$ does not imply the existence of $f'(t)$ almost everywhere. When this happens it is said that the Banach space X has the *Radon–Nikodym property*. For instance, every reflexive Banach space has the Radon–Nikodym property. Now, there are important Banach spaces like $L^1(\Omega)$, $L^\infty(\Omega)$ or $C(\overline{\Omega})$ without the Radon–Nikodym property.

As we mentioned before, our aim is to study evolution problems of the form:

$$\begin{cases} u'(t) + Au(t) \ni f(t) & \text{on } t \in (0,T),\\ u(0) = x, \end{cases} \tag{A.7}$$

where $f : (0,T) \to X$ and A is an operator in X. A problem of the form (A.7) is called an *abstract Cauchy problem*, and it will be denoted by $(\mathrm{CP})_{x,f}$. In the homogeneous case, that is, for $f = 0$, we will write $(\mathrm{CP})_x$ instead $(\mathrm{CP})_{x,0}$.

In principle, one natural notion of solution for $(\mathrm{CP})_{x,f}$ is the classical one, that is, a function u satisfying:

$$\begin{cases} u \in C([0,T];X) \cap C^1(]0,T[;X), \\ u'(t) + Au(t) \ni f(t) \ \ \forall\, t \in]0,T[, \\ u(0) = x. \end{cases} \tag{A.8}$$

In fact, this is a common notion of solution in the classical theory of ordinary differential equation (i.e., for $X = \mathbb{R}^N$) when A and f are continuous. But as soon as discontinuities arise, the notion of classical solution turns out to be too restrictive as may be illustrated by the following example.

Example A.2. Let $X = \mathbb{R}$, $f = 0$, $x = 1$ and A the Heaviside function

$$A(r) = \begin{cases} 1 & \text{if } \ r > 0, \\ 0 & \text{if } \ r \leq 0. \end{cases}$$

Then, $(\mathrm{CP})_{x,f}$ becomes

$$\begin{cases} u'(t) = -1 \quad \text{if } \ u(t) > 0, \\ u'(t) = 0 \quad \text{if } \ u(t) \leq 0, \\ u(0) = 1. \end{cases} \tag{A.9}$$

The solution of problem (A.9) is given by

$$u(t) = \begin{cases} 1 - t & \text{if } \ 0 \leq t \leq 1, \\ 0 & \text{if } \ t \geq 1. \end{cases}$$

But u is not a classical solution since is not differentiable at $t = 1$.

This example motivates the following weaker notion of solution for $(\mathrm{CP})_{x,f}$.

Definition A.3. A function u is called a *strong solution* of $(\mathrm{CP})_{x,f}$ if

$$\begin{cases} u \in C([0,T];X) \cap W^{1,1}_{loc}(]0,T[;X), \\ u' + Au(t) \ni f(t) \ \text{ a.e. } t \in]0,T[, \\ u(0) = x. \end{cases} \tag{A.10}$$

Clearly, the previous example is covered by this notion of solution. However, it is still not sufficient in general, as the following simple example due to G. Webb ([206]) shows.

Example A.4. Consider the problem

$$\begin{cases} w_t - w_x + w^+ = 0 & \text{on } [0,+\infty[\times\mathbb{R}, \\ w(0,x) = u_0(x) & x \in \mathbb{R}. \end{cases} \tag{A.11}$$

We are interested in solving (A.11) in the space $X = C_0(\mathbb{R})$. To this end, we define the operator A in X by $Au := -u' + u^+$ with domain

$$D(A) := \left\{ u \in C^1(\mathbb{R}) \ : \ u, u' \in C_0(\mathbb{R}) \right\}.$$

We rewrite the problem (A.11) as an evolution problem in X:

$$\begin{cases} u'(t) + Au(t) = 0 & \text{in } [0,+\infty[, \\ u(0) = u_0. \end{cases} \tag{A.12}$$

Observe that this is a semi-linear problem with $A = A_0 + F$, being $A_0 u = -u'$ and $F(u) = u^+$. Then, since $-A_0$ is the infinitesimal generator of a C_0-semigroup $(S(t))_{t\geq 0}$ in X and F is Lipschitz continuous, it is well known that for every $u_0 \in X$ there is a unique solution of (A.12) given by the classical Duhamel formula

$$u(t) = S(t)u_0 - \int_0^t S(t-s)F(u(s))\, ds \qquad \forall\, t \geq 0.$$

Nevertheless, u need not be a strong solution of problem (A.12), even if $u_0 \in D(A)$. In fact: Let $u_0 \in X$ such that there exists $x_0 \in \mathbb{R}$ satisfying: $u_0(x) > 0$ if $x > x_0$ and $u_0(x) < 0$ if $x < x_0$. Then, using the classical method of the characteristics, it is not difficult to see that the solution of (A.11) is given by

$$w(t,x) = \begin{cases} e^{-t}u_0(x+t) & \text{if } x+t > x_0, \\ u_0(x+t) & \text{if } x+t \leq x_0. \end{cases}$$

From which it follows that if $u_0'(x_0) \neq 0$, this solution is not a strong solution.

Consequently, we need to introduce a more general concept of solution for $(\mathrm{CP})_{x,f}$. The more adequate notion of solution for $(\mathrm{CP})_{x,f}$ in general Banach spaces is the concept of mild solution, introduced by M.G. Crandall and T.M. Liggett in [85] and Ph. Bénilan in [40], which is studied in the next section.

A.3 Mild Solutions

Let A be an operator in X and $f \in L^1(a,b;X)$. Roughly speaking a mild solution of the problem

$$u' + Au \ni f \qquad \text{on } [a,b] \tag{A.13}$$

is a continuous function $u \in C([a,b];X)$ which is the uniform limit of solutions of time-discretized problems, given by the implicit Euler scheme of the form

$$\frac{v(t_i) - v(t_{i-1})}{t_i - t_{i-1}} + Av(t_i) \ni f_i,$$

where f_i are approximations of f when $|t_i - t_{i-1}| \to 0$. So the underlying idea of the notion of mild solution is simple and from the point of view of numerical analysis, even classical. Formally, the definition is as follows.

Definition A.5. Let $\epsilon > 0$. An *ϵ-discretization* of $u' + Au \ni f$ on $[a,b]$ consists of a partition $t_0 < t_1 < \cdots < t_N$ and a finite sequence $f_1, f_2, \ldots, f_N$ of elements of X such that

$$\begin{aligned} &a \le t_0 < t_1 < \cdots < t_N \le b, \quad \text{with} \\ &t_i - t_{i-1} \le \epsilon, \ i = 1, \ldots, N, \quad t_0 - a \le \epsilon \ \text{ and } \ b - t_N \le \epsilon. \end{aligned} \tag{A.14}$$

$$\sum_{i=1}^{N} \int_{t_{i-1}}^{t_i} \|f(s) - f_i\| \, ds \le \epsilon. \tag{A.15}$$

We will denote this discretization by $D_A(t_0, \ldots, t_N; f_1, \ldots, f_N)$.

A *solution of the discretization* $D_A(t_0, \ldots, t_N; f_1, \ldots, f_N)$ is a piecewise constant function $v : [t_0, t_N] \to X$ whose values $v(t_0) = v_0$, $v(t) = v_i$ for $t \in]t_{i-1}, t_i]$, $i = 1, \ldots, N$ satisfy

$$\frac{v_i - v_{i-1}}{t_i - t_{i-1}} + Av_i \ni f_i, \qquad i = 1, \ldots, N. \tag{A.16}$$

A *mild solution* of $u' + Au \ni f$ on $[a,b]$ is a continuous function $u \in C([a,b];X)$ such that, for each $\epsilon > 0$ there is $D_A(t_0, \ldots, t_N; f_1, \ldots, f_N)$, an ϵ-discretization of $u' + Au \ni f$ on $[a,b]$ which has a solution v satisfying

$$\|u(t) - v(t)\| \le \epsilon \qquad \text{for } \ t_0 \le t \le t_N.$$

It is easy to see that if u is a mild solution of $u' + Au \ni f$ on $[a,b]$ and $[c,d] \subset [a,b]$, then $u_{|[c,d]}$ is a mild solution of $u' + Au \ni f$ on $[c,d]$. Therefore, the following definition is consistent.

Definition A.6. Let I an interval of $\mathbb{R}$, and $f \in L^1_{loc}(I;X)$. A *mild solution* of $u' + Au \ni f$ on I is a function $u \in C(I;X)$ whose restriction to each compact subinterval $[a,b]$ of I is a mild solution of $u' + Au \ni f$ on $[a,b]$.

In the next result we will see that mild solutions generalize the concept of the strong solutions.

Theorem A.7. *Let $f \in L^1_{loc}(I;X)$ and u be a strong solution of $u' + Au \ni f$ on I. Then u is a mild solution of $u' + Au \ni f$ on I.*

The heart of the proof of the above theorem is the following result concerning the approximation of Bochner integrals by Riemann sums in a strong sense.

Lemma A.8. *Let Y be a Banach space, $g \in L^1(a,b;Y)$ and K be a subset of $[a,b]$ such that $[a,b] \setminus K$ has measure zero. Then, given $\delta > 0$, there is a partition $a = t_0 < t_1 < \cdots < t_N \leq b$ satisfying:*

$$t_i \in K \text{ and } t_i \text{ is a Lebesgue point of } g \text{ for all } i = 1, \ldots, N. \tag{A.17}$$

$$b - t_N < \delta \quad \text{and} \quad t_i - t_{i-1} < \delta, \quad i = 1, \ldots, N. \tag{A.18}$$

$$\sum_{i=1}^{N} \int_{t_{i-1}}^{t_i} \|g(t) - g(t_i)\| \, dt < \delta. \tag{A.19}$$

The converse of Theorem A.7 is false; mild solutions need not be strong solutions. One counterexample is given by the equation of Example A.4.

The next result collects some of the properties of mild solutions.

Theorem A.9. *Let A be an operator in X and $f \in L^1_{loc}(I;X)$. Then:*

(i) *If u is a mild solution of $u' + Au \ni f$ on I, then $u(t) \in \overline{D(A)}$ for all $t \in I$.*

(ii) *Let I_1, I_2 be subintervals of I with $I \subset \overline{I_1 \cup I_2}$. If $u \in C(I;X)$ is a mild solution of $u' + Au \ni f$ on I_1 and on I_2, then u is a mild solution of $u' + Au \ni f$ on I.*

(iii) *Let $\overline{A}$ be the closure of the operator A. Then, u is a mild solution of $u'+Au \ni f$ on I if and only if u is a mild solution of $u' + \overline{A}u \ni f$ on I.*

(iv) *Let $\{u_n\} \subset C(I;X)$, $\{f_n\} \subset L^1_{loc}(I;X)$ and u_n be a mild solution of $u_n' + Au_n \ni f_n$ on I. Assume $u \in C(I;X)$, $f \in L^1_{loc}(I;X)$ and for each compact subinterval $[a,b]$ of I,*

$$\lim_{n\to\infty} \left(\int_a^b \|f_n(t) - f(t)\| \, dt + \sup_{a \leq t \leq b} \|u_n(t) - u(t)\| \right) = 0,$$

then u is a mild solution of $u' + Au \ni f$ on I.

Definition A.10. Let D be a subset of X. A family of mappings $S(t) : D \to D$, $(t \geq 0)$ satisfying:

$$S(t+s)x = S(t)S(s)x \qquad \text{for all} \;\; t, s \geq 0, \;\; x \in D, \tag{A.20}$$

$$\lim_{t\to 0} S(t)x = x \qquad \text{for} \;\; x \in D, \tag{A.21}$$

is called a *strongly continuous semigroup* on D.

One may now associate to every operator A in X a strongly continuous semigroup $(S^A(t))_{t\geq 0}$ by the following definition:

$$D\left(S^A\right) := \left\{x \in X \ : \ \begin{array}{l} \exists! \text{ mild solution } u_x \text{ of } u' + Au \ni 0 \\ \text{on } \ (0, +\infty) \text{ with } u_x(0) = x \end{array}\right\}.$$

For $t \geq 0$ and $x \in D(S^A)$, we set

$$S^A(t)x := u_x(t).$$

It is an immediate consequence of the properties of mild solutions that, in fact, $\left(S^A(t)\right)_{t\geq 0}$ is a strongly continuous semigroup on $D(S^A)$.

In the linear case, that is, if $S(t) \in \mathcal{L}(X)$, the strongly continuous semigroups are called C_0-semigroups. In this situation, each C_0-semigroup $(S(t))_{t\geq 0}$ has associated its *infinitesimal generator* B defined by

$$Bx := \lim_{t\to 0} \frac{S(t)x - x}{t} \qquad \text{for } \ x \in D(B)$$

and

$$D(B) := \left\{x \in X \ : \ \exists \ \lim_{t\to 0} \frac{S(t)x - x}{t}\right\}.$$

In the linear case it is well known that

"*$-A$ is the infinitesimal generator of a C_0-semigroup $\left(S(t)\right)_{t\geq 0}$ of a bounded linear operator on X, if and only if A is linear, closed and $D(S^A) = X$, and then $S^A(t) = S(t)$ for all $t \geq 0$.*"

This motivates the development of a nonlinear semigroup theory analogous to the classical linear one. We will see that in the nonlinear case the situation is very different from the linear one, and has more difficulties.

A.4 Accretive Operators

We are going to introduce now the class of operators for which we could obtain existence and uniqueness results of mild solutions.

The existence of mild solutions requires, as we pointed out before, the existence of solutions of discretized equations of the form

$$\frac{x_i - x_{i-1}}{t_i - t_{i-1}} + Ax_i \ni f_i, \qquad i = 1, \ldots, N$$

or equivalently

$$x_i + (t_i - t_{i-1})Ax_i \ni (t_i - t_{i-1})f_i + x_{i-1}, \qquad i = 1, \ldots, N. \tag{A.22}$$

Then, to solve (A.22) we need that the inverse of the operator $(I + \lambda A)$ be a singlevalued operator. Operators satisfying this property are the following:

Definition A.11. An operator A in X is *accretive* if

$$\|x - \hat{x}\| \leq \|x - \hat{x} + \lambda(y - \hat{y})\|, \quad \text{whenever } \lambda > 0 \text{ and } (x, y), (\hat{x}, \hat{y}) \in A.$$

Note that A is accretive if and only if for $\lambda > 0$ and $z \in X$, $x + \lambda y = z$ has at most one solution $(x, y) \in A$ and the relations $x + \lambda y = z$, $(x, y) \in A$, $\hat{x} + \lambda\hat{y} = \hat{z}$, $(\hat{x}, \hat{y}) \in A$ imply

$$\|x - \hat{x}\| = \left\|(I + \lambda A)^{-1}z - (I + \lambda A)^{-1}\hat{z}\right\| \leq \|z - \hat{z}\|.$$

Therefore, we have

"A is accretive if and only if $(I + \lambda A)^{-1}$ is a singlevalued nonexpansive map for $\lambda \geq 0$"

In case A is accretive, we denote $J_\lambda^A = (I + \lambda A)^{-1}$ and we call J_λ^A the *resolvent* of A. Note that $D(J_\lambda^A) = R(I + \lambda A)$.

It is easy to see that if β is an operator in $\mathbb{R}$, then β is accretive if and only if $(y - \hat{y})(x - \hat{x}) \geq 0$ for all $(x, y), (\hat{x}, \hat{y}) \in \beta$. Thus, if β is univalued, then β is accretive if and only if β is nondecreasing. We have the following examples of accretive operators in $\mathbb{R}$:

$$\mathrm{sign}_0(r) := \begin{cases} 1 & \text{if } r > 0, \\ 0 & \text{if } r = 0, \\ -1 & \text{if } r < 0, \end{cases}$$

and

$$\mathrm{sign}(r) := \begin{cases} 1 & \text{if } r > 0, \\ [-1, 1] & \text{if } r = 0, \\ -1 & \text{if } r < 0. \end{cases}$$

In order to verify accretivity of a given operator, it is useful to take into account alternative characterizations of this property. To do that we need to introduce the bracket and the duality map.

For each $\lambda \neq 0$ define $[\cdot, \cdot]_\lambda : X \times X \to \mathbb{R}$ by

$$[x, y]_\lambda := \frac{\|x + \lambda y\| - \|x\|}{\lambda}.$$

For fixed $(x, y) \in X \times X$, $\lambda \mapsto [x, y]_\lambda$ is nondecreasing for $\lambda > 0$. Indeed, if $\lambda \geq \mu > 0$, then

$$\|x + \mu y\| = \left\|\left(1 - \frac{\mu}{\lambda}\right)x + \frac{\mu}{\lambda}(x + \lambda y)\right\| \leq \left(1 - \frac{\mu}{\lambda}\right)\|x\| + \frac{\mu}{\lambda}\|x + \lambda y\|,$$

from which it follows that $[x,y]_\mu \le [x,y]_\lambda$. Therefore for every $(x,y) \in X \times X$ we can define:

$$[x,y] := \lim_{\lambda \downarrow 0} [x,y]_\lambda = \inf_{\lambda > 0} [x,y]_\lambda .$$

The number $[x,y]$ is the right-hand derivative of the norm of x in the direction y. In the next proposition we collect some of the useful properties of the bracket $[\cdot,\cdot]$.

Proposition A.12. *If $x,y,z \in X$ and $\alpha,\beta \in \mathbb{R}$, then*

(i) $[\cdot,\cdot] : X \times X \to \mathbb{R}$ *is upper-semi-continuous,*

(ii) $[\alpha x, \beta y] = |\beta| [x,y]$ *if* $\alpha \cdot \beta > 0$,

(iii) $[x, \alpha x + y] = \alpha \|x\| + [x,y]$,

(iv) $[x,y] \ge 0$ *if and only if* $\|x + \lambda y\| \ge \|x\|$ *for* $\lambda \ge 0$,

(v) $|[x,y]| \le \|y\|$ *and* $[0,y] = \|y\|$,

(vi) $[x,y] \ge -[x,-y]$,

(vii) $[x, y+z] \le [x,y] + [x,z]$.

(viii) *Let* $u :]a,b[\to \mathbb{R}$ *and* $t_0 \in]a,b[$, *such that* u *is differentiable at* t_0, *then* $t \mapsto \|u(t)\|$ *is differentiable at* t_0 *if and only if* $[u(t_0), u'(t_0)] = -[u(t_0), -u'(t_0)]$. *In this case*

$$\frac{d}{dt} \|u(t)\|_{|t=t_0} = [u(t_0), u'(t_0)] .$$

As a consequence of (iv) of the above proposition we have the following characterization of accretive operators.

Corollary A.13. *An operator A in X is accretive if and only if $[x - \hat{x}, y - \hat{y}] \ge 0$ whenever $(x,y), (\hat{x},\hat{y}) \in A$.*

In some concrete Banach spaces the bracket $[\cdot,\cdot]$ can be computed explicitly. We give some examples.

Example A.14. Suppose $(H, (\ |\ ))$ is a Hilbert space. Then for $x,y \in H$,

$$\big(\|x+\lambda y\| - \|x\|\big)\big(\|x+\lambda y\| + \|x\|\big) = \|x+\lambda y\|^2 - \|x\|^2 = 2\lambda (x|y) + \lambda^2 \|y\|^2 .$$

Dividing this equality by λ yields

$$\big(\|x+\lambda y\| + \|x\|\big)[x,y]_\lambda = 2(x|y) + \lambda \|y\|^2 ,$$

and, thus, we find

$$\|x\| [x,y] = (x|y).$$

Then, by Corollary A.13, it follows that: An operator A in H is accretive if and only if

$$(x - \hat{x}|y - \hat{y}) \geq 0 \qquad \text{for all } (x,y), (\hat{x},\hat{y}) \in A. \tag{A.23}$$

An operator in a Hilbert space satisfying (A.23) is called *monotone* and therefore in Hilbert spaces monotone and accretive operators coincide.

Example A.15. Let $X = L^p(\Omega)$ where $1 < p < \infty$. By the convexity of the map $t \mapsto |t|^p$, and applying the dominated convergence theorem, it is easy to see that

$$[f,g] = \|f\|_p^{1-p} \int_\Omega g|f|^{p-1} \operatorname{sign}_0(f).$$

In the case $p = 1$, i.e., for $X = L^1(\Omega)$, we have

$$[f,g] = \int_\Omega g \operatorname{sing}_0(f) + \int_{\{f=0\}} |g|.$$

The formulas for the bracket given in the above examples are very useful to prove that a concrete operator is accretive. Another useful tool to study the accretivity of concrete operators is the *duality map* $J : X \to 2^{X^*}$, defined as

$$J(x) := \{x^* \in X^* \ : \ \|x^*\| \leq 1, \ \langle x, x^* \rangle = \|x\|\}.$$

By the Hahn–Banach theorem, we have $J(x) \neq \emptyset$ for every $x \in X$.

Given $x^* \in J(x)$, since $\|x^*\| \leq 1$,

$$|\langle x^*, x + \lambda y \rangle| \leq \|x + \lambda y\|$$

and

$$\langle x^*, y \rangle = \frac{1}{\lambda} (\langle x^*, x + \lambda y \rangle - \|x\|) \leq [x,y]_\lambda.$$

Hence

$$\langle x^*, y \rangle \leq [x,y] \qquad \forall \ x^* \in J(x).$$

On the other hand, if $V = LIN\{x,y\}$ and we define $\xi^* \in V^*$ by

$$\langle \xi^*, \alpha x + \beta y \rangle := \alpha \|x\| + \beta [x,y],$$

then, by the Hahn–Banach theorem, there exists $x^* \in X^*$ such that $x^*|_V = \xi^*$, so

$$\langle x^*, x \rangle = -\|x\| \quad \text{and} \quad \langle x^*, y \rangle = [x,y].$$

Moreover, it is not so difficult to see that $\|x^*\| \leq 1$, therefore $x^* \in J(x)$. Consequently, we have the following result.

Proposition A.16. *For $x, y \in X$,*

$$[x,y] = \max_{x^* \in J(x)} \langle x^*, y \rangle.$$

As a consequence of the above proposition and Corollary A.13, we have the following characterization of accretive operators.

Corollary A.17. *An operator A in X is accretive if and only if, for (x,y), $(\hat{x},\hat{y}) \in A$, there exists $x^* \in J(x-\hat{x})$ such that*

$$\langle x^*, y-\hat{y}\rangle \geq 0.$$

Example A.18. Let $X = L^p(\Omega)$ where $1 < p < \infty$, then by Hölder inequality we have

$$J(f) = \operatorname{sign}_0(f)|f|^{p-1}\|f\|_p^{1-p}.$$

In $L^1(\Omega)$, we have

$$J(f) = \operatorname{sign}(f) = \{g \in L^\infty(\Omega) \ : \ |g| \leq 1, \ \ gf = |f| \ \text{ a.e.}\}.$$

Given $w \in \mathbb{R}$, we define:

$$\mathcal{A}(w) := \{A \subset X \times X \ : \ A + wI \ \text{ is accretive}\}.$$

Proposition A.19. *Let A be an operator in X. The following statements are equivalent:*

(i) $A \in \mathcal{A}(w)$.

(ii) $(1-\lambda w)\|x-\hat{x}\| \leq \|x-\hat{x}+\lambda(y-\hat{y})\| \ \ \forall\ \lambda < 0,\ (x,y),(\hat{x},\hat{y}) \in A$.

(iii) $[x-\hat{x}, y-\hat{y}] + w\|x-\hat{x}\| \geq 0$.

(iv) *For $\lambda > 0$, $\lambda w < 1$, $J_\lambda^A = (I+\lambda A)^{-1}$ is Lipschitz continuous with Lipschitz constant $\dfrac{1}{1-\lambda w}$.*

(v) *For $(x,y),(\hat{x},\hat{y}) \in A$, there exists $x^* \in J(x-\hat{x})$ such that*

$$\langle x^*, y-\hat{y}\rangle + w\|x-\hat{x}\| \geq 0.$$

We have that accretivity implies uniqueness of the strong solutions. More precisely we have:

Theorem A.20. *Let $f, \hat{f} \in L^1(0,T;X)$, $A \in \mathcal{A}(w)$ and $u, \hat{u}$ strong solutions of $u' + Au \ni f$, $\hat{u}' + A\hat{u} \ni \hat{f}$, respectively, on $[0,T]$. Then*

$$\begin{aligned}\|u(t)-\hat{u}(t)\| &\leq e^{wt}\|u(0)-\hat{u}(0)\| + \int_0^t e^{w(t-s)}\left[u(s)-\hat{u}(s), f(s)-\hat{f}(s)\right] ds \\ &\leq e^{wt}\|u(0)-\hat{u}(0)\| + \int_0^t e^{w(t-s)}\|f(s)-\hat{f}(s)\|\, ds\end{aligned}$$

for $t \in [0,T]$.

In particular, the strong solutions of $(CP)_{x,f}$ are unique.

Proof. For simplicity, we suppose $w = 0$, i.e., A is accretive. Since u and $\hat{u}$ are differentiable a.e. in $]0,T[$, by (viii) of Proposition A.12, we have

$$\begin{aligned}\frac{d}{dt}\|u(t)-\hat{u}(t)\| &= -\left[u(t)-\hat{u}(t), \hat{u}'(t)-u'(t)\right] \\ &= -\left[u(t)-\hat{u}(t), (f(t)-u'(t))-(\hat{f}(t)-\hat{u}'(t))+(\hat{f}(t)-f(t))\right]\end{aligned}$$

for almost all $t \in]0,T[$.

Moreover, for almost all $t \in]0,T[$, $(u(t), f(t)-u'(t)) \in A$ and $(\hat{u}(t), \hat{f}(t)-\hat{u}'(t)) \in A$. Then, by Corollary A.13 and (vi), (vii) of Proposition A.12, we get

$$\begin{aligned}&\left[u(t)-\hat{u}(t), (f(t)-u'(t))-(\hat{f}(t)-\hat{u}'(t))+(\hat{f}(t)-f(t))\right] \\ &\geq \left[u(t)-\hat{u}(t), (f(t)-u'(t))-(\hat{f}(t)-\hat{u}'(t))\right]-\left[u(t)-\hat{u}(t), f(t)-\hat{f}(t)\right] \\ &\geq -\left[u(t)-\hat{u}(t), f(t)-\hat{f}(t)\right].\end{aligned}$$

Hence

$$\frac{d}{dt}\|u(t)-\hat{u}(t)\| \leq \left[u(t)-\hat{u}(t), f(t)-\hat{f}(t)\right].$$

From here, applying Gronwall's inequality we obtain

$$\begin{aligned}\|u(t)-\hat{u}(t)\| &\leq \|u(0)-\hat{u}(0)\| + \int_0^t \left[u(s)-\hat{u}(s), f(s)-\hat{f}(s)\right] ds \\ &\leq \|u(0)-\hat{u}(0)\| + \int_0^t \|f(s)-\hat{f}(s)\|\, ds.\end{aligned}$$

□

We have seen that accretivity of operator A implies uniqueness of the solution x_i of the discretized equation

$$\frac{x_i - x_{i-1}}{t_i - t_{i-1}} + Ax_i \ni f_i, \qquad i = 1, \ldots, N$$

which, if they exist, are given by

$$x_i = J^A_{(t_i - t_{i-1})}\left((t_i - t_{i-1})f_i + x_i\right) \qquad i = 1, \ldots, N.$$

This formula indicates that apart from accretivity one should expect a range condition (i.e., a condition on $R(I + \lambda A) = D(J^A_\lambda)$) to hold in order to get existence of a solution as well. This motivates the following definition.

Definition A.21. An operator A is called m-*accretive* in X if and only if A is accretive and $R(I + \lambda A) = X$ for all $\lambda > 0$.

Applying the Banach fixed point theorem it is not hard to see that if A is accretive, then A is m-accretive if there exists $\lambda > 0$ such that $R(I + \lambda A) = X$.

It is easy to see that each m-accretive operator A in X is *maximal accretive* in the sense that every accretive extension of A coincides with A. In general, the converse is not true, but it is true in Hilbert spaces due to the following classical result of G. Minty [154]:

Minty Theorem. *Let H be a Hilbert space and A an accretive operator in H. Then, A is m-accretive if and only if A is maximal monotone.*

One of the most important examples of a maximal monotone operator in Hilbert spaces comes from optimization theory; they are the subdifferentials of convex functions which we introduce next.

Let $(H, (\ |\ ))$ be a Hilbert space and $\varphi : H \to]-\infty, +\infty]$. We denote

$$D(\varphi) = \{x \in H \ : \ \varphi(x) \neq +\infty\} \qquad \text{(effective domain)}.$$

We say that φ is *proper* if $D(\varphi) \neq \emptyset$.

φ is *convex* if $\varphi(\alpha x + (1-\alpha)y) \leq \alpha\varphi(x) + (1-\alpha)\varphi(y)$ for all $\alpha \in [0,1]$ and $x, y \in H$.

Some of the properties of φ are reflected in its epigraph:

$$\operatorname{epi}(\varphi) := \{(x, r) \in H \times \mathbb{R} \ : \ r \geq \varphi(x)\}.$$

For instance, φ is convex if and only if $\operatorname{epi}(\varphi)$ is a convex subset of H; and φ is lower-semi-continuous if and only if $\operatorname{epi}(\varphi)$ is closed.

The *subdifferential* $\partial\varphi$ of φ is the operator defined by

$$w \in \partial\varphi(z) \iff \varphi(x) \geq \varphi(z) + (w|x - z) \qquad \forall\, x \in H.$$

Observe that

$$0 \in \partial\varphi(z) \iff \varphi(x) \geq \varphi(z) \ \forall\, x \in H \iff \varphi(z) = \min_{x \in D(\varphi)} \varphi(x).$$

Therefore, we have that $0 \in \partial\varphi(z)$ is the *Euler equation* of the variational problem

$$\varphi(z) = \min_{x \in D(\varphi)} \varphi(x).$$

If $(z, w), (\hat{z}, \hat{w}) \in \partial\varphi$, then $\varphi(z) \geq \varphi(\hat{z}) + (\hat{w}|z - \hat{z})$ and $\varphi(\hat{z}) \geq \varphi(z) + (\hat{w}|\hat{z} - z)$. Adding these inequalities we get

$$(w - \hat{w}|z - \hat{z}) \geq 0.$$

Thus, $\partial\varphi$ is a monotone operator. Now, if φ is convex, lower-semi-continuous and proper, it can be proved that $\partial\varphi$ is maximal monotone and $\overline{D(\partial\varphi)} = \overline{D(\varphi)}$ (see, [58], [32]).

As we mentioned in the linear case, the existence and uniqueness of mild solutions is equivalent to the requirement that $-A$ be the infinitesimal generator of a C_0-semigroup. Now, there are classical results connecting this fact with the m-accretivity of the operator A, for instance:

Lumer–Phillips Theorem. *$-A$ is the infinitesimal generator of a C_0-semigroup $(S(t))_{t\geq 0}$ of linear contractions on X if and only if A is linear, m-accretive and $\overline{D(A)} = X$. Moreover, in this case*

$$S(t)x = \lim_{n\to\infty}\left(I + \frac{t}{n}A\right)^{-n} x.$$

A first extension to the nonlinear case of this type of results has been given by Y. Komura in [142].

Komura Theorem. (i) *Let A be a maximal monotone operator in the Hilbert space H. Then $\overline{D(A)}$ is a closed convex subset of H and $D(S^A) = \overline{D(A)}$.*

(ii) *Given some closed convex set $C \subset H$ and a strongly continuous semigroup of contractions $(S(t))_{t\geq 0}$ on C, then there exists a unique maximal monotone operator A in H such that $\overline{D(A)} = C$ and $S^A(t) = S(t)$ for all $t \geq 0$.*

This result has been extended to some Banach spaces with good geometrical properties, but it turns out to be false in general Banach spaces. The good extension to nonlinear operators in general Banach spaces was done by Crandall–Liggett ([85]) and Ph. Bénilan ([40]) at the beginning of the 1970s. In the next section we give an outline of this theory.

A.5 Existence and Uniqueness Theorem

Suppose A is an operator in X and $f \in L^1(0,T;X)$. Consider the abstract Cauchy problem

$$(CP)_{x_0,f}\begin{cases} u'(t) + Au(t) \ni f(t) & \text{on } t \in (0,T),\\ u(0) = x. \end{cases}$$

Definition A.22. An *ϵ-approximate solution* of $(\mathrm{CP})_{x_0,f}$ is a solution v of an ϵ-discretization $D_A(0 = t_0,\ldots,t_N,f_1,\ldots,f_N)$ of $u' + Au \ni f$ on $[0,T]$ with $\|v(0) - x_0\| < \epsilon$.

It follows from this definition that u is a mild solution of $(\mathrm{CP})_{x_0,f}$ on $[0,T]$ if and only if $u \in C([0,T];X)$ and for each $\epsilon > 0$ there is an ϵ-approximate solution v of $(\mathrm{CP})_{x_0,f}$ such that $\|u(t) - v(t)\| < \epsilon$ on the domain of v.

Definition A.23. Suppose that for each $\epsilon > 0$ there are ϵ-approximate solutions of $(\mathrm{CP})_{x_0,f}$ on $[0,T]$. We say that the *ϵ-approximate solutions converge* on $[0,T]$

as $\epsilon \downarrow 0$ to $u \in C([0,T];X)$ if there exists a function $\psi : [0,+\infty[\to [0,+\infty[$ with $\lim_{\epsilon\downarrow 0}\psi(\epsilon) = 0$ such that $\|u(t) - v(t)\| \leq \psi(\epsilon)$ whenever $\epsilon > 0$, v is an ϵ-approximate solution of $(\mathrm{CP})_{x_0,f}$ on $[0,T]$ and t is in the domain of v.

Theorem A.24. *Suppose that $A \in \mathcal{A}(w)$, $f \in L^1(0,T;X)$ and $x_0 \in \overline{D(A)}$. If the problem $(CP)_{x_0,f}$ has an ϵ-approximate solution on $[0,T]$ for every $\epsilon > 0$, then it has a unique mild solution on $[0,T]$ to which the ϵ-approximate solutions of $(CP)_{x_0,f}$ converge as $\epsilon \downarrow 0$.*

This theorem was given by Ph. Bénilan in his Thesis ([40]) as an extension of the Crandall–Liggett theorem (which corresponds to $f = 0$). We also have the following result.

Theorem A.25. *Let A be an accretive operator in X and let u be a mild solution of $u' + Au \ni 0$ on $[0,T]$. Then:*

(i) *If v is an ϵ-approximate solution of $u' + Au \ni 0$ on $[0,T]$ with $[0,s]$ in its domain, $0 \leq t \leq T$, and $(x,y) \in A$, then*

$$\|u(t) - v(s)\| \leq 2\|u(0) - x\| + \|y\||t-s| \qquad 0 \leq s,t \leq T. \tag{A.24}$$

(ii) *If $\hat{u}$ is a mild solution of $\hat{u}' + A\hat{u} \ni 0$ on $[0,T]$, then*

$$\|u(t) - \hat{u}(t)\| \leq \|u(0) - \hat{u}(0)\| \qquad 0 \leq t \leq T. \tag{A.25}$$

Theorem A.24 tells us that, for accretive operators to have existence and uniqueness of mild solutions, it is enough to have existence of ϵ-approximate solutions for each $\epsilon > 0$. Now, we have seen this is the case for m-accretive operators, consequently we have the following result

Theorem A.26. *Let A be an operator in X, $f \in L^1(0,T;X)$ and $x_0 \in \overline{D(A)}$. If $A + wI$ is m-accretive, then the problem*

$$u' + Au \ni f \quad \text{on} \quad [0,T], \qquad u(0) = x_0$$

has a unique mild solution u on $[0,T]$.

Recall that

$$D(S^A) := \{x \in X \ : \ \exists! \text{ mild solution } u_x \text{ of } u' + Au \ni 0$$
$$\text{on } (0,+\infty) \text{ with } u_x(0) = x\},$$

and for $t \geq 0$ and $x \in D(S^A)$, $S^A(t)x := u_x(t)$. From now on, we denote $S^A(t)$ by e^{-tA}, and we call $(e^{-tA})_{t\geq 0}$ the *semigroup generated* by $-A$.

As a consequence of Theorem A.25, if A is accretive, then $(e^{-tA})_{t\geq 0}$ is a contraction semigroup, i.e.,

$$\|e^{-tA}x - e^{-tA}\hat{x}\| \leq \|x - \hat{x}\| \qquad \forall\, x,\hat{x} \in D(S^A),\ \forall\, t \geq 0.$$

Moreover, by the properties of mild solutions, it is easy to see that $D(S^A)$ is closed and, by Theorem A.25, we have that the map

$$(t,x) \mapsto e^{-tA}x \quad \text{is continuous in } [0,+\infty[\times D(S^A).$$

As a consequence of Theorem A.26 we have that if A is m-accretive in X, then $D(S^A) = \overline{D(A)}$ and $(e^{-tA})_{t\geq 0}$ is a contraction semigroup in $\overline{D(A)}$.

Let us see now that in the homogeneous case we can weaken the m-accretivity of the operator and get an explicit representation of the mild solution. Suppose for the moment that A is m-accretive. Let $\lambda > 0$ and v be a solution of the discretization $D_A(0,\lambda,2\lambda,\ldots,N\lambda;0,\ldots,0)$ satisfying $v(0) = x_0$. Due to the fact that the discretization has a constant step size λ, the difference equation for v is equivalent to

$$\begin{cases} v(t) = x_0 \quad \text{for} \quad -\lambda < t \leq 0, \\ \dfrac{v(t)-v(t-\lambda)}{\lambda} + Av(t) \ni 0 \quad \text{for} \quad 0 < t \leq N\lambda. \end{cases} \tag{A.26}$$

Moreover, $v(k\lambda) = J_\lambda v((k-1)\lambda)$ or, iterating

$$v(k\lambda) = J_\lambda^k v(0) = J_\lambda^k x_0.$$

Then in order to solve (A.26) we only need that $\overline{D(A)} \subset D(J_\lambda)$ for $\lambda > 0$ and of course the accretivity of the operator A.

Definition A.27. An accretive operator A satisfies the *range condition* if $\overline{D(A)} \subset R(I+\lambda A)$ for all $\lambda > 0$.

Theorem A.28. (Crandall–Liggett Theorem) *If A is accretive and satisfies the range condition, then $-A$ generates a semigroup of contractions $(e^{-tA})_{t\geq 0}$ on $\overline{D(A)}$ and:*

(i) *For $x_0 \in \overline{D(A)}$ and $0 \leq t < \infty$,*

$$\lim_{\lambda\downarrow 0, k\lambda \to t} J_\lambda^k x_0 = e^{-tA}x_0$$

holds uniformly for t on compact subintervals of $[0,\infty[$.

(ii) *If $x_0 \in \overline{D(A)}$, $t > 0$ and $n \in \mathbb{N}$, then*

$$\left\| J_{t/n}^n x_0 - e^{-tA}x_0 \right\| \leq \frac{t}{\sqrt{n}}\|y\| + 2\|x_0 - x\| \tag{A.27}$$

for every $(x,y) \in A$.

From either (i) or (ii) of the last theorem we deduce

$$e^{-tA}x = \lim_{n\to\infty}\left(I + \frac{t}{n}A\right)^{-n} x \quad \text{for } x \in \overline{D(A)}. \tag{A.28}$$

This representation of the semigroup $(e^{-tA})_{t\geq 0}$ is called the *exponential formula* by analogy with the formula $\lim_{n\to\infty}(1+\frac{t}{n}a)^{-n} = e^{-ta}$ for $a \in \mathbb{C}$.

Observe the analogy of (A.28) with the exponential formula given by the Lumer–Phillips theorem for the linear case. Now, there are strong differences between the linear and nonlinear cases. For instance, in the linear case, $-A$ is the infinitesimal generator of the C_0-semigroup $(e^{-tA})_{t\geq 0}$, and in the nonlinear case there are examples of operators A satisfying the assumptions of Crandall–Liggett's theorem, such that the domain of the infinitesimal generator of the semigroup $(e^{-tA})_{t\geq 0}$ is empty ([85]).

Let us give now an example of how to apply Crandall–Liggett's theorem.

Example A.29. Consider the nonlinear partial differential equation

$$\begin{cases} u_t(t,x) = \Delta\varphi(u(t,x)), & (t,x) \in]0,\infty[\times\Omega, \\ \varphi(u(t,x)) = 0, & (t,x) \in]0,\infty[\times\partial\Omega, \\ u(0,x) = u_0(x), & x \in \Omega, \end{cases} \tag{A.29}$$

where $\varphi : \mathbb{R} \to \mathbb{R}$ is a nondecreasing function and Ω is a smooth domain in $\mathbb{R}^N$. This equation is called the *Filtration Equation* and different elections of φ correspond to equations that appear in applications. For instance, if $\varphi(r) = |r|^m \mathrm{sign}_0(r)$, we have, for $m > 1$, the *Porous Medium Equation*, which appears in the study of a gas flow through a porous medium (see [195]); moreover, this equation also appears in models for population dynamics (Gurtin and McCamy). The case $0 < m < 1$ occurs in the theory of plasma, and in this case the equation is called the *Fast Diffusion Equation.*

To simplify the discussion we will assume that $\varphi \in C(\mathbb{R}) \cap C^1(\mathbb{R}\setminus\{0\})$, $\varphi(0) = 0$ and $\varphi'(s) > 0$ for $s \neq 0$.

Associated to the problem (A.29) we consider the operator A in $L^1(\Omega)$ defined by

$$\begin{aligned} D(A) &:= \left\{u \in L^1(\Omega) \,:\, \varphi(u) \in W_0^{1,1}(\Omega),\ \Delta\varphi(u) \in L^1(\Omega)\right\}, \\ Au &:= -\Delta\varphi(u) \quad \text{for } u \in D(A). \end{aligned}$$

We rewrite problem (A.29) as the abstract Cauchy problem

$$\begin{cases} u'(t) + Au(t) = 0 & t \in]0,+\infty[, \\ u(0) = u_0. \end{cases} \tag{A.30}$$

Since, $\{u \in L^1(\Omega) : \varphi(u) \in D(\Delta)\} \subset D(A)$, where

$$D(\Delta) = \left\{v \in W_0^{1,1}(\Omega) : \Delta v \in L^1(\Omega)\right\},$$

we have $\overline{D(A)} = L^1(\Omega)$. Therefore, if we prove that A is m-accretive in $L^1(\Omega)$, for each $u_0 \in L^1(\Omega)$, $e^{-tA}u_0$ solves problem (A.29) in the mild sense, i.e., $e^{-tA}u_0$ is the unique mild solution of (A.30). Let us see that A is m-accretive in $L^1(\Omega)$. To see the accretivity of A we need to show that

$$0 \leq [u - \hat{u}, Au - A\hat{u}] = \int_\Omega (Au - A\hat{u})\mathrm{sign}_0(u - \hat{u}) + \int_{\{u=\hat{u}\}} |Au - A\hat{u}|. \quad \text{(A.31)}$$

To this goal, choose $p_n \in C^1(\mathbb{R})$ with the properties: $p_n(0) = 0$, $|p_n(s)| \leq 1$, $p_n'(s) \geq 0$, $\lim_{n\to\infty} p_n(s) = \mathrm{sign}_0(s)$ for all $s \in \mathbb{R}$ (for example, $p_n(s) = \dfrac{ns}{n|s|+1}$, $s \in \mathbb{R}$). Applying Green's formula we have

$$\begin{aligned}\int_\Omega (Au - A\hat{u})p_n\left(\varphi(u) - \varphi(\hat{u})\right) &= -\int_\Omega \Delta\left(\varphi(u) - \varphi(\hat{u})\right)p_n\left(\varphi(u) - \varphi(\hat{u})\right)\\ &= \int_\Omega \nabla\left(\varphi(u) - \varphi(\hat{u})\right)\cdot\nabla\left(p_n\left(\varphi(u) - \varphi(\hat{u})\right)\right)\\ &= \int_\Omega p_n'\left(\varphi(u) - \varphi(\hat{u})\right)\left|\nabla\left(\varphi(u) - \varphi(\hat{u})\right)\right|^2 \geq 0.\end{aligned}$$

Then, letting $n \to +\infty$, we obtain

$$\int_\Omega (Au - A\hat{u})\mathrm{sign}_0\left(\varphi(u) - \varphi(\hat{u})\right) \geq 0.$$

Now, since φ is increasing,

$$\mathrm{sign}_0\left(\varphi(u) - \varphi(\hat{u})\right) = \mathrm{sign}_0(u - \hat{u}).$$

Hence, we get

$$\int_\Omega (Au - A\hat{u})\mathrm{sign}_0(u - \hat{u}) \geq 0$$

and consequently, (A.31) holds.

It remains to prove that for each $f \in L^1(\Omega)$ there exists a (unique) $u \in D(A)$, such that

$$u - \Delta\varphi(u) = f. \quad \text{(A.32)}$$

The proof of (A.32) is more complicated than the proof of the accretivity and is a consequence of a result due to H. Brezis and W. Strauss ([59]).

A.6 Regularity of Mild Solutions

As we have already pointed out, mild solutions may not satisfy any additional regularity properties, in general, they can not be interpreted as a solution of the Cauchy problem in a pointwise sense, that they are not strong solutions.

Nevertheless, the question arises naturally whether under certain additional assumptions one may obtain more regularity of mild solutions. This will be done now. We do emphasize, before this, that even in applications one does not want to be limited to strong solutions, since there are important partial differential equations which simply do not have strong solutions.

A basic fact is the following consistence between the accretivity of A and the differentiability of mild solutions of $u' + Au \ni f$.

Theorem A.30. *Let A be an accretive operator in X, $f \in L^1(0,T;X)$ and u be a mild solution of $u' + Au \ni f$ on $[0,T]$. If u has a right derivative $\frac{d^+u}{dt}(\tau)$ at $\tau \in]0,T[$ and*

$$\lim_{h\downarrow 0} \frac{1}{h} \int_\tau^{\tau+h} \|f(t) - f(\tau)\| \, dt = 0,$$

that is, τ is a right Lebesgue point of f, then the operator $\hat{A}$ given by

$$\hat{A}x = Ax \qquad \text{for } x \neq u(\tau)$$

$$\hat{A}u(\tau) = Au(\tau) \cup \left\{ f(\tau) - \frac{d^+u}{dt}(\tau) \right\}$$

is accretive.

Since every m-accretive operator is maximal accretive, as a consequence of the above theorem we have the following result.

Corollary A.31. *Suppose A is an m-accretive operator in X, $f \in L^1(0,T;X)$ and u is a mild solution of $u' + Au \ni f$ on $[0,T]$. Then,*

(i) *if u is differentiable at $t \in]0,T[$ and t is a right Lebesgue point of f, then*

$$u'(t) + Au(t) \ni f(t).$$

(ii) *If $u \in W^{1,1}(0,T;X)$, then u is a strong solution of $u' + Au \ni f$ on $[0,T]$.*

Then, the problem is: When is a mild solution in $W^{1,1}(0,T;X)$?

We denote by $BV(0,T;X)$ the subspace of functions in $L^1(0,T;X)$ which are of *bounded variation*, i.e., $f \in BV(0,T;X)$ if $f \in L^1(0,T;X)$ and

$$\operatorname{Var}(f,T) := \limsup_{h\downarrow 0} \int_0^{T-h} \frac{\|f(\tau+h) - f(\tau)\|}{h} \, d\tau < +\infty.$$

The principal conditions guaranteeing that a mild solution is in $W^{1,1}(0,T;X)$ are given by the following result.

Proposition A.32. *Let A be an accretive operator in X, $f \in BV(0,T;X)$ and $x \in D(A)$. If u is a mild solution of $(CP)_{x,f}$ on $[0,T]$, then u is locally Lipschitz continuous on $[0,T[$. Moreover, if X has the Radon–Nikodym property, then $u \in W^{1,1}(0,T;X)$ and consequently u is a strong solution of $(CP)_{x,f}$ on $[0,T]$.*

In the case that the operator is the subdifferential of a convex lower semi-continuous function in a Hilbert space, we have good regularity. More precisely, we have the following result.

Theorem A.33. *Let H be a Hilbert space and $\varphi : H \to]-\infty,+\infty]$ a proper, convex and lower semi-continuous function such that $\operatorname{Min}\varphi = 0$, and let $K := \{v \in H : \varphi(v) = 0\}$. Assume $f \in L^2(0,T;H)$ and $u_0 \in \overline{D(\partial\varphi)}$, then the mild solution $u(t)$ of*

$$u' + \partial\varphi(u) \ni f \quad \text{on } [0,T], \quad u(0) = u_0$$

is a strong solution and we have the following estimates:

$$\|u'(t)\|_{L^2(\delta,T;H)} \leq \|f\|_{L^2(0,T;H)} + \frac{1}{\sqrt{2\delta}}\int_0^\delta \|f(t)\|\,dt + \frac{1}{\sqrt{2\delta}}\operatorname{dist}(u_0,K) \tag{A.33}$$

for $0 < \delta < T$.

$$\left(\int_0^T \|u'(t)\|^2 t\,dt\right)^{\frac{1}{2}} \leq \left(\int_0^T \|f(t)\|^2 t\,dt\right)^{\frac{1}{2}} + \frac{1}{\sqrt{2}}\int_0^T \|f(t)\|^2\,dt + \frac{1}{\sqrt{2}}\operatorname{dist}(u_0,K). \tag{A.34}$$

Moreover, for almost all $t \in [0,T]$, we have

$$\frac{d}{dt}\varphi(u(t)) = (h|u'(t)) \qquad \forall\, h \in \partial\varphi(u(t)). \tag{A.35}$$

In the homogeneous case, i.e., $f = 0$, we have

$$\|u'(t)\|_{L^\infty(\delta,T;H)} \leq \frac{1}{\delta}\|u_0\| \quad \text{for } 0 < \delta < T. \tag{A.36}$$

A.7 Completely Accretive Operators

Many nonlinear semigroups that appear in applications are also order-preserving and contractions in every L^p. Ph. Bénilan and M. Crandall introduced in [44] a class of operators, named completely accretive, for which the semigroup generated by the Crandall–Ligget exponential formula enjoys these properties. In this section we outline some of the main points given in [44].

Let $(\Omega, \mathcal{B}, \mu)$ be a σ-finite measure space and let $M(\Omega)$ denote the space of measurable functions from Ω into $\mathbb{R}$. We denote by $L(\Omega)$ the space

$$L(\Omega) := L^1(\Omega) + L^\infty(\Omega);$$

$L(\Omega)$ is exactly the subset of $M(\Omega)$ on which the functional

$$\|u\|_{1+\infty} := \inf\{\|f\|_1 + \|g\|_\infty \ : f, g \in M(\Omega), \ f + g = u\}$$

is finite and $L(\Omega)$ equipped with $\| \ \|_{1+\infty}$ is a Banach space.

Let

$$\begin{aligned} L_0(\Omega) &:= \{u \in L(\Omega) \ : \ \mu((\{|u| > k\}) < \infty \text{ for } k > 0\} \\ &= \left\{u \in M(\Omega) \ : \ \int_\Omega (|u| - k)^+ < \infty \ \text{ for } k > 0\right\}. \end{aligned}$$

$L_0(\Omega)$ is a closed subspace of $L(\Omega)$; in fact, it is the closure in $L(\Omega)$ of the linear span of the set of characteristic functions of sets of finite measure. Hereafter, $L_0(\Omega)$ carries the norm $\| \ \|_{1+\infty}$, it is then a Banach space. With the natural pairing $\langle u, v\rangle = \int_\Omega uv$, the dual space of $L_0(\Omega)$ is isometrically isomorphic to

$$L^{1\cap\infty}(\Omega) := L^1(\Omega) \cap L^\infty(\Omega),$$

when in $L^{1\cap\infty}(\Omega)$ is given the norm

$$\|u\|_{1\cap\infty} := \max\{\|u\|_1, \|u\|_\infty\}.$$

Given $u, v \in \mathcal{M}(\Omega)$, we shall write

$$u \ll v \quad \text{if and only if} \int_\Omega j(u)dx \le \int_\Omega j(v)dx \tag{A.37}$$

for all $j \in J_0$, where

$$J_0 = \{j : \mathbb{R} \to [0, \infty], \text{ convex, l.s.c., } j(0) = 0\} \tag{A.38}$$

(l.s.c. is an abbreviation for lower semi-continuous function).

Definition A.34. A functional $N : M(\Omega) \to]-\infty, +\infty]$ is *normal* if $N(u) \le N(v)$ whenever $u \ll v$.

A map $S : D(S) \subset M(\Omega) \to M(\Omega)$ is a *complete contraction* if it is an N-contraction for every normal functional N, i.e., if

$$N(Su - Sv) \le N(u - v) \quad \text{for } u, v \in D(S).$$

A Banach space $(X, \| \|_X)$, with $X \subset M(\Omega)$ is a *normal Banach space* if it has the property

$$u \in X,\ v \in M(\Omega),\ v \ll u\ \Rightarrow\ v \in X \ \text{ and } \ \|v\|_X \leq \|u\|_X. \tag{A.39}$$

Simple examples of normal Banach spaces are: $L^p(\Omega)$, $1 \leq p \leq \infty$ and $L(\Omega)$, $L_0(\Omega)$, $L^{1\cap\infty}(\Omega)$.

Proposition A.35. *Let $S : D(S) \subset M(\Omega) \to M(\Omega)$ and assume*

$$u, v \in D(S) \text{ and } k \geq 0\ \Rightarrow\ \ u \wedge (v + k) \text{ or } v \vee (u - k) \in D(S). \tag{A.40}$$

Then S is a complete contraction if and only if it is order-preserving and a contraction for $\| \|_1$ and $\| \|_\infty$.

Definition A.36. Let A be an operator in $M(\Omega)$. We shall say that A is *completely accretive* if

$$u - \hat{u} \ll u - \hat{u} + \lambda(v - \hat{v}) \quad \text{for all } \lambda > 0 \text{ and all } (u, v), (\hat{u}, \hat{v}) \in A. \tag{A.41}$$

In other words, A is completely accretive if

$$N(u - \hat{u}) \leq N\left(u - \hat{u} + \lambda(v - \hat{v})\right) \tag{A.42}$$

for all $\lambda > 0$, all $(u, v), (\hat{u}, \hat{v}) \in A$ and every normal functional N in $M(\Omega)$.

Let

$$P_0 = \left\{p \in C^\infty(\mathbb{R}) : 0 \leq p' \leq 1,\ \operatorname{supp}(p') \text{ is compact and } 0 \notin \operatorname{supp}(p)\right\}.$$

The following result, which is a generalization of one due to H. Brezis and W. Strauss ([59]), provides a very useful characterization of the complete accretivity.

Proposition A.37. *Let $u \in L_0(\Omega)$, $v \in L(\Omega)$. Then,*

$$u \ll u + \lambda v \quad \forall\, \lambda > 0 \iff \int_\Omega p(u)v \geq 0 \quad \forall\, p \in P_0.$$

Observe that if $\mu(\Omega) < \infty$, then $L_0(\Omega) = L(\Omega) = L^1(\Omega)$. Consequently, from the above proposition we get the following characterization.

Corollary A.38. *Assume that $\mu(\Omega) < \infty$. If $A \subseteq L^1(\Omega) \times L^1(\Omega)$, then A is completely accretive if and only if*

$$\int_\Omega p(u - \hat{u})(v - \hat{v}) \geq 0 \quad \textit{for any} \ \ p \in P_0,\ (u, v), (\hat{u}, \hat{v}) \in A. \tag{A.43}$$

Proposition A.39. *Let $u \in L_0(\Omega)$. Then $\{v \in M(\Omega)\ : v \ll u\}$ is a weakly sequentially compact subset of $L_0(\Omega)$.*

Definition A.40. Let X be a linear subspace of $M(\Omega)$. An operator A in X is m-*completely accretive* in X if A is completely accretive and $R(I+\lambda A)=X$ for $\lambda>0$

Remark A.41. The above definition does not require X to be a Banach space and so does not require A to be m-accretive in any Banach space. However, if A is completely accretive, then it is accretive in $L(\Omega)$ and if A is m-completely accretive in a subspace X of $L(\Omega)$, then the closure $\overline{A}$ of A in $L(\Omega)$ is completely accretive and m-accretive in the closure $\overline{X}$ of X in $L(\Omega)$. We also note that if A is completely accretive in a subspace X of $M(\Omega)$ and $R(I+\lambda A)=X$ for some $\lambda>0$, the only completely accretive operator B in X which extends A is A.

Proposition A.42. *Let X be a normal Banach space, $X\subset L_0(\Omega)$, and A be a completely accretive operator in X. Then, if there exists $\lambda>0$ for which $R(I+\lambda A)$ is dense in $L_0(\Omega)$, then the operator $A^X:=\overline{A}\cap(X\times X)$ is the unique m-completely accretive extension of A in X.*

Definition A.43. Let A be an operator in $L_0(\Omega)$. Then A° is the restriction of A defined by

$$v\in A^\circ u \iff v\in Au \text{ and } v\ll w\ \forall\, w\in Au.$$

In the case X is a normal Banach space and A is m-completely accretive in X, by Crandall–Ligget's theorem, A generates a contraction semigroup in X given by the exponential formula

$$e^{-tA}u_0 = X-\lim_{n\to\infty}\left(I+\frac{t}{n}A\right)^{-n}u_0 \quad \text{for any } u_0\in\overline{D(A)}^X.$$

Now, since $\overline{A}$ is m-completely accretive in $\overline{X}$ endowed with the norm of $L(\Omega)$, we also may consider the semigroup $e^{-t\overline{A}}$ on $\overline{D(A)}$. We have the following relation between these two semigroups.

Proposition A.44. *Let X be a normal Banach space and A an m-completely accretive operator in X. Then, we have*

(i) *e^{-tA} is a complete contraction for $t\geq 0$.*

(ii) *e^{-tA} is the restriction of $e^{-t\overline{A}}$ to $\overline{D(A)}^X$ and $e^{-t\overline{A}}$ is the closure of e^{-tA} in $L(\Omega)$.*

(iii) *$e^{-t\overline{A}}(\overline{D(A)}\cap X)\subset\overline{D(A)}\cap X$.*

As a consequence of (iii) of the above proposition, if we denote by $S^A(t)$ the restriction of $e^{-t\overline{A}}$ to $\overline{D(A)}\cap X$, we have $S^A(t)$ is given by the exponential formula

$$S^A(t)u = L(\Omega)-\lim_{n\to\infty}\left(I+\frac{t}{n}A\right)^{-n}u \quad \text{for } u\in\overline{D(A)}\cap X.$$

Theorem A.45. *Let X be a normal Banach space with $X \subset L_0(\Omega)$ and A an m-completely accretive operator in X. Then, we have*

(i) $D(A) = \left\{u \in \overline{D(A)} \cap X \ : \exists v \in X \ s.t. \ \dfrac{S^A(t)u - u}{t} \ll v \text{ for small } t > 0\right\}.$

(ii) $S^A(t)D(A) \subset D(A)$ *for* $t > 0$.

(iii) *If $u \in D(A)$, then*

$$\frac{u - S^A(t)u}{t} \ll v \quad \text{for } t > 0 \quad \text{and} \quad v \in Au$$

and

$$L(\Omega) - \lim_{t \to 0} \frac{S^A(t)u - u}{t} = -A^\circ u.$$

Corollary A.46. *Assume that $\mu(\Omega) < \infty$. If $A \subseteq L^1(\Omega) \times L^1(\Omega)$, is an m-completely accretive operator in $L^1(\Omega)$, then for every $u_0 \in D(A)$, the mild solution $u(t) = e^{-tA}u_0$ of the problem*

$$\frac{du}{dt} + Au \ni 0, \quad u(0) = u_0 \tag{A.44}$$

is a strong solution.

The following result is a variant of the regularizing effect of the homogeneous evolution equation obtained in [43] in the m-completely accretive case.

Theorem A.47. *In addition to the assumptions of Theorem* A.45, *assume that A is positively homogeneous of degree $0 < m \neq 1$, i.e., $A(\lambda u) = \lambda^m Au$ for $u \in D(A)$. Then for $u \in \overline{D(A)} \cap X$ and $t > 0$, we have $S^A(t)u \in D(A)$ and*

$$|A^\circ S^A(t)u| \leq 2\frac{|u|}{|m-1|t}.$$

To finish we summarize the following results about the completely accretive subdifferentials. Let X be a linear subspace of $M(\Omega)$ and $\Phi : X \to]-\infty, +\infty]$. We define the operator $\partial_X \Phi$ in X by

$$\begin{cases} v \in \partial_X \Phi(u) \iff u \in D(\Phi), \ v \in X \text{ and} \\ \Phi(w) - \Phi(u) \geq \displaystyle\int_\Omega (w-u)v \quad \text{for} \quad w \in X \quad \text{with} \quad (w-u)v \in L^1(\Omega). \end{cases}$$

For example, if $X \subset L^2(\Omega)$ and $D(\Phi) \neq \emptyset$, then $\partial_X \Phi$ coincides with the subdifferential in $L^2(\Omega)$ of the extension $\hat{\Phi}$ of Φ to $L^2(\Omega)$ which is $+\infty$ on $L^2(\Omega) \setminus X$.

Lemma A.48. *Let $X \subset L_0(\Omega)$ be a normal Banach space and $\Phi : X \to]-\infty, +\infty]$. Assume that*

$$\Phi(u + p(\hat{u} - u)) + \Phi(\hat{u} - p(\hat{u} - u)) \leq \Phi(u) + \Phi(\hat{u}) \tag{A.45}$$

holds for $u, \hat{u} \in X$ and $p \in P_0$. Then $\partial_X \Phi$ is completely accretive.

As one expects from the classical Hilbert space theory, in order to get the range condition for the operator $\partial_X \Phi$, one needs lower-semi-continuity of the functional Φ. If $\Phi : X \to]-\infty, +\infty]$ and X is a Banach space, we will denote by Φ^X the functional defined by

$$\Phi^X(u) := \liminf_{r \downarrow 0} \{\Phi(w) \ : \ w \in X \text{ and } \|w - u\| \leq r\}.$$

Φ^X is the l.s.c. envelope of Φ. It is clear that Φ^X is a l.s.c. functional on X, $\Phi^X \leq \Phi$, $D(\Phi^X) \subset \overline{D(\Phi)}^X$ and for $u \in X$, $\Phi^X(u) = \Phi(u)$ if and only if Φ is l.s.c. in X at the point u.

Lemma A.49. *Let X be a normal Banach space with $X \subset L_0(\Omega)$, $\Phi : X \to]-\infty, +\infty]$ and Φ^X be the l.s.c. envelope of Φ in X. If $L^1(\Omega) \cap L^\infty(\Omega)$ is dense in X and (A.45) holds, then $\partial_X \phi^X$ is an extension of $\partial_X \phi$.*

Theorem A.50. *Let X be a normal Banach space with $L^1(\Omega) \cap L^\infty(\Omega)$ dense in X and $\Phi : X \to]-\infty, +\infty]$. Assume that (A.45) holds, $0 \in \partial_X \Phi(0)$ and Φ is l.s.c. for the topology of $X + L^2(\Omega)$. Then the closure in X of $\partial_X \Phi$ is m-completely accretive in X.*

Appendix B

Functions of Bounded Variation

Due to the linear growth condition on the Lagrangians associated with the problems we study in this monograph, the natural energy space to study them is the space of functions of bounded variation. In this appendix we collect some basic results of the theory of functions of bounded variation. For more information we refer the reader to [10], [25], [110], [122], [209]

B.1 Definitions

Throughout this chapter, Ω denotes an open subset of $\mathbb{R}^N$.

Definition B.1. A function $u \in L^1(\Omega)$ whose partial derivatives in the sense of distributions are measures with finite total variation in Ω is called a *function of bounded variation.* The vector space of functions of bounded variation in Ω is denoted by $BV(\Omega)$. Thus $u \in BV(\Omega)$ if and only if $u \in L^1(\Omega)$ and there are Radon measures $\mu_1, \ldots, \mu_N$ with finite total mass in Ω such that

$$\int_\Omega u \frac{\partial \varphi}{\partial x_i} dx = -\int_\Omega \varphi d\mu_i \quad \forall \varphi \in C_0^\infty(\Omega),\ i = 1, \ldots, N.$$

If $u \in BV(\Omega)$, the total variation of the measure Du is

$$\|Du\| = \sup \left\{ \int_\Omega u \operatorname{div}(\phi)\, dx : \phi \in C_0^\infty(\Omega, \mathbb{R}^N),\ |\phi(x)| \leq 1 \text{ for } x \in \Omega \right\}.$$

The space $BV(\Omega)$, endowed with the norm

$$\|u\|_{BV} = \|u\|_1 + \|Du\|,$$

is a Banach space. If $u \in BV(\Omega)$, the total variation $\|Du\|$ may be regarded as a measure, whose value on an open set $U \subseteq \Omega$ is

$$\|Du\|(U) = \sup \left\{ \int_U u \operatorname{div}(\phi)\, dx : \phi \in C_0^\infty(U, \mathbb{R}^N),\ |\phi(x)| \leq 1 \text{ for } x \in U \right\}.$$

We also use

$$\int_U \|Du\|$$

to denote $\|Du\|(U)$.

For $u \in BV(\Omega)$, the gradient Du is a Radon measure that decomposes into its absolutely continuous and singular parts

$$Du = D^a u + D^s u.$$

Then $D^a u = \nabla u \, \mathcal{L}^N$ where ∇u is the Radon–Nikodym derivative of the measure Du with respect to the Lebesgue measure $\mathcal{L}^N$. There is also the polar decomposition $D^s u = \overrightarrow{D^s u} |D^s u|$ where $|D^s u|$ is the total variation measure of $D^s u$.

The total variation is lower semi-continuous. More concretely, we have the following result.

Theorem B.2. *Suppose that $u_i \in BV(\Omega)$, $i = 1, 2, \ldots$, and $u_i \to u$ in $L^1_{loc}(\Omega)$. Then*

$$\|Du\|(\Omega) \leq \liminf_{i \to \infty} \|Du_i\|(\Omega).$$

We say that $u \in L^1_{loc}(\Omega)$ is *locally of bounded variation* if $\varphi u \in BV(\Omega)$ for any $\varphi \in C_0^\infty(\Omega)$. We denote by $BV_{loc}(\Omega)$ the space of functions which are locally of bounded variation.

Here and in what follows we shall denote by $\mathcal{H}^\alpha$ the Hausdorff measure of dimension α in $\mathbb{R}^N$. In particular, $\mathcal{H}^{N-1}$ denotes the $(N-1)$-dimensional Hausdorff measure and $\mathcal{H}^N$, the N-dimensional Hausdorff measure, coincides with the (outer) Lebesgue measure in $\mathbb{R}^N$. Given any Borel set $B \subseteq \mathbb{R}^N$ with $\mathcal{H}^\alpha(B) < \infty$, we denote by $\mathcal{H}^\alpha \llcorner B$ the finite Borel measure $\chi_B \mathcal{H}^\alpha$, i.e., $\mathcal{H}^\alpha \llcorner B(C) = \mathcal{H}^\alpha(B \cap C)$ for any Borel set $C \subseteq \mathbb{R}^N$. We recall that

$$\lim_{r \to 0^+} \frac{\mathcal{H}^k(B \cap B(x,r))}{r^k} = 0 \qquad \text{for } \mathcal{H}^k\text{-a.e. } x \in \mathbb{R}^N \setminus B \tag{B.1}$$

holds whenever $B \subseteq \mathbb{R}^N$ is a Borel set with finite k-dimensional Hausdorff measure (see for instance §2.3 of [110]).

B.2 Approximation by Smooth Functions

Theorem B.3. *Assume that $u \in BV(\Omega)$. There exists a sequence of functions $u_i \in C^\infty(\Omega) \cap BV(\Omega)$ such that*

(i) *$u_i \to u$ in $L^1(\Omega)$;*

(ii) *$\|Du_i\|(\Omega) \to \|Du\|(\Omega)$ as $i \to \infty$.*

Moreover,

(iii) *if* $u \in BV(\Omega) \cap L^q(\Omega)$, $q < \infty$, *we can find functions* u_i *such that* $u_i \in L^q(\Omega)$ *and* $u_i \to u$ *in* $L^q(\Omega)$;

(iv) *if* $u \in BV(\Omega) \cap L^\infty(\Omega)$, *we can find* u_i *such that* $\|u_i\|_\infty \leq \|u\|_\infty$ *and* $u_i \to u$ *in* $L^\infty(\Omega)$*-weakly*.*

Finally,

(v) *if* $\partial\Omega$ *is Lipschitz continuous one can find* u_i *such that*

$$u_i|_{\partial\Omega} = u|_{\partial\Omega} \qquad \text{for all} \quad i.$$

Theorem B.4. *Assume that* $u \in BV(\Omega)$. *There exists a sequence of functions* $u_i \in C^\infty(\Omega) \cap BV(\Omega)$ *such that*

(i) $u_i \to u$ *in* $L^1(\Omega)$;

(ii) *if* $U \subset\subset \Omega$ *is such that* $\|Du\|(\partial U) = 0$, *then*

$$\lim_{i\to\infty} \|Du_i\|(U) = \|Du\|(U).$$

Moreover, if $u \in L^q(\Omega)$, $1 \leq q < \infty$ *or* $u \in L^\infty(\Omega)$, *one can find* u_i *satisfying* (iii) *or* (iv), *respectively, of the above result.*

Definition B.5. Let $u_i, u \in BV(\Omega)$, $i = 1, 2, \ldots$. We say that u_i *strictly converges* to u in $BV(\Omega)$ if both conditions (i), (ii) of Theorem B.3 hold.

Definition B.6. Let $u_i, u \in BV(\Omega)$, $i = 1, 2, \ldots$. We say that u_i weakly* converges to u in $BV(\Omega)$ if $u_i \to u$ in $L^1_{loc}(\Omega)$ and Du_i weakly* converges to Du as measures in Ω.

Remark B.7. $BV(\Omega)$ is the dual of a separable space and, at least for sufficiently regular domains, the convergence of Definition B.6 coincides with the weak* convergence in the usual sense. The predual of $BV(\Omega)$ can be described as a quotient space $\frac{E}{F}$ where $E = C_0(\Omega)^{N+1}$, $C_0(\Omega)$ being the space of functions vanishing at the boundary of Ω, i.e., the closure of $C_0^\infty(\Omega)$ with respect to the uniform norm, and F being the closure in E of the space

$$\{(\phi_0, \phi_1, \ldots, \phi_N) : \phi_i \in C_0^\infty(\Omega), \text{ and } \phi_0 = \operatorname{div}(\phi_1, \ldots, \phi_N)\}.$$

Indeed, if $S : BV(\Omega) \to E^*$ is the map defined by

$$S(w) := \left(w\,dx, \frac{\partial w}{\partial x_1}, \ldots, \frac{\partial w}{\partial x_N}\right),$$

then $S(BV(\Omega))$ is isomorphic to $(\frac{E}{F})^*$.

Proposition B.8. *If $u_i, u \in BV(\Omega)$. Then $u_i \to u$ weakly* in $BV(\Omega)$ if and only if $\{u_i\}$ is bounded in $BV(\Omega)$ and converges to u in $L^1_{loc}(\Omega)$. Moreover, if*

$$\|Du_i\|(\Omega) \to \|Du\|(\Omega) \quad \text{as } i \to \infty,$$

and we consider the measures

$$\mu_i(B) = \int_{B\cap\Omega} Du_i, \quad \mu(B) = \int_{B\cap\Omega} Du,$$

for all Borel sets $B \subset \mathbb{R}^N$, then $\mu_i \rightharpoonup \mu$ weakly as (vector valued) Radon measures in $\mathbb{R}^N$.*

Theorem B.9. *If $(u_k) \subseteq BV(\Omega)$ strictly converges to u and $f : \mathbb{R}^N \to \mathbb{R}$ is continuous and 1-positively homogeneous, we have*

$$\lim_{k\to\infty} \int_\Omega \phi f\left(\frac{Du_k}{\|Du_k\|}\right) d\|Du_k\| = \int_\Omega \phi f\left(\frac{Du}{\|Du\|}\right) d\|Du\|$$

for any bounded continuous function $\phi : \Omega \to \mathbb{R}$. As a consequence

$$f\left(\frac{Du_k}{\|Du_k\|}\right)\|Du_k\| \quad \text{weakly}^* \text{ converge in } \Omega \text{ to } f\left(\frac{Du}{\|Du\|}\right)\|Du\|.$$

In particular, $\|Du_k\| \to \|Du\|$ weakly in Ω.*

B.3 Traces and Extensions

Assume that Ω is open and bounded with $\partial\Omega$ Lipschitz. We observe that since $\partial\Omega$ is Lipschitz, the outer unit normal ν exists $\mathcal{H}^{N-1}$ a.e. on $\partial\Omega$.

Theorem B.10. *Assume that Ω is open and bounded, with $\partial\Omega$ Lipschitz. There exists a bounded linear mapping*

$$T : BV(\Omega) \to L^1(\partial\Omega, \mathcal{H}^{N-1})$$

such that

$$\int_\Omega u \operatorname{div}(\varphi)\, dx = -\int_\Omega \varphi \cdot dDu + \int_{\partial\Omega} \varphi \cdot \nu Tu \, d\mathcal{H}^{N-1}$$

for all $u \in BV(\Omega)$ and $\varphi \in C^1(\mathbb{R}^N, \mathbb{R}^N)$. Moreover, for any $u \in BV(\Omega)$ and for $\mathcal{H}^{N-1}$ a.e. $x \in \partial\Omega$, we have

$$\lim_{r\to+} r^{-N} \int_{B(x,r)\cap\Omega} |u - Tu(x)|\, dy = 0.$$

Theorem B.11. *Let Ω be an open bounded set, with $\partial\Omega$ Lipschitz. Then the trace operator $u \to Tu$ is continuous between $BV(\Omega)$, endowed with the topology induced by the strict convergence, and $L^1(\partial\Omega, \mathcal{H}^{N-1} \llcorner \partial\Omega)$.*

Theorem B.12. *Assume that Ω is open and bounded, with $\partial\Omega$ Lipschitz. Let $u_1 \in BV(\Omega)$, $u_2 \in BV(\mathbb{R}^N \setminus \overline{\Omega})$. We define*

$$v(x) = \begin{cases} u_1(x) & \text{if } x \in \Omega, \\ u_2(x) & \text{if } x \in \mathbb{R}^N \setminus \overline{\Omega}. \end{cases}$$

Then $v \in BV(\mathbb{R}^N)$ and

$$\|Dv\|(\mathbb{R}^N) = \|Du_1\|(\Omega) + \|Du_2\|(\mathbb{R}^N \setminus \overline{\Omega}) + \int_{\partial\Omega} |Tu_1 - Tu_2| d\mathcal{H}^{N-1}.$$

In particular, if

$$Eu(x) = \begin{cases} u(x) & \text{if } x \in \Omega, \\ 0 & \text{if } x \in \mathbb{R}^N \setminus \overline{\Omega}, \end{cases}$$

then $Eu \in BV(\mathbb{R}^N)$ provided $u \in BV(\Omega)$.

B.4 Sets of Finite Perimeter and the Coarea Formula

Definition B.13. An $\mathcal{L}^N$ measurable subset E of $\mathbb{R}^N$ has *finite perimeter* in Ω if $\chi_E \in BV(\Omega)$. The perimeter of E in Ω is $P(E, \Omega) = \|D\chi_E\|(\Omega)$.

We shall denote the measure $\|D\chi_E\|$ by $\|\partial E\|$ and $P(E, \mathbb{R}^N)$ by $\mathrm{Per}(E)$.

Theorem B.14. *Let E be a set of finite perimeter in Ω and let $D\chi_E = \nu_E \|D\chi_E\|$ be the polar decomposition of $D\chi_E$. Then the generalized Gauss–Green formula holds*

$$\int_E \operatorname{div}(\varphi)\, dx = -\int_\Omega \langle \nu_E, \varphi \rangle d\|D\chi_E\|$$

for all $\varphi \in C_0^1(\Omega, \mathbb{R}^N)$.

Theorem B.15 (Coarea formula for BV-functions). *Let $u \in BV(\Omega)$. Then*

(i) *$E_{u,t} := \{x \in \Omega : u(x) > t\}$ has finite perimeter for $\mathcal{L}^1$* a.e. *$t \in \mathbb{R}$ and*

(ii) $\|Du\|(\Omega) = \int_{-\infty}^{\infty} P(E_{u,t}, \Omega) dt.$

(iii) *Conversely, if $u \in L^1(\Omega)$ and*

$$\int_{-\infty}^{\infty} P(E_{u,t}, \Omega) dt < \infty,$$

then $u \in BV(\Omega)$.

We need to consider the truncations $T_{a,b}$, $a < b$ (see Section 3.6)

Proposition B.16. *If $u \in BV(\Omega)$ and $f : \mathbb{R} \to \mathbb{R}$ is a Lipschitz function, then $f(u) \in BV(\Omega)$. In particular, $T_{a,b}(u) \in BV(\Omega)$ and we have*

$$\|DT_{a,b}(u)\|(\Omega) = \int_a^b P(E_{u,t}, \Omega)dt.$$

The next proposition follows from Theorem B.15 and Proposition B.16.

Proposition B.17. *If $u \in BV(\Omega)$, then $T_k(u), G_k(u) \in BV(\Omega)$ where $G_k(r) = r - T_k(r)$, $k \geq 0, r \in \mathbb{R}$. Moreover,*

$$\|Du\|(\Omega) = \|DT_k(u)\|(\Omega) + \|DG_k(u)\|(\Omega)$$

for any $k \geq 0$.

B.5 Some Isoperimetric Inequalities

Theorem B.18 (Sobolev inequality). *There exists a constant $C > 0$ such that*

$$\|u\|_{L^{N/N-1}(\mathbb{R}^N)} \leq C\|Du\|(\mathbb{R}^N)$$

for all $u \in BV(\mathbb{R}^N)$.

If $u \in L^1(\Omega)$, the mean value of u in Ω is

$$u_\Omega = \frac{1}{\mathcal{L}^N(\Omega)} \int_\Omega u(x)\, dx.$$

Theorem B.19 (Poincaré inequality). *Let Ω be open and bounded with $\partial\Omega$ Lipschitz. Suppose that Ω is connected. Then*

$$\int_\Omega |u - u_\Omega|\, dx \leq C\|Du\|(\Omega) \quad \forall u \in BV(\Omega)$$

for some constant C depending only on Ω.

Theorem B.20 (Isoperimetric Inequality). *Let $N > 1$. For any set E of finite perimeter in $\mathbb{R}^N$ either E or $\mathbb{R}^N \setminus E$ has finite Lebesgue measure and*

$$\min\left\{\mathcal{L}^N(E), \mathcal{L}^N(\mathbb{R}^N \setminus E)\right\} \leq C[\operatorname{Per}(E)]^{\frac{N}{N-1}}$$

for some dimensional constant C.

Theorem B.21 (Embedding Theorem). *Let Ω be open and bounded, with $\partial\Omega$ Lipschitz. Then the embedding $BV(\Omega) \to L^{N/N-1}(\Omega)$ is continuous and $BV(\Omega) \to L^p(\Omega)$ is compact for all $1 \leq p < \frac{N}{N-1}$.*

The continuity of the embedding of Theorem B.21 and Theorem B.19 imply the following *Sobolev–Poincaré inequality*

$$\|u - u_\Omega\|_p \leq C\|Du\|(\Omega) \quad \forall u \in BV(\Omega),\ 1 \leq p \leq \frac{N}{N-1} \tag{B.2}$$

for some constant C depending only on Ω.

B.6 The Reduced Boundary

In this section we assume that E is a set of finite perimeter in $\mathbb{R}^N$.

Definition B.22. Let $x \in \mathbb{R}^N$. We say that $x \in \partial^* E$, the *reduced boundary* of E, if

(i) $\|D\chi_E\|(B(x,r)) > 0$ for all $r > 0$,

(ii) $\displaystyle\lim_{r\to 0+} \frac{1}{\mathcal{L}^N(B(x,r))} \int_{B(x,r)} \nu_E d\|\partial E\| = \nu_E(x)$, and

(iii) $|\nu_E(x)| = 1$.

According to the properties of the Radon–Nykodym derivatives, we have

$$\|\partial E\|(\mathbb{R}^N \setminus \partial^* E) = 0.$$

Definition B.23. For each $x \in \partial^* E$, we define the hyperplane

$$H(x) = \left\{y \in \mathbb{R}^N : \nu_E(x) \cdot (y - x) = 0\right\}$$

and the half-spaces

$$H^+(x) = \left\{y \in \mathbb{R}^N : \nu_E(x) \cdot (y - x) \geq 0\right\},$$

$$H^-(x) = \left\{y \in \mathbb{R}^N : \nu_E(x) \cdot (y - x) \leq 0\right\}.$$

Proposition B.24. *Assume $x \in \partial^* E$. Then*

$$\lim_{r\to 0+} \frac{\mathcal{L}^N(B(x,r) \cap E \cap H^+(x))}{r^N} = 0, \tag{B.3}$$

$$\lim_{r\to 0+} \frac{\mathcal{L}^N((B(x,r) \setminus E) \cap H^-(x))}{r^N} = 0, \tag{B.4}$$

and

$$\lim_{r\to 0+} \frac{\|\partial E\|(B(x,r))}{\omega_{N-1} r^{N-1}} = 1, \tag{B.5}$$

where ω_{N-1} denotes the volume of the unit ball in $\mathbb{R}^{N-1}$.

Definition B.25. A unit vector $\nu_E(x)$ for which (B.3) and (B.4) hold is called a *measure theoretic unit outer normal* to E at x.

Theorem B.26 (Structure theorem for sets of finite perimeter). *Assume that E has locally finite perimeter in $\mathbb{R}^N$.*

(i) *Then*

$$\partial^* E = \bigcup_{k=1}^{\infty} K_k \cup N,$$

where

$$\|\partial E\|(N) = 0$$

and K_k is a compact set of a C^1-hypersurface S_k, $k = 1, 2, \ldots$.

(ii) *Furthermore, $\nu_E|_{S_k}$ is normal to S_k, $k = 1, 2, \ldots$, and*

(iii) $\|\partial E\| = \mathcal{H}^{N-1} \llcorner \partial^* E$.

For a Lebesgue measurable subset $E \subseteq \mathbb{R}^N$ and a point $x \in \mathbb{R}^N$, the upper and lower densities of E at x are respectively defined by

$$\overline{D}(E, x) := \limsup_{r \to 0^+} \frac{\mathcal{L}^N(E \cap B(x, r))}{\mathcal{L}^N(B(x, r))},$$

$$\underline{D}(E, x) := \liminf_{r \to 0^+} \frac{\mathcal{L}^N(E \cap B(x, r))}{\mathcal{L}^N(B(x, r))}.$$

If the upper and lower densities are equal, their common value will be called the density of x at E and it will be denoted by $D(x, E)$. We shall use the word measurable to mean Lebesgue measurable.

Using densities we can define the essential interior $\mathring{E}^{\mathrm{M}}$, the essential closure $\overline{E}^{\mathrm{M}}$ and the essential boundary $\partial^{\mathrm{M}} E$ of a measurable set E as follows:

$$\mathring{E}^{\mathrm{M}} := \{x : D(x, E) = 1\}, \qquad \overline{E}^{\mathrm{M}} := \{x : \overline{D}(x, E) > 0\} \tag{B.6}$$

$$\partial^{\mathrm{M}} E := \overline{E}^{\mathrm{M}} \cap \overline{\mathbb{R}^N \setminus E}^{\mathrm{M}} = \{x : \overline{D}(x, E) > 0,\ \overline{D}(x, \mathbb{R}^N \setminus E) > 0\}. \tag{B.7}$$

Notice also that by the Lebesgue differentiation theorem the symmetric difference $\mathring{E}^{\mathrm{M}} \Delta E$ is Lebesgue negligible, hence the measure theoretic interior of $\mathring{E}^{\mathrm{M}}$ is $\mathring{E}^{\mathrm{M}}$ (in this sense $\mathring{E}^{\mathrm{M}}$ is essentially open), and also that

$$\partial^{\mathrm{M}} E = \mathbb{R}^N \setminus \left(\mathring{E}^{\mathrm{M}} \cup \overset{\circ}{\overline{\mathbb{R}^N \setminus E}}{}^{\mathrm{M}} \right).$$

Proposition B.27. *We have $\partial^* E \subseteq \partial^M E$ and*

$$\mathcal{H}^{N-1}\left(\partial^M E \setminus \partial^* E\right) = 0.$$

Theorem B.28. *Let $E \subseteq \mathbb{R}^N$ a set of locally finite perimeter. Then*

$$\mathcal{H}^{N-1}\left(\partial^M E \cap K\right) < \infty$$

for each compact set $K \subseteq \mathbb{R}^N$. Furthermore, for $\mathcal{H}^{N-1}$ a.e. $x \in \partial^M E$, there is a unique measure theoretic unit outer normal $\nu_E(x)$ such that

$$\int_E \operatorname{div}(\varphi)\, dx = \int_{\partial^M E} \varphi \cdot \nu_E \, d\mathcal{H}^{N-1}$$

for all $\varphi \in C_0^1(\mathbb{R}^N, \mathbb{R}^N)$.

B.7 Connected Components of Sets of Finite Perimeter

This section reviews some results on the decomposition of sets of finite perimeter into connected components, and we shall follow [9].

To simplify, the Lebesgue measure of a Lebesgue measurable set $E \subseteq \mathbb{R}^N$ will be denoted by $|E|$. Given A, $B \subseteq \mathbb{R}^N$, we shall write $E_1 = E_2$ (mod $\mathcal{H}^\alpha$) if $\mathcal{H}^\alpha(E_1 \Delta E_2) = 0$, where $E_1 \Delta E_2 = (E_1 \setminus E_2) \cup (E_2 \setminus E_1)$ is the symmetric difference of E_1 and E_2. We will use an analogous notation for the inclusion and in some cases, in order to simplify the notation, the equivalence or inclusion (mod $\mathcal{H}^N$) will be tacitly understood.

Let $E \subseteq \mathbb{R}^N$ be a set with finite perimeter. We say that E is *decomposable* if there exists a partition (A, B) of E such that $\mathrm{Per}(E) = \mathrm{Per}(A) + \mathrm{Per}(B)$ and both $|A|$ and $|B|$ are strictly positive. We say that E is *indecomposable* if it is not decomposable; notice that the properties of being decomposable or indecomposable are invariant (mod $\mathcal{H}^N$) and that, according to our definition, any Lebesgue negligible set is indecomposable. It was proved in [9] that any connected open set $\Omega \subseteq \mathbb{R}^N$ satisfying $\mathcal{H}^{N-1}(\partial^{\mathrm{M}}\Omega) < \infty$ is indecomposable.

The following decomposition theorem was proved in [9]; a similar decomposition result for integer currents is stated in 4.2.25 of [113]. This result has also been used in G. Dolzmann and S. Müller ([101]) and B. Kirchheim ([139]) to prove Liouville type theorems for a class of partial differential inclusions.

Theorem B.29 (Decomposition theorem). *Let E be a set with finite perimeter in $\mathbb{R}^N$. Then there exists a unique finite or countable family of pairwise disjoint indecomposable sets $\{E_i\}_{i\in I}$ such that $|E_i| > 0$ and $\mathrm{Per}(E) = \sum_i \mathrm{Per}(E_i)$. Moreover*

$$\mathcal{H}^{N-1}\left(\overset{\circ}{E}{}^{\mathrm{M}} \setminus \bigcup_{i\in I} \overset{\circ}{E}{}^{\mathrm{M}}_i\right) = 0 \tag{B.8}$$

and the E_i's are maximal indecomposable sets, i.e. any indecomposable set $F \subseteq E$ is contained (mod $\mathcal{H}^N$) *in some set E_i.*

Definition B.30 (M-connected components). In view of the previous theorem, we call the sets E_i the *M-connected components* of E and denote this family by $\mathcal{CC}^M$.

Notice that $\mathcal{CC}^M(E) = \emptyset$ whenever E is Lebesgue negligible and by the results in [9] we have

$$\partial^{\mathrm{M}} F \subset \partial^{\mathrm{M}} E \quad (\mathrm{mod}\ \mathcal{H}^{N-1}) \qquad \text{for any } F \in \mathcal{CC}^M(E). \tag{B.9}$$

The family $\mathcal{CC}^M(A)$ coincides with the family of connected components of A for any sufficiently regular open set A; moreover for any Lipschitz function $u : \mathbb{R}^N \to \mathbb{R}$ almost every upper level set $\{u > \lambda\}$ has this (weak) regularity property. In general an open indecomposable set needs not be connected: for instance a disk without a diameter is disconnected but indecomposable.

Theorem B.31. *Let $A \subseteq \mathbb{R}^N$ be an open set such that $\mathcal{H}^{N-1}(\partial A) = \mathcal{H}^{N-1}(\partial^{\mathrm{M}} A)$. Then $\mathcal{CC}^M(A)$ coincides with the family of connected components of A.*

B.7.1 Holes, saturation, simple sets

The decomposition theorem leads to reasonably good definitions of "hole" and "saturation" for a set of finite perimeter. These concepts permit us to recover a canonical decomposition of the measure theoretic boundary.

Definition B.32 (Holes, saturation). Let E be an indecomposable set. We call any M-connected component of $\mathbb{R}^N \setminus E$ with finite measure a *hole* of E. We define the *saturation* of E, denoted by $\mathrm{sat}(E)$, as the union of E and its holes. In the general case when E has finite perimeter, we define

$$\mathrm{sat}(E) := \bigcup_{i \in I} \mathrm{sat}(E_i), \qquad \text{where} \qquad \mathcal{CC}^M(E) = \{E_i\}_{i \in I}.$$

We call E *saturated* if $\mathrm{sat}(E) = E$.

Definition B.33 (Simple sets). Any indecomposable and saturated subset of $\mathbb{R}^N$ will be called *simple.*

Notice that the only simple set with infinite measure is $\mathbb{R}^N$ and that the saturation of any indecomposable set E is simple (actually, the smallest simple set containing E) ([9]).

B.7.2 Description of sets of finite perimeter in terms of their boundary

In general a decomposition in M-connected components does not lead directly to a canonical decomposition of the boundary. This goal can be achieved by looking to the saturations and to the holes of all M-connected components of E.

Definition B.34 (Exterior). If $E \subseteq \mathbb{R}^N$ has finite perimeter and $|E| < \infty$, we call the unique (mod $\mathcal{H}^N$) M-connected component of $\mathbb{R}^N \setminus E$ with infinite measure the *exterior* of E, denoted by $\mathrm{ext}(E)$.

Notice that the notion of exterior makes sense only if $|E| < \infty$, due to the fact that $\mathbb{R}^N \setminus E$ has finite measure if $\mathrm{Per}(E) < \infty$ and $|E| = \infty$.

Definition B.35 (Jordan boundary). We say that a set J is a Jordan boundary if there is a simple set E such that $J = \partial^{\mathrm{M}} E$ (mod $\mathcal{H}^{N-1}$).

According to [9], the simple set E associated to a Jordan boundary J is unique. In this sense, J can also be thought as an *oriented* set, with the orientation induced by the generalized inner normal to E. We shall write $\mathrm{int}(J) = E$ and $\mathrm{ext}(J) = \mathbb{R}^N \setminus E$; notice that $\mathrm{ext}(J) = \mathrm{ext}(E)$.

In order to simplify the following statements we enlarge the class of Jordan boundaries by introducing a *formal* Jordan boundary J_∞ whose interior is $\mathbb{R}^N$ and a *formal* Jordan boundary J_0 whose interior is empty; we also set $\mathcal{H}^{N-1}(J_\infty) = \mathcal{H}^{N-1}(J_0) = 0$ and denote by $\mathcal{S}$ this extended class of Jordan boundaries. This permits us to consider at the same time sets with finite and infinite measure and we can always assume that the list of components (or holes of the components) is infinite, possibly adding to it infinitely many $\mathrm{int}(J_0)$.

Proposition B.36. *Let E be indecomposable and let $\{Y_i\}_{i\in I}$ be its holes. Then*

$$E = \mathrm{sat}(E) \setminus \bigcup_{i\in I} Y_i = \mathrm{sat}(E) \cap \bigcap_{i\in I} \mathrm{ext}(Y_i) \tag{B.10}$$

and

$$\mathrm{Per}(E) = \mathrm{Per}(\mathrm{sat}(E)) + \sum_{i\in I} \mathrm{Per}(Y_i). \tag{B.11}$$

Conversely, let F be simple and let $\{G_i\}_{i\in I}$ be indecomposable sets such that

$$E = F \setminus \bigcup_{i\in I} G_i \tag{B.12}$$

and

$$\mathrm{Per}(E) = \mathrm{Per}(F) + \sum_{i\in I} \mathrm{Per}(G_i). \tag{B.13}$$

Then, $F = \mathrm{sat}(E)$ and $\{G_i\}_{i\in I}$ are the holes of E.

Theorem B.37 (Decomposition of $\partial^{\mathrm{M}} E$ in Jordan boundaries). *Let $E \subseteq \mathbb{R}^N$ be a set of finite perimeter. Then, there is a unique decomposition of $\partial^{\mathrm{M}} E$ into Jordan boundaries $\{J_i^+, J_k^- : i, k \in \mathbb{N}\} \subseteq \mathcal{S}$, such that:*

(i) *Given $\mathrm{int}(J_i^+)$, $\mathrm{int}(J_k^+)$, $i \neq k$, they are either disjoint or one is contained in the other; given $\mathrm{int}(J_i^-)$, $\mathrm{int}(J_k^-)$, $i \neq k$, they are either disjoint or one is contained in the other. Each $\mathrm{int}(J_i^-)$ is contained in one of the $\mathrm{int}(J_k^+)$.*

(ii) $\mathrm{Per}(E) = \sum_i \mathcal{H}^{N-1}(J_i^+) + \sum_k \mathcal{H}^{N-1}(J_k^-)$.

(iii) *If $\mathrm{int}(J_i^+) \subseteq \mathrm{int}(J_j^+)$, $i \neq j$, then there is some Jordan boundary J_k^- such that $\mathrm{int}(J_i^+) \subseteq \mathrm{int}(J_k^-) \subseteq \mathrm{int}(J_j^+)$. Similarly, if $\mathrm{int}(J_i^-) \subseteq \mathrm{int}(J_j^-)$, $i \neq j$, then there is some Jordan boundary J_k^+ such that $\mathrm{int}(J_i^-) \subseteq \mathrm{int}(J_k^+) \subseteq \mathrm{int}(J_j^-)$.*

(iv) *Setting $L_j = \{i : \mathrm{int}(J_i^-) \subseteq \mathrm{int}(J_j^+)\}$, the sets $Y_j = \mathrm{int}(J_j^+) \setminus \bigcup_{i\in L_j} \mathrm{int}(J_i^-)$ are pairwise disjoint, indecomposable and $E = \bigcup_j Y_j$.*

Theorem B.38. *Let $\{J_i^+, J_k^- : i, k \in \mathbb{N}\} \subset \mathcal{S}$ satisfy the conditions* (i), (iii) *of Theorem* B.37 *and assume* (ii′) *each two different Jordan boundaries of the system* $\{J_i^+, J_k^- : i,\, k \geq 0\}$ *are disjoint* (mod $\mathcal{H}^{N-1}$).

(iv′) $\sum_i \mathrm{Per}(J_i^+) + \sum_k \mathrm{Per}(J_k^-) < \infty$.

Let $E = \bigcup_j Y_j$, *where*

$$Y_j := \mathrm{int}(J_j^+) \setminus \bigcup_{i \in L_j} \mathrm{int}(J_i^-).$$

Then, E *is a set of finite perimeter and* $\partial^{\mathrm{M}} E = \bigcup_i J_i^+ \cup \bigcup_k J_k^-$ (mod $\mathcal{H}^{N-1}$).

B.7.3 Indecomposability and Jordan curves in the plane

We say that $\Gamma \subseteq \mathbb{R}^2$ is a *Jordan curve* if $\Gamma = \gamma([a,b])$ for some a, $b \in \mathbb{R}$ (with $a < b$) and some continuous map γ, one-to-one on $[a,b)$ and such that $\gamma(a) = \gamma(b)$. In a more geometric language, Γ can be viewed as the image of a continuous and one-to-one map defined on the unit circle $\mathbf{S}^1$. According to the celebrated Jordan curve theorem any Jordan curve Γ splits $\mathbb{R}^2 \setminus \Gamma$ in exactly two connected components, a bounded one and an unbounded one, whose common boundary is Γ. As for Jordan boundaries, these components will be respectively denoted by $\mathrm{int}(\Gamma)$ and $\mathrm{ext}(\Gamma)$.

Theorem B.39. *Let* E *be a subset of* $\mathbb{R}^2$ *of finite perimeter. Then, there is a unique decomposition of* $\partial^{\mathrm{M}} E$ *into rectifiable Jordan curves*

$$\{C_i^+, C_k^- : i,\, k \in \mathbb{N}\} \subset \mathcal{S},$$

such that:

(i) *Given* $\mathrm{int}(C_i^+)$, $\mathrm{int}(C_k^+)$, $i \neq k$, *they are either disjoint or one is contained in the other; given* $\mathrm{int}(C_i^-)$, $\mathrm{int}(C_k^-)$, $i \neq k$, *they are either disjoint or one is contained in the other. Each* $\mathrm{int}(C_i^-)$ *is contained in one of the* $\mathrm{int}(C_k^+)$.

(ii) $\mathrm{Per}(E) = \sum_i \mathcal{H}^1(C_i^+) + \sum_k \mathcal{H}^1(C_k^-)$.

(iii) *If* $\mathrm{int}(C_i^+) \subseteq \mathrm{int}(C_j^+)$, $i \neq j$, *then there is some rectifiable Jordan curve* C_k^- *such that* $\mathrm{int}(C_i^+) \subseteq \mathrm{int}(C_k^-) \subseteq \mathrm{int}(C_j^+)$. *Similarly, if* $\mathrm{int}(C_i^-) \subseteq \mathrm{int}(C_j^-)$, $i \neq j$, *then there is some rectifiable Jordan curve* C_k^+ *such that* $\mathrm{int}(C_i^-) \subseteq \mathrm{int}(C_k^+) \subseteq \mathrm{int}(C_j^-)$.

(iv) *Setting* $L_j = \{i : \mathrm{int}(C_i^-) \subseteq \mathrm{int}(C_j^+)\}$, *the sets* $Y_j = \mathrm{int}(C_j^+) \setminus \bigcup_{i \in L_j} \mathrm{int}(C_i^-)$ *are pairwise disjoint, indecomposable and* $E = \bigcup_j Y_j$.

The next result characterizes the M-connected components (or, better, suitable representatives in the equivalence class (mod $\mathcal{H}^2$)), by the classical topological property of connectedness by arcs. For that, we need another definition of boundary which, more than ∂^{M}, is suitable for the analysis of connected components. For any set E with finite perimeter in $\mathbb{R}^N$ let us define

$$\partial^S E := \left\{ x \in \mathbb{R}^N : \limsup_{r \to 0+} \frac{\mathcal{H}^{N-1}(\partial^{\mathrm{M}} E \cap B(x,r))}{r^{N-1}} > 0 \right\}.$$

Notice that the relative isoperimetric inequality, together with a continuity argument, gives that $\partial^{\mathrm{M}} E \subset \partial^{\mathrm{s}} E$ (see [9]); however (B.1) guarantees that $\mathcal{H}^{N-1}(\partial^{\mathrm{s}} E \setminus \partial^{\mathrm{M}} E) = 0$, hence $P(E) = \mathcal{H}^{N-1}(\partial^{\mathrm{s}} E)$ still holds.

Theorem B.40 (Indecomposability and connectedness by arcs). *Let $E \subset \mathbb{R}^2$ be a set of finite perimeter and let $\{E_i\}_{i \in I} = CC^M(E)$. Then, $\mathring{E}^{\mathrm{M}} \setminus \partial^{\mathrm{s}} E$ is the disjoint union of $\mathring{E}_i^{\mathrm{M}} \setminus \partial^{\mathrm{s}} E$ and $x,\, y \in \mathring{E}^{\mathrm{M}} \setminus \partial^{\mathrm{s}} E$ belong to the same M-connected component E_i of E if and only if there exists a rectifiable curve Γ joining x to y contained in $\mathring{E}^{\mathrm{M}} \setminus \partial^{\mathrm{s}} E$. Moreover, for any $\delta > 0$, Γ can be chosen so that*

$$\mathcal{H}^1(\Gamma) \leq \|x - y\| + \mathrm{Per}(E_i) + \delta.$$

In particular the sets $\mathring{E}_i^{\mathrm{M}} \setminus \partial^{\mathrm{s}} E$ are connected.

Appendix C

Pairings Between Measures and Bounded Functions

In this appendix we give some of the main points of the results about pairing between measures and bounded functions given by G. Anzellotti in [25] (see also [141]).

C.1 Trace of the Normal Component of Certain Vector Fields

It is well known that summability conditions on the divergence of a vector field z in Ω yield trace properties for the normal component of z on $\partial\Omega$. In this section we define a function $[z,\nu] \in L^\infty(\partial\Omega)$ which is associated to any vector field $z \in L^\infty(\Omega, \mathbb{R}^N)$ such that $\mathrm{div}(z)$ is a bounded measure in Ω.

Let Ω be an open set in $\mathbb{R}^N$, $N \geq 2$, and $1 \leq p \leq N$, $\frac{N}{N-1} \leq q \leq \infty$. We shall consider the following spaces:

$$
\begin{aligned}
BV(\Omega)_q &:= BV(\Omega) \cap L^q(\Omega) \\
BV(\Omega)_c &:= BV(\Omega) \cap L^\infty(\Omega) \cap C(\Omega) \\
X(\Omega)_p &:= \{z \in L^\infty(\Omega, \mathbb{R}^N) \,:\, \mathrm{div}(z) \in L^p(\Omega)\} \\
X(\Omega)_\mu &:= \{z \in L^\infty(\Omega, \mathbb{R}^N) \,:\, \mathrm{div}(z) \text{ is a bounded measure in } \Omega\}.
\end{aligned}
$$

In the next theorem we define a pairing $\langle z, u\rangle_{\partial\Omega}$, for $z \in X(\Omega)_\mu$ and $u \in BV(\Omega)_c$. We need the following result, which can be easily obtained by the same technique that Gagliardo uses in [114] in proving his extension theorem $L^1(\partial\Omega) \to W^{1,1}(\Omega)$.

Lemma C.1. *Let Ω be a bounded open set in $\mathbb{R}^N$ with Lipschitz boundary. Then, for any given function $u \in L^1(\partial\Omega)$ and for any given $\epsilon > 0$ there exists a function $w \in W^{1,1}(\Omega) \cap C(\Omega)$ such that*

$$w|_{\partial\Omega} = u,$$

$$\int_\Omega |\nabla w|\, dx \leq \int_{\partial\Omega} |u|\, d\mathcal{H}^{N-1} + \epsilon,$$

$$w(x) = 0 \quad \text{if} \quad \operatorname{dist}(x, \partial\Omega) > \epsilon.$$

Moreover, for any fixed $1 \leq q < \infty$, one can find the function w such that

$$\|w\|_q \leq \epsilon.$$

Finally, if one has also $u \in L^\infty(\partial\Omega)$, one can find w such that

$$\|w\|_\infty \leq \|u\|_\infty.$$

Theorem C.2. *Assume that $\Omega \subset \mathbb{R}^N$ is an open bounded set with Lipschitz boundary $\partial\Omega$. Denote by $\nu(x)$ the outward unit normal to $\partial\Omega$. Then there exists a bilinear map $\langle z, u\rangle_{\partial\Omega} : X(\Omega)_\mu \times BV(\Omega)_c \to \mathbb{R}$ such that*

$$\langle z, u\rangle_{\partial\Omega} = \int_{\partial\Omega} u(x) z(x) \cdot \nu(x)\, d\mathcal{H}^{N-1} \qquad \text{if} \quad z \in C^1(\Omega, \mathbb{R}^N), \tag{C.1}$$

$$|\langle z, u\rangle_{\partial\Omega}| \leq \|z\|_\infty \int_{\partial\Omega} |u(x)|\, d\mathcal{H}^{N-1} \qquad \text{for all} \quad z, u. \tag{C.2}$$

Proof. For $u \in BV(\Omega)_c \cap W^{1,1}(\Omega)$ and $z \in X(\Omega)_\mu$, we define

$$\langle z, u\rangle_{\partial\Omega} := \int_\Omega u \operatorname{div}(z)\, dx + \int_\Omega z \cdot \nabla u\, dx.$$

We remark that if $u, v \in BV(\Omega)_c \cap W^{1,1}(\Omega)$ and $u = v$ on $\partial\Omega$, then one has

$$\langle z, u\rangle_{\partial\Omega} = \langle z, v\rangle_{\partial\Omega} \qquad \text{for all} \quad z \in X(\Omega)_\mu.$$

In fact, by standard techniques in Sobolev spaces theory, we can find a sequence of functions $g_i \in \mathcal{D}(\Omega)$ such that, for all $z \in X(\Omega)_\mu$, one has

$$\begin{aligned}\langle z, u - v\rangle_{\partial\Omega} &= \int_\Omega (u - v) \operatorname{div}(z)\, dx + \int_\Omega z \cdot \nabla(u - v)\, dx \\ &= \lim_{i\to\infty} \left(\int_\Omega g_i \operatorname{div}(z)\, dx + \int_\Omega z \cdot \nabla g_i\, dx \right) = 0.\end{aligned}$$

Now, we define $\langle z, u\rangle_{\partial\Omega}$ for all $u \in BV(\Omega)_c$ by setting

$$\langle z, u\rangle_{\partial\Omega} = \langle z, w\rangle_{\partial\Omega},$$

where w is any function in $BV(\Omega)_c \cap W^{1,1}(\Omega)$ such that $u = w$ on $\partial\Omega$. This is a valid definition, in view of the preceding remark and because of Lemma C.1.

To prove (C.2), we take a sequence $u_n \in BV(\Omega)_c \cap C^\infty(\Omega)$ converging to u as in Theorem B.3 and we get

$$|\langle z, u\rangle_{\partial\Omega}| = |\langle z, u_n\rangle_{\partial\Omega}| \leq \left|\int_\Omega u_n \operatorname{div}(z)\, dx\right| + \|z\|_\infty \int_\Omega |\nabla u_n|\, dx$$

for all z and for all n. Hence, taking the limit when $n \to \infty$ we have

$$|\langle z, u\rangle_{\partial\Omega}| \leq \left|\int_\Omega u \operatorname{div}(z)\, dx\right| + \|z\|_\infty \int_\Omega \|Du\|.$$

Now, for a fixed $\epsilon > 0$ we consider a function w as in Lemma C.1. Then

$$|\langle z, u\rangle_{\partial\Omega}| = |\langle z, w\rangle_{\partial\Omega}| \leq \|w\|_\infty \int_{\Omega\setminus\Omega_\epsilon} |\operatorname{div}(z)| + \|z\|_\infty \left(\int_{\partial\Omega} |u|\, dx + \epsilon\right),$$

where $\Omega_\epsilon = \{x \in \Omega \ : \ \operatorname{dist}(x, \partial\Omega) > \epsilon\}$. Since $\operatorname{div}(z)$ is a measure of bounded total variation in Ω,

$$\lim_{\epsilon\to 0^+} \int_{\Omega\setminus\Omega_\epsilon} |\operatorname{div}(z)|\, dx = 0.$$

Consequently, (C.2) holds. □

Theorem C.3. *Let Ω be as in Theorem C.2. Then there exists a linear operator $\gamma : X(\Omega)_\mu \to L^\infty(\partial\Omega)$ such that*

$$\|\gamma(z)\|_\infty \leq \|z\|_\infty, \tag{C.3}$$

$$\langle z, u\rangle_{\partial\Omega} = \int_{\partial\Omega} \gamma(z)(x) u(x)\, d\mathcal{H}^{N-1} \quad \text{for all } \ u \in BV(\Omega)_c, \tag{C.4}$$

$$\gamma(z)(x) = z(x)\cdot\nu(x) \qquad \text{for all } \ x \in \partial\Omega \ \text{ if } \ z \in C^1(\overline{\Omega}, \mathbb{R}^N). \tag{C.5}$$

The function $\gamma(z)$ is a weakly defined trace on $\partial\Omega$ of the normal component of z. We shall denote $\gamma(z)$ by $[z, \nu]$.

Proof. Take a fixed $z \in X(\Omega)_\mu$. Consider the functional $F : L^\infty(\partial\Omega) \to \mathbb{R}$ defined by

$$F(u) := \langle z, w\rangle_{\partial\Omega},$$

where $w \in BV(\Omega)_c$ is such that $w|_{\partial\Omega} = u$. By estimate (C.2),

$$|F(u)| \leq \|z\|_\infty \, \|u\|_1.$$

Hence there exists a function $\gamma(z) \in L^\infty(\partial\Omega)$ such that

$$F(u) = \int_{\partial\Omega} \gamma(z)(x) u(x)\, d\mathcal{H}^{N-1}$$

and the result follows. □

Obviously, $X(\Omega)_p \subset X(\Omega)_\mu$ for all $p \geq 1$ and the trace $[z, \nu]$ is defined for all $z \in X(\Omega)_p$.

C.2 The Measure (z, Du)

Approximating by smooth functions and applying Green's formula, the following result can be deduced easily.

Proposition C.4. *Let Ω be as in Theorem C.2 and $1 \leq p \leq \infty$. Then, for all $z \in X(\Omega)_p$ and $u \in W^{1,1}(\Omega) \cap L^{p'}(\Omega)$, one has*

$$\int_\Omega u \operatorname{div}(z)\, dx + \int_\Omega z \cdot \nabla u\, dx = \int_{\partial\Omega} [z, \nu] u \, d\mathcal{H}^{N-1}. \tag{C.6}$$

In the sequel we shall consider pairs (z, u) such that one of the following conditions holds

$$\begin{cases} a)\ u \in BV(\Omega)_{p'},\ z \in X(\Omega)_p \quad \text{and} \quad 1 < p \leq N; \\ b)\ u \in BV(\Omega)_\infty,\ z \in X(\Omega)_1; \\ c)\ u \in BV(\Omega)_c,\ z \in X(\Omega)_\mu. \end{cases} \tag{C.7}$$

Definition C.5. Let z, u be such that one of the conditions C.7 holds. Then we define a functional $(z, Du) : \mathcal{D}(\Omega) \to \mathbb{R}$ as

$$\langle (z, Du), \varphi \rangle := -\int_\Omega u\varphi \operatorname{div}(z)\, dx - \int_\Omega uz \cdot \nabla\varphi\, dx.$$

Theorem C.6. *For all open sets $U \subset \Omega$ and for all functions $\varphi \in \mathcal{D}(U)$, one has*

$$|\langle (z, Du), \varphi \rangle| \leq \sup \|\varphi\|_\infty \, \|z\|_{L^\infty(U)} \int_U \|Du\|, \tag{C.8}$$

hence (z, Du) is a Radon measure in Ω.

Proof. Take a sequence $u_n \in C^\infty(\Omega)$ converging to u as in Theorem B.4. Take $\varphi \in \mathcal{D}(U)$ and consider an open set V such that $\operatorname{supp}(\varphi) \subset V \subset\subset U$. Then

$$|\langle (z, Du_n), \varphi \rangle| \leq \sup \|\varphi\|_\infty \, \|z\|_{L^\infty(U)} \int_V \|Du_n\| \quad \text{for all } \ n \in \mathbb{N}.$$

From here, taking the limit as $n \to \infty$, the result follows. □

We shall denote by $|(z, Du)|$ the measure total variation of (z, Du) and by $\int_B |(z, Du)|$, $\int_B (z, Du)$ the values of these measures on every Borel set $B \subset \Omega$.

As a consequence of the above theorem, the following result holds.

Corollary C.7. *The measures (z, Du), $|(z, Du)|$ are absolutely continuous with respect to the measure $\|Du\|$ and*

$$\left| \int_B (z, Du) \right| \leq \int_B |(z, Du)| \leq \|z\|_{L^\infty(U)} \int_B \|Du\|$$

for all Borel sets B and for all open sets U such that $B \subset U \subset \Omega$. Moreover, by the Radon–Nikodym theorem, there exists a $\|Du\|$-measurable function

$$\theta(z, Du, \cdot) : \Omega \to \mathbb{R}$$

such that

$$\int_B (z, Du) = \int_B \theta(z, Du, x) \, \|Du\| \qquad \text{for all Borel sets } \; B \subset \Omega$$

and

$$\|\theta(z, Du, \cdot)\|_{L^\infty(\Omega, \|Du\|)} \leq \|z\|_\infty .$$

Assume u, z satisfy one of the conditions (C.7). By writing

$$z \cdot D^s u := (z, Du) - (z \cdot \nabla u) \, d\mathcal{L}^N ,$$

we have that $z \cdot D^s u$ is a bounded measure. Furthermore, with an approximation argument to the one used in the proof of Theorem C.6, we have that $z \cdot D^s u$ is absolutely continuous with respect to $\|D^s u\|$ (and, thus, it is a singular measure respect to $\mathcal{L}^N$), and

$$|z \cdot D^s u| \leq \|z\|_\infty |D^s u|. \tag{C.9}$$

Lemma C.8. *Assume u, z satisfy one of the conditions* (C.7). *Let $u_n \in C^\infty(\Omega) \cap BV(\Omega)$ converging to u as in Theorem* B.3. *Then we have*

$$\int_\Omega z \cdot \nabla u_n \, dx \to \int_\Omega (z, Du).$$

Proof. For a given $\epsilon > 0$, we take an open set $U \subset\subset \Omega$ such that

$$\int_{\Omega \setminus U} \|Du\| < \epsilon.$$

Let $\varphi \in \mathcal{D}(\Omega)$ be such that $\varphi(x) = 1$ in U and $0 \leq \varphi \leq 1$ in Ω. Then

$$\left| \int_\Omega (z, Du_n) - \int_\Omega (z, Du) \right|$$

$$\leq |\langle (z, Du_n), \varphi \rangle - \langle (z, Du), \varphi \rangle| + \int_\Omega |(z, Du_n)|(1 - \varphi) + \int_\Omega |(z, Du)|(1 - \varphi).$$

Since

$$\lim_{n \to \infty} \langle (z, Du_n), \varphi \rangle = \langle (z, Du), \varphi \rangle,$$

$$\limsup_{n \to \infty} \int_\Omega |(z, Du_n)|(1 - \varphi) \leq \|z\|_\infty \limsup_{n \to \infty} \int_{\Omega \setminus U} \|Du_n\| < \epsilon \, \|z\|_\infty ,$$

$$\int_\Omega |(z, Du)|(1 - \varphi) \leq \epsilon \, \|z\|_\infty$$

and ϵ is arbitrary, the lemma follows. □

We give now the expected *Green's formula* relating the function $[z,\nu]$ and the measure (z,Du).

Theorem C.9. *Let Ω be a bounded open set in $\mathbb{R}^N$ with Lipschitz boundary and let z, u be such that one of the conditions* (C.7) *holds, then we have*

$$\int_\Omega u\,\mathrm{div}(z)\,dx + \int_\Omega (z,Du) = \int_{\partial\Omega} [z,\nu]u\,d\mathcal{H}^{N-1}. \tag{C.10}$$

Proof. Take a sequence of functions $u_n \in C^\infty(\Omega)\cap BV(\Omega)$ converging to u as in Theorem B.3. Then, by Lemma C.8 and Proposition C.4, we have

$$\begin{aligned}\int_\Omega u\,\mathrm{div}(z)\,dx + \int_\Omega (z,Du) &= \lim_{n\to\infty}\left(\int_\Omega u_n\,\mathrm{div}(z)\,dx + \int_\Omega z\cdot\nabla u_n\,dx\right)\\ &= \lim_{n\to\infty}\int_{\partial\Omega}[z,\nu]u_n\,d\mathcal{H}^{N-1} = \int_{\partial\Omega}[z,\nu]u\,d\mathcal{H}^{N-1}.\end{aligned}$$ □

Remark C.10. Observe that with a similar proof to that of Theorem C.9, in the case $\Omega = \mathbb{R}^N$, the following integration by parts formula, for z and w satisfying one of the conditions (C.7), holds:

$$\int_{\mathbb{R}^N} w\;\mathrm{div}(z)\;dx + \int_{\mathbb{R}^N}(z,Dw) = 0. \tag{C.11}$$

In particular, if Ω is bounded and has finite perimeter in $\mathbb{R}^N$, from (C.11) and Corollary C.7 it follows:

$$\int_\Omega \mathrm{div}(z)\,dx = \int_{\mathbb{R}^N}(z,-D\chi_\Omega) = \int_{\partial^*\Omega}\theta(z,-D\chi_\Omega,x)\;d\mathcal{H}^{N-1}. \tag{C.12}$$

Notice also that if $z_1,z_2\in X(\mathbb{R}^N)_p$ and $z_1 = z_2$ almost everywhere on Ω, then $\theta(z_1,-D\chi_\Omega,x) = \theta(z_2,-D\chi_\Omega,x)$ for $\mathcal{H}^{N-1}$-almost every $x\in\partial^*\Omega$.

If Ω is a bounded open set with Lipschitz boundary, then (C.12) has a meaning also if z is defined only on Ω and not on the whole of $\mathbb{R}^N$, precisely when $z\in L^\infty(\Omega;\mathbb{R}^N)$ with $\mathrm{div}(z)\in L^N(\Omega)$. In this case we mean that $\theta(z,-D\chi_\Omega,\cdot)$ coincides with $[z,\nu]$.

Remark C.11. Let $\Omega\subset\mathbb{R}^2$ be a bounded open set with Lipschitz boundary, and let $z_{\mathrm{inn}}\in L^\infty(\Omega;\mathbb{R}^2)$ with $\mathrm{div}(z_{\mathrm{inn}})\in L^2_{\mathrm{loc}}(\Omega)$, and $z_{\mathrm{out}}\in L^\infty(\mathbb{R}^2\setminus\overline{\Omega};\mathbb{R}^2)$ with $\mathrm{div}(z_{\mathrm{out}})\in L^2_{\mathrm{loc}}(\mathbb{R}^2\setminus\overline{\Omega})$. Assume that

$$\theta(z_{\mathrm{inn}},-D\chi_\Omega,x) = -\theta(z_{\mathrm{out}},-D\chi_{\mathbb{R}^2\setminus\overline{\Omega}},x)\qquad \text{for } \mathcal{H}^1-\text{a.e } x\in\partial\Omega.$$

Then if we define $z := z_{\mathrm{inn}}$ on Ω and $z := z_{\mathrm{out}}$ on $\mathbb{R}^2\setminus\overline{\Omega}$, we have $z\in L^\infty(\mathbb{R}^2;\mathbb{R}^2)$ and $\mathrm{div}(z)\in L^2_{\mathrm{loc}}(\mathbb{R}^2)$.

C.3 Representation of the Radon–Nikodym Derivative $\theta(z, Du, \cdot)$

This section is devoted to the problem of whether or not one can write

$$\theta(z, Du, x) = z(x) \cdot \frac{Du}{\|Du\|}(x) \tag{C.13}$$

where $\frac{Du}{\|Du\|}$ is the density function of the measure Du with respect to the measure $\|Du\|$.

For the sake of simplicity, we shall assume throughout this section that $z \in X(\Omega)_N$ and $u \in BV(\Omega)$, but it is clear that analogous results can be obtained for pairs (z, u) satisfying any of the conditions (C.7). First we have the following continuity result.

Proposition C.12. *Assume that*

$$z_n \rightharpoonup z \qquad \text{in } \ L^\infty(U) - \text{weak}^*, \tag{C.14}$$

$$\operatorname{div}(z_n) \rightharpoonup \operatorname{div}(z) \qquad \text{in } \ L^N(U) - \text{weak} \tag{C.15}$$

for all open sets $U \subset\subset \Omega$. Then, for all $u \in BV(\Omega)$, we have

$$(z_n, Du) \to (z, Du) \qquad \text{as measures in } \ \Omega \tag{C.16}$$

and

$$\theta(z_n, Du, \cdot) \rightharpoonup \theta(z, Du, \cdot) \quad \text{in } \ L^\infty(U) - \text{weak}^* \text{ for all } \ U \subset\subset \Omega. \tag{C.17}$$

Proof. By (C.14), for all $U \subset\subset \Omega$,

$$\sup_{n \in \mathbb{N}} \|z_n\|_{L^\infty(U)} = c(U) < +\infty.$$

Moreover,

$$\int_U |(z_n, Du)| \le \|z_n\|_{L^\infty(U)} \int_U \|Du\|.$$

Hence, it is sufficient to check the weak convergence (C.16) on $\mathcal{D}(\Omega)$ functions. Now, if $\varphi \in \mathcal{D}(\Omega)$, we have

$$\langle (z_n, Du), \varphi \rangle = -\int_\Omega u\varphi \operatorname{div}(z_n)\, dx - \int_\Omega u z_n \cdot \nabla\varphi\, dx \to \langle (z, Du), \varphi \rangle$$

and (C.16) is proved.

By Corollary C.7, we have

$$\|\theta(z_n, Du, \cdot)\|_{L^\infty(U, \|Du\|)} \le \|z_n\|_{L^\infty(U)} \le c(U).$$

Hence the convergence (C.17) has to be checked only on $C_c(\Omega)$ functions, now this is a consequence of (C.16). □

Using mollifiers it is easy to get the following result.

Lemma C.13. *For every function* $z \in X(\Omega)_N$, *there exists a sequence of functions* $z_n \in C^\infty(\Omega, \mathbb{R}^N) \cap L^\infty(\Omega, \mathbb{R}^N)$ *such that*

$$\|z_n\|_\infty \leq \|z\|_\infty \qquad \text{for all } \; n \in \mathbb{N},$$

$$z_n \rightharpoonup z \quad \text{in } \; L^\infty(\Omega, \mathbb{R}^N) - \text{weak}^* \;\text{ and in }\; L^p_{loc}(\Omega, \mathbb{R}^N) \;\text{ for }\; 1 \leq p < \infty,$$

$z_n(x) \to z(x)$ at every Lebesgue point x of z, and uniformly in sets of uniformly continuity for z,

$$\operatorname{div}(z_n) \to \operatorname{div}(z) \quad \text{in } \; L^N_{loc}(\Omega).$$

Now we give the representation result for $\theta(z, Du, \cdot)$.

Theorem C.14. *Assume that* $z \in X(\Omega)_N$ *and* $u \in BV(\Omega)$. *Then, we have*

$$\theta(z, Du, x) = z(x) \cdot \frac{Du}{\|Du\|}(x), \qquad \|D^a u\| - \text{a.e. in } \; \Omega. \tag{C.18}$$

Moreover, if $z \in C(\Omega, \mathbb{R}^N)$, *we have*

$$\theta(z, Du, x) = z(x) \cdot \frac{Du}{\|Du\|}(x), \qquad \|Du\| - \text{a.e. in } \; \Omega, \tag{C.19}$$

and consequently,

$$z \cdot D^s u = (z \cdot \overrightarrow{D^s u})\, |D^s u|. \tag{C.20}$$

Proof. Suppose first that $z \in C(\Omega, \mathbb{R}^N)$. (C.19) is equivalent to

$$\langle (z, Du), \varphi \rangle = \int_\Omega \varphi z \, Du \qquad \forall\, \varphi \in \mathcal{D}(\Omega). \tag{C.21}$$

Now, (C.21) is true by definition if $z \in C^1(\Omega, \mathbb{R}^N)$. If $z \in C(\Omega, \mathbb{R}^N)$, we take a sequence z_n as in Lemma C.13, and by Proposition C.12, for any $\varphi \in \mathcal{D}(\Omega)$, we have

$$\langle (z, Du), \varphi \rangle = \lim_{n\to\infty} \langle (z_n, Du), \varphi \rangle = \lim_{n\to\infty} \int_\Omega \varphi z_n \, Du = \int_\Omega \varphi z \, Du,$$

where, in the last step, we have used the fact that z_n converges uniformly to z on $\operatorname{supp}(\varphi)$.

Let us see now (C.18). This equality is equivalent to

$$\int_B \theta(z, Du, x) |\nabla u(x)| \, dx = \int_B z(x) \cdot \nabla u(x) \, dx \tag{C.22}$$

for all Borel sets $B \subset \Omega$. Let E^a and E^s be two disjoint Borel sets such that $E^a \cup E^s = \Omega$ and

$$\int_{E^s} \|D^a u\| = \int_{E^a} \|D^s u\| = 0.$$

Let $\epsilon > 0$ be fixed. Then, there exists a compact set $K \subset E^s$ such that

$$\int_{E^s \setminus K} \|D^s u\| < \epsilon. \tag{C.23}$$

Given a compact set $B_0 \subset E^a$, we can find an open set U with regular boundary, such that

$$B_0 \subset U \subset \Omega \setminus K, \qquad \int_{U \setminus B_0} \|Du\| < \epsilon$$

and by (C.23) it follows that

$$\int_U \|D^s u\| < \epsilon.$$

Take now a sequence $u_n \in C^\infty(U) \cap BV(U)$ approximating u as in Theorem B.3. Then, by Lemma C.8, it follows that

$$\begin{aligned}
&\left| \int_U \theta(z, Du, x)\, Du - \int_U z(x) \cdot \nabla u(x)\, dx \right| \\
&= \lim_{n \to \infty} \left| \int_U z(x) \cdot \nabla u_n(x)\, dx - \int_U z(x) \cdot \nabla u(x)\, dx \right| \\
&\leq \|z\|_\infty \lim_{n \to \infty} \int_U |\nabla u_n(x) - \nabla u(x)|\, dx \leq \|z\|_\infty \int_U \|D^s u\| \leq \epsilon \|z\|_\infty.
\end{aligned}$$

On the other hand, we have

$$\left| \int_U z(x) \cdot \nabla u(x)\, dx - \int_{B_0} z(x) \cdot \nabla u(x)\, dx \right| \leq \|z\|_\infty \int_{U \setminus B_0} \|Du\| \leq \epsilon \|z\|_\infty$$

and by Corollary C.7, we also have

$$\left| \int_U \theta(z, Du, x)\, \|Du\| - \int_{B_0} \theta(z, Du, x)\, \|Du\| \right| \leq \|z\|_\infty \int_{U \setminus B_0} \|Du\| \leq \epsilon \|z\|_\infty.$$

Therefore, we obtain that

$$\left| \int_{B_0} \theta(z, Du, x)\, \|Du\| - \int_{B_0} z(x) \cdot \nabla u(x)\, dx \right| \leq 3\epsilon \|z\|_\infty.$$

Hence (C.22) is proved for all compact sets $B_0 \subset E^a$. From which it follows, having in mind the regularity of the Radon measures, that (C.22) holds for all Borel subsets of Ω. $\square$

For later use we recall that by the coarea formula (Theorem B.15), if $u \in BV(\Omega)$ and $E_{u,t} := \{x \in \Omega : u(x) > t\}$, we have that

$$\frac{Du}{\|Du\|}(x) = \frac{D\chi_{E_{u,t}}}{\|D\chi_{E_{u,t}}\|}(x), \qquad \|D\chi_{E_{u,t}}\| - \text{a.e. in } \Omega$$

for $\mathcal{L}^1$ a.e. $t \in \mathbb{R}$.

In the next result we link the measure (z, Du) with the measure $(z, D\chi_{E_{u,t}})$.

Theorem C.15. *If $z \in X(\Omega)_N$ and $u \in BV(\Omega)$, then we have:*

(i) *For all functions $\varphi \in C_c(\Omega)$, the function $t \mapsto \langle (z, D\chi_{E_{u,t}}), \varphi\rangle$ is $\mathcal{L}^1$-measurable and*

$$\langle (z, Du), \varphi\rangle = \int_{-\infty}^{+\infty} \langle (z, D\chi_{E_{u,t}}), \varphi\rangle \, dt.$$

(ii) *For all Borel sets $B \subset \Omega$, the function $t \mapsto \int_B (z, D\chi_{E_{u,t}})$ is $\mathcal{L}^1$-measurable and*

$$\int_B (z, Du) = \int_{-\infty}^{+\infty} \left(\int_B (z, D\chi_{E_{u,t}}) \right) dt.$$

(iii) $\theta(z, Du, x) = \theta(z, D\chi_{E_{u,t}}, x)$ $\|D\chi_{E_{u,t}}\|$-a.e. *in Ω for $\mathcal{L}^1$-almost all $t \in \mathbb{R}$.*

Proof. (i) Take a sequence $z_n \in C^\infty(\Omega, \mathbb{R}^N) \cap L^\infty(\Omega, \mathbb{R}^N)$ converging to z as in Lemma C.13. By the coarea formula we have

$$\begin{aligned} \langle (z_n, Du), \varphi\rangle &= \int_\Omega z_n(x) \cdot \frac{Du}{\|Du\|}(x)\varphi(x) \, \|Du\| \\ &= \int_{-\infty}^{+\infty} \left(\int_\Omega z_n(x) \cdot \frac{D\chi_{E_{u,t}}}{\|D\chi_{E_{u,t}}\|}(x)\varphi(x) \, \|D\chi_{E_{u,t}}\| \right) dt \\ &= \int_{-\infty}^{+\infty} \langle (z_n, D\chi_{E_{u,t}}), \varphi\rangle \, dt. \end{aligned} \tag{C.24}$$

Since

$$|\langle (z_n, D\chi_{E_{u,t}}), \varphi\rangle| \leq \|z\|_\infty \|\varphi\|_\infty \int_\Omega \|D\chi_{E_{u,t}}\|, \qquad \forall \, n \in \mathbb{N},$$

having in mind Proposition C.12, by the dominated convergence theorem, taking the limit in (C.24) we get (i).

We shall prove (ii) after (iii). Let us prove (iii). For $a, b \in \mathbb{R}$, $a < b$, let $v = T_{a,b}(u)$ be. Then,

$$D\chi_{E_{u,t}} = D\chi_{E_{v,t}} \quad \text{and} \quad \frac{D\chi_{E_{u,t}}}{\|D\chi_{E_{u,t}}\|} = \frac{D\chi_{E_{v,t}}}{\|D\chi_{E_{v,t}}\|}, \quad \text{if } \; a \leq t < b$$

and

$$D\chi_{E_{v,t}} = 0 \qquad \text{if } \; t \geq b \; \text{ or } \; t < a,$$

from which it follows that

$$\frac{Du}{\|Du\|}(x) = \frac{D\chi_{E_{u,t}}}{\|D\chi_{E_{u,t}}\|}(x) = \frac{D\chi_{E_{v,t}}}{\|D\chi_{E_{v,t}}\|}(x) = \frac{Dv}{\|Dv\|}(x),$$

$\|D\chi_{E_{v,t}}\|$-a.e in Ω for $\mathcal{L}^1$-almost all $t \in \mathbb{R}$. Hence,

$$\frac{Du}{\|Du\|}(x) = \frac{Dv}{\|Dv\|}(x), \qquad \|Dv\| - \text{a.e in } \; \Omega.$$

From this, taking again the sequence z_n of the first part, we get

$$\theta(z_n, Du, x) = z_n(x) \cdot \frac{Du}{\|Du\|}(x) = \theta(z_n Dv, x), \qquad \|Dv\| - \text{a.e. in } \; \Omega, \;\; \forall \, n \in \mathbb{N}.$$

Then taking limit as $n \to \infty$, by the uniqueness of the limit in the $L^\infty(\Omega, \|Dv\|)$-weak* topology, we get

$$\theta(z, Du, x) = \theta(z, Dv, x), \qquad \|Dv\| - \text{a.e. in } \; \Omega. \tag{C.25}$$

Now, using statement (i) for v, we have, for a fixed $\varphi \in \mathcal{D}(\Omega)$,

$$\langle (z, Dv), \varphi \rangle = \int_{-\infty}^{+\infty} \langle (z, D\chi_{E_{v,t}}), \varphi \rangle \, dt.$$

From this, using (C.25) and the coarea formula, we obtain that

$$\int_a^b \left(\int_\Omega \theta(z, Du, x)\varphi(x) \, \|D\chi_{E_{v,t}}\| \right) dt$$
$$= \int_a^b \left(\int_\Omega \theta(z, D\chi_{E_{v,t}}, x)\varphi(x) \, \|D\chi_{E_{v,t}}\| \right) dt$$

and this implies that

$$\int_\Omega \theta(z, Du, x)\varphi(x) \, \|D\chi_{E_{v,t}}\| = \int_\Omega \theta(z, D\chi_{E_{v,t}}, x)\varphi(x) \, \|D\chi_{E_{v,t}}\|$$

for $\mathcal{L}^1$-almost all $t \in \mathbb{R}$. Then by a density argument, we finish the proof of (iii).

Finally, (ii) is a consequence of (iii) since

$$\int_b (z, du) = \int_B \theta(z, Du, x) \, \|Du\| = \int_{-\infty}^{+\infty} \left(\int_B \theta(z, Du, x) \, \|D\chi_{E_{u,t}}\| \right) dt$$
$$= \int_{-\infty}^{+\infty} \left(\int_B \theta(z, D\chi_{E_{u,t}}, x) \, \|D\chi_{E_{u,t}}\| \right) dt = \int_{-\infty}^{+\infty} \left(\int_B (z, D\chi_{E_{u,t}}) \right) dt.$$

□

Corollary C.16. *Assume that $z \in X(\Omega)_N$ and $u \in BV(\Omega)$. If $f : \mathbb{R} \to \mathbb{R}$ is a Lipschitz continuous increasing function, then*

$$\theta(z, D(f \circ u), x) = \theta(z, Du, x), \qquad \|Du\| - \text{a.e. in } \Omega. \tag{C.26}$$

Proof. Observe first that

$$E_{u,t} = \{x \in \Omega \,:\, u(x) > t\} = \{x \in \Omega \,:\, (f \circ u)(x) > f(t)\} = E_{fou,f(t)}.$$

Hence, for almost all $t \in \mathbb{R}$, we have

$$D\chi_{E_{u,t}} = D\chi_{E_{fou,f(t)}}.$$

Therefore,

$$\theta(z, Du, x) = \theta(z, D\chi_{E_{u,t}}, x) = \theta(z, D\chi_{E_{fou,f(t)}}, x) = \theta(z, D(f \circ u), x),$$

$\|D\chi_{E_{u,t}}\|$-a.e. in Ω for $\mathcal{L}^1$-almost all $t \in \mathbb{R}$, and consequently (C.26) follows. □

Bibliography

[1] R. Acar and C.R. Vogel, *Analysis of Total Variation Penalty Methods for Ill-Posed Problems,* Inverse Problems, 10 (1994), pp. 1217–1229.

[2] R.A, Adams, *Sobolev spaces*, Academic Press, New York, 1975.

[3] F. Alter, *Dualité et sous-différentielle pour une fonctionnelle convexe homogène positive.* Unpublished notes.

[4] F. Alter, V. Caselles and A. Chambolle, *Evolution of Convex Sets by the Minimizing Total Variation Flow,* Preprint 2002.

[5] L. Alvarez, Y. Gousseau and J.M. Morel, *The size of objects in natural images,* Advances in Imaging and Electron Physics, **111** (1999).

[6] L. Alvarez, F. Guichard, P.L. Lions, and J.M. Morel, *Axioms and fundamental equations of image processing*, Arch. Rational Mechanics and Anal. **16** (1993), pp. 200–257.

[7] L. Alvarez, P.L. Lions, and J.M. Morel, *Image selective smoothing and edge detection by nonlinear diffusion*, SIAM J. Numer. Anal. **29** (1992), pp. 845–866.

[8] L. Ambrosio, *Corso introduttivo alla teoria geometrica della misura ed alle superfici minime*, Quad. Scuola Norm. Sup. Pisa, Pantograph, Genova, 1997.

[9] L. Ambrosio, V. Caselles, S. Masnou and J.M. Morel, *Connected Components of Sets of Finite Perimeter and Applications to Image Processing*, J. European Math. Soc., **3** (2001), 39–92.

[10] L. Ambrosio, N. Fusco and D. Pallara, *Functions of Bounded Variation and Free Discontinuity Problems*, Oxford Mathematical Monographs, 2000.

[11] L. Ambrosio and E. Paolini, *Partial regularity for quasiminimizers of perimeter,* Ricerche Mat. **48** (1999), 167–186.

[12] F. Andreu, C. Ballester, V. Caselles and J.M. Mazón, *Minimizing Total Variation Flow*, C. R. Acad. Sci. Paris **331** (2000), 867–872.

[13] F. Andreu, C. Ballester, V. Caselles and J.M. Mazón, *Minimizing Total Variation Flow*, Diff. and Int. Eq. **14** (2001), 321–360.

[14] F. Andreu, C. Ballester, V. Caselles and J.M. Mazón, *The Dirichlet problem for the total variational flow*, J. Funct. Anal. **180** (2001), 347–403.

[15] F. Andreu, V. Caselles, J.I. Diaz, and J.M. Mazón, *Qualitative properties of the total variation flow*, J. Funct. Analysis **188** (2002), 516–547.

[16] F. Andreu, V. Caselles, and J.M. Mazón, *A parabolic quasilinear problem for linear growth functionals*, Rev. Mat. Iberoamericana **18** (2002), 135–185.

[17] F. Andreu, V. Caselles, and J.M. Mazón, *Existence and uniqueness of solution for a parabolic quasilinear problem for linear growth functionals with L^1 data*, Math. Ann. **322** (2002), 139–206.

[18] F. Andreu, V. Caselles, and J.M. Mazón, *The Cauchy Problem for Linear Growth Functionals with*, J. Evol. Equat. **3** (2003), 39–65.

[19] F. Andreu, V. Caselles, J.M. Mazón and S. Moll, *The Minimizing Total Variation Flow with Measure Initial Conditions*, To appear in Comm. Contemporary Math.

[20] F. Andreu, J.M. Mazón, S. Segura and J. Toledo, *Quasilinear Elliptic and Parabolic Equations in L^1 with Nonlinear Boundary Conditions*, Adv. in Math. Sci. and Appl. **7** (1997), 183–213.

[21] F. Andreu, J.M. Mazón, S. Segura and J. Toledo, *Existence and Uniqueness for a degenerate Parabolic Equation with L^1-data*, Trans. Amer. Math. Soc. **315** (1999), 285–306.

[22] H.C. Andrews and B.R. Hunt, *Digital Signal Processing,* Tech. Englewood Cliffs, NJ, Prentice Hall, 1977.

[23] S.N. Antonsev and J.I. Diaz, *New results on space and time localization of solutions of nonlinear elliptic or parabolic equations via energy methods*, Soviet math. Dokl. **303** (1988), 524–528.

[24] S.N. Antonsev, J.I. Díaz and S.I. Shmarev, *Energy methods for free boundary problems*, Birkhäuser, Boston, Basel, 2001.

[25] G. Anzellotti, *Pairings Between Measures and Bounded Functions and Compensated Compactness*, Ann. di Matematica Pura ed Appl. IV (135) (1983), 293–318.

[26] G. Anzellotti, *The Euler equation for functionals with linear growth*, Trans. Amer. Math. Soc. **290** (1985), 483–500.

[27] G. Anzellotti, *BV solutions of quasilinear PDEs in divergent form*, Commun. in Partial Differential Equations **12** (1987), 77–122.

[28] G. Aubert and P. Kornprobst, *Mathematical Problems in Image Processing*, Springer Verlag, 2002.

[29] C. Ballester, M. Bertalmio, V. Caselles, G. Sapiro and J. Verdera, *Filling-in by joint interpolation of vector fields and gray levels,* IEEE Trans. Image Process. **10**(8) (2001), 1200–1211.

[30] C. Ballester, V. Caselles, and J. Verdera, *Disocclusion by joint interpolation of vector fields and gray levels,* To appear in Multiscale Modeling and Simulation.

[31] A. Bamberger, *Etude d'une equation doublement non lineaire*, J. Functional Analysis **24** (1977), 148–155.

[32] V. Barbu, *Nonlinear Semigroups and Differential Equations in Banach Spaces,* Noordhoff International Publisher, 1976.

[33] G. Barles, H.M. Soner and P. Souganidis, *Front propagation and phase field theory,* J. Control Optim **31** (1993), 439–469.

[34] G. Bellettini, V. Caselles and M. Novaga, *The Total Variation Flow in* $\mathbb{R}^N$, J. Diff. Equations **184** (2002), 475–525.

[35] G. Bellettini, V. Caselles and M. Novaga, *Explicit solutions of the eigenvalue problem* $-div\frac{Du}{|Du|} = u$, Preprint 2003.

[36] G. Bellettini, M. Novaga, and M. Paolini, *On a crystalline variational problem, part I: first variation and global* L^∞*-regularity,* Arch. Rational Mech. Anal. **157** (2001), 165–191.

[37] G. Bellettini, M. Novaga, and M. Paolini, *On a crystalline variational problem, part II: BV-regularity and structure of minimizers on facets,* Arch. Rational Mech. Anal. **157** (2001), 193–217.

[38] G. Bellettini, M. Novaga, and M. Paolini, *Characterization of facet breaking for nonsmooth mean curvature flow in the convex case,* Interfaces and Free Boundaries, **3** (2001), 415–446.

[39] G. Bellettini and M. Paolini, *Anisotropic motion by mean curvature in the context of Finsle geometry,* Hokkaido Mathematical J. **25** (1996), 537–566.

[40] Ph. Bénilan, *Equations d'évolution dans un espace de Banach quelconque et applications,* Thése Orsay, 1972.

[41] Ph. Bénilan, L. Boccardo, T. Gallouet, R. Gariepy, M. Pierre and J.L Vazquez, *An L^1-Theory of Existence and Uniqueness of Solutions of Nonlinear Elliptic Equations*, Ann. Scuola Normale Superiore di Pisa, IV, Vol. XXII (1995), 241–273.

[42] Ph. Bénilan and F. Bouhsiss, *Une remarque sur l'unicité des solutions pour l'opérateur de Serrin*, C. R. Acad. Sci. Paris **324** (1997), 611–616.

[43] Ph. Bénilan and M.G. Crandall, *Regularizing Effects in Homogeneous Equations*, in Contributions to Analysis and Geometry, D.N. Clark et al. editors, John Hopkins University Press, 1981, pp. 23–39.

[44] Ph. Bénilan and M.G. Crandall, *Completely Accretive Operators*, in Semigroups Theory and Evolution Equations, Ph. Clement et al. editors, Marcel Dekker, 1991, pp. 41–76.

[45] Ph. Bénilan, M.G. Crandall and A. Pazy, *Evolution Equations Governed by Accretive Operators*, Forthcoming.

[46] S. Bernstein, *Sur les équations du calcul des variations*, Ann. Sci. Ecole Norm. Sup. **29** (1912), 431–485.

[47] J.G. Berryman and C.J. Holland, *Stability of the Separable Solution for Fast Diffusion*, Arch. Rational. Mech. Anal. **74** (1980), 279–288.

[48] M. Bertalmio, V. Caselles, B. Rougé and A. Solé, *TV based image restoration with local constraints*, Journal of Scientific Computing **19**(1–3) (2003), 95–122.

[49] D. Blanchard and F. Murat, *Renormalized solutions of nonlinear parabolic problems with L^1-data: existence and uniqueness*, Proc. Roy Soc. Edinburgh Sect A. **127** (1997), 1137–1152.

[50] L. Blanc-Féraud, P. Charbonnier, G. Aubert and M. Barlaud, *Nonlinear image processing: modeling and fast algorithm for regularization with edge detection*, Proceedings of the International Conference on Image Processing, 1995, Vol. 1 , 1995, pp. 474–477.

[51] P. Blomgren and T.F. Chan, *Color TV: total variation methods for restoration of vector-valued images*, IEEE Transactions on Image Processing, **7**(3) (1998), 304–309.

[52] P. Blomgren, T. Chan, P. Mulet and C.K. Wong, *Total Variation Image Restoration: Numerical Methods and Extensions,* Proceedings Int. Conf. Image Processing, ICIP 97, Vol. III (1997), pp. 384–387.

[53] P. Blomgren, T. F. Chan, and P. Mulet, *Extensions to Total Variation Denoising*, Technical report, UCLA Cam report 97-42, Department of Mathematics, UCLA, Los Angeles, CA, September 1997.

[54] L. Boccardo T. Gallouët, *Nonlinear elliptic equations with right-hand side measures*, Comm. in Partial Diff. Equat. **17** (1992), 641–655.

[55] L. Boccardo D. Giachetti, J.I. Diaz and F. Murat, *Existence and regularity of renormalized solutions for some elliptic problems involving derivatives of nonlinear term*, J. Diff. Equat. **106** (1993), 215–237.

[56] E. Bombieri, E. de Giorgi and M. Miranda, *Una maggiorazione a priori relative alle ipersuperfici minimalli non parametriche*, Arch. Rat. Mech. Anal. **32** (1965), 255–267.

[57] A. Bressan, *Hyperbolic Systems of Conservation Laws: The One-Dimensional Cauchy Problem*, Oxford University Press, 2000.

[58] H. Brezis, *Operateurs Maximaux Monotones*, North Holland, Amsterdam, 1973.

[59] H. Brezis and W. Strauss, *Semilinear elliptic equations in L^1*, J. Math. Soc. Japan **25** (1973), 565–590.

[60] R.E. Bruck, *Asymmptotic convergence of nonlinear contraction semi-groups in Hilbert spaces*, J. Functional Anal. **18** (1975), 15–26.

[61] J. Carrillo, *On the uniquenes of solution of the evolution dam problem*, Nonlinear Anal. **22** (1994), 573–607.

[62] E. Casas, K. Kunisch, and C. Pola, *Regularization by functions of bounded variation and applications to image enhancement*, Appl. Math. Optim. **40** (1999), no. 2, 229–257.

[63] K.R. Castleman, *Digital Image Processing,* Prentice Hall 1996.

[64] A. Chambolle and P.L. Lions, *Image Recovery via Total Variation Minimization and Related Problems*, Numer. Math. **76** (1997), 167–188.

[65] A. Chambolle, R.A. de Vore, N. Lee and B.J. Lucier, *Nonlinear Wavelet image Processing: Variational Problems, Compression and Noise Removal Through Wavelet Shrinkage,* IEEE Transactions on Image Processing, Vol. 7 (1998), pp. 319–335.

[66] A. Chambolle, *An algorithm for Total Variation Minimization and Applications,* Preprint 2002.

[67] T.F. Chan, R. H. Chan, and H. M. Zhou, *A continuation method for total variation denoising problems,* Proceedings of the SPIE Conference on Advanced Signal Processing Algorithms (SPIE Code No. 2563), July, 1995.

[68] T.F. Chan, G. Golub and P. Mulet, *A nonlinear primal-dual method for total variation-based image restoration.* ICAOS '96 (Paris, 1996), 241–252, Lecture Notes in Control and Inform. Sci., 219, Springer, London, 1996.

[69] T.F. Chan and P. Mulet, *Iterative methods for total variation image restoration*, Iterative methods in scientific computing (Hong Kong, 1995), 359–381, Springer, Singapore, 1997.

[70] T.F. Chan and P. Mulet, *On the convergence of the lagged diffusivity fixed point method in total variation image restoration*, SIAM J. Numer. Anal. **36**, (1999), 354–367.

[71] T.F. Chan, G.H. Golub and P. Mulet, *A Nonlinear Primal-Dual Method for Total Variation Based Image Restoration*, SIAM J. Sci. Comput. **20** (1999), 1964–1977.

[72] T.F. Chan, G.H. Golub and P. Mulet, *Total Variation Image Restoration: Numerical Methods and Extensions*, Proceedings International Conference on Image Processing, ICIP-97, October 26-29, 1997, Santa Barbara, California, Vol. III, pp. 384–387.

[73] T.F. Chan and C.K. Wong, *Total variation blind deconvolution*, IEEE Transactions on Image Processing, **7**(3) (1998), 370–375.

[74] T.F. Chan, A. Marquina and P. Mulet, *High-order total variation-based image restoration*, SIAM J. Sci. Comput. **22** (2000), 503–516.

[75] T.F. Chan, and J. Shen, *Variational restoration of nonflat image features: models and algorithms*, SIAM J. Appl. Math. **61** (2000), 1338–1361.

[76] T.F. Chan, S. Osher, and J. Shen, *The digital TV filter and nonlinear denoising*, IEEE Transactions on Image Processing, **10**(2) (2001), 231–241.

[77] P. Charbonnier, L. Blanc-Feraud, G. Aubert and M. Barlaud, *Deterministic Edge-Preserving Regularization in Computed Imaging*, IEEE Transactions on Image Processing, **6**(2) (1997), 298–311.

[78] P. Charbonnier, L. Blanc-Féraud, G. Aubert and M. Barlaud, *Two deterministic half-quadratic regularization algorithms for computed imaging*, Proceedings of IEEE International Conference on Image Processing, 1994, ICIP-94, Vol. 2 (1994), pp. 168–172.

[79] G. Chavent and K. Kunisch, *Regularization of linear least squares problems by total bounded variation*, ESAIM Control Optim. Calc. Var. **2** (1997), 359–376.

[80] P.G. Ciarlet, *Introduction to Numerical Linear Algebra and Optimization*, Cambridge Univ. Press, 1988.

[81] A. Cohen, W. Dahmen, I. Daubechies, and R. De Vore, *Harmonic analysis and the space $BV(R^2)$*, Preprint, 2000.

[82] A. Cohen, R. De Vore, P. Petrushev, and H. Xu, *Nonlinear approximation and the space BV.* Amer. J. Math. **121** (1999), 587–628.

[83] M.G. Crandall, *Semigroups of Nonlinear Transformations in Banach Spaces,* In Contribution to Nonlinear Functional Analysis (E.H. Zarantonello, ed.) Academic Press, New York 1971, 149–179.

[84] M.G. Crandall, *Nonlinear Semigroups and Evolution Governed by Accretive Operators,* In Proceeding of Symposium in Pure Mat., Part I (F. Browder, ed.) A.M.S., Providence 1986, 305–338.

[85] M.G. Crandall and T.M. Liggett, *Generation of Semigroups of Nonlinear Transformations on General Banach Spaces,* Amer. J. Math. **93** (1971), 265–298.

[86] E. De Giorgi, L. Ambrosio, *Un nuovo tipo di funzionale del calcolo delle variazioni,* Atti Accad. Naz. Lincei Rend. Cl. Sci. Mat Fis. Natur. s. 8 **82** (1988), 199–210.

[87] E. De Giorgi, M. Carriero, and A. Leaci, *Existence theorem for a minimum problem with free discontinuity set,* Arch. Rational Mech. Anal. **108**(3) (1989), 195–218.

[88] M. C. Delfour and J. P. Zolesio, *Shape Analysis via Oriented Distance Functions,* J. Functional Anal. **123** (1994), 129–201.

[89] F. Demengel, *Functions locally almost 1-harmonic,* Prépublication de l'Université Cergy Pontoise. No. 31, 2001.

[90] F. Demengel and R. Temam, *Convex Functions of a Measure and Applications,* Indiana Univ. Math. J. **33** (1984), 673–709.

[91] G. Demoment, *Image reconstruction and restoration: Overview of common estimation structures and problems,* IEEE Transactions on Acoustics, Speech and Signal Processing, Vol. 37, 12 (1989), pp. 2024–2036.

[92] R. De Vore and B.J. Lucier, *Fast wavelet techniques for near optimal image compression.* IEEE Military Communications Conference, 1992.

[93] J.I. Díaz and M.A. Herrero, *Propertiés de support compact pour certaines équations elliptiques et paraboliques non linéaires,* C. R. Acad. Sci. Paris, **286**, (1978), 815–817.

[94] J.I. Díaz and M.A. Herrero, *Estimates on the support of the solutions of some nonlinear elliptic and parabolic problems,* Proceedings of the Royal Society of Edinburgh, **89A**, (1981), 249–258.

[95] J.I. Díaz and A. Liñán, *Movimiento de descarga de gases en conductos largos: modelización y estudio de una ecuación doblemente no lineal.* In the book *Reunión Matemática en Honor de A.Dou* (J.I.Díaz y J.M.Vegas eds.) Universidad Complutense de Madrid. 1989, 95–119.

[96] F. Dibos and G. Koepfler, *Global total variation minimization,* SIAM J. Numer. Anal. **37** (2000), 646–664.

[97] U. Dierkes, S. Hildebrandt, A Küster and O. Wohlrab, *Minimal Surfaces,* Vol I, II, Springer-Verlag, 1992.

[98] J. Diestel and J.J. Uhl, *Vector Measures,* Math. Surveys **15**, Amer. Math. Soc., Providence, 1977.

[99] R.J. DiPerna and P.L. Lions, *On the Cauchy problem for Boltzmann equation: global existence and weak stability,* Ann. Math. **130** (1989), 321–366.

[100] D.C. Dobson, and C.R. Vogel, *Convergence of an iterative method for total variation denoising,* SIAM J. Numer. Anal. **34** (1997), 1779–1791.

[101] G. Dolzmann and S. Müller, *Microstructures with finite surface energy: the two-well problem,* Arch. Rational Mech. Anal. **132** (1995), 101–141.

[102] D. Donoho, *Denoising via soft-threholding,* IEEE Transactions Inf. Theory **41** (1995), 613–627, .

[103] D. Donoho, *Nonlinear solution of linear inverse problems by wavelet-vaguelette decomposition,* Applied and Computational Harmonic Analysis **2** (1995), 101–126, .

[104] D. Donoho and I.M. Johnstone, *Ideal spatial adaptation by wavelet shrinkage,* Biometrika, 81(3), (1994), pp. 425–455.

[105] D. Donoho, I. Johnstone, G. Kerkyacharian, and D. Picard, *Wavelet shrinkage: Asymptopia,* Journal R. Stat. Soc. B **57** (1995), 301–369.

[106] D.L. Donoho, and I.M. Johnstone, *Minimax estimation via wavelet shrinkage,* Ann. Statist. **26**(3) (1998), 879–921.

[107] D.L. Donoho, M. Vetterli, R.A. DeVore and I. Daubechies, *Data compression and harmonic analysis. Information theory: 1948–1998,* IEEE Trans. Inform. Theory **44**(6) (1998), 2435–2476.

[108] S. Durand, F. Malgouyres and B. Rougé, *Image Deblurring, Spectrum Interpolation and Application to Satellite Imaging,* ESAIM Control Optim. Calc. Var. **5** (2000), 445–475.

[109] I. Ekeland and R. Temam, *Convex Analysis and Variational Problems,* North Holland, Amsterdam, 1976.

[110] L.C. Evans and R.F. Gariepy, *Measure Theory and Fine Properties of Functions*, Studies in Advanced Math., CRC Press, 1992.

[111] L.C. Evans and J. Spruck, *Motion of level sets by mean curvature I*, J. Diff. Geometry **33** (1991), 635–681.

[112] L.C. Evans and J. Spruck, *Motion of level sets by mean curvature II*, Trans. Amer. math. Soc, **330** (1992), 321–332.

[113] H. Federer, *Geometric Measure Theory*, Springer Verlag, 1969.

[114] E. Gagliardo, *Caratterizzazione delle trace sulla frontiera relative ad alcune classi di funzioni in n variabili*, Rend. Sem. mat. Padova **27** (1957), 284–305.

[115] G. Gagneux and M. Madaune-Tort, *Unicité des solutions faibles d'équations de diffusion-convection*, C. R. Acad. Sc. Paris Sér I Math. **318** (1994), 919–924.

[116] D. Geman and G. Reynolds, *Constrained Image Restoration and Recovery of Discontinuities*, IEEE Trans. Pattern Anal. Machine Intell., 14 (1992), 367–383.

[117] D. Geman and C. Yang, *Nonlinear image recovery with half-quadratic regularization*, IEEE Trans. Image Processing, **4**(7) (1995), 932–946.

[118] M. Giaquinta, G. Modica and J. Soucek, *Functionals with linear growth in the calculus of variations I*, Comment. Math. Univ. Carolinae **20** (1979), 143–156.

[119] M. Giaquinta, G. Modica and J. Soucek, *Cartesian Currents in the Calculus of Variations II, Variational Integrals*, Springer Verlag, 1997.

[120] M-H. Giga and Y. Giga, *Evolving graph by singular nonlocal weighted curvature*, Arch. Rational Mech. Anal. **141** (1998), 117–198.

[121] M-H. Giga and Y. Giga, *Stability for evolving graph by singular nonlocal weighted curvature*, Comm. in Partial Differential equations **24** (1999), 109–184.

[122] E. Giusti, *Boundary Value Problems for Non-Parametric Surfaces of Prescribed Mean Curvature*, Ann. Scuola Normale Sup. di Pisa (4) **3** (1976), 501–548.

[123] E. Giusti, *On the equation of surfaces of prescribed mean curvature. Existence and uniqueness without boundary conditions*, Invent. Math. **46** (1978), 111–137.

[124] E. Giusti, *Minimal Surface and Functions of Bounded variation*, Birkhäuser, Basel, 1984.

[125] G.H. Golub, T. F. Chan and P. Mulet. *A nonlinear primal-dual method for total variation based image restoration,* SIAM J. Sci. Computing **20** (1999), 1964–1977.

[126] Y. Gousseau, and J.M. Morel, *Are natural images of bounded variation?,* SIAM J. Math. Anal. **33** (2001), 634–648.

[127] C.W. Groetsch, *The Theory of Tikhonov Regularization for Fredholm Integral Equations of the First Kind,* Pitman, Boston, 1984.

[128] H. Jenkins and J. Serrin, *The Dirichlet problem for minimal surface equation in higher dimensions,* J. Reine Angew. Math. **229** (1968), 170–187.

[129] R. Hardt and X. Zhou, *An evolution problem for linear growth functionals,* Commun. Partial Differential Equations **19** (1994), 1879–1907.

[130] E. Hecht and A. Zajac, *Optica,* Addison Wesley Iberoamericana, 1986.

[131] M. Herrero and J.L. Vazquez, *Asymptotic behaviour of the solutions of a strongly nonlinear parabolic problem,* Annales Fac. Sci. Toulouse **3** (1981), 113–127.

[132] M. Herrero and J.L. Vazquez, *On the propagation properties of a nonlinear degenerate parabolic equation,* Comm. in Part. Diff. Equations **7** (1982), 1381–1402.

[133] L. Hörmander, *The Analysis of Linear Partial Differential Operators I,* Springer-Verlag, 1983.

[134] L. Hörmander, *Notions of Convexity.* Birkhäuser, Boston, 1994.

[135] B.R. Hunt, *The Application of Constrained Least Square Estimation to Image Restoration by Digital Computer,* IEEE Trans. Comp., Vol. C-22 (1973), pp. 805–812.

[136] K. Ito and K. Kunisch, *An active set strategy based on the augmented Lagrangian formulation for image restoration,* M2AN Math. Model. Numer. Anal. **33** (1999), no. 1, 1–21.

[137] T. Kailath, *A view of three decades of linear filtering theory,* IEEE Trans. on Information Theory, Vol. 20, No. 2 (1974).

[138] D. Kinderlehrer and G. Stampacchia, *An Introduction to Variational Inequalities and their Applications,* Academic Press, 1980.

[139] B. Kirchheim, *Lipschitz minimizers of the 3-well problem having gradients of bounded variation,* Forthcoming.

[140] R. Kobayashi and Y. Giga, *Equations with Singular Diffusivity,* Journ. Statistical Physics **95** (1999), 1187–1220.

[141] R. Kohn and R. Temam, *Dual space of stress and strains with application to Hencky plasticity,* Appl. Math. Optim. **10** (1983), 1–35.

[142] Y. Komura, *Nonlinear Semigroups in Hilbert Spaces.* J. Math. Soc. Japan **19** (1967), 493–507.

[143] S.N. Kruzhkov, *First order quasilinear equations in several independent variables,* Math. USSR-Sb. **10** (1970), 217–243.

[144] J. Leray and J.L. Lions, *Quelques reésultats de Visik sur le problémes elliptiques non linéaires par le méthodes de Minty-Browder,* Bul. Soc. Math. France **93** (1965), 97–107.

[145] A. Lichnewski and R. Temam, *Pseudosolutions of the Time Dependent Minimal Surface Problem,* Journal of Differential Equations **30** (1978), 340–364.

[146] P.L. Lions, S. Osher and L. Rudin, *Denoising and Deblurring using Constrained Nonlinear Partial Differential Equations,* Tech. Repport, Cognitech Inc., Santa Monica, CA, 1992, submitted to SINUM.

[147] P.L. Lions and F. Murat, *Renormalized solutions of nonlinear elliptic equation,* (unpublished paper)

[148] F. Malgouyres and F. Guichard, *Edge direction preserving image zooming: a mathematical and numerical analysis,* SIAM J. Numer. Anal. **39** (2001), 1–37.

[149] F. Malgouyres, *A framework for image deblurring using wavelet packet bases,* Appl. Comput. Harmon. Anal. **12** (2002), 309–331.

[150] F. Malgouyres, *A unified framework for image restoration,* Preprint 2001.

[151] S. Mallat, *A wavelet tour of signal processing,* Academic Press, Inc., San Diego, CA, 1998.

[152] A. Marquina and S. Osher, *Explicit algorithms for a new time dependent model based on level set motion for nonlinear deblurring and noise removal,* SIAM J. Sci. Comput. **22** (2000), 387–405.

[153] Y. Meyer, *Oscillating patterns in image processing and nonlinear evolution equations,* The fifteenth Dean Jacqueline B. Lewis memorial lectures. University Lecture Series, 22. American Mathematical Society, Providence, RI, 2001.

[154] G. Minty, *Monotone (nonlinear) operators in Hilbert space,* Duke Math. **29** (1962), 341–346.

[155] L. Moisan, *Extrapolation de spectre et variation totale ponderée,* Proceedings of GRETSI, 2001.

[156] J.M. Morel and S. Solimini, *Variational Methods in Image Processing,* Birkhäuser, Boston, 1994.

[157] C.B. Jr. Morrey, *Multiple Integrals in the Calculus of Variations*, Springer-Verlag, 1966.

[158] D. Mumford and J. Shah, *Optimal approximations by Piecewise Smooth Functions and Associated Variational Problems,* Communications on Pure and Applied Mathematics, **17** (1989), 577–685.

[159] D. Mumford and B. Gidas, *Stochastic Models for Generic Images*, Quart. Appl. Math. **59**(1) (2001), 85–111.

[160] F. Murat, *Soluciones renormalizadas de EDP elípticas no lineales*, (Technical report R93023, Laboratoire d'Analyse Numérique, Paris VI, 1993).

[161] S.G. Nash and A. Sofer, *Linear and Nonlinear Programming*, McGraw-Hill, 1996.

[162] M. Nikolova, *Local strong homogeneity of a regularized estimator,* SIAM J. Appl. Math. **61** (2000), 633–658.

[163] F. Oru, *Le role des oscillations dans quelques problèmes d'analyse non-linéaire.* PhD thesis, Thèse CMLA, ENS-Cachan, France, 1998.

[164] S. Osher and J.A. Sethian, *Fronts propagating with curvature-dependent speed: Algorithms base on Hamilton–Jacobi formulations* J. of Comp. Phys. **79** (1988), 12–49.

[165] S. Osher and L. Vese, *Numerical Methods for $p-$ harmonic Flows and Applications to Image Processing,* CAM report 01-22, UCLA, 2001.

[166] S.J. Osher, A. Solé and L. Vese, *Image decomposition and restoration using total variation minimization and the H^{-1} norm,* Preprint 2002.

[167] D.L. Phillips, *A Technique for the Numerical Solution of Certain Integral Equations of the First Kind,* Journal of the ACM., 9 (1962), pp. 84–97.

[168] A. Prignet, *Existence and uniqueness of entropy solutions of parabolic problems with L^1 data,* Nonlinear Analysis TMA **28** (1997), 1943–1954.

[169] Yu. G. Reshetnyak, *Weak convergence of completely additive vector functions on a set*, Sibirsk. Mat. Z. **9** (1968), 1386–1394. (Translated)

[170] R.T. Rockafellar, *Duality and stability in extremum problem involvimg convex functions*, Pacific J. Math. **21** (1967), 167–187.

[171] J.G. Rosen, *The Gradient Projection Method for Nonlinear Programming. Part II, Nonlinear Constraints,* J. Soc. Indust. Appl. Math. 9 (1961), 514–532.

[172] B. Rougé, *Théorie de l'echantillonage et satellites d'observation de la terre,* Analyse de Fourier et traitement d'images, Journées X-UPS 1998.

[173] L. Rudin, *Images, Numerical Analysis of Singularities and Shock Filters,* Ph. D. Thesis, Caltech 1987.

[174] L. Rudin and S. Osher, *Total Variation based Image Restoration with Free Local Constraints*, Proc. of the IEEE ICIP-94, vol. 1, Austin, TX, 1994, pp. 31–35.

[175] L. Rudin, S. Osher and E. Fatemi, *Nonlinear Total Variation based Noise Removal Algorithms*, Physica D.**60** (1992), 259–268.

[176] C. Samson, L. Blanc-Feraud, G. Aubert and J.A. Zerubia, *A variational model for image classification and restoration*, IEEE Transactions on Pattern Analysis and Machine Intelligence, **22**(5) (2000), 460–472.

[177] G. Sapiro and V. Caselles, *Histogram Modification via Differential Equations*, Journal of Differential Equations **135** (1997), 238–268.

[178] G. Sapiro and A. Tannenbaum, *On Affine Plane Curve Evolution*, Journal of Functional Analysis, **119** (1994), 79–120.

[179] L. Schwartz, *Analyse IV. Applications a la théorie de la mesure*, Hermann, 1993.

[180] J. Serra, *Image analysis and mathematical morphology*, Academic Press, 1982.

[181] J. Serra, *Image analysis and mathematical morphology. Volume 2: Theoretical Advances*, Academic Press, 1988.

[182] J. Serrin, *A new definition of the integral for non-parametric problems in the Clculus of Variations*, Acta Math., **102** (1959), 23–32.

[183] J. Serrin, *Pathological solutions of elliptic differential equations*, Ann. Scuola Norm. Sup. Pisa, **18** (1964), 385–387.

[184] A. Sommerfeld, *Optics,* Academic Press, 1992.

[185] D. Strong, P. Blomgren and T. Chan, *Spatially Adaptative Local Feature Driven Total Variation Minimizing Image Restoration,* CAM Report 97-32, UCLA, 1997.

[186] D. Strong and T. Chan, *Exact Solutions to Total Variation Regularization Problems*, CAM Report 96-41, UCLA, October 1996.

[187] D. Strong and T.F. Chan, *Spatially and Adaptative Total Variation Based Regularization and Anisotropic Diffusion in Image Processing,* CAM Report 96-46, UCLA, 1996.

[188] B. Tang, G. Sapiro and V. Caselles, *Diffusion of general data on non-flat manifolds via harmonic maps theory: The direction diffusion case,* Int. J. Computer Vision, 36(2), pp. 149–161, February 2000.

[189] B. Tang, G. Sapiro and V. Caselles, *Color Image Enhancement via Chromaticity Diffusion,* IEEE Transactions on Image Processing, vol. 10(5), pp. 701–707, 2001.

[190] S. Teboul, L. Blanc-Féraud, G. Aubert and M. Barlaud, *Variational Approach for Edge-Preserving Regularization using Coupled PDE's,* IEEE Transactions on Image Processing, **7**(3) (1998), 387–397.

[191] R. Temam, *Solution Généralisées de Certain Equations du Type Hypersurfaces Minima,* Arch Rat. Mech. Anal. **44** (1971), 121–156.

[192] R. Temam, *On the Continuity of the Trace of Vector Functions with Bounded Deformation,* Appl. Analysis **11** (1981), 291–302.

[193] A.N. Tikhonov and V.Y. Arsemin, *Solutions of Ill-Posed problems,* John Wiley, New York, 1977.

[194] S. Twomey, *The Application of Numerical Filtering to the Solution of Integral Equations Encountered in Indirect Sensing Measurements,* J. Franklin Inst. **297** (1965), 95–109.

[195] J.L. Vazquez, *An introduction to the mathematical theory of porous medium equation,* Shape optimization and free boundaries (Montreal. PQ, 1990), 347–389. NATO Adv. Sci. Inst. Ser. C. Math. Phy. Sci. 380, Kluwer Acad. Publ., Dordrecht. 1992.

[196] J.L. Vazquez, *Entropy solutions and the uniqueness problem for nonlinear second-order elliptic equation,* Nonlinear partial differential equations (Fès. 1994), 179–203, Pitman Res. Notes Math. Ser. 343. Longman. Harlow, 1996.

[197] L. Veron, *Effects regularisantes des semi-groupes non linéaires dans les espaces de Banach,* Ann. Fac. Sci. Toulouse **1** (1979), 171–200.

[198] L. Vese, *A Study in the BV Space of a Denoising-Deblurring Variational Problem,* Appl. Math. Optim. **44** (2001), 131–161.

[199] L. Vese and S.J. Osher, *Modelling Textures with Total Variation minimization and oscillating patterns in image processing*, CAM report 02-19, UCLA, 2002.

[200] C. Vogel, Book in preparation.

[201] C.R. Vogel and M.E. Oman, *Iterative Methods for Total Variation Denoising*, SIAM J. Sci. Computing, **17** (1996), 227–238.

[202] C.R. Vogel and M. E. Oman, *Fast Total Variation Based Image Reconstruction*, Proceedings of the 1995 ASME Design Engineering Conferences, Vol. 3, pp. 1009–1015, 1995.

[203] C.R. Vogel and M.E. Oman, *Fast numerical methods for total variation minimization in image reconstruction*, Proceedings of SPIE 1995, San Diego, Advanced Signal Processing Algorithms, Vol. 2563, edited by F. T. Luk.

[204] C.R. Vogel and M.E. Oman, *Fast, robust total variation-based reconstruction of noisy, blurred images*, IEEE Trans. Image Process., vol. 7, no. 7, pp. 813–824, July 1998.

[205] C.R. Vogel, *A multigrid method for total variation-based image denoising*, in Computation and Control IV, K. Bowers and J. Lund, editors, Progess in Systems and Control Theory, 20, Birkhauser, Boston, 1995.

[206] G.F. Webb, *Continuous nonlinear perturbations of linear accretive operators in Banach spaces,* J. Funct. Anal. **10** (1972), 191–203.

[207] L.P. Yaroslavsky and M. Eden, *Fundamentals of digital optics*, Birkhäuser, Boston, 1996.

[208] X. Zhou, *An Evolution Problem for Plastic Antiplanar Shear*, Appl. Math. Optm. **25** (1992), 263–285.

[209] W. P. Ziemer, *Weakly Differentiable Functions*, GTM 120, Springer Verlag, 1989.

Index

Progress in Mathematics

Previously published titles

PM 217: **Thangavelu, S.**, An Introduction to the Uncertainty Principle. Hardy's Theorem on Lie Groups (2003). ISBN 0-8176-4330-3

PM 216: **Borovik, A. / Gelfand, M.I. / White, N.**, Coxeter Matroids (2003). ISBN 0-8176-3764-8

PM 215: **Arone, G. / Hubbuck, J. / Levi, R. / Weiss, M.** (Eds.), Categorical Decomposition Techniques in Algebraic Topology. International Conference in Algebraic Topology, Isle of Skye, Scotland, June 2001 (2003). ISBN 3-7643-0400-6

PM 214: **Dungey, N. / Elst, A.F.M. ter / Robinson, D.**, Analysis on Lie Groups with Polynomial Growth (2003). ISBN 0-8176-3225-5

PM 213: **Duval, C. / Guieu, L. / Ovsienko, V.**, The Orbit Method in Geometry and Physics. In Honor of A.A. Kirillov (2003). ISBN 0-8176-4232-3

PM 212: **Lubotzky, A. / Segal, D.**, Subgroup Growth (2003). ISBN 3-7643-6989-2

PM 211: **Hernández-Lerma, O. / Lasserre, J.B.**, Markov Chains and Invariant Probabilities (2003). ISBN 3-7643-7000-9

PM 210: **Joseph, A. / Melnikov, A. / Rentschler, R.** (Eds.), Studies in Memory of Issai Schur (2003). ISBN 0-8176-4208-0

PM 209: **Unterberger, A.**, Automorphic Pseudodifferential Analysis and Higher-Level Weyl Calculi (2002). ISBN 3-7643-6909-4

PM 208: **Turaev, V.**, Torsions of 3-dimensional Manifolds (2002). ISBN 3-7643-6911-6

PM 207: **Long, Y.**, Index Theory for Symplectic Paths with Applications (2002). ISBN 3-7643-6647-8

PM 206: **Snaith, V. P.**, Algebraic K-Groups as Galois Modules (2002). ISBN 3-7643-6717-2

PM 205: **Bouwknegt, P. / Wu, S.** (Eds.), Geometric Analysis and Applications to Quantum Field Theory (2002). ISBN 0-8176-4287-0

PM 204: **Kumar, S.**, Kac-Moody Groups, their Flag Varieties and Representation Theory (2002). ISBN 0-8176-4227-7

PM 203: **Blair, D.E.**, Riemannian Geometry of Contact and Symplectic Manifolds (2002). ISBN 0-8176-4261-7

PM 202: **Casacuberta, C. / Miró-Roig, R.M. / Verdera, J. / Xambó-Descamps, S.** (Eds.), European Congress of Mathematics. Barcelona, July 10-14, 2000. Volume II (2001). ISBN 3-7643-6418-1

PM 201: **Casacuberta, C. / Miró-Roig, R.M. / Verdera, J. / Xambó-Descamps, S.** (Eds.), European Congress of Mathematics. Barcelona, July 10-14, 2000. Volume I (2001). ISBN 3-7643-6417-3

PM 200: **Golubitsky, M. / Stewart, I.**, The Symmetry Perspective. From Equilibrium to Chaos in Phase Space and Physical Space (2002). ISBN 3-7643-6609-5

PM 199: **Peyre, E. / Tschinkel, Y.** (Eds.), Rational Points on Algebraic Varieties (2001). ISBN 3-7643-6612-5

PM 198: **Landsman, N.P. / Pflaum, M. / Schlichenmaier, M.** (Eds.), Quantization of Singular Symplectic Quotients (2001). ISBN 3-7643-6608-7

PM 197: **Cherix, P.-A. / Cowling, M. / Jolissaint, P.** / ., Groups with the Haagerup Property. Gromov's a-T-menability (2001). ISBN 3-7643-6598-6

PM 196: **Aguadé, J. / Broto, C. / Casacuberta, C.** (Eds.) Cohomological Methods in Homotopy Theory (2001) ISBN 3-7643-6588-9

PM 195: **Faber, C. / van der Geer, G. / Oort, F.** (Eds.) Moduli of Abelian Varieties (2001) ISBN 3-7643-6517-X

PM 194: **Juhl, A.** Cohomological Theory of Dynamical Zeta Functions (2001) ISBN 3-7643-6405-X

PM 193: **Knudson, K.P.** Homology of Linear Groups (2001) ISBN 3-7643-6415-7

PM 192: **Dehornoy, P.** Braids and Self-Distributivity (2000) ISBN 3-7643-6343-6

PM 191: **Kashiwara, M. / Miwa, T.** (Eds.) Physical Combinatorics (2000) ISBN 0-8176-4175-0

PM 190: **Essen van den, A.** Polynomial Automorphisms and the Jacobian Conjecture (2000) ISBN 3-7643-6350-9

PM 189: **André, Y. / Baldassarri, F.** De Rham Cohomology of Differential Modules on Algebraic Varieties (2001) ISBN 3-7643-6348-7

PM 188: **Dolbeault, P. / Iordan, A. / Henkin, G. / Skoda, H.** (Eds.) Complex Analysis and Geometry. International Conference in Honor of Pierre Lelong (2000) ISBN 3-7643-6352-5

PM 187: **Bost, J.-B. / Loeser, F. / Raynaud, M.** (Eds.) Courbes semi-stables et groupe fondamental en géométrie algébrique. Luminy, Décembre 1998 (2000) ISBN 3-7643-6308-8

PM 186: **Kleinert, E.** Units in Skew Fields (2000) ISBN 3-7643-6293-6

PM 185: **Faraut, J. / Kaneyuki, S. / Korányi, A. / Lu, Q.** Analysis and Geometry on Complex Homogeneous Domains (2000) ISBN 0-8176-4138-6

PM 184: **du Sautoy, M. / Segal, D. / Shalev, A.** (Eds.) New Horizons in pro-p Groups ((p kursiv)) (2000) ISBN 0-8176-4171-8

PM 183: **Kapovich, M.** Hyperbolic Manifolds and Discrete Groups (2000) ISBN 0-8176-3904-7

PM 182: **Billey, S. / Lakshmibai, V.** Singular Loci of Schubert Varieties (2000) ISBN 0-8176-4092-4

PM 181: **Hauser, H. / Lipmann, J. / Oort, F. / Quirós, A.** (Eds.) Resolution of Singularities. A research textbook in tribute to Oscar Zariski Based on the courses given at the Working Week in Obergurgl, Austria, September 7-14, 1997 (2000) ISBN 3-7643-6178-6

PM 180: **Paternain, G.P.** Geodesic Flows (1999) ISBN 0-8176-4144-0

PM179: **Morales Ruiz, J.J.** Differential Galois Theory and Non-Integrability of Hamiltonian Systems. (Sunyer Prize 1998) (1999) ISBN 3-7643-6078-X

PM 178: **Puig, L.** On the Local Structure of Morita and Rickard Equivalences between Brauer Blocks (1999) ISBN 3-7643-6156-5

PM 177: **Birkenhake, C. / Lange, H.** Complex Tori (1999) ISBN 0-8176-4103-3

PM 176: **Bass, H. / Lubotsky, A.** Tree Lattices (2000) ISBN 0-8176-4120-3

PM 175: **Bañuelos, R. / Moore, C.N.** Probabilistic Behavior of Harmonic Functions (1999) ISBN 3-7643-6062-3

PM 174: **Goerss, P.G. / Jardine, J.F.** Simplicial Homotopy Theory (1999) ISBN 3-7643-6064-X

PM 173: **Dräxler, P. / Michler, G. / Ringel, C.M.** (Eds.) Computational Methods for Representations of Groups and Algebras. Euroconference in Essen (Germany), April 1-5, 1997 (1999) ISBN 3-7643-6063-1

PM 172: **Brylinski, J.-L. / Brylinski, R. / Nistor, V. / Tsygan, B.** et al. (Eds.) Advances in Geometry (1998) ISBN 0-8176-4044-4

PM 171: **Reznikov, A. / Schappacher, N.** (Eds.) Regulators in Analysis, Geometry and Number Theory (1999) ISBN 0-8176-4115-7

PM 170: **Paterson, A.** Groupoids, Inverse Semigroups and their Operator Algebras (1998) ISBN 0-8176-4051-7

PM 169: **Balog, A. / Katona, G.O.H. / Sza'sz, D. / Recski, A.** (Eds.) European Congress of Mathematics, Budapest, July 22-26, 1996. Volume II (1998) ISBN 3-7643-5498-4

PM 168: **Balog, A. / Katona, G.O.H. / Sza'sz, D. / Recski, A.** (Eds.) European Congress of Mathematics, Budapest, July 22-26, 1996. Volume I (1998) ISBN 3-7643-5497-6

PM 167: **Facchini, A.** Module Theory. Endomorphism Rings and Direct Sum Decompositions in some Classes of Modules (1998) ISBN 3-7643-5908-0

PM 166: **Elias, J. / Giral, J.M. / Miró-Roig, R.M. / Zarzuela, S.** (Eds.) Six Lectures on Commutative Algebra (1998) ISBN 3-7643-5951-X

PM 165: **Migliore, J.C.** Introduction to Liaison Theory and Deficiency Modules (1998) ISBN 0-8176-4027-4

PM 164: **Roussarie, R.** Bifurcations of Planar Vector Fields and Hilbert's Sixteenth Problem (1998) ISBN 3-7643-5900-5

PM 163: **Berndt, R. / Schmidt, R.** Elements of the Representation Theory of the Jacobi Group (1998) ISBN 3-7643-5922-6

PM 162: **Arnold, V.I. / Greuel, G.M. / Steenbrink, J.H.M.** (Eds.) Singularities. The Brieskorn Anniversary Volume (1998) ISBN 3-7643-5913-7

PM 161: **Sagan, B.E. / Stanley, R.P.** Mathematical Essays in Honor of Gian-Carlo Rota (1998) ISBN 0-8176-3872-5

PM 160: **Kashiwara, M. / Matsuo, A. / Saito, K. / Satake, I.** (Eds.) Topological Field Theory, Primitive Forms and Related Topics (1998) ISBN 0-8176-3975-6

PM 159: **Thangavelu, S.** Harmonic Analysis on the Heisenberg Group (1998) ISBN 0-8176-4050-9